MW00598047

Encyclopedia of Electronic Circuits

VOLUME 4

Patent notice

Purchasers and other users of this book are advised that several projects described herein could be proprietary devices covered by letters patent owned or applied for. Their inclusion in this book does not, by implication or otherwise, grant any license under such patents or patent rights for commercial use. No one participating in the preparation or publication of this book assumes responsibility for any liability resulting from unlicensed use of information contained herein.

Encyclopedia of Electronic Circuits

VOLUME 4

Rudolf F. Graf & William Sheets

TAB BOOKS

Blue Ridge Summit, PA

FIRST EDITION
FIRST PRINTING

© 1992 by **Rudolf F. Graf and William Sheets**.
Published by TAB Books.
TAB Books is a division of McGraw-Hill, Inc.

Printed in the United States of America. All rights reserved. The publisher takes no
responsibility for the use of any of the materials or methods described in this book,
nor for the products thereof.

Library of Congress Cataloging-in-Publication Data

(Revised for volume 4)

Graf, Rudolf F.
    Encyclopedia of electronic circuits.

    Vol. 4: By Rudolf F. Graf and William Sheets.
    Includes bibliographical references and indexes.
    1. Electronic circuits—Encyclopedias.   I. Sheets,
William.   II. Title.
TK7867.G66   1985   621.3815′.3   84-26772
ISBN 0-8306-0938-5 (v. 1)
ISBN 0-8306-1938-0 (pbk. : v. 1)
ISBN 0-8306-3138-0 (pbk. : v. 2)
ISBN 0-8306-3138-0 (v. 2)
ISBN 0-8306-3348-0 (pbk. : v. 3)
ISBN 0-8306-7348-2 (v. 3)
ISBN 0-8306-3895-4 (pbk. : v. 4)
ISBN 0-8306-3896-2 (v. 4)

TAB Books offers software for sale. For information and a catalog, please contact
TAB Software Department, Blue Ridge Summit, PA 17294-0850.

Acquisitions Editor: Roland S. Phelps
Technical Editor: Andrew Yoder
Director of Production: Katherine G. Brown
Cover Design: Holberg Design, York, PA   TAB1

Contents

Introduction

Volume 4 of *Encyclopedia of Electronic Circuits* contains many new, not previously covered circuits, organized into 104 chapters. Circuit titles are listed at the beginning of each chapter, for references. Most of these circuits appeared in publications since 1988 and should be very useful for obtaining new ideas for research and development, or simply to fill a need for a specific circuit idea or application. Those wishing to develop their own circuits will find this book indispensable as a source of ideas, to see how others have solved a problem or approach a design, and to obtain a starting point toward a new design.

A brief explanation accompanies almost every entry. Those that have been omitted are either repetitive, obvious, or too involved to describe in few words. In this case, the reader should consult the original sources (as listed in the back of the book).

We also wish to extend our sincere thanks to Ms. Loretta Gonsalves for her fine work at the word processor. Her skill and cooperation contributed much to the successful completion of this book.

Rudolf Graf and William Sheets

Introduction

1

Active Antennas

The sources of the following circuits are contained in the Sources section, which begins on page 660. The figure number in the box of each circuit correlates to the entry in the Sources section.

ACTIVE ANTENNA WITH GAIN

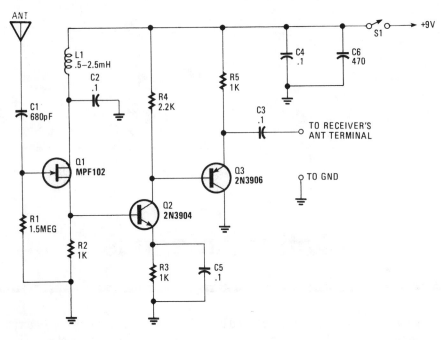

POPULAR ELECTRONICS

Fig. 1-1

The signal booster, built around a few transistors and support components, offers an RF gain of about 12 to 18 dB (from about 100 kHz to over 30 MHz).

The RF signal is direct-coupled from Q1's source terminal to the base of Q2, which is configured as a voltage amplifier. The output of Q2 is then direct-coupled to the base of Q3 (configured as an emitter-follower amplifier). Transistor Q3 is used to match and isolate the gain stage from the receiver's RF-input circuitry.

Inductor L1 is used to keep any power source noise from reaching the FET (Q1) and any value of RF choke from 0.5 to 2.5 mH will do. The value of R2 sets the Q2 bias at about 2 V. If the voltage is less than 2 V, increase the value of R2 to 1.5 kΩ. To go below 100 kHz (to the bottom of the RF spectrum), increase the value of C1 to 0.002 µF. The antenna is a short pull-up type (42″ to 86″ long).

ACTIVE ANTENNA I

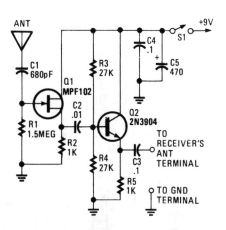

This circuit is designed to make a short pull-up antenna perform like a long wire antenna, while offering no voltage gain. The circuit boosts the receiver's performance only if the signal at the antenna is of sufficient level to begin with.

This circuit takes a short pull-up antenna that has a high output impedance and couples it to the receiver's low input impedance through a two-transistor impedance-matching network. Transistor Q1's high input impedance and high-frequency characteristics make it a good match for the short antenna, and Q2's low output impedance is a close match for the receiver's input. This circuit is usable over the range from 100 kHz to 30 MHz.

POPULAR ELECTRONICS   *Fig. 1-2*

ACTIVE ANTENNA II

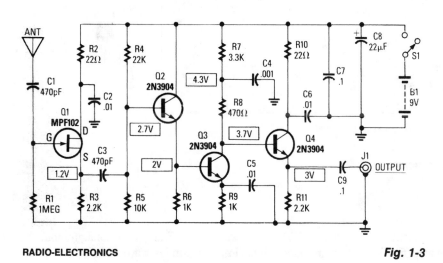

RADIO-ELECTRONICS   *Fig. 1-3*

This circuit provides 14- to 20-dB gain at frequencies from 10 kHz to 30 MHz. The antenna length can be anything between 5 and 10 feet. A 102-inch CB whip is excellent for this purpose.

WIDEBAND ACTIVE ROD ANTENNA

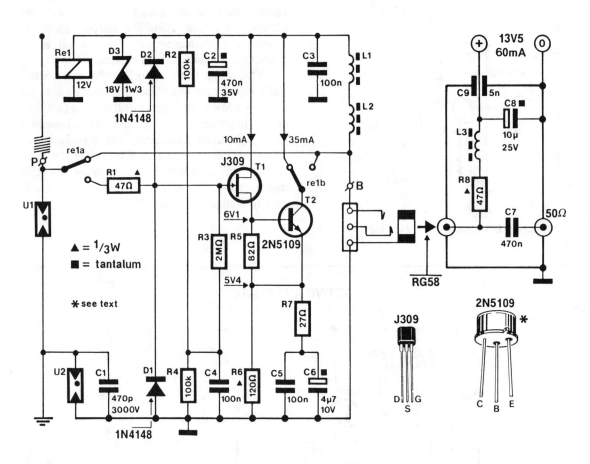

Fig. 1-4

A J309 Siliconix FET feeds a 2N5109 in a wideband RF amplifier configuration. A relay is used to bypass the amplifier in the transmit mode (if desired). A 2-m ⁵/₈-wave whip is used as the active antenna element. The amplifier is fed dc via the coax cable, which makes the use of only a single coax lead for both signal and power. U1 is a surge arrester for electrostatic discharge protection.

2

Analog-to-Digital Converter

The source of the following circuit is contained in the Sources section, which begins on page 660. The figure number in the box correlates to the entry in the Sources section.

A/D Board

A/D BOARD

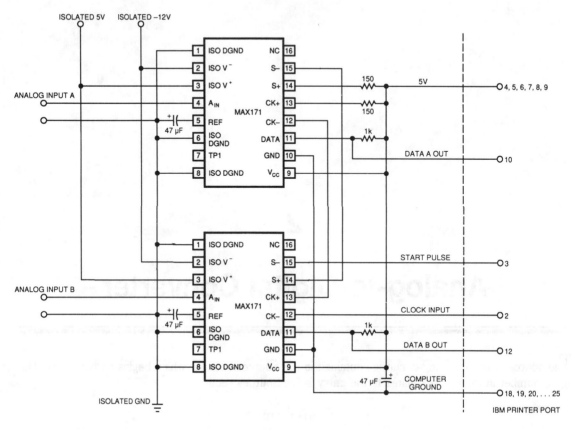

EDN

Fig. 2-1

An IBM PC can operate the two 12-bit A/D converters in Fig. 1 via its printer port. The converters' serial outputs use only two of the printer port's eight data lines (DATA A OUT, DATA B OUT). Because the IBM PC's printer port supplies no power, interface software running on the PC programs the six unused data lines high. Busing these data lines provides power for the digital portion of the A/D converters. (The converters have internal optoisolators. Consequently, you must provide isolated supplies for their analog sides.)

Although the converters can execute 12-bit conversions in 6 μs, the slow software-driven approach used in this Design Idea stretches conversion periods out to about 100 μs (depending on your PC's clock speed).

The circuit takes advantage of the converters' optoisolator inputs to put their clock and start inputs in series. Therefore, the converters operate synchronously.

The accompanying software starts the conversions, issues clock pulses, reads the data bits as they become available, and stores them in memory. The listing is too long to reproduce here; you can obtain it from the EDN BBS (617–558–4241, 2400, 8, N, 1).

3

Annunciators

The sources of the following circuits are contained in the Sources section, which begins on page 660. The figure number in the box of each circuit correlates to the entry in the Sources section.

Electronic Door Buzzer
Door Buzzer
SCR Circuit with Self-Interrupting Load
Electronic Bell
Two-Door Annunciator

ELECTRONIC DOOR BUZZER

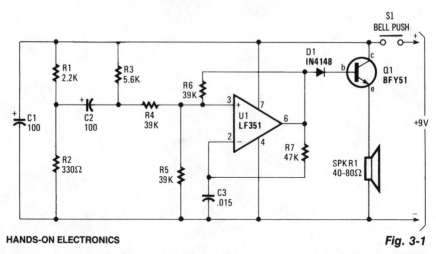

HANDS-ON ELECTRONICS

Fig. 3-1

When S1 is depressed, an initial positive voltage is placed on C2 and the noninverting terminal of U1. The circuit oscillates at a low frequency. As C2 charges up through R3, a rapid increase in frequency of oscillation results, producing (at SPKR1) a rapidly rising pitched sound. This sound is easily recognized over ambient noise.

DOOR BUZZER

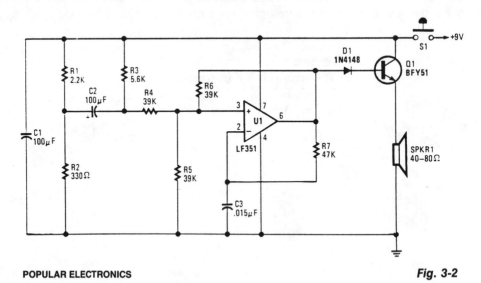

POPULAR ELECTRONICS

Fig. 3-2

An LF357 functions as a swept-tone oscillator, driving Q1 and SPKR1. A 9-Vdc supply is required.

SCR CIRCUIT WITH SELF-INTERRUPTING LOAD

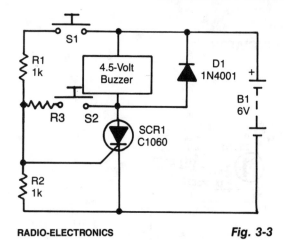

RADIO-ELECTRONICS

Fig. 3-3

A self-interrupting device connected to a voltage source functions as a switch that repeatedly opens and closes; therefore, the circuit does not latch in the normal way, so the alarm operates only as long as S1 is closed. Because of the inductive nature of that type of load, a damping diode (D1) must be wired across it.

The circuit can be modified to provide a self-latching action simply by wiring a 470-Ω resistor in parallel with the alarm. The circuit latches because the anode current of the SCR does not fall to zero when the alarm self-interrupts, but to a value that is determined by the value of the R3. The circuit can be unlatched by pressing S2, thereby enabling the anode current to fall to zero when the alarm self-interrupts.

ELECTRONIC BELL

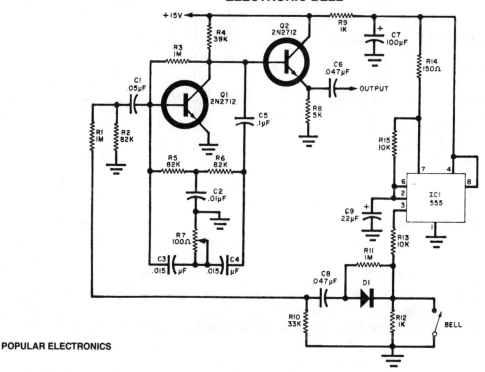

POPULAR ELECTRONICS

Fig. 3-4

A 555 timer pulses twin-T oscillator Q1. Q2 acts as an output buffer. R7 adjusts the frequency of oscillator Q1.

TWO-DOOR ANNUNCIATOR

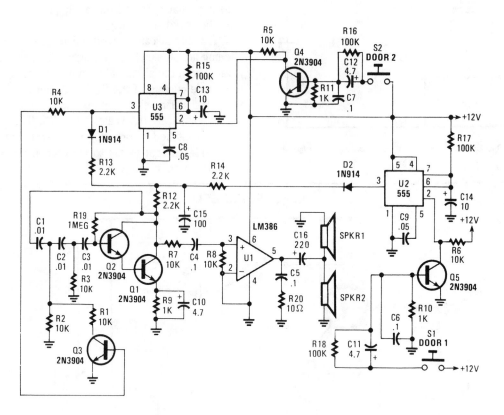

POPULAR ELECTRONICS

Fig. 3-4

When the pushbuttons at either door are depressed, this circuit generates a different tone for each door. Tones are generated by phase-shift oscillator Q1/Q2. Q3 provides tone frequency change by changing the phase-shift network. U2 and U3 are timers for the tones and Q4/Q5 interface the timers with the pushbuttons.

4

Antenna Circuits

The sources of the following circuits are contained in the Sources section, which begins on page 660. The figure number in the box of each circuit correlates to the entry in the Sources section.

Loop Antenna for 3.5 MHz
1-to 30-MHz Antenna Tuner

LOOP ANTENNA FOR 3.5 MHz

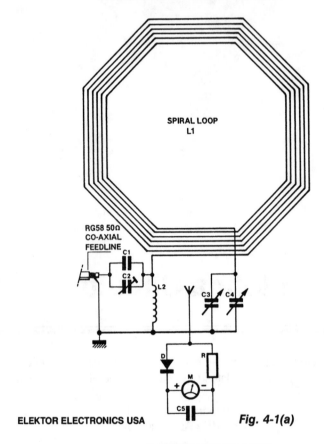

SPIRAL LOOP
L1

RG58 50Ω
CO-AXIAL
FEEDLINE

ELEKTOR ELECTRONICS USA

Fig. 4-1(a)

COMPONENTS LIST

C1 = 3 750 pF 500 V silver-mica capacitor.
C2 = 100 pF preset capacitor (Jackson C803).
C3 = 75 pF variable capacitor (Jackson C809), plus knob.
C4 = 12.7 pF variable capacitor (Jackson C16), plus knob.
C5 = 22 nF mica capacitor.
M = 250 µA f.s.d. 40 × 40 mm moving coil meter (Maplin LB808).
D = HF silicon diode.
R = 1 kΩ resistor (see text).
L1 = 5 1/8 turns of PVC covered stranded 7/0.2 mm wire. Outside diameter: 1.2 mm, 1 kV/1.5 A rating (see text).
L2 = 13 turns 16SWG tinned wire, 1 inch internal diameter.
Feedline = 48 inch RG58 coaxial cable, plus plug to suit transmitter.
Box = ABS box type MB3, 118 × 96 × 45 mm. Maplin ref. LH22.
Terminal blocks = qty. 4 12-way 2 amp terminal block. Maplin ref. FE78.
Spacers = qty. 3 insulated spacer type M3, 30 mm long, Maplin ref. FS40T.
Spokes = qty. 4 8-foot lengths of 5/8 × 1/4 inch molded hardwood (DIY store).
Vertical support = 23 × 0.8 × 0.8 inch wood (DIY store).
Wood base = 12 × 8 × 0.5 inch plywood or similar.
2 1/2 inch steel support bracket.

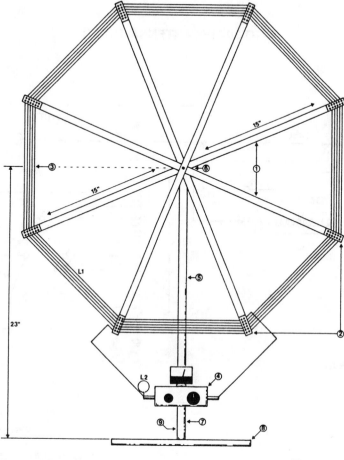

ELEKTOR ELECTRONICS USA   *Fig. 4-1(b)*

1. 4 lengths molded hardwood 30″ × 5/8 × 1/4″. Varnished. 2BA holes drilled in the centre. Glued and bolted together.
2. 8 off 6-way 2-amp polythene terminal blocks used as insulated wire spacers.
3. 5 1/8 turns of PVC stranded wire (for specs see components list).
4. See Fig. 3.
5. Wood vertical support 23″ × 0.8″ × 0.8″, wood stained.
6. 2″ × 2BA bolt.
7. Box front vertical support, 4 1/2″ × 1/2″ × 3/4″, wood stained.
8. Wood base 12″ × 8″ × 1/2″ (for similar), wood stained.
9. 2 1/2 steel support bracket behind wood vertical support.
10. Drilled and secured with glue and c/s wood screws.
Note: ″ = inch = 2.54 cm.

Suitable for receiving or transmitting (10 W or less) on the 80-m band, this loop antenna might be helpful when an outside antenna is not possible.

1-TO 30-MHz ANTENNA TUNER

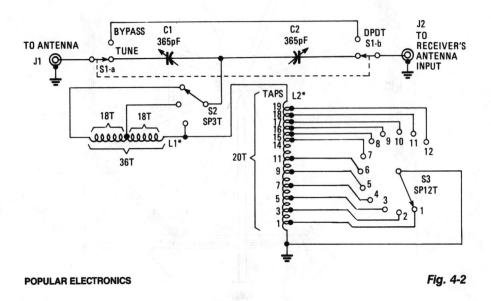

POPULAR ELECTRONICS

Fig. 4-2

L1 = 36T #18 enamel wire
on 2″ PVC SCH 40 pipe.

L2 = 20T #18 enamel (as in L1)
tapped as shown C1/C2.

365-pF variable capacitor,
receiving type.

This tuner will match a random length wire antenna to a receiver or low-power transmitter (≤ 25 W) for optimum signal transfer.

5

Audio Effects Circuits

The sources of the following circuits are contained in the Sources section, which begins on page 660. The figure number in the box of each circuit correlates to the entry in the Sources section.

Audio-Frequency Doubler
Audio Fader
Audio Equalizer
Vocal Eliminator
Voltage-Controlled Amplifier
Analog Delay Line (Echo and Reverb)
Musical Envelope Generator and Modulator
Audio Ditherizing Circuit for Digital Audio Use
Derived Center-Channel Stereo System
Low-Distortion Amplifier/Compressor

AUDIO-FREQUENCY DOUBLER

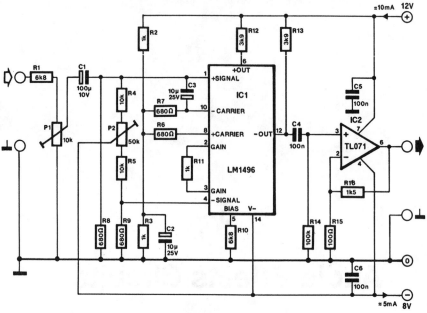

HANDS-ON ELECTRONICS

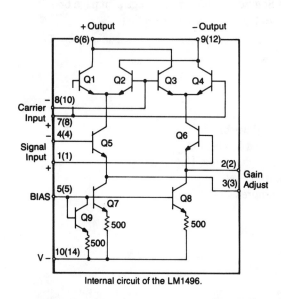

+ Output   − Output

Internal circuit of the LM1496.

ELEKTOR ELECTRONICS

Fig. 5-1

Often the frequency of a signal must be doubled: modulator/demodulator chip LM1496 is an ideal basis for this.

From trigonometry it is well known that:

$$2\sin x\cos x = \sin 2x$$

and:

$$\sin^2 = 1 - x\cos 2x.$$

These equations indicate that the product of two pure sinusoidal signals of the same frequency is one signal of double that frequency. The purity of the original signals is important: composite signals would give rise to all sorts of undesired products.

The LM1496 can only process signals that are not greater than 25 mV: above that level, serious distortion will occur. The design is therefore provided with a potential divider at its input. This addition makes it possible, for instance, to arrange for a 500-mV input signal to result in a signal of only 25 mV at the input of the LM1496.

AUDIO-FREQUENCY DOUBLER (*Cont.*)

To provide a sufficiently high output signal, the output of IC1 is magnified by op amp IC2, which is connected as a noninverting amplifier. Because the output of IC1 contains a dc component of about 8 V, the coupling between the two stages must be via a capacitor, C4.

With values of R15 and R16 as shown, IC2 gives an amplification of 16 (24 dB). The overall amplification of the circuit depends on the level of the input signal: with an input of 1.2 V, the amplification is unity; when the input drops to 0.1 V, the amplification is just 0.1. The value of the input resistors has been fixed at 680 Ω: this value gives a reasonable compromise between the requirements for a high input impedance and a low noise level.

To ensure good suppression of the input signal at the output, the voltages at pin 1 and pin 4 of IC1 must be absolutely identical to P4. It is possible, with the aid of a spectrum analyzer, to suppress the fundamental (input) frequency by 60 to 70 dB.

The output signal at pin 12 is distorted easily, because the IC is not really designed for this kind of operation. The distortion depends on the level of the input signal. At a frequency of 1 kHz and an input level of 100 mV, the distortion is about 0.6%; when the input level is raised to 500 mV, the distortion increases to 2.3%, and when the input level is 1 V, the distortion is 6%. The signal-to-noise ratio under these conditions varies between 60 and 80 dB.

The circuit draws a current of 10 mA from the positive supply line and 5 mA from the negative line. The phase shift between the input and output signals is about 45° (output lags). Finally, although the normal output is taken from pin 12, a similar output that is shifted by 180° (with respect to that at pin 12), is available at pin 6.

AUDIO FADER

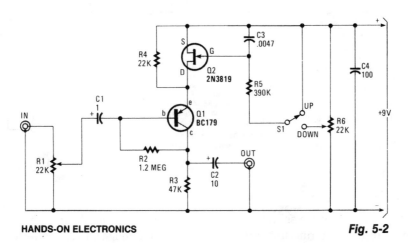

HANDS-ON ELECTRONICS   *Fig. 5-2*

In this circuit, Q1 is a simple amplifier that has its gain controlled by a variable emitter resistance supplied by FET Q2. In the up position of S1, C3 discharges through R5 and the gain of Q1 decreases because Q2 is driven toward cut-off. In the down position, Q2 conducts more, depending on the setting of R6, which causes a gain increase. By varying R5 or C3, various fade rates can be obtained.

AUDIO EQUALIZER

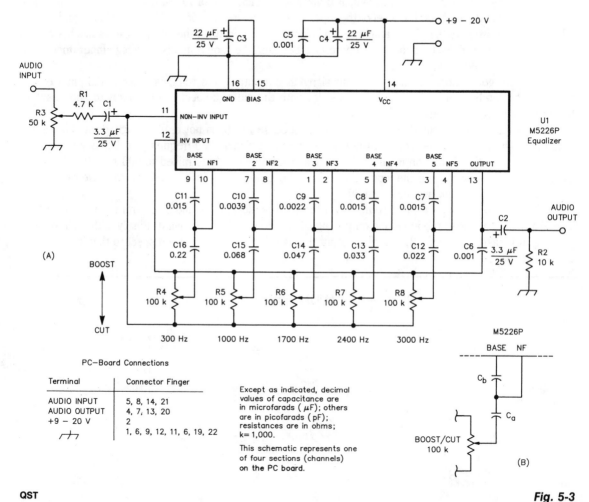

Except as indicated, decimal values of capacitance are in microfarads (µF); others are in picofarads (pF); resistances are in ohms; k= 1,000.

This schematic represents one of four sections (channels) on the PC board.

PC—Board Connections

Terminal	Connector Finger
AUDIO INPUT	5, 8, 14, 21
AUDIO OUTPUT	4, 7, 13, 20
+9 – 20 V	2
⏚	1, 6, 9, 12, 11, 6, 19, 22

Fig. 5-3

Designed for communications use, this equalizer circuit uses a Mitsubishi M5226P audio equalizer IC to adjust frequency response. It runs from a 9 to 20 V supply. C6 through C16 are polyester film capacitors of ± 5% tolerance.

VOCAL ELIMINATOR

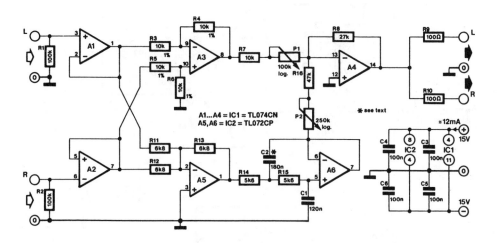

ELEKTOR ELECTRONICS

Fig. 5-4

Otherwise properly mixed sounds often suffer from a predominant solo voice (which might, of course, be the intention). If such a voice needs to be suppressed, the present circuit will do the job admirably.

The circuit is based on the fact that solo voices are invariably situated "at the center" of the stereo recordings that are to be mixed. Thus, voice levels in the left- and right-hand channels are about equal. Arithmetically, therefore, left minus right equals zero; that is, a mono signal without voice.

There is, however, a problem: the sound levels of bass instruments, more particularly the double basses, are also just about the same in the two channels. On the one hand low-frequency sounds are virtually nondirectional and on the other hand, the recording engineers purposely use these frequencies to give a balance between the two channels.

However, the bass instruments can be recovered by adding those appearing in the left + right signal to the left – right signal. The whole procedure is easily followed in the circuit diagram. The incoming stereo signal is buffered by A1 and A2. The buffered signal is then fed to differential amplifier A3 and subsequently to summing amplifier A5. The latter is followed by a low-pass filter formed by A6. You can choose between a first-order and a second-order filter by respectively omitting or fitting C2. Listen to what sounds best.

The low-frequency signal and the difference signal are applied to summing amplifier A4. The balance between the two is set by P1 and P2 to individual taste.

You have noticed that the circuit does not contain input or output capacitors. If you wish, output capacitors can be added without detriment. However, adding input capacitors is not advisable, because the consequent phase shift would adversely affect the circuit operation.

VOLTAGE-CONTROLLED AMPLIFIER

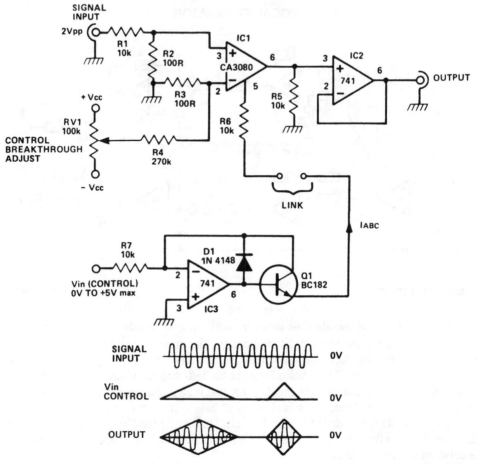

Fig. 5-5

The CA3080 can be used as a gain controlling device. The input signal is attenuated by R1/R2 so that a 20-mVpp signal is applied to the input terminals. If this voltage is much larger, significant distortion will occur at the output. In fact, this distortion is put to good use in the triangle-to-sine wave converter.

The gain of the circuit is controlled by the magnitude of the current IABC. This current flows into the CA3080 at pin 5, which is held at one diode voltage drop above the $-V_{CC}$ rail. The gain of the CA3080 is "linearly" proportional to the magnitude of the IABC current over a range of 0.1 μA to 1 mA. Thus, by controlling IABC, you can control the signal level at the output. The output is a current output, which has to be "dumped" into a resistive load (R5) to produce a voltage output. The output impedance at IC1 pin 6 is 10 kΩ (R5), but this is "unloaded" by the voltage follower (IC2) to produce a low output impedance.

The circuit around IC3 is a precision voltage-to-current converter and this can be used to generate IABC. When V_{in} (control) is positive, it linearly controls the gain of the circuit. When it is negative, IABC is zero and so the gain is zero.

ANALOG DELAY LINE (ECHO AND REVERB)

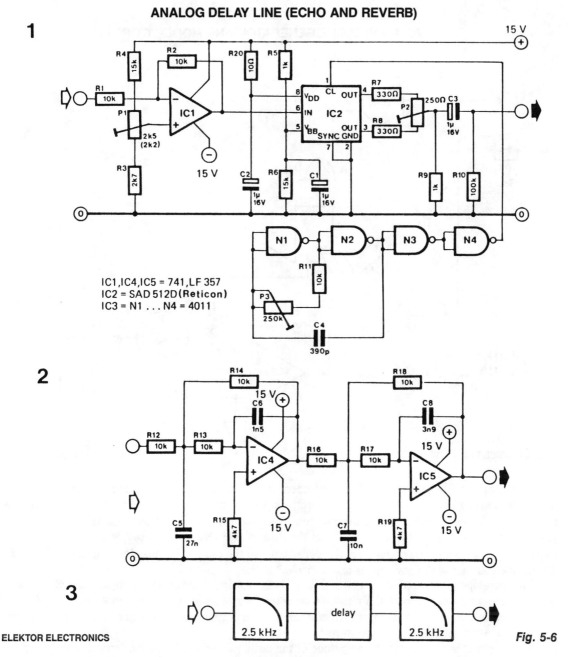

IC1,IC4,IC5 = 741, LF 357
IC2 = SAD 512D (Reticon)
IC3 = N1 ... N4 = 4011

Fig. 5-6

This circuit uses an SAD 512D (Reticon) chip, which is a 512-stage analog shift register. By varying the clock frequency between 5 and 50 kHz, delay time can be set between 51.2 and 5.12 ms. The clock frequency must be at least twice the highest audio frequency.

MUSICAL ENVELOPE GENERATOR AND MODULATOR

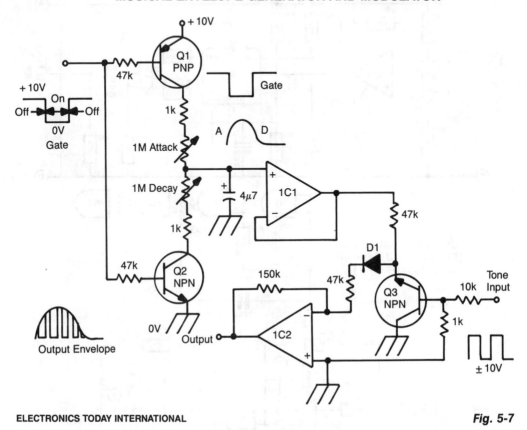

ELECTRONICS TODAY INTERNATIONAL

Fig. 5-7

A gate voltage is applied to initiate the proceedings. When the gate voltage is in the ON state, Q1 is turned on, and capacitor C is charged up via the attack pot in series with the 1-kΩ resistor. By varying this pot, the attack time constant can be manipulated. A fast attack gives a percussive sound, a slow attack gives the effect of "backward" sounds. When the gate voltage returns to its OFF state, Q2 is turned on and the capacitor is then discharged via the decay pot and the other 1-kΩ resistor to ground. Thus, the decay time constant of the envelope is also variable.

This envelope is buffered by IC1, a high-impedance voltage follower and is applied to Q3, which is being used as a transistor chopper. A musical tone in the form of a square wave is connected to the base of Q3. This turns the transistor on or off. Thus, the envelope is chopped up at regular intervals, which are determined by the pitch of the square wave.

The resultant waveform has the amplitude of the envelope and the harmonic structure of the square wave. IC2 is used as a virtual earth amplifier to buffer the signal and D1 ensures that the envelope dies away at the end of a note.

AUDIO DITHERIZING CIRCUIT FOR DIGITAL AUDIO USE

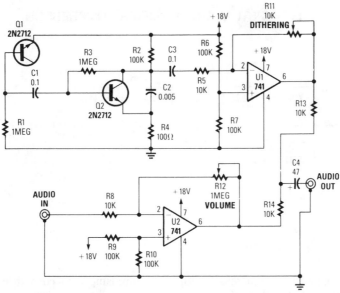

POPULAR ELECTRONICS

Fig. 5-8

By adding a small amount of noise to a signal to be digitized (about 0.7 bit):

$$V_{\text{NOISE}} \approx 0.7 \left(\frac{V_{\text{INPUT}}\text{p-p}}{2^N} \right)$$

where: $n = \#$ of bits. For example, 8 bits and 2 V p-p would be 0.0055 V.

This circuit uses a transistor (Q1) and an amplifier (Q2 and U1) to generate the noise signal. R11 controls the noise injection and R12 controls the gain of the system.

DERIVED CENTER-CHANNEL STEREO SYSTEM

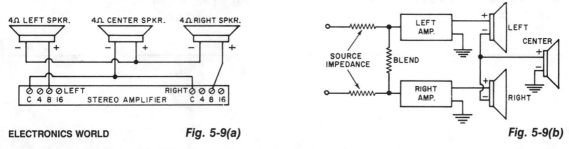

ELECTRONICS WORLD

Fig. 5-9(a)

Fig. 5-9(b)

A simple method of deriving a center or third channel without the use of an extra transformer or amplifier. (a) 4-Ω speakers are connected to 8-Ω amplifier taps. 8 and 16-Ω speakers connect to 16-Ω taps. (b) By blending the inputs it is possible to cancel out undesired crosstalk.

LOW-DISTORTION AMPLIFIER/COMPRESSOR

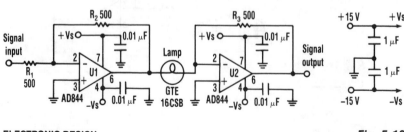

ELECTRONIC DESIGN

Fig. 5-10

Designers can build a 15-dB compressor with a miniature lamp and a current-feedback amplifier. The circuit possesses extremely low distortion at frequencies above the lamp's thermal time constant. This means that distortion is negligible from audio frequencies to beyond 10 MHz. There's also relatively little change in phase versus gain compared to other automatic gain-control circuits. Lastly, the circuit has many instrumentation, audio, and high-frequency applications as a result of its low distortion and small phase change.

The AD844 op amp is a perfect fit for this application because it's a current-feedback amplifier. Each stage of the circuit, U2, lamp, and feedback resistor compresses an ac signal by over 15 dB (see the figure). Cascading a number of stages delivers higher compression ranges.

Op amp U1 operates as a unity-gain buffer to drive the input to the compressor. However, U1 is optional if a low-impedance signal source is used. The lamp's resistance will increase with temperature, which reduces the ratio of resistor R3 to the resistance of the lamp. This ratio reduces the gain of U2. The lamp's cold resistance should be greater than the input resistance of U2 (more than 50 Ω) for proper operation. The lamp's resistance will change slightly for low input levels. Therefore, the ratio of R3 to the resistance of the lamp and the gain of U2 stays high.

6

Audio Scramblers

The sources of the following circuits are contained in the Sources section, which begins on page 661. The figure number in the box of each circuit correlates to the entry in the Sources section.

Voice Scrambler/Descrambler
Voice Scrambler/Disguiser Circuit

VOICE SCRAMBLER/DESCRAMBLER

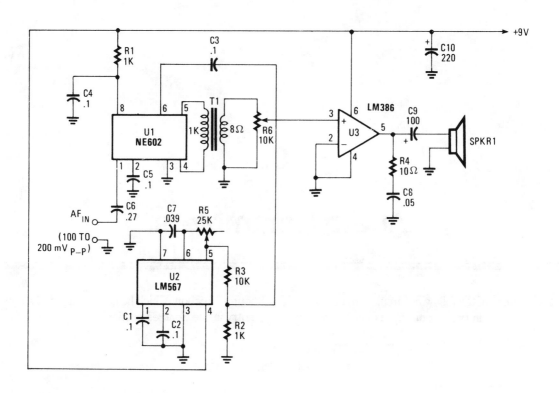

Fig. 6-1

This circuit uses an NE602 as an inversion mixer. U2 is set to run at about 2.5 to 3.5 kHz. U3 drives a loudspeaker. Because speech inversion scrambling is its own inverse, the circuit will also descramble.

VOICE SCRAMBLER/DISGUISER CIRCUIT

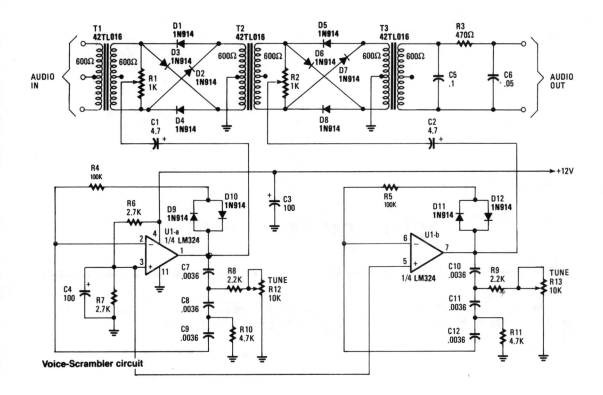

Voice-Scrambler circuit

Fig. 6-2

This circuit uses two balanced modulators to produce a DSB signal and then reinsert the carrier, except the carrier now has a different frequency. This causes an input signal to be distorted. A voice signal will be recognizable with this circuit, but the original speakers' voice will not be identifiable with correct adjustments.

Two LM324 op amps act as oscillators that are tuneable from 2 to 3.5 kHz. The frequencies are set with R12 and R13. T1, T2, and T3 are 600 Ω CT/600 Ω audio transformers—available from Mouser Electronics, Inc.

7

Audio Power Amplifiers

The sources of the following circuits are contained in the Sources section, which begins on page 661. The figure number in the box of each circuit correlates to the entry in the Sources section.

12-V/20-W Stereo Amplifier
General-Purpose 5-W Audio Amplifier with ac Power Supply
Bull Horn
Receiver Audio Circuit
Audio Amplifier
8-W Audio Amplifier
Simple Op Amp Audio Amplifier

12-V/20-W STEREO AMPLIFIER

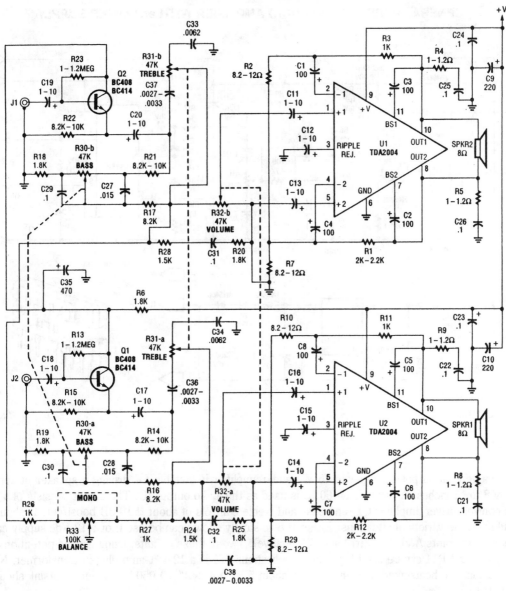

Fig. 7-1

This amplifier delivers 20 W per channel. Input sensitivity is about 300 mV into 47 kΩ. Notice that a bridged output is used, so the speakers are operated with both wires above ground. A +12-V supply is used. U1 and U2 must be heatsinked.

GENERAL-PURPOSE 5-W AUDIO AMPLIFIER WITH ac POWER SUPPLY

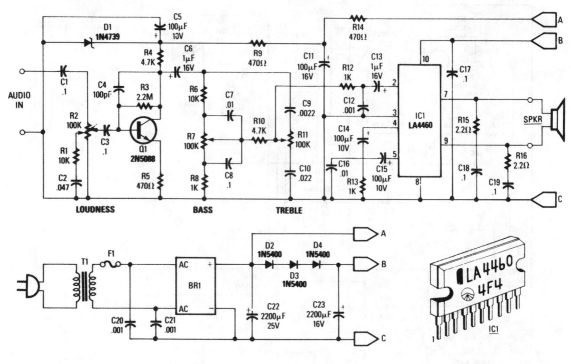

RADIO-ELECTRONICS

Fig. 7-2

This general-purpose low-power (5 W) audio amplifier is suitable for driving a speaker of approximately 8 to 12 inches. A Sanyo LA4460 IC is used as the audio output IC. The circuit consists of a loudness control, driver amplifier Q1, and bass and treble controls of about ±10 dB boost/cut. It should be useful in a wide variety of situations. Either the ac supply shown can be used, or a 12 Vdc supply can be connected to points A&B (positive) and C (negative). Two of these circuits, using ganged potentiometers at R2, R7, and R11 can be used for stereo applications. T1 is a 12-V 1-amp plug-in transformer. Notice that IC1 must be heatsinked. Power output is about 5 W. A 4″×2″×0.050″ aluminum heatsink should be adequate.

BULL HORN

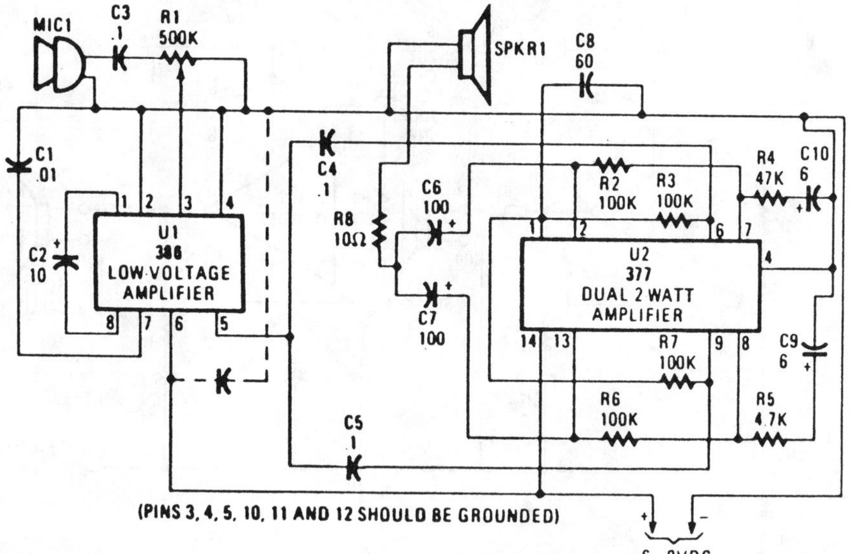

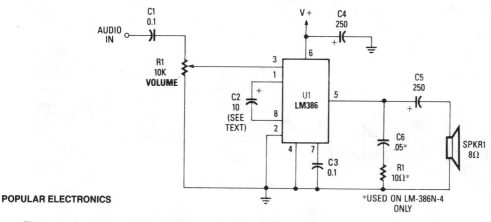

(PINS 3, 4, 5, 10, 11 AND 12 SHOULD BE GROUNDED)

WELS' THINK TANK

6 -9VDC SOURCE

Fig. 7-3

This bull horn uses U1 as a driver stage and U2 as an output driver. U1 is set up for a gain of 200. The microphone should have about 200-mVpp output. The two sections of U2 produce about 4-W of output power. Use shielded cable for all audio leads. Power is a 6-to 9-V battery or other source.

RECEIVER AUDIO CIRCUIT

POPULAR ELECTRONICS

*USED ON LM-386N-4 ONLY

Fig. 7-4

This simple receiver AF amplifier can supply several hundred milliwatts to an 8-Ω speaker. The gain is about 200X. If high gain is not needed, C2 can be deleted and a gain of 20 will be obtained. R1 and C6 are musts, otherwise ultrasonic (30 to 60 kHz) oscillations might occur. C6 can be 0.1 μF on all LM386N versions for protection against these oscillations. The supply voltage is typically 6 to 12 V. No heatsink is necessary, but good grounding is a must.

AUDIO AMPLIFIER

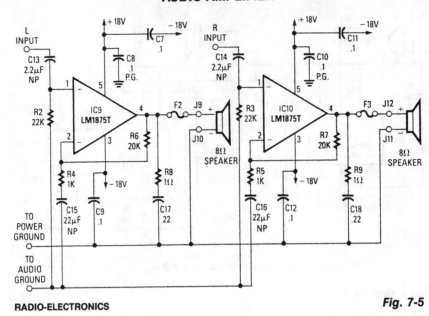

RADIO-ELECTRONICS

Fig. 7-5

This amplifier will deliver around 20 W to an 8-Ω speaker.

8-W AUDIO AMPLIFIER

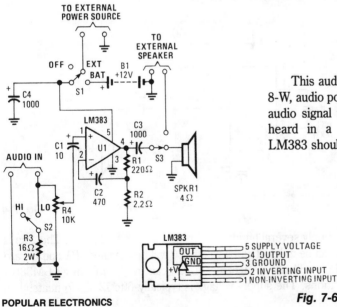

This audio power amp (built around an LM383 8-W, audio power amplifier) can be used to boost an audio signal to a sufficient level so that it can be heard in a high-noise environment. Note that LM383 should be heatsinked.

LM383	
OUT	5 SUPPLY VOLTAGE
GND	4 OUTPUT
+V	3 GROUND
	2 INVERTING INPUT
	1 NON-INVERTING INPUT

POPULAR ELECTRONICS

Fig. 7-6

SIMPLE OP AMP AUDIO AMPLIFIER

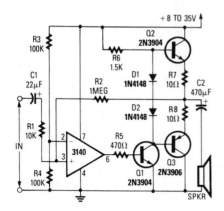

RADIO-ELECTRONICS　　　　*Fig. 7-7*

A CA3140 drives a complementary output stage Q1, Q2, and Q3. Output power depends on supply voltage and limits on dissipations of Q2 and Q3, but it can be 1 or 2 W with a higher impedance speaker and a 30-V supply.

8

Audio Signal Amplifiers

The sources of the following circuits are contained in the Sources section, which begins on page 661. The figure number in the box of each circuit correlates to the entry in the Sources section.

PREAMPLIFIER FOR MAGNETIC PHONO CARTRIDGES

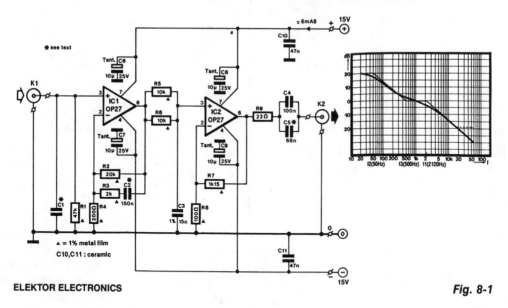

ELEKTOR ELECTRONICS

Fig. 8-1

This amplifier is intended to be added to preamplifiers that have no phono input. Such a phono input is required for normal record players with a dynamic pick-up, of which millions are still around. Moreover, the amplifier does not only bring the output of the pick-up to line level, it also adds the correction to the frequency response (according to RIAA requirements).

When recording gramophone records, the frequency characteristic is lifted at the high end. This lift must be countered in the playback (pre)amplifier. The corrections to the frequency response characteristic are according to a norm set by the Record Industries Association of America (RIAA) and also by the IEC.

The corrective curve provided by the amplifier is shown in the graph (bold line). The thin line shows the ideal corrective curve. The sharp bends in this at 50 and 500 Hz are nearly obtained in the practical curve by network R3/C2; just above 2 kHz is approached in practice by filter R5/R6/C3. The arrangement of R3/C2 in the feedback loop of IC1 gives noticeably better results than the usual (passive) filter approach.

Circuit IC1 provides a dc amplification of 40 dB, which drops to about 20 dB when the frequency rises above 500 Hz. To minimize the (resistor) noise and the load of the op amp at higher frequencies, the value of R3 is a compromise. The associated polystyrene capacitor, C2, should have a tolerance of 1 to 2%.

To raise the 2-mV output of the dynamic pick-up to line level at 1 kHz, linear amplifier IC2 has been added. This stage has a gain of 22 dB, so a signal of 250 mV is available at its output.

Capacitors C4/C5 at the output, in conjunction with the input impedance of the following preamplifier, form a high-pass filter with a cut-off frequency of 20 Hz; this serves to suppress any rumble or other low frequency noise. The value of C1 is normally given in the instruction booklet of the dynamic pick-up.

The power supply for the amplifier must be of good quality. Particularly, the transformer should be class A1 with a small stray magnetic field.

When the amplifier is built into the record player (best), the power supply should not be included unless it is very well screened; otherwise, hum is unavoidable.

35

SIMPLE TAPE PLAYBACK AMPLIFIER

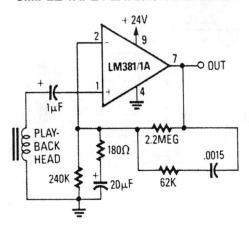

RADIO-ELECTRONICS **Fig. 8-2**

This circuit uses an LM381/1A as a tape preamp. The feedback network includes NAB Equalization.

LOW-NOISE PHONO PREAMP

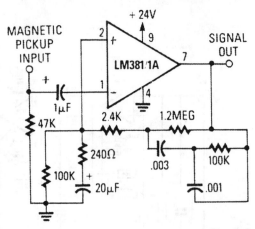

RADIO-ELECTRONICS **Fig. 8-3**

This circuit uses an LM381/1A as a low-noise phono preamp. The feedback network provides RIAA compensation.

SIMPLE 40-dB GAIN AMPLIFIER

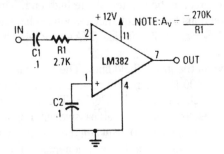

RADIO-ELECTRONICS **Fig. 8-4**

An LM382 low-noise preamp is used here to obtain a 40-dB gain amplifier, using only the IC and three peripheral components.

ULTRA-LOW-NOISE MAGNETIC PHONO PREAMP

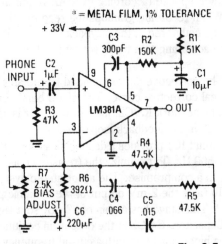

RADIO-ELECTRONICS **Fig. 8-5**

This phono preamp uses an LM381/1A in a circuit that includes RIAA equalization. Adjust R7 for a voltage that is equal to half of the supply voltage (≈ 16.5 V).

IMPEDANCE-MATCHING PREAMP

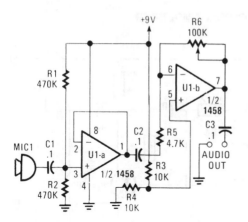

POPULAR ELECTRONICS   **Fig. 8-6**

This circuit will match a crystal microphone to a device that requires a low-impedance dynamic microphone.

LOW-NOISE AMPLIFIER

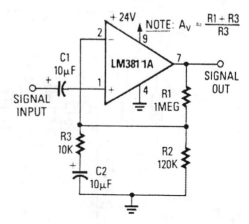

$$\text{NOTE: } A_V \approx \frac{R1 + R3}{R3}$$

RADIO-ELECTRONICS   **Fig. 8-7**

This low-noise LM381/1A noninverting amplifier has a gain of 100.

LOW-NOISE 1 000 × PREAMP

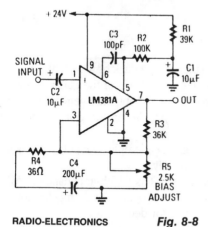

RADIO-ELECTRONICS   **Fig. 8-8**

An LM381A is used here as a low-noise preamp with a gain of approximately 1 000×. Adjust R5 for 12 V at pin 7, assuming a 24-V supply.

SIMPLE MICROPHONE PREAMP

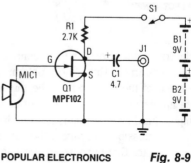

POPULAR ELECTRONICS   **Fig. 8-9**

This preamp uses a small dynamic microphone coupled to the gate of Q1. R1 is a load resistor. Audio is taken out between the negative side of C1 and ground. Output will be between 10 and 100 mVpp, depending on the microphone.

ELECTRIC GUITAR MATCHING AMPLIFIER

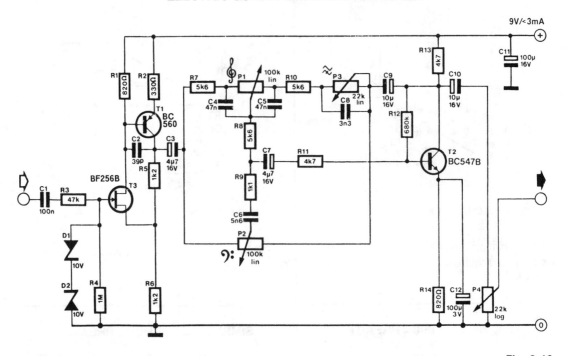

Fig. 8-10

An electric guitar often has to be connected to a mixing panel, a tape deck or a portable studio. As far as cabling is concerned, that is no problem, but matching the high impedance of the guitar element to the low impedance of the line input of the mixing panel or tape deck is a problem. Even the so-called high impedance inputs of those units are not suitable for the guitar output. When the guitar is connected to such an input, hardly any signal is left for the panel or deck to process.

It would be possible to connect the guitar to the (high impedance) microphone input, but it is normally far too sensitive for that purpose; guitar clipping occurs all too readily.

The matching amplifier presented here solves those problems: it has a high-impedance (1 MΩ) input that can withstand voltages of over 200 V. The output impedance is reasonably low. Amplification is ×2 (6 dB). Dual tone control, presence control, and volume control are provided.

The circuit can handle input levels of up to 3 V. Above that level distortion increases, but that is, of course, a good thing with guitar music. Real clipping of the input signal does not occur until much higher levels than are obtainable from a guitar are applied. Power is supplied by a 9-V (PP3) battery from which the circuit draws a current that does not exceed 3 mA.

UNIVERSAL AUDIO LINE AMPLIFIER

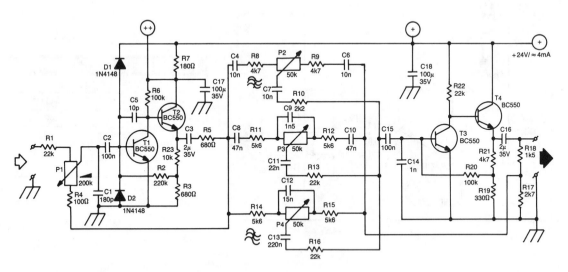

ELEKTOR ELECTRONICS

Fig. 8-11

A line amplifier is always a useful unit to have around, whether it is for matching a line signal or raising its level somewhat. This might be needed during a recording session or with a public-address system. Furthermore, a line mixer can be constructed from a number of these amplifiers. The input of the amplifier is high-voltage proof. The output impedance is low.

The circuit is a conventional design: two dc-coupled stages of amplification are separated by a three-fold Baxandall tone control system. The volume control at the input is conspicuous by having its ''cold'' side connected, not to ground, but to the output of the first amplifier. Because the signal there is out of phase with the input signal, the amplifier obtains negative feedback via P1. The amplification is therefore inversely proportional to the magnitude of the input signal. Thus, it is possible for the amplifier to accept a wide range of input levels. It is quite possible to input a signal taken directly from the loudspeaker terminals of a power amplifier.

The supply voltage is 24 V. At that voltage, the amplifier draws a current of about 4 mA. If several amplifiers are used in conjunction (as, for instance, in a mixer panel), the various supplies (+ and + + in the diagram) can be interlinked. Capacitors C17 and C18, and resistor R7 don't need to be duplicated in that case.

CD4049 AMPLIFIER

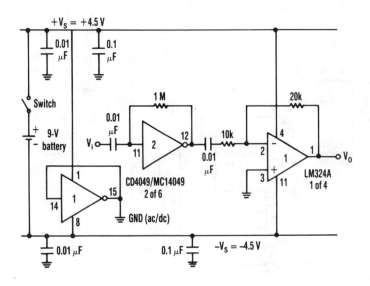

INVERTER CHARACTERISTICS

	V supply	
	9V	13.6V
Av	30 V/V	40 V/V
f(–3dB)	2.5 MHz	3.5 MHz nom.
Ioh	–1.25 mA	–3.0 mA min.
Iol	8.0 mA	20.0 mA min.

ELECTRONIC DESIGN

Fig. 8-12

When an inverter is biased with one resistor from its input to output in the range of 100 kΩ to 10 MΩ and is capacitor coupled, it exhibits amplifier characteristics (see the table).

Furthermore, when a split power-supply bus is needed and only one battery is used, the inverter can be configured to supply a pseudo-dc ground of relatively low impedance, coincident with the ac ground (see the figure). Depending on the magnitude of the dc ground return currents, anywhere from one inverter to several in parallel are sufficient. Also, the supply buses must be capacitor bypassed.

The configured input-to-input shorted inverter now acts as a voltage regulator that sinks and sources current. In this configuration, the inverter is forced to operate at the midpoint of its transfer characteristic. This divides the battery potential into two equal parts—as referenced to the defined dc ground by virtue of its internal gain and physical structure. Op amps such as the LM324A, can be powered from one battery while being referenced to the dc ground that is generated by the inverter. This novel technique surpasses the use of discrete resistors for battery potential dividing. It can be employed in other applications where individual component savings and improved design performance are needed.

LOW-NOISE AUDIO PREAMP

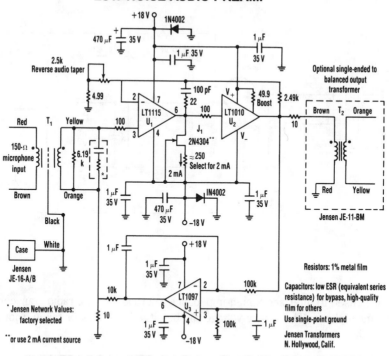

ELECTRONIC DESIGN

Fig. 8-13

A low-noise LT1115 (Linear Technology, Inc.) op amp is coupled to a class-A buffer amplifier to produce a variable gain (12-to-50 dB) microphone preamp. THD is less than 0.01% from 80 Hz to over 20 kHz. The transformers must be properly grounded and shielded.

LOW-IMPEDANCE MICROPHONE PREAMP

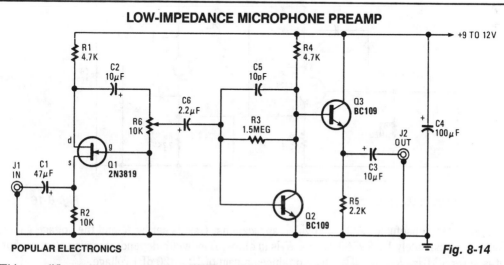

POPULAR ELECTRONICS

Fig. 8-14

This amplifier uses a common-gate FET amplifier to match a low-Z microphone.

41

MICROPHONE PREAMP

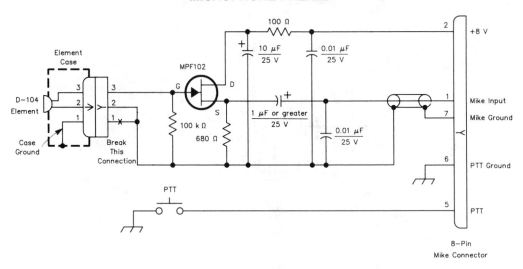

QST

Fig. 8-15

This circuit is used to interface a high-impedance microphone to a radio transceiver that requires a low-impedance microphone. The supply voltage can be either a battery or taken from the transceiver the circuit is used with.

GENERAL-PURPOSE PREAMP

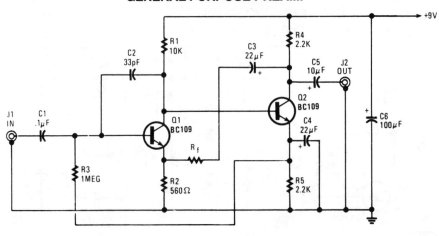

POPULAR ELECTRONICS

Fig. 8-16

This amplifier is useful for audio and video applications. Gain is set by R_f and the voltage gain of this amplifier is approximately $1 + R_f/560$, where R_f is in ohms. Bandwidth depends on gain selected, but typically it is several MHz. $R_f = 5.1 \text{ k}\Omega$, which produces a gain of $10 \times$ (20 dB) voltage.

9

Automotive Instrumentation Circuits

The sources of the following circuits are contained in the Sources section, which begins on page 661. The figure number in the box of each circuit correlates to the entry in the Sources section.

Digital Oil-Pressure Gauge
Water Temperature Gauge
Automotive Electrical Tester
Digital Vacuum Gauge
Digital Fuel Gauge
Analog Expanded-Scale Meter for Autos
Digital Pressure Gauge
Voltage Gauge
Digital Miscellaneous Temperature Gauge

DIGITAL OIL-PRESSURE GAUGE

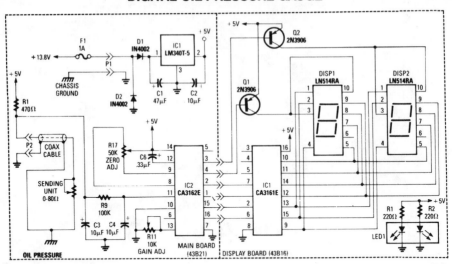

RADIO-ELECTRONICS

Fig. 9-1

This gauge uses a sensor in conjunction with R1 to develop a dc voltage proportional to oil pressure. IC1 and IC2 form a two-digit DVM. Q1 and Q2 are display selectors for the multiplexed display. IC1 provides the necessary +5 V to the circuitry. Calibration is via R11 and zero adjust via R17.

WATER TEMPERATURE GAUGE

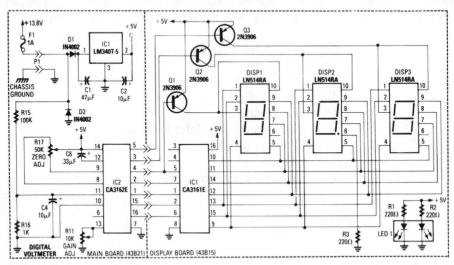

RADIO-ELECTRONICS

Fig. 9-2

This gauge is similar to the miscellaneous temperature gauge, except that a thermostat is used as a sensing element.

AUTOMOTIVE ELECTRICAL TESTER

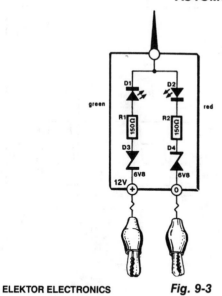

This little tester is useful for checking vehicle electrical circuits. Two LEDs indicate whether one of the clips is connected to the positive supply line (red) or to ground (green).

The unit is powered by the vehicle battery. It is advisable to terminate the unit into two insulated heavy-duty crocodile clips. These enable connection to be made directly to the battery or to terminals on the fuse box. It is also possible to terminate it into a suitable connector that fits into the cigarette lighter socket. If a sharp needle is soldered to one of the terminals, it is possible to check insulated wiring—but only those that carry 12 V. Although the needle pierces the insulation, it does not damage it.

ELEKTOR ELECTRONICS   *Fig. 9-3*

DIGITAL VACUUM GAUGE

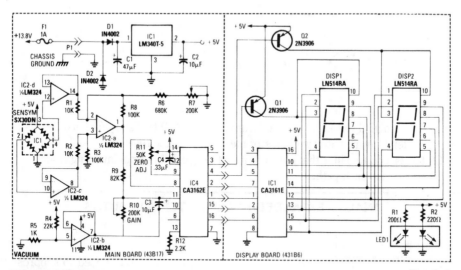

RADIO-ELECTRONICS   *Fig. 9-4*

A bridge circuit is used to produce a signal from the output of vacuum sensor IC1. IC2b provides about a 0.2 V offset for IC4, the A/D converter. IC2b and d are voltage followers that drive differential amp IC2a. The output of this circuit is used to drive IC4 and IC1, the display drivers.

DIGITAL FUEL GAUGE

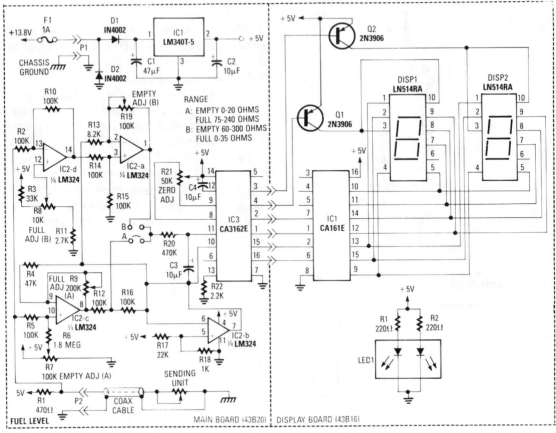

RADIO-ELECTRONICS

Fig. 9-5

This circuit uses a digital voltmeter (formed from IC1 and IC3) to display fuel quantity as a percentage of a full tank. In order to work with two kinds of fuel sensors, low resistance = full. Where higher resistance = full, IC2 forms a dc amplifier that has both inverting (path A) or noninverting (path B) outputs, and calibration adjustments for each path.

ANALOG EXPANDED-SCALE METER FOR AUTOS

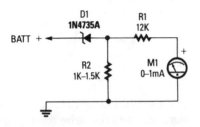

Zener diode D1 is used to suppress the first 6 V of the scale, which gives a meter reading of 6 to 8 V—useful for automotive electrical system monitoring.

POPULAR ELECTRONICS

Fig. 9-6

DIGITAL PRESSURE GAUGE

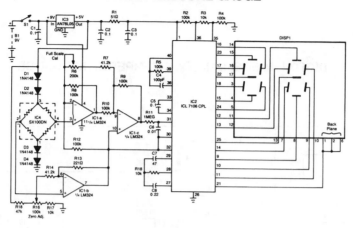

RADIO-ELECTRONICS

Fig. 9-7

Using an intersil ICL 7106 A/D converter chip and an LED display module, this gauge uses a Sensym Corp. pressure transducer SX100pn (100 psi full scale) in a Wheatstone bridge configuration to drive an op amp (IC1a, b, c) translator circuit that supplies a dc voltage to IC2 that is proportional to pressure. R6 sets the gain of IC1A (full-scale sensitivity) and R16 supplies a zero adjustment. IC3 provides regulated +5 V to power the circuit.

VOLTAGE GAUGE

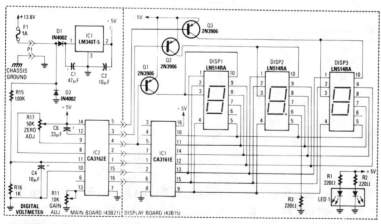

RADIO-ELECTRONICS

Fig. 9-8

This circuit uses an RCA CA3162E (IC2) A/D converter. This converter has 12-bit output (BCD) which is sent to display decoder driver IC1. +5 V is obtained from IC1. R17 adjusts to 0 and R11 should be set to produce correct calibration of gauge unit. Displays are common cathode types. No limiting resistors are necessary because the output drivers are constant current. R15 and R16 sample the applied voltage (usually 8 to 18 V). LED1 is used to illuminate the gauge legend (Volts, Temp, etc.).

DIGITAL MISCELLANEOUS TEMPERATURE GAUGE

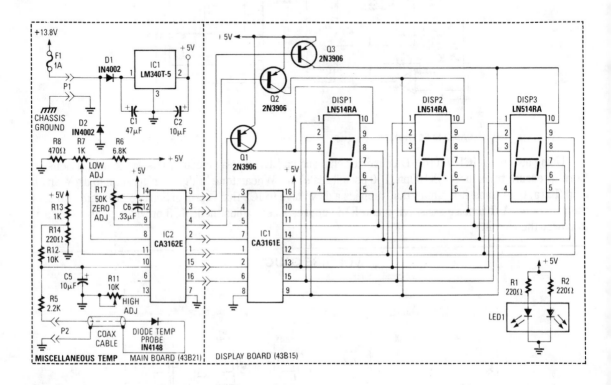

RADIO-ELECTRONICS

Fig. 9-9

A diode (IN4148) is used as a temperature sensor. IC2 is an A/D converter with BCD output. A reference voltage set by R7 is applied to the positive input of IC2. As the temperature increases, the voltage across the temperature sensor resistance decreases. This increases the differential input voltage across pins 10 and 11 of IC2. R3 adjusts low calibration. R17 zeros the A/D converter and R7 adjusts high calibration.

10

Automotive Security Circuits

The sources of the following circuits are contained in the Sources section, which begins on page 662. The figure number in the box of each circuit correlates to the entry in the Sources section.

AUTOMATIC ARMING AUTO ALARM

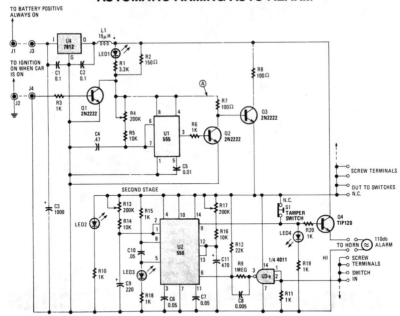

POPULAR ELECTRONICS

Fig. 10-1

The circuit automatically turns on when the car is turned off. It gives you a variable time to get out and lock up, and also provides a variable time delay to get in and start the car.

The 555 oscillator/timers are always powered down when the car is on. That keeps the alarm from going off while you're driving. As soon as the car is turned off, Q2 switches off and shunts power to U1. When that happens, U1 immediately sends its output high, keeping Q3 on, and thereby prevents power from returning to U2.

Transistor Q2 also sends power to Q3's collector to be used only when U1 has completed its timing cycle. When U1 has finished, it turns Q3 off, which in turn activates Q4, and sends power to the balance of the circuit. That timing period was the time needed to get out of the car. LED1 indicates that the system is disarmed and LED2 indicates that the system is armed.

At this point, U2 waits for a trigger pulse from the car's door switches or dome light. A positive impulse at the 4011's input sends a negative trigger pulse to the first stages of U2, which is connected as a cascading timer. The first stage's output becomes high for a time to allow the car to be turned on.

If that does not happen, the first stage's output lowers, which sends a low trigger pulse to the second stage. The second stage then sends its output high, turning on Q5, which sounds the alarm for a given time. Once that time has elapsed, the alarm is shut off by a low output to Q5 and the system is reset. If the car door is closed or a second door opened while the alarm is sounding, the first stage retriggers and prepares to extend the ON-time of the alarm.

The cascading or counting action continues until the car is left alone. You can add a switch on the positive supply rail at J3 to override and silence the alarm, if (for example) you plan to work on the car. Switch S1 is a normally closed type that is built into the case of the alarm; S1 is pushed to the open position when the case is mounted flush with a surface. Any attempt to remove the alarm will sound the alarm.

BACKUP BEEPER

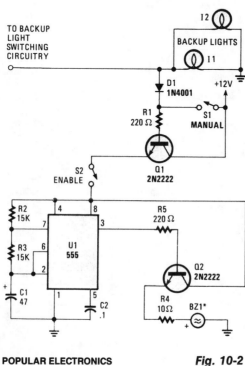

POPULAR ELECTRONICS   *Fig. 10-2*

When the vehicle's backup lights kick on, or when the manual switch (S1) is closed, a small current is fed to the base of Q1. Transistor Q1 allows current to flow through it and, if the enable switch (S2) is closed, it sends 12 V to U1, a 555 timer. Timer U1 sends high pulses that last 0.977 13 s and low signals that last 0.488 565 s to the base of Q2. When U1 switches Q2 on, it sends 12 V to BZ1, a piezoelectric buzzer. For best results, the buzzer should be mounted under the vehicle—somewhere where people around the car can hear the warning beeps.

AUTO TURN-OFF ALARM WITH 8-MINUTE DELAY

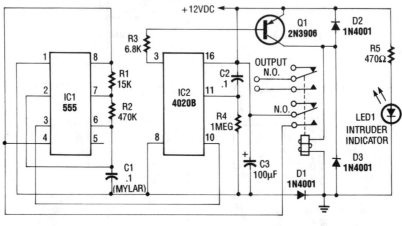

RADIO-ELECTRONICS

Fig. 10-3

This circuit uses a NE555 timer and CD4020B. When +12 Vdc is applied to the circuit, the output of IC2 is set low via C2, which turns on the relay, and IC1, a pulse generator. IC1 pulses counter IC2. After 8192 clocks, IC2 output (pin 3) goes high, cuts off Q2, and completes the cycle.

AUTO ALARM

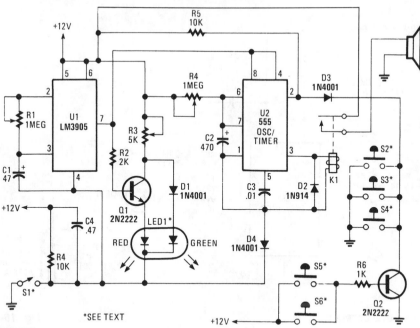

POPULAR ELECTRONICS

*SEE TEXT

Fig. 10-4

S1 is external key switch. The alarm allows a 0- to 45-s delay after S1 is operated before the circuit is armed. During this period, LED1 lights green. After this delay, LED1 lights red, which indicates that the circuit is armed. Then, sensors S2 through S4 -(NO) or S5 through S6 (NO) pull pin 2 of U2 low, which activates K1 and sounds the alarm. The alarm sounds for a duration determined by R4 and C2. After this time, K1 releases and the circuit is again ready. Manual reset is via the key switch, S1.

AUTO IGNITION CUT-OFF

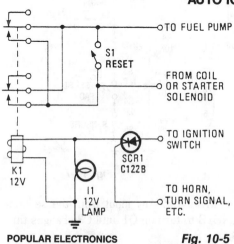

Using an SCR/relay combination, this circuit can be made to cut off ignition, unless a positive voltage is applied to the gate of the SCR. This is useful as an anti-theft device, because depending on hook-up, the car will not start unless a certain accessory or a hidden switch is closed.

POPULAR ELECTRONICS

Fig. 10-5

CAR ALARM WITH HORN AS LOUDSPEAKER

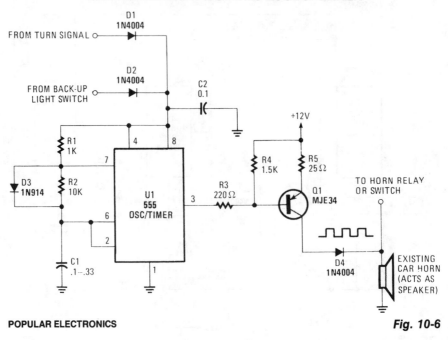

POPULAR ELECTRONICS

Fig. 10-6

An auto horn will work as a speaker of limited audio-frequency range. This circuit uses a 555 timer as an oscillator to drive an MJE34 transistor, which in turn drives the horn. Normal horn operation is ensured by blocking diode D4.

AUTOMATIC TURN-OFF ALARM WITH DELAY

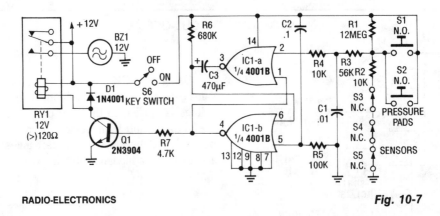

RADIO-ELECTRONICS

Fig. 10-7

In this circuit, IC1A and IC1B act as a monostable multivibrator. Any input from the sensors S1 through S5 forces IC1A to produce logic low, which causes IC1B to turn on Q1 until C3 changes through R6. This action resets the latch formed by IC1A and IC1B.

SINGLE-IC ALARM

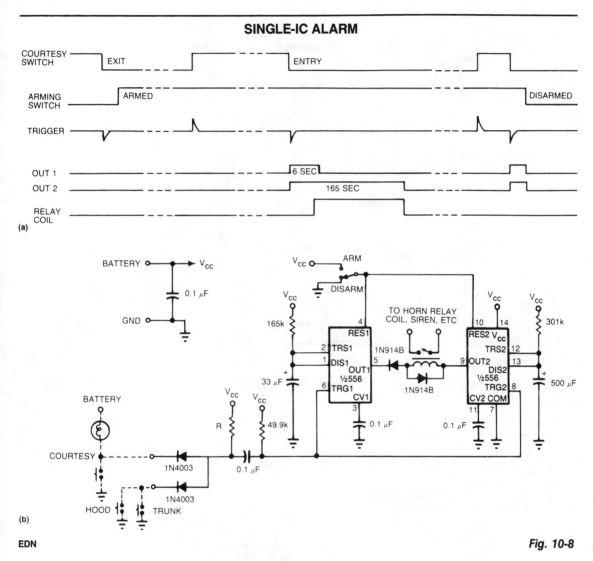

(a)

(b)

EDN

Fig. 10-8

With a single IC, you can build a simple, reliable auto burglar alarm or a similar alarm. See (a) for the timing information for the alarm circuit in (b).

When you leave your vehicle, flip the arming switch and close the door behind you to arm the device. Subsequent opening of an entrance triggers both timers. After the expiration of the entry delay timer, the alarm sounds for a time that is determined by the second timer.

The value of R should be less than 1 kΩ. If you use an incandescent lamp instead of a resistor, you get an extra function—an open-entrance indicator. By keeping the resistance low, you avoid false tripping should water collect under the hood.

If your door switch connects the courtesy light to 12 V rather than to ground, use a single transistor as an inverter at the input. Although this circuit's simplicity has its drawbacks, you can add more features, such as no-entry delays for the hood and trunk, and retripping when doors remain open.

LOW-CURRENT SIMPLE CMOS ALARM

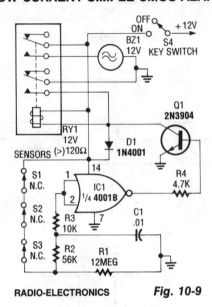

RADIO-ELECTRONICS

Fig. 10-9

This CMOS-aided alarm draws only 1 μA standby current. An open sensor allows IC1 to bias Q1 on, activating RY1.

BACK-UP ALARM

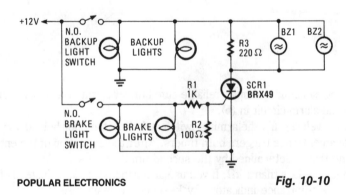

POPULAR ELECTRONICS

Fig. 10-10

The brake lights of the automobile trigger this circuit on and off. This saves the annoyance of the alarm when it is not needed.

11

Automotive Light Circuits

The sources of the following circuits are contained in the Sources section, which begins on page 662. The figure number in the box of each circuit correlates to the entry in the Sources section.

Car Lights Monitor
Rear Fog Light Controller with Delay
Interior Convenience Light
Third Brake Light
Lights-On Warning
Car Headlight Control
Night Safety Light for Autos
Lights-On Reminder for Autos
Automobile Lights-On Reminder

CAR LIGHTS MONITOR

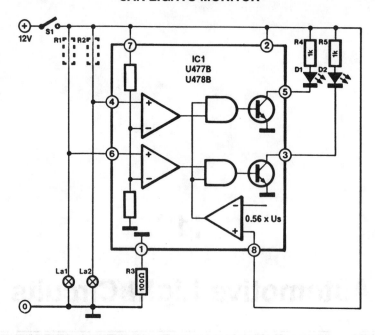

ELEKTOR ELECTRONICS

Fig. 11-1

This circuit is for the purpose of monitoring automotive lighting. Two special ICs are available from Telefunken that are designed to measure the current through a light bulb. In practice, detecting whether a current flows through a bulb or not is a most suitable way of determining whether the bulb still works.

If a small resistance is connected in series with a bulb, a small voltage drop will develop across it when the bulb lights (R1 and R2 in the diagram). Each IC can cope with only two bulbs, so three or four ICs are needed per car. The junction of the bulb and resistor is connected to one of the inputs (pin 4 or pin 6) of the IC. The potential across the resistor is compared in the IC with an internal reference voltage. Depending on which of the two ICs is used, the voltage drop must be about 16 mV (U477B) or 100 mV (U478B). This voltage drop is so small that it will not affect the brightness of the relevant bulb.

The value of the series resistor is determined quite easily. If, for instance, it is in series with the brake light (normally 21 W), the current through the bulb, assuming that the vehicle has a 12-V battery, is $21 \div 12 = 1.75$ A. The resistance must then be of $16 \div 1.75 = 9$ mΩ (U477B) or $100 \div 1.75 = 57$ mΩ (U478B).

These resistors can be made from a length of resistance wire (available from most electrical retailers). Failing that, standard circuit wire of 0.7 mm diameter can be used. This has a specific resistance of about 100 mΩ per meter. However, in most cars, the existing wiring will have sufficient resistance to serve as series resistor.

LEDs can be connected to the outputs of the IC (pins 3 and 5). These will only light if the relevant car light fails to work properly.

REAR FOG LIGHT CONTROLLER WITH DELAY

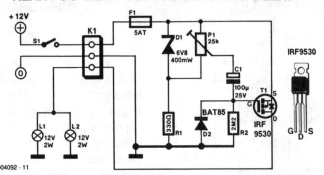

ELEKTOR ELECTRONICS   904092 - 11

Fig. 11-2

We assume that most of our readers are thoughtful drivers who do not switch on their rear fog lights when closely followed by other traffic. Following drivers (for an instant) will think that you are braking (although they have seen no reason for your doing so) and thus slam on their brakes as well. This could cause a very dangerous situation. Avoid a potentially dangerous action and install the rear fog light delay circuit, presented here.

Switch S1 is the on/off control for rear fog lights L1 and L2. As soon as this switch is closed, the gate-source voltage (Vgs) of MOSFET T1 will become more and more negative. Thus, the IC will conduct harder and harder. This in turn causes the brightness of the lights to gradually become brighter. Maximum brightness is reached after a delay of about 20 seconds, which is determined by time constant R2/C1.

The gate of T1 can be given a bias by preset P1. This provides compensation for the initial period after the lights are switched on and the lamps do not light, because they need hundreds of milliamperes before they can do so. With P1 set correctly, the lamps will light, albeit weakly, and immediately the control switch is closed. The gate potential is then equal to the voltage at the wiper of P1 (remember that C1 is then still discharged).

Although the dissipation of T1 is a maximum during the transitional period (between switch on and the lamps lighting brightly), the heatsink required is calculated on the basis of the dissipation when the lamps light brightly.

INTERIOR CONVENIENCE LIGHT

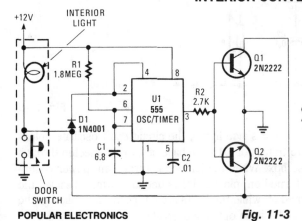

POPULAR ELECTRONICS

Fig. 11-3

When the door is closed, this circuit keeps the dome light on for a period determined by R1/C1. The time is approximately $(1.1)R_1 \times C_1$.

THIRD BRAKE LIGHT

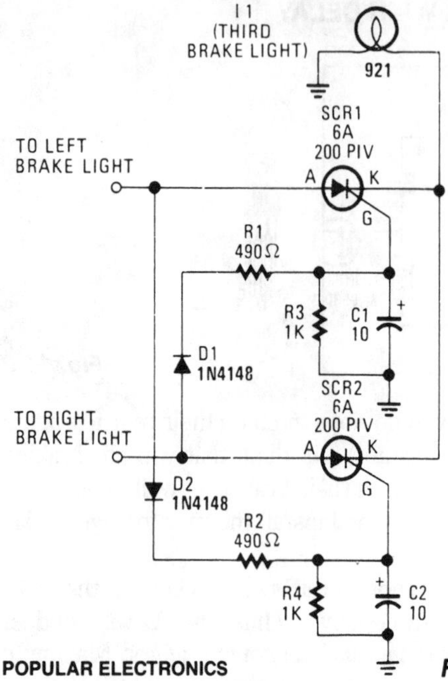

The circuit is designed to light (via SCR1 and SCR2) only when both the left and right brake lights are activated. The circuit operates on 12-V negative-ground systems. When the brake pedal is depressed, 12 V is applied to the left and right brake lights. The gates of the two SCRs are triggered and current flows through the SCRs, which turns on the third brake light.

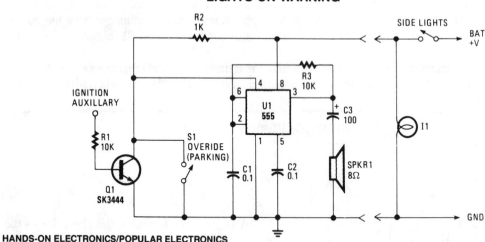

POPULAR ELECTRONICS

Fig. 11-4

LIGHTS-ON WARNING

HANDS-ON ELECTRONICS/POPULAR ELECTRONICS

Fig. 11-5

Because power for the circuit is obtained from the car's side lights, the circuit can't oscillate unless the lights are on. The reset pin on the 555 connects to transistor Q1. The base of Q1 is connected through R1 to the ignition auxiliary terminal on the car's fuse box. When the ignition is turned on, power is supplied to the base of Q1, which turns it on. With Q1 turned on, pin 4 of U1 is tied low, which disables the oscillator and inhibits the alarm. If the ignition is turned off while the lights are on, power is applied to the 555 and Q1 is turned off, and the alarm starts. Switch S1 is an optional override.

CAR HEADLIGHT CONTROL

It is annoying to realize that you have left your car headlights on only to find that the battery is flat. One possible way to prevent this is with the present control.

The circuit does not provide a warning, but an action: when you switch off the ignition, relay Re1 is de-energized and the headlights are switched off—unless you deliberately decide otherwise. That decision is made possible by switch S1, which, when operated, triggers silicon-controlled rectifier Th1 so that Re1 is energized. Notice that this is possible only when the ignition switch, S2, is off. Otherwise, the voltage across Th1 is so low, owing to shunt diode D1, that it cannot be triggered. However, the headlights should not normally be switched on when the ignition is off; in most cases S1 will be used only rarely and the switch can then be omitted altogether. The relay should be a standard 12-V car type with contacts that can switch up to 25 A.

ELEKTOR ELECTRONICS   **Fig. 11-6**

NIGHT SAFETY LIGHT FOR AUTOS

POPULAR ELECTRONICS   **Fig. 11-7**

This circuit turns on the brake lights of a parked car when the headlights of an oncoming car are detected, warning the driver of the oncoming car about the parked vehicle. LDR4 is the sensor. LDR1 disables the circuit by causing U1 gate input to be pulled high during daylight hours, causing pin 2 of U1a to become low, disabling it and the circuit.

LIGHTS-ON REMINDER FOR AUTOS

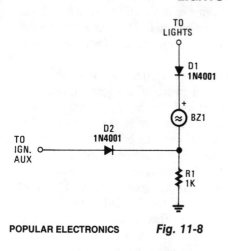

When ignition is off, BZ1 will sound if the headlights are on. With the ignition on, BZ1 receives no voltage.

POPULAR ELECTRONICS   *Fig. 11-8*

AUTOMOBILE LIGHTS-ON REMINDER

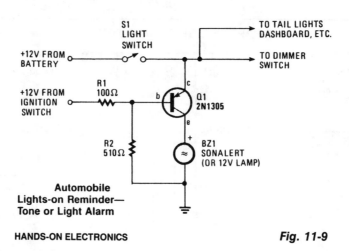

Automobile
Lights-on Reminder—
Tone or Light Alarm

HANDS-ON ELECTRONICS   *Fig. 11-9*

The circuit can be used to give a visible or an audible warning that the headlights are on. It uses a 2N1305 transistor as a switch to turn on a Sonalert tone generator or a small 12-V lamp. Operating current for the transistor is supplied from the wire that feeds the headlights. When the ignition is on, the transistor is biased off and the alarm is not activated. Turning off the ignition while the lights are on sets off the alarm.

12

Automotive Electronic Circuits

The sources of the following circuits are contained in the Sources section, which begins on page 662. The figure number in the box of each circuit correlates to the entry in the Sources section.

Solid-State Windshield-Wiper Delay
Electronic Ignition
High-Power Car Audio Amplifier
Windshield Wiper Interval Controller
Voltage Regulator for Cars and Motorcycles

SOLID-STATE WINDSHIELD-WIPER DELAY

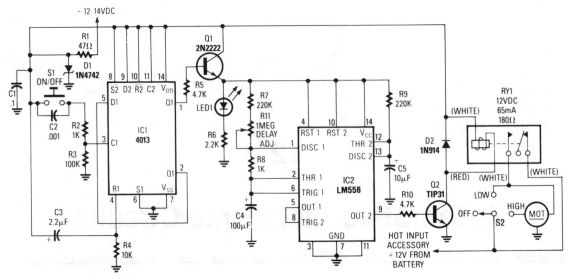

RADIO-ELECTRONICS

Fig. 12-1

In the wiper-delay schematic shown, with the ignition on, D1 maintains regulated +12 Vdc. When S1 closes, C1 bypasses transients and passes this +12 Vdc to divider R2-R3, producing a TTL high at pin 3 of IC1, a 4013 CMOS dual leading-edge triggered D-type flip-flop. Filter R4/C3 keeps IC1 from triggering erroneously when the ignition is on. When S1 is pressed, output Q1 (pin 1) of IC1 latches high, turning on Q1, which conducts via R5, turning on IC2; LED1 indicates power, and R6 sets the current. Because IC2 depends on Q1 for power, IC2 stays off until Q1 turns on.

The left half of IC2 is an astable, with its delay set by R7, R8, R11, and C4. The right half of IC2 is a monostable, with its pulse duration set by R9 and C5. With the values used, you might expect R11 to vary the delay from about 15 to 84 seconds, with a 2.42-second monostable pulse operating the wiper blades on each cycle. However, the actual delay will range between 2 to 18 s, with a 1-s monostable pulse on each cycle. That discrepancy stems from the fact that IC2 is being fed from the emitter of Q1, rather than directly from the regulated +12-V supply. Transistor Q1 acts as an active current source, charges and discharges C4 faster than it ordinarily would.

The astable output (OUT1, pin 5) is tied to TRIG2 (pin 8). When OUT2 (pin 9) becomes high, Q2 is biased via R10 and current flows through RY1, with D2 dissipating back-emf when RY1 shuts off.

ELECTRONIC IGNITION

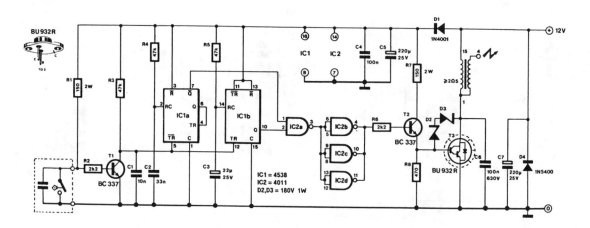

ELEKTOR ELECTRONICS

Fig. 12-2

This electronic ignition circuit is intended to be inserted into a car's conventional ignition system. In effect, it replaces the original 12-V switching circuit in the primary winding of the coil by one generating more than 100 V. It thereby converts a current circuit, which is upset by lead and stray resistance, into a voltage circuit that is much more efficient.

The pulses emanating from the contact breaker, shown at the extreme lower left-hand side of the diagram, are applied to transistor T1 and subsequently differentiated by R3/C1. This causes a negligible ignition delay. The current through the contact-breaker points is determined by the value of R1. This value has been chosen to ensure that the points remain clean.

Transistor T1 is followed by two monostables, IC1A and IC1B, which are both triggered by the output pulses of T1. However, whereas IC1A is triggered by the trailing edge, IC1B is triggered by the leading edge.

Monostable IC1A passes a pulse of about 1.5 ms (determined by R4/C2) to NAND gate IC2A. This gate switches off high-voltage Darlington T3 via gates IC2B, IC2C and IC2D, and driver T2, for the duration of the pulse. Gate IC2 ensures that T3 is switched on only when the engine is running, to prevent a current of some amperes flowing through the ignition coil.

As long as pulses emanate from the contact breaker, IC1B is triggered and its Q output remains logic high. The mono time of this stage is about 1 s and is determined by R5/C3.

Darlington T3 is switched on via T2 and IC2A through IC2D as long as IC1A does not pass an ignition pulse. When the engine is not running, the Q output of IC2B goes low after 1 s and this causes T2 and T3 to be switched off. The two series-connected 180-V zener diodes protect the collector of the BU932R against too high of a voltage. The Darlington must be fitted on a suitable heatsink.

HIGH-POWER CAR AUDIO AMPLIFIER

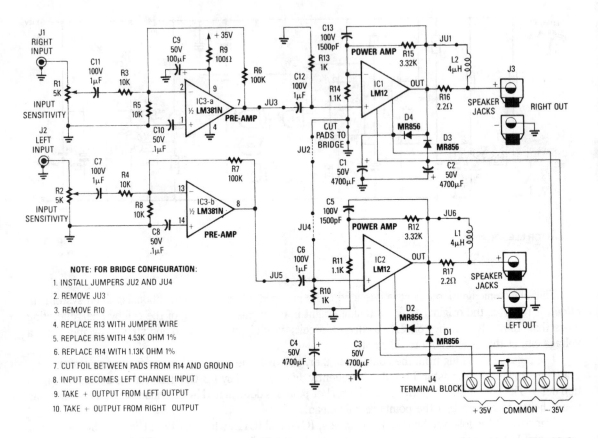

NOTE: FOR BRIDGE CONFIGURATION:

1. INSTALL JUMPERS JU2 AND JU4
2. REMOVE JU3
3. REMOVE R10
4. REPLACE R13 WITH JUMPER WIRE
5. REPLACE R15 WITH 4.53K OHM 1%
6. REPLACE R14 WITH 1.13K OHM 1%
7. CUT FOIL BETWEEN PADS FROM R14 AND GROUND
8. INPUT BECOMES LEFT CHANNEL INPUT
9. TAKE + OUTPUT FROM LEFT OUTPUT
10. TAKE + OUTPUT FROM RIGHT OUTPUT

RADIO-ELECTRONICS

Fig. 12-3

This stereo amp will supply 60 W rms into 8 Ω or 100 W rms into 4 Ω. Notice that the LM12C1 line level (about 300 mV) into 5 KΩ. A ±35-V supply is required for full power output. Power can be obtained from a dc-dc converter.

WINDSHIELD WIPER INTERVAL CONTROLLER

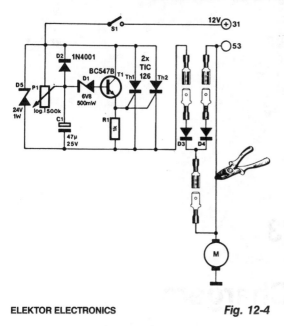

ELEKTOR ELECTRONICS

Fig. 12-4

The windshield wiper interval circuit presented here is very compact and is noteworthy for its use of two thyristors, instead of a relay. It has only two connections and operates without any problems—even in conjunction with multistage wiper circuits.

The connecting wire between the wiper motor and terminal 53 is cut and new connections are made (as shown in the diagram). When the interval switch, S1, is closed, capacitor C1 charges via P1 and the wiper motor. After a time set with P1, transistor T1 switches on and triggers the thyristors. The wiper motor is then energized via the thyristors and D3 and sets the wipers into motion. At the same time, C1 discharges via D2 and the thyristors.

After a short time, the wiper stop switch connects terminal 53 to the +12-V line so that the wiper motor is energized via D4. The thyristors are switched off because the voltage drop across D3 plus Th1/Th2 is then greater than that across D4. When the wipers reach the end of their travel again, the stop switch connects terminal 53 to ground and enables C1 to charge again.

VOLTAGE REGULATOR FOR CARS AND MOTORCYCLES

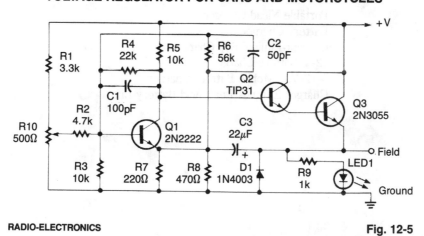

RADIO-ELECTRONICS

Fig. 12-5

This regulator circuit can be used on an alternator that has one field terminal grounded. When +V (input) gets too high, Q1 conducts, and the base of Q2 is driven toward ground, reducing the voltage fed to Q3. This lowers the voltage fed to the field of the alternator.

13

Battery Chargers

The sources of the following circuits are contained in the Sources section, which begins on page 662. The figure number in the box of each circuit correlates to the entry in the Sources section.

PORTABLE NICAD CHARGER

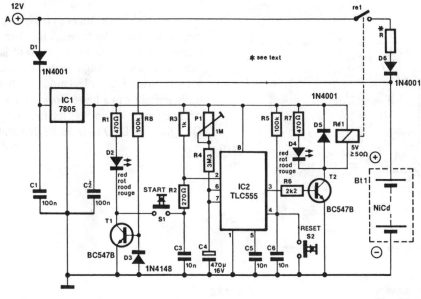

ELEKTOR ELECTRONICS

Fig. 13-1

The portable charger is intended primarily to give model enthusiasts the opportunity of charging their Nicad batteries from a car battery out in the open. The supply voltage for the circuit is regulated by IC1.

When the circuit is connected to the car battery, D2 lights only if the Nicad to be charged has been connected with correct polarity. For that purpose, the + terminal of the Nicad battery is connected to the base of T1 via R8. Because even a discharged battery provides some voltage, T1 is switched on and D2 lights.

Only if the polarity is correct will the pressing of the start switch, S1, have any effect. If so, the collector voltage of T1 is virtually zero so that monostable IC2 is triggered by S1. The output, pin 3, of this CMOS timer then becomes high, T2 is switched on and relay Re1 is energized. Charging of the Nicad battery, via R5 and D6, then begins and charging indicator D4 lights. During the charging, C4 is charged slowly via P1 and R4. The value of these components determines the mono time of IC2 and thus the charging period of the Nicad battery. With values as shown in the diagram, that period can be set with P1 to between 26 and 33 min. Notice that this time is affected by the leakage current of C4; use a good-quality capacitor here. The charging can be interrupted with reset switch S2.

The charging current through the Nicad battery is determined by the value of R, which can be calculated:

$$R = \left\{ \frac{12 - (0.7 + 1.3x \text{ no. of cells})}{I_c} \right\} [\Omega]$$

I_c is the charging current, which is here because the chosen charging period is twice the nominal value of the capacity of the Nicad battery. Resistor R must be able to dissipate a power of $I_c^2 R$ W. Finally, make sure that the Nicad battery is suitable for fast charging; never charge for longer than half an hour!

BATTERY CHARGER

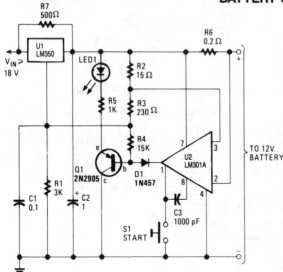

This high-performance charger quickly charges gelled lead-acid batteries, and turns off at full charge. At first, the charge current is held at 2 A, but as battery voltage rises, current decreases. When current falls to 150 mA, the charger automatically switches to a lower float voltage to keep from overcharging. When you hit full charge, transistor Q1 lights the LED to indicate that status.

HANDS-ON ELECTRONICS   *Fig. 13-2*

CURRENT-LIMITED 6-V CHARGER

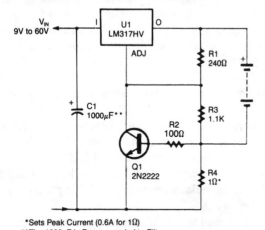

*Sets Peak Current (0.6A for 1Ω)
**The 1000μF is Recommended to Filter
  Out Input Transients.

POPULAR ELECTRONICS   *Fig. 13-3*

An LM317HV regulator is used as a current-limited charger. If current through R4 exceeds 0.6 A, Q1 is biased on, which pulls the ADJ terminal of the LM317 HV to ground and reduces the battery-charging current.

12-V BATTERY CHARGER

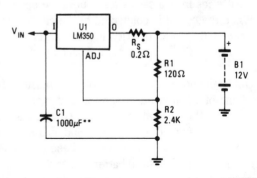

*R_S—SETS OUTPUT IMPEDANCE OF CHARGER $Z_{OUT} = R_S \left(1 + \frac{R2}{R1}\right)$
USE OF R_S ALLOWS LOW CHARGING RATES WITH
FULLY CHARGED BATTERY.

**THE 1000μF IS RECOMMENDED TO FILTER OUT
  INPUT TRANSIENTS.

POPULAR ELECTRONICS   *Fig. 13-4*

This simple charger uses an LM350 regulator as a battery charger.

+12-Vdc MOBILE BATTERY CHARGER

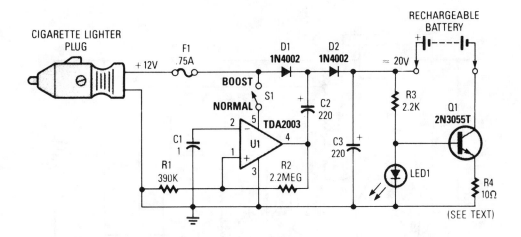

Cell Size	Amp/hr Rate	R4 value (14 hr rate)
N	150 mA	120 ohms @ .25-watt
AA	500 mA	47 ohms @ .5 watt
C	1500 mA	12 ohms @ .5 watt
D	1500 mA	12 ohms @ .5 watt
D	4000 mA	3.3 ohms @ 2 watt
(High capacity)		

POPULAR ELECTRONICS   *Fig. 13-5*

This circuit provides up to 20 V output from a 12-V automotive supply, to enable constant current charging of Nicad battery assemblies up to about 18 V total. V1 forms a square-wave oscillator, D1 and D2, coupling this square wave to the 12-V battery supply to obtain over 20 Vdc. If this is not needed, S1 is left open. Q1 forms a current regulator to determine the charging rate of the rechargeable battery. R4 is selected from the table or it can be switched with a rotary selector switch.

CHARGER EXTENDS LEAD-ACID BATTERY LIFE

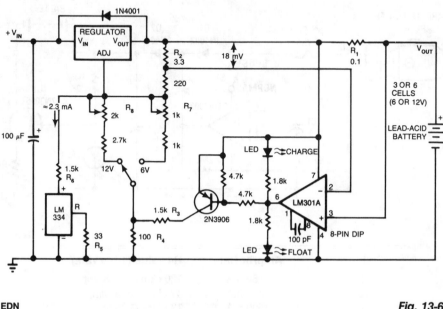

EDN

Fig. 13-6

The circuit furnishes an initial charging voltage of 2.5 V per cell at 25°C to rapidly charge a battery. The charging current decreases as the battery charges, and when the current drops to 180 mA, the charging circuit reduces the output voltage to 2.35 V per cell, floating the battery in a fully charged state. This lower voltage prevents the battery from overcharging, which would shorten its life. The LM301A compares the voltage drop across R1 with an 18-mV reference set by R2. The comparator's output controls the voltage regulator, forcing it to produce the lower float voltage when the battery-charging current passing through R1 drops below 180 mA. the 150-mV difference between the charge and float voltages is set by the ratio of R3 to R4. The LEDs show the state of the circuit.

14

Battery Monitors

The sources of the following circuits are contained in the Sources section, which begins on page 662. The figure number in the box of each circuit correlates to the entry in the Sources section.

INTERNAL RESISTANCE BATTERY TESTER

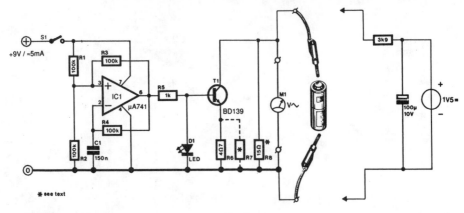

ELEKTOR ELECTRONICS

Fig. 14-1

A designer often needs to know the value of the internal resistance of a battery. Quite a few testers give a relative indication of the value, but this is seldom in ohms. The present tester can, in principle, provide that information.

The basic idea behind it is to load the battery with a varying current, so as to cause an alternating-voltage drop across the internal resistance that can be measured at the battery terminals. Provided that current variations are regular and constant, the voltage drop is directly proportional to the internal resistance.

Choose the variation of the current carefully to read the value of the internal resistance directly on the scale of an ac voltmeter.

The load current is varied with the aid of a current source, T1 in the diagram, which is switched on and off by square-wave generator IC1. The chosen switching frequency of 50 Hz ensures that the ac component at the battery terminals can be measured by a standard ac voltmeter (universal meter).

The battery is loaded constantly by R8, which has a value of 1.5 Ω for 1.5-V batteries, shunted by the ac voltmeter. The indicated voltage times 10 is the value of the internal resistance of the battery. When the battery under test is flat or if the supply battery is flat, no current flows and the meter will read zero. It would then appear as if the battery under test is an ideal type—without internal resistance.

A flat supply battery is indicated if D1 does not light. You can ascertain that the battery under test is flat by measuring the direct voltage across its terminals. The load must be left connected, of course, otherwise the emf is measured and this may well be 1.5 V—even if the battery is flat.

The tester is calibrated with the aid of the auxiliary circuit (shown at the extreme right in the circuit diagram). The 1.5-V supply and electrolytic capacitor form a virtually ideal voltage source, of which the 3.9-Ω resistor forms the internal resistance. With this source connected across the output terminals of the tester, a suitable value should be ascertained for R7. That value is found when the ac voltmeter shows 0.39 V. Notice that this procedure is not the same for all measuring instruments: the alternate use of the digital and a moving coil meter, for instance, is not feasible.

The tester is intended for 1.5-V batteries. The load current is fairly high: about 100 mA through R8 and around 170 mA through T1. For 9-V batteries that current is too high: the current should then be reduced by taking greater values for R6 through R8.

BATTERY-SAVING DISCONNECT SWITCH

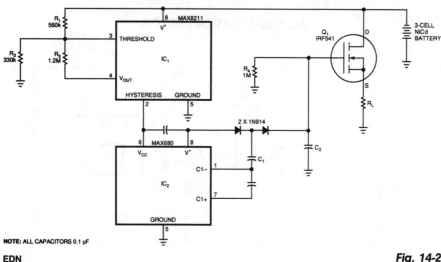

Fig. 14-2

At a predetermined level of declining terminal voltage, the circuit disconnects the battery from the load and halts potentially destructive battery discharge. Q1, a high-side, floating-source MOSFET, acts as the switch. The overall circuit draws about 500 μA when the switch is closed and about 8 μA when it's open.

The values of R_1, R_2, and R_3 set the upper and lower voltage thresholds, V_U and V_L, according to the relationships

$$R_1 = R_2 \left(\frac{V_L}{1.15} \right) - 1$$

and

$$R_3 = 1.15 \left(\frac{R_1}{V_U - V_L} \right)$$

For the circuit to start, $V+$ must exceed V_U. The voltage detector IC1 then powers IC2, but only while $V+$ remains above V_L. Otherwise, IC2 loses its power, removes gate drive from Q1, and turns it off. IC2 is a dual charge-pump inverter that normally converts 5 V to ± 10 V. Capacitors C1, C2, and the two associated diodes form a voltage tripler that generates a gate drive for Q1 that is approximately equal to two times the battery voltage.

With the values in the schematic, the circuit disconnects 3-cell Nicad battery from its load when $V+$ reaches a V_L of 3.1 V. Approximately 0.5 V of hysteresis prevents the switch from turning immediately when the circuit removes the load; $V+$ must first return to V_U, which is 3.6 V. The gate drive declines as the battery voltage declines, cause the ON-resistance of Q1 to reach a maximum of approximately 0.1 Ω, just before $V+$ reaches its 3.1-V threshold. A 300-mA load current at that time will cause a 30-mV drop across the disconnect switch. The drop will be 2 to 3 mV less for higher battery voltages. Resistor R4 ensures that Q1 can adequately turn off by providing a discharge path for C2.

LOW-BATTERY DETECTOR

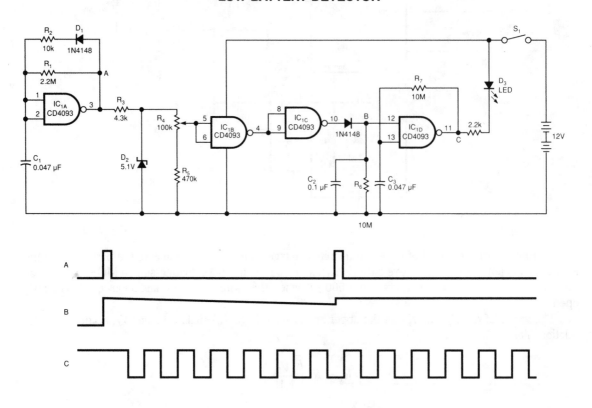

EDN

Fig. 14-3

The battery low-voltage detector uses a CD4093 Schmitt trigger and a capacitor that acts as a 1-bit dynamic RAM. The circuit conserves power by using a periodic test method. IC1A, C1, R1, R2, and D1 generate a narrow, positive pulse at point A.

D2, R4, and R5 regulate and divide the signal at A. Thus, the input of IC1B is independent of the power supply. Because the threshold voltage of the Schmitt trigger depends on the power supply, the threshold voltage will drop if the power-supply voltage drops. When the threshold voltage is lower than the input voltage, IC1B will become low, and IC1C's output will become high.

Capacitor C2 stores the results of the periodic test. The time constant C2 and R6 set is 1 s, and the test period is approximately 0.1 s. When point B is high, which implies that the battery is low, IC1D, C3, and R7 generate a square waveform, which lights D3. You can adjust the detected voltage level by adjusting R4. You can test different battery voltages by changing the voltage level of D2.

BATTERY-TEMPERATURE SENSING NICAD CHARGER

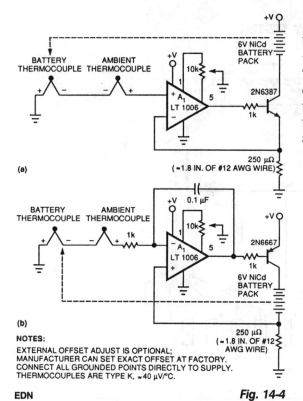

Two simple circuits permit Nicad charging of a battery based on temperature differences between the battery pack and the ambient temperature. This method has the advantage of allowing fast charging because the circuit senses the temperature rise that occurs after charging is complete and the battery under charge is producing heat, not accumulating charge.

NOTES:
EXTERNAL OFFSET ADJUST IS OPTIONAL; MANUFACTURER CAN SET EXACT OFFSET AT FACTORY. CONNECT ALL GROUNDED POINTS DIRECTLY TO SUPPLY. THERMOCOUPLES ARE TYPE K, ≈40 µV/°C.

EDN

Fig. 14-4

BATTERY-VOLTAGE MEASURING REGULATOR

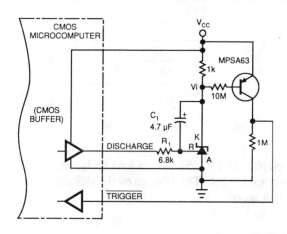

This circuit allows a microprocessor system to measure its own battery voltage. A Texas Instrument TI431 precision shunt regulator acts as a precision reference and integrator/amplifier, measuring its own supply via voltage-dependent charge/discharge time intervals. Notice that you must write a short control and voltage calculation software routine for your system.

EDN

Fig. 14-5

BATTERY TESTER

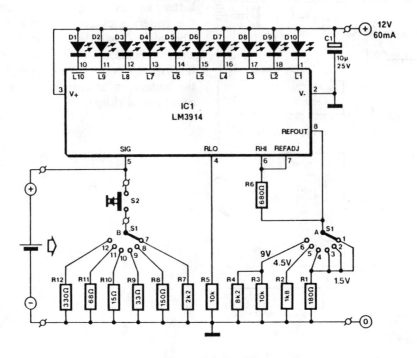

Range	LED									
	D1	D2	D3	D4	D5	D6	D7	D8	D9	D10
	red	orange		green						
1.5 V	0.86	0.96	1.04	1.13	1.21	1.29	1.38	1.46	1.55	1.63
4.5 V	2.58	2.83	3.05	3.31	3.57	3.82	4.07	4.33	4.57	4.82
9.0 V	5.3	5.8	6.3	6.9	7.4	7.9	8.5	9.0	9.5	10.2

ELEKTOR ELECTRONICS

Fig. 14-6

This battery tester makes use of an LM3914 bar-graph driver IC. S1 selects load on battery under test and programs the voltage range. S2 loads the battery under test. The table gives the calibration factors for the tester. LEDs D1 through D10 are used as indicators.

NICAD BATTERY TESTER

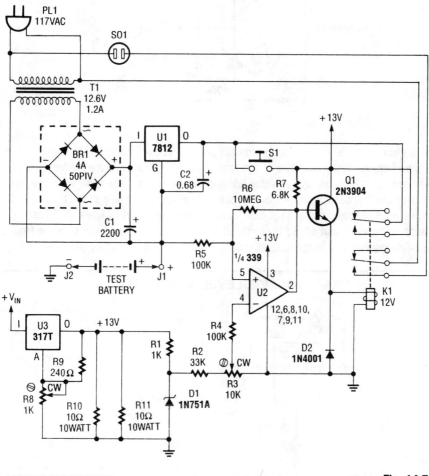

POPULAR ELECTRONICS

Fig. 14-7

This Nicad battery tester discharges the test battery at a rate of 500 mA. When the endpoint of 1 V (determined by setting of R3) is resolved, pin 2 of U2 becomes low, deactivating Q1 and disconnecting the test battery from the circuit. Power for U3 comes from the 12-V regulator in series with the battery being tested. A clock or timer can be plugged into S1 to indicate the time it takes to discharge the battery under test.

LOW-BATTERY INDICATOR

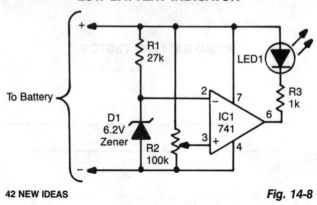

Fig. 14-8

The sensing circuit consists of a 741 op amp set up as a voltage comparator, using a zener diode as a voltage reference. The op amp is inserted as a bridge between two resistance ladders, one which contains the zener reference, and the other a high-value linear potentiometer. When the voltage at the wiper of the potentiometer drops below the voltage set by the zener, the output of the op amp becomes low; that turns on the LED connected between it and V_{CC}. The circuit can be adapted to work with battery-powered circuits that require between 6 and 18 V; the only changes needed would be a lower-voltage zener and a smaller current-limiting resistor in the case of voltage below 9 V, and a larger resistor for higher voltages.

VOLTAGE-LEVEL INDICATOR

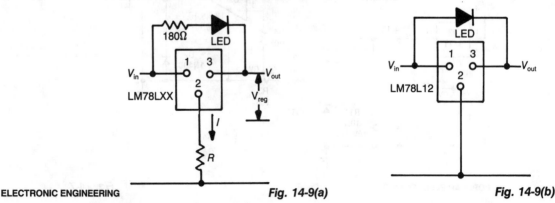

Fig. 14-9(a)   Fig. 14-9(b)

Three-terminal regulator device (LM78LXX) has $V_{out} = V_{in}$ until the input rises 1.5 to 2 V above the output when the regulated voltage $V_{reg} = XX$ is obtained. A differential of 1.5 V between input and output is necessary to light the LED. Thus, the LED lights when V_{in} rises above $V_{reg} + IR + 1.5$ V, where I is typically 6 mA (a zener diode could be used in place of R). For input voltages much higher than necessary to light the LED, a current-limiting resistor in series might be necessary. A useful automotive application is shown in Fig. 14-9(b). The circuit indicates when battery voltage is above 13.5 V which indicates (in conjunction with an ammeter) whether the alternator/regulator/battery system is operating correctly. With the engine off, the battery voltage drops to 12 V and the LED extinguishes. The circuit requires no calibration.

15

Bridge Circuits

The sources of the following circuits are contained in the Sources section, which begins on page 663. The figure number in the box of each circuit correlates to the entry in the Sources section.

Air-Flow Sensing Thermistor Bridge
Bridge Circuit With One Power Supply
Inductance Bridge

AIR-FLOW SENSING THERMISTOR BRIDGE

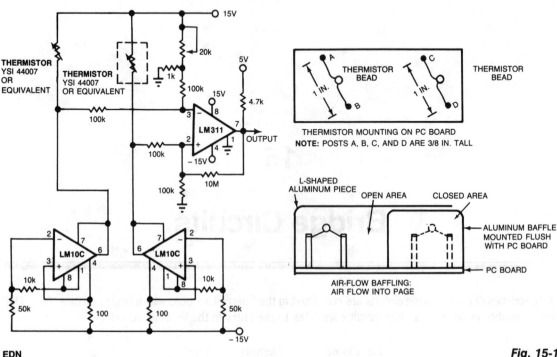

THERMISTOR MOUNTING ON PC BOARD
NOTE: POSTS A, B, C, AND D ARE 3/8 IN. TALL

AIR-FLOW BAFFLING:
AIR FLOW INTO PAGE

EDN

Fig. 15-1

Using the thermistor-bridge circuit, you can detect system-cooling air losses caused by filter or inlet blockage or fan failure. One thermistor is mounted directly in the air flow; the other is baffled. The exposed thermistor senses the temperature in the cooling system; the baffled thermistor senses the ambient temperature in still air. As long as the thermistors are at different temperatures, the bridge stays unbalanced and the circuit produces a logical high, indicating that the cooling system is working. If the air flow stops, the exposed thermistor will reach ambient temperature, the bridge will become balanced, and the circuit will indicate ventilation-system failure by producing a logical low.

The bridge circuit's matched thermistors are biased by matched-current sources. Two LM10C operational amplifiers act as constant-current sources, and an LM311 comparator senses the difference between the voltage drops across the thermistors, producing the logical high when the bridge is unbalanced and the logical low when the bridge is balanced. Use a 20-kΩ potentiometer to set the comparator's threshold; this setting determines the minimum air flow that will cause the circuit to produce a logical high.

BRIDGE CIRCUIT WITH ONE POWER SUPPLY

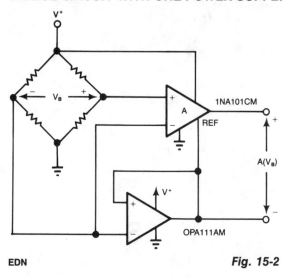

EDN

Fig. 15-2

For systems with only one power supply, two op amps act as instrumentation and buffer amps. The OPA111AM buffers the reference mode of the bridge and applies that voltage to the instrumentation amps REF terminal. Output is taken between the amplifier outputs to exclude the fixed output offset.

The additional op amp creates a bridge error of $I_B \times R/2$, where I_B = bias current of op amp and R is the resistance of one leg of the bridge.

INDUCTANCE BRIDGE

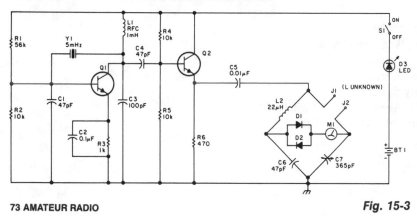

73 AMATEUR RADIO

Fig. 15-3

This bridge will measure inductances from about 1 to 30 μH at a test frequency of 5 mHz. A 365-pF AM-type tuning capacitor is used as a variable element. The circuit should be constructed in a metal enclosure. Calibration can be done on known inductors or by plotting a curve of the capacitance of the 365-pF capacitor versus rotation and calculating the inductance from this. The range of measurement can be charged by using a different frequency crystal and/or variation of L2 and C6.

16

Burglar Alarms

The sources of the following circuits are contained in the Sources section, which begins on page 663. The figure number in the box of each circuit correlates to the entry in the Sources section.

SELF-LATCHING BURGLAR ALARM

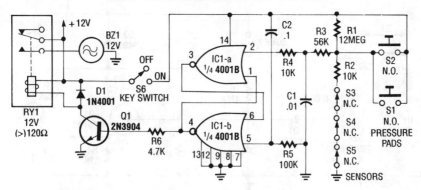

A SIMPLE SELF-LATCHING BURGLAR ALARM.

RADIO-ELECTRONICS

Fig. 16-1

This alarm uses IC1A and IC1B as a latch. When sensors S1 through S5 activate, IC1A turns on and forces IC1B to cut off. Q1 drives RY1.

BURGLAR ALARM WITH TIMED SHUTOFF

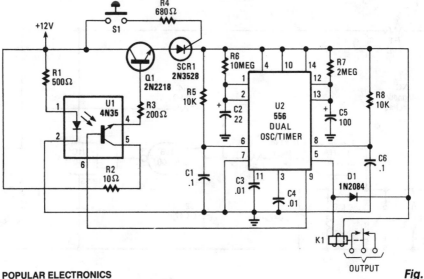

POPULAR ELECTRONICS

Fig. 16-2

When S1 (sensor) is closed, power is applied to U2, a dual timer. After a time determined by C2, C1 is energized after a predetermined time determined by the value of C5, pin 9 of U2 becomes low, switching off the transistor in the optoisolator, cutting anode current of SCR1 and de-energizing K1. The system is now reset. Notice that $(R_6 \times C_2)$ is less than $(R_7 \times C_5)$. The ON time is approximately given by:

$$(R_7 \times C_5) - (R_6 \times C_2) = t_{ON}$$

SIMPLE BURGLAR ALARM

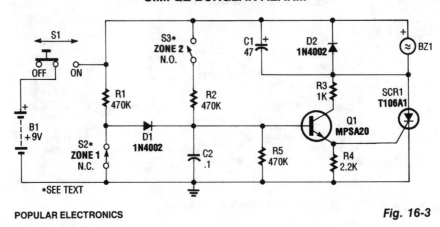

POPULAR ELECTRONICS

Fig. 16-3

A simple circuit using either NO or NC sensors uses an RC delay circuit (R2/C2 or R1/C2) to drive emitter-follower Q1, switching SCR1 and buzzer (or bell) BZ1. S1 is used for activation and reset.

SIMPLE BURGLAR ALARM

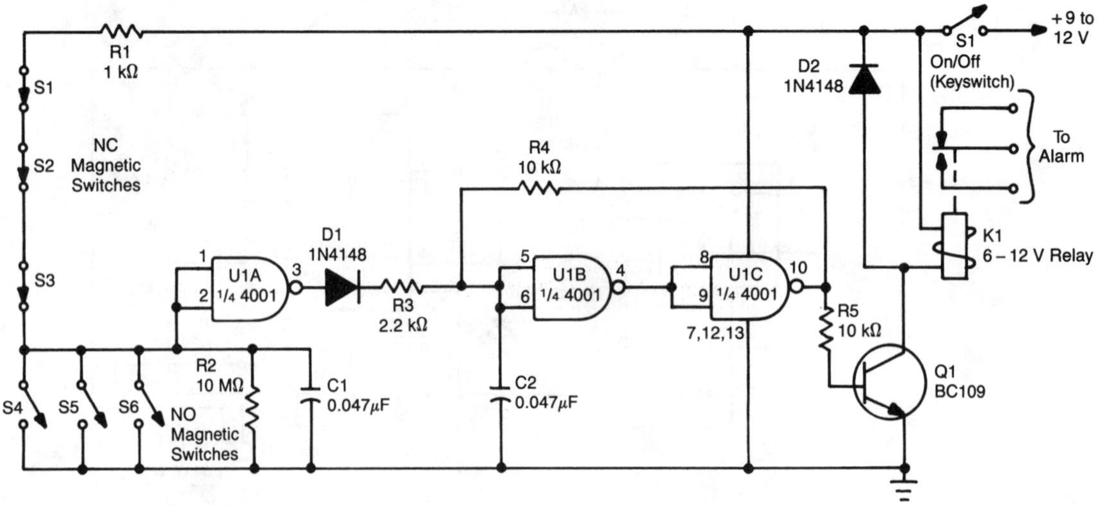

POPULAR ELECTRONICS

Fig. 16-4

Using one IC and a driver transistor, this simple alarm uses either NO or NC sensors. When a sensor operates, the input to U1A goes low, causing U1A to go high, U1B low, and U1C high. This biases Q1 ON and activates relay K1. On/off is via keyswitch S1.

HOME SECURITY SYSTEM

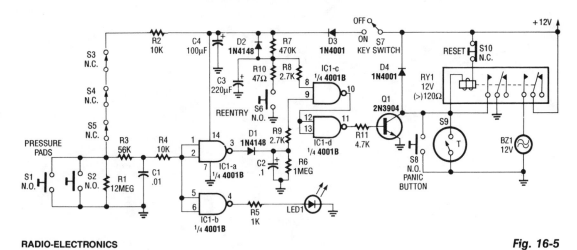

RADIO-ELECTRONICS

Fig. 16-5

This alarm circuit activates when S1 through S5 are activated. This lights LED1 and activates Q1 via IC1C and IC1D. RY1 is wired to self latch. S10 is used to reset. When key switch S1 is activated or when re-entry buttons at S6 are depressed, IC1C is deactivated until RC network R7/C3 charges.

SIMPLE BURGLAR ALARM
WITH NC SWITCHES

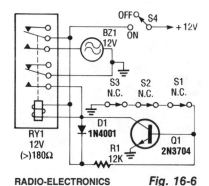

RADIO-ELECTRONICS   Fig. 16-6

This relay draws 1 mA of idling current. Q1 detects open switch and energizes RY1.

BURGLAR ALARM WITH NC
AND NO SWITCHES

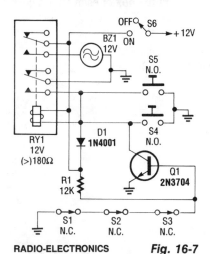

RADIO-ELECTRONICS   Fig. 16-7

This circuit uses both NC and NO sensors. Series NC sensors allow Q1 to activate RY1. NO sensors directly activate RY1.

17

Buffers

The sources of the following circuits are contained in the Sources section, which begins on page 663. The figure number in the box of each circuit correlates to the entry in the Sources section.

Oscillator Buffers
Precision-Increasing Buffer
Inverting Bistable Buffer

OSCILLATOR BUFFERS

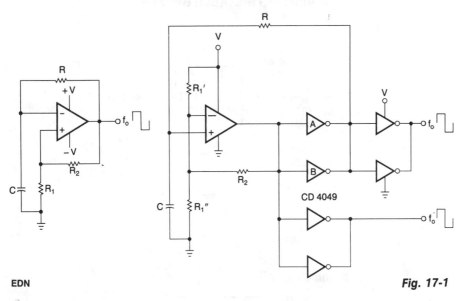

EDN

Fig. 17-1

CMOS buffers added to an op amp oscillator improve performance, largely as a result of nonsymmetry and variability of the op amp's output saturation voltages.

PRECISION-INCREASING BUFFER

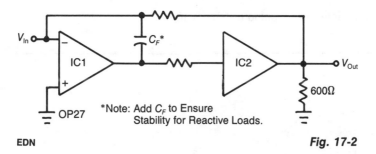

*Note: Add C_F to Ensure Stability for Reactive Loads.

EDN

Fig. 17-2

Adding an unity-gain buffer to your analog circuit can increase its precision. For example, by itself, the op amp IC1 exhibits a maximum dV_{OS}/dT of 1.8 $\mu V/°C$ and can drive a 600-Ω load. Under these conditions, IC1 would dissipate 94 mW incrementally. Thus, the op amp's O_{JA} of 150°C/W would change its V_{OS} by 25 μV.

The buffer, IC2, will isolate IC1 from the load and eliminate the change in power dissipation in IC1, thereby achieving IC1's minimum, rated offset-voltage drive. The loop gain of IC1 essentially eliminates the offset of the buffer. Almost any unity-gain buffer will work, provided that it exhibits a 3-dB bandwidth that is at least 5 times the gain-bandwidth product of the op amp.

INVERTING BISTABLE BUFFER

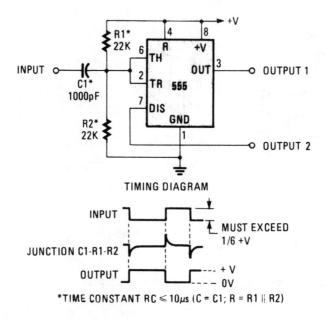

TIMING DIAGRAM

*TIME CONSTANT RC ≤ 10μs (C = C1; R = R1 ‖ R2)

Fig. 17-3

This circuit uses a 555 timer as a flip-flop bistable buffer.

18

Carrier-Current Circuits

The sources of the following circuits are contained in the Sources section, which begins on page 663. The figure number in the box of each circuit correlates to the entry in the Sources section.

Carrier-Current Transmitter for Data Transmission
Carrier-Current Receiver for Data Transmission

CARRIER CURRENT TRANSMITTER FOR DATA TRANSMISSION

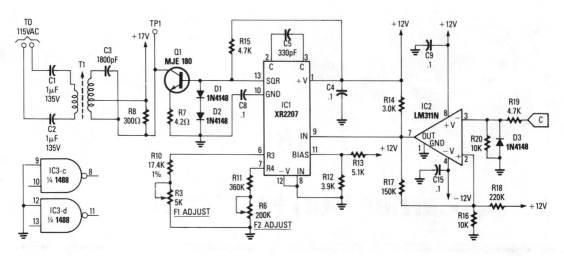

RADIO-ELECTRONICS

Fig. 18-1

$$f_1 = \frac{1}{R_3 C_1}$$

$$f_2 = f_1 + \Delta f_1$$

where:

$$\Delta f_1 = \frac{1}{R_4 C_1}$$

In this circuit, data at input C is amplified by IC2 and then fed to modulator IC1. IC1 generates two frequencies, depending on the values of C_5, R_{10}, R_3, R_{11}, and R_6. The frequency f, is generated if pin 9 IC1 is low and f_2 if pin 9 IC1 is high. A square wave appears at pin 13 of IC1 and is fed to Q1, an amplifier stage, that is coupled via tuned transformer T, to the ac line via C1 and C2. Notice that, for safety reasons, C1 and C2 must be specifically rated for the ac line voltage.

CARRIER-CURRENT RECEIVER FOR DATA TRANSMISSION

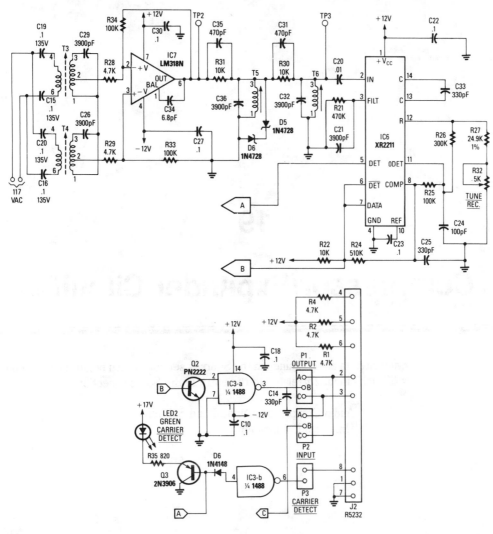

Fig. 18-2

Receive frequency: $f_0 = \dfrac{1}{(R_{27} + R_{32})\, C_{33}}$

This receiver consists of an input network amplifier IC7 FSK PLL detector ICG, and output amplifier/interface Q2, Q3, IC3A and IC3B, a 1488 Quad RS232 line driver of the carrier-current signal. Tuned amplifier IC7 amplifies this signal and drives PLL detector IC8. The values shown in the circuit are suitable for operation in the 100-kHz range. Recovered data at pins 5, 6, 7 is fed to the output amplifier/interface circuit (Fig. 6). This circuit is also used with the carrier-current data transmitter to form a pair.

19

Compressor/Expander Circuits

The sources of the following circuits are contained in the Sources section, which begins on page 663. The figure number in the box of each circuit correlates to the entry in the Sources section.

Audio Compressor/Audio-Band Splitter
Universal Compander

AUDIO COMPRESSOR/AUDIO-BAND SPLITTER

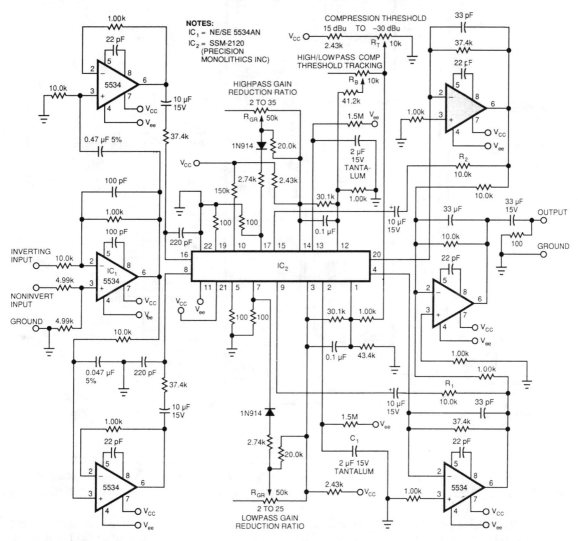

Fig. 19-1

This 2-band compressor splits the audio into high and low frequencies and allows independent adjustment of each. Two active filters drive the two halves of dual voltage controlled amplifier/rectifier IC. Each section has a dynamic range greater than 100 dB. Compression gain slopes are adjustable from 2 to 25 for both audio bands. R_B adjusts the threshold amplitude between the two bands. RK1 and R2/C2 control the compressor attack times (10 kΩ and 2 μF, respectively), while the 1.5-MΩ resistor in the integrator circuit controls the release line.

UNIVERSAL COMPANDER

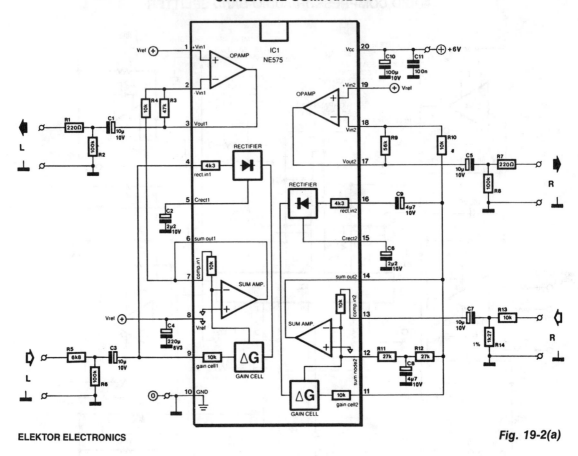

Fig. 19-2(a)

Signetics' type NE575 compander IC is intended primarily for use with battery power supplies of 3 to 7 V (max. 8 V). It draws a current of 3.5 mA at 3 V and 5 mA at 7 V. The compander process (compression at the input, expansion at the output) significantly improves the signal-to-noise ratio in a communications link.

The IC contains two almost identical circuits, of which one (pins 1 to 9) is arranged as an expander. The other (pins 11 to 19) can be used as expander, compressor or automatic load control (ALC), depending on the externally connected circuit. For the compressor function, the inverting output of the internal summing amplifier is brought out to pin 12. This is not the case in the expander section, where a reference voltage is available at pin 8. This pin is interlinked to pins 1 and 19 to enable the setting of the dc operating point of the op amps.

The op amp in the expander section, pins 1 through 3, serves as output buffer in the compressor section, pins 17 through 19 as the input buffer. The IC has a relatively high output sensitivity and is evidently intended for processing small signals (microphone output level). A signal of 100 mV, for instance, is amplified by 1 only. The present circuit caters to larger input signals (line level); its maximum input level is 1.5 Vrms.

UNIVERSAL COMPANDER (*Cont.*)

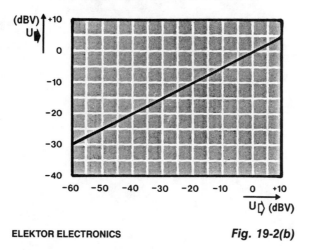

ELEKTOR ELECTRONICS

Fig. 19-2(b)

With a 1-V input into R13, a potential of about 500 mV exists between compressor output R7 and expander input R5. The compression characteristic is shown in Fig. 19-2(b). The signal range is reduced by about one half at the output, which is doubled in the expander. Thus, the range after compression and expansion is the same again, but that is not necessarily the case with the input and output level. The compander can be arranged to provide a constant attenuation or amplification. With the circuit values as shown in the diagram, the input and output levels are the same. The prototype had an overall gain of 0.5 dB when the expander input was connected directly to the compressor output.

To allow acceptance of high input levels, R13, R14, and the compressor input resistance form a 10:1 attenuator. At the expander input, R5 and the expander input impedance of about 3 kΩ form a potential divider. If the compander is to be used with smaller signals, the attenuation can be reduced as appropriate. If the input level lies below 100 mV, R5, R13 and R14 can be omitted.

The compander covers the frequency range of 20 Hz to 20 kHz, the overall distortion is less than 1%, and the signal-to-noise ratio is about 80 dB.

20

Computer-Related Circuits

The sources of the following circuits are contained in the Sources section, which begins on page 663. The figure number in the box of each circuit correlates to the entry in the Sources section.

DUAL 8051s EXECUTE IN LOCK-STEP

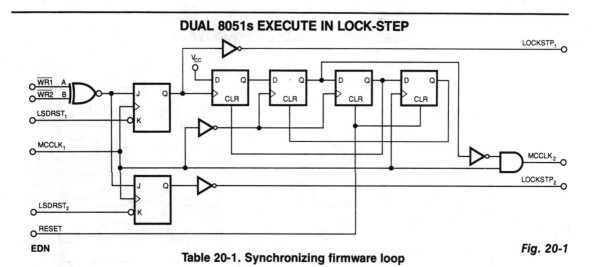

EDN

Fig. 20-1

Table 20-1. Synchronizing firmware loop

SYNCUP:	MOV	P3, #0FFH	;Activate internal port pull-ups.
;	MOV	DPTR, #SAFLOC	;Location for safely writing values so that WR* can be generated.
;	MOV	A, #SAFVAL	;Value to be written into location DPTR is pointing to. Use value that will not cause adverse effects.
; ; ;			
WAIT:	CLR	P1.0	;Toggle the JK flip flop lock-step-reset (K input).
	SETB	P1.0	
	MOVX	@DPTR, A	;Generate WR*.
	JNB	P3.5, WAIT	;If signal is high (LOCKSTP1) or LOCKSTP2) then lock step has been achieved. Otherwise, try again.
; ;			
; ;	Program will now execute synchronously on each microcontroller.		

This hardware-software combination deletes clocks from the slave μP until both μPs synchronize. The firmware loop causes each μP to generate a WR signal once per loop. The circuit exclusive-ORs the two WR signals to produce a miss-compare pulse. The miss-compare pulse latches into the two JK flip-flops via outputs LOCKSTP1 and LOCKSTP2. A high on these signals indicates that the μPs are in lock-step, causing both μPs' programs' execution to exit the firmware loop. If you use discrete components, you'll probably want to use the Q output of the JK flip-flop and delete the circuit's inverters.

The listing uses the μPs' ports 1 and 3. You cannot use a memory-mapped location for the lock-step-detect clear (K input) because this scheme would generate additional WR signals. You could apply this idea to other μPs, perhaps using their RD signals. This way, generating an RD signal to activate the lock-step-detect clear would not affect the synchronization inputs.

RESET PROTECTION FOR COMPUTERS

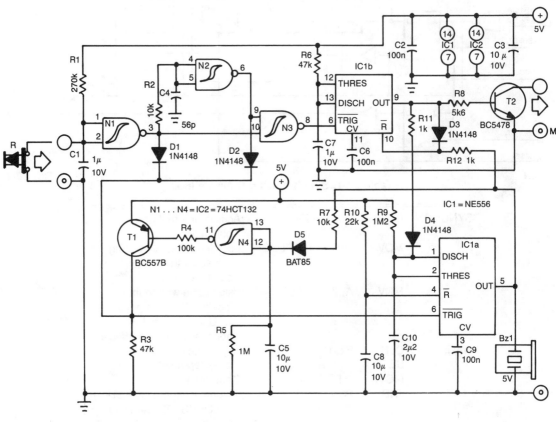

Fig. 20-2

This protection circuit is inserted between the reset switch and the motherboard. The earth connection of the computer must be linked to terminal M of the protection circuit. The protection circuit can draw its power from the computer supply.

When the circuit has been fitted, operation of the reset switch will not immediately restart the computer. Instead, a buzzer will sound to alert you to the reset operation. The buzzer is actuated for 4 s by monostable IC1A, which is triggered by the reset switch. During these 4 s, the output, pin 5, of IC1A ensures that the reset function, pin 10, of IC1B is disabled. When the reset switch is operated again, monostable IC1B will be triggered and this starts the reset procedure. Transistor T2 is then switched on for 0.5 s and the buzzer is deactuated via R11 and D4.

The circuit around T1 and N4 ensures that IC1A can accept trigger pulses again 10 s after the mono time of IC1B has lapsed. This arrangement prevents, for example, children operating the reset switch.

3 μP I/O LINE PROTECTORS

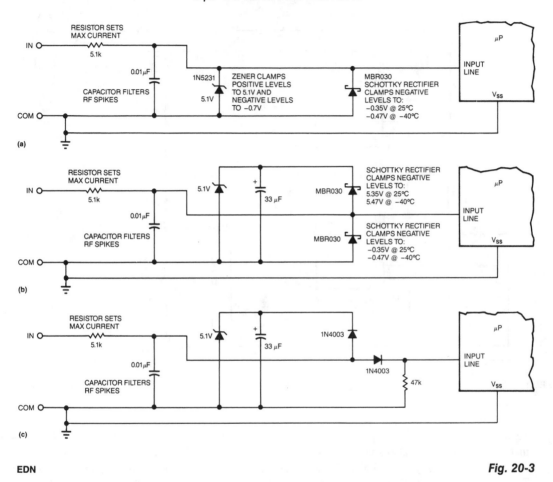

EDN

Fig. 20-3

In Fig. 20-3(a), a 5.1-V zener diode clamps positive-going transients, and a Schottky rectifier clamps negative-going transients. The Schottky rectifier has problems at both ends of the temperature scale. At 125°C (257°F), its leakage current can reach 50 μA when the input line is at 5 V. This leakage is not a big deal unless the input resistor has a value of 100 kΩ or more. More troubling, at temperatures below −40°C (−40°F), the Schottky rectifier's forward voltage rises to about 0.47 V, which is perilously close to the −0.50-V max spec that most HCMOS-type μP's inputs can tolerate.

The third circuit, Fig. 20-3(c), uses two regular silicon rectifiers. One rectifier is connected in series with the input line, thereby isolating the μP's inputs from negative-going voltage spikes. The other rectifier is in series with a 5.1-V zener, which clamps positive-going transients.

DATA LINE REMOTE SHORT SENSOR

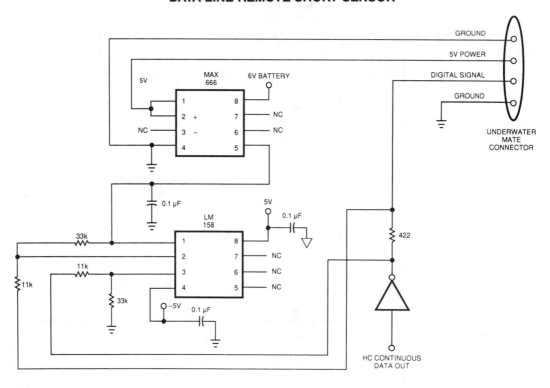

EDN

Fig. 20-4

Sensing short circuits in equipment that performs under water is especially critical, but Fig. 20-4's wet-mate connector design also suits other remote short-circuit sensing applications. Because of the limits imposed by the battery and voltage levels, the circuit uses the data line to sense short circuits.

The differential voltage that develops across the 422-Ω resistor in the data line drives a low-bandwidth op amp, which amplifies and filters the differential signal. The resistor values produce a gain of 3. The op amp's output controls the voltage regulator's shutdown pin.

To operate correctly, the circuit must have a continuous stream of digital data. Under normal conditions, and using high-speed CMOS logic, the data source sinks less than 10 μA. This normal operation generates about −3 mV across the sense resistor. The op amp's output will be slightly negative, producing a solid ON signal to the voltage-regulator chip. When a short occurs, the resistor and op amp together produce an average of 2.4 Vdc. This voltage provides a solid OFF to the voltage-regulator chip.

The peak signal-line current is about 12 mA (5-V data divided by 422 Ω), which HCMOS logic can handle. The addition of the resistor and op amp only changes the rise time to about 40 ns and doesn't cause any problems with the 2.5-MHz data rate. When the short is no longer present, the voltage regulator chip turns on again. You can use the same circuit with any TTL on/off-type voltage-regulator IC.

Z-80 BUS MONITOR/DEBUGGER

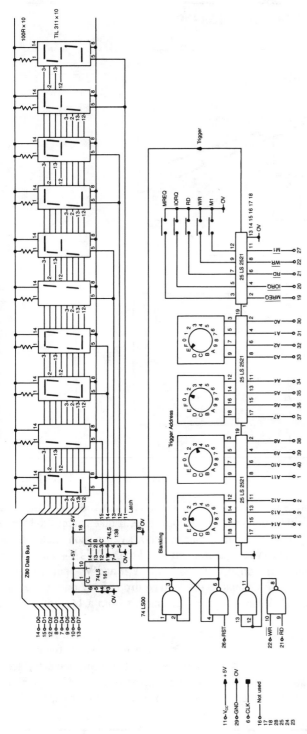

Fig. 20-5

ELECTRONIC ENGINEERING

Getting microprocessor designs to work is notoriously difficult when both the software and hardware are new. The usual approach is to run test routines that address memory and I/O, but do not rely on their correct functioning. However, miswiring in any part of the circuit usually leads to a misleading jumble of signals that might require a logic analyzer to interpret.

This simple circuit will trace the program execution and help point to the problems. Although the circuit shows connections for a Z-80, the circuit can very easily be adapted for any 8-bit microprocessor or with additional circuitry for CPUs of any bus width.

The circuit consists of a 5-byte hexadecimal display and comparator, which are wired to a 40-pin IC test clip. The test clip sits over the microprocessor (in this case a Z-80), where it gets power and all the required signals. The address bus and control lines are fed to the comparator, where (by means of switches) a trigger condition can be set. Following the trigger, the next 5 occurrences of either RD or WR will latch the contents of the data bus into the 5 hex displays, each in turn.

For example, select address 0000 M1 Rd and reset the CPU. The displays will show the very first instruction fetch, followed by its data and any consequent action. Even details, such as stack writes and subroutine addresses, are included. To trace longer portions of a program the address switches can be incremented to follow the execution path.

103

RS-232C LINE-DRIVEN CMOS CIRCUITS

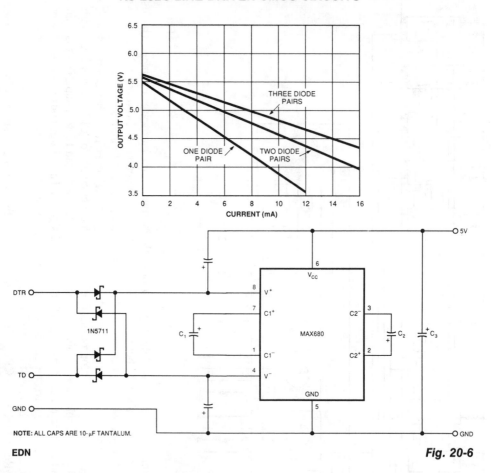

NOTE: ALL CAPS ARE 10-μF TANTALUM.

EDN

Fig. 20-6

The circuit illustrates a way to power CMOS ICs from RS-232C lines. The MAX680 is normally used to generate a voltage equal to $\pm 2 \times V_{CC}$. This circuit does exactly the opposite. It takes in ± 10.5 to ± 12 V on the DTR and TD lines and puts out a 5.25- to 6-V signal. A pair of Schottky or silicon diodes rectifies each RS-232C line. The resultant energy is stored by the capacitors attached to the IC's V+ and V− pins. C1 or both C1 and C2 then reverse-charge pump the energy stored at the V+ and V− pins to C3. The input source current of the MAX680 is approximately equal to the voltage drop of any one of the diodes that is divided by the series resistance of 160 Ω. When you drive this circuit from a 1488 driver with a ± 12-V supply, it can deliver 5 V at 3 mA.

To increase the output current, you can use as many as three sets of diodes on each RS-242 line to provide 5 V at 8 mA. The more diodes you use, the lower the source resistance: R_S equals the inversion of the sum of the diodes' conductances. If your circuit requires even more output current, you can place two MAX680s in parallel, tie their V+ and V− capacitors together, and use separate C1 and C2 capacitors for each ship. If you do connect the devices in parallel, make sure not to exceed the power capability of the RS-232C lines.

BIT GRABBER

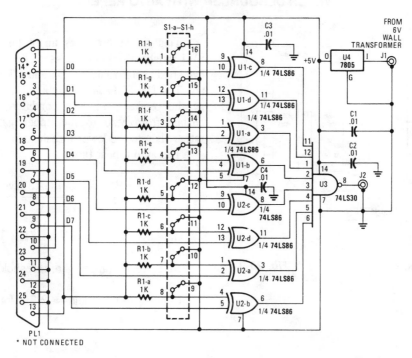

POPULAR ELECTRONICS

Fig. 20-7

When the user set character (D_0 through D_7) from a computer matches the character programmed on S1A through S1H, the output from J2 becomes low. This device can be used as a test aid to check printer cable or as a control circuit for interfacing with a computer.

SWITCH DEBOUNCER

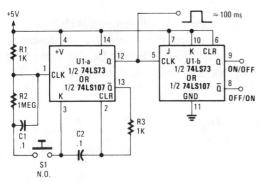

Using a 7473 JK flip-flop U1A connected as a monostable to drive U1B, as a switch debouncer. The circuit is self-clearing during power up. A 100-ms pulse is available at pin 12 U1A.

POPULAR ELECTRONICS

Fig. 20-8

SWITCH DEBOUNCER WITH AUTO REPEAT

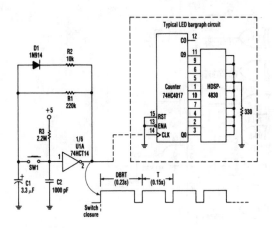

ELECTRONIC DESIGN

Fig. 20-9

This circuit produces an output pulse when SW1 (pushbutton) is depressed. It also becomes a hysteresis gate oscillator. D1 and R2 add asymmetry. The DBRT (*delay before repeat time*) is caused by the oscillator start-up conditions: C1 has to change from zero to the upper gate threshold rather than to the lower threshold.

The auto repeat time:

$$t = \frac{(R_1 + R_2)(C_1 + C_2)(\text{Gate Hysteresis})}{V_{\text{SUPPLY}}}$$

Gate hysteresis ≈ 1 V for 74HCT14 gate

DBRT = 0.7T (upper gate threshold hysteresis)

Upper gate threshold ≈ 2.3 V for HCT14

$$R_1 << R_3$$
$$R_2 << R_1$$

PRINTER ERROR ALARM

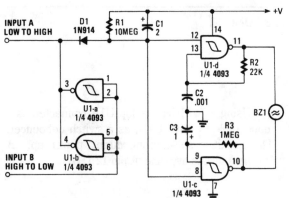

When a printer is shut down, this alarm sounds an alarm. The input can be either a high-to-low or low-to-high transition. This can be a logic level that corresponds with the printer being on or off. The oscillator produces an interrupted (on-off) tone.

POPULAR ELECTRONICS

Fig. 20-10

CHILD-PROOF RESET SWITCH

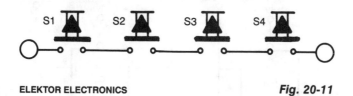

ELEKTOR ELECTRONICS   *Fig. 20-11*

The reset switch on a computer is very important. If an operating instruction threatens to wreck the internal management of a computer, the reset button is often the only way of avoiding a possible disaster. On the other hand, it also could cause a disaster.

It is particularly important that children or pets cannot inadvertently operate the control. The circuit proposed here should put an end to your worries in this respect. Instead of one reset switch, it is necessary to press four switches simultaneously. The chances of this happening via accident, child, or pet are negligible.

The four switches are placed in positions that make it impossible to operate them all with one hand. Instead, two of them can be operated with the fingers of one hand and the other two with the fingers on the other hand. As shown, the four switches are connected in series and are intended to replace the existing switch.

XOR GATE

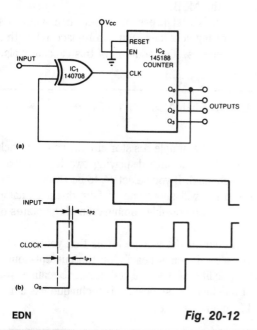

Inverting the negative-going input transactions allows the counter to count both positive- and negative-going edges. The XOR gate transforms the input signal into a series of short pulses whose width is equal to the sum of the counter and gate propagation delays.

EDN   *Fig. 20-12*

FLIP-FLOP DEBOUNCER SWITCH

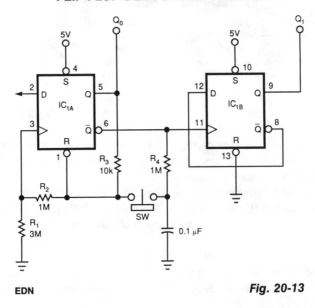

EDN

Fig. 20-13

Although this circuit uses a 74HC74, any CMOS variant of this flip-flop will work. IC1A acts as a true/complement buffer. R1 and R2 ensure that IC1A comes out of reset before the clock's edge occurs. R3 applies IC1A's logic state to pins 1 and 3. When the switch closes, the next logic state stored on the capacitor transfers to the flip-flop's reset and clock inputs. Releasing the switch lets the capacitor charge to the next state via R4. IC1A's output is the LSB; IC1B's output is the MSB.

Notice that the counter's state advances when the switch is first pressed, rather than when it's released; the latter is the case with many other switch-debouncing schemes. You can replace R1 with a 22-pF capacitor to reduce the circuit's sensitivity to parasitic effects. The addition of this capacitor also lets you lower the magnitude of R2 and R3 by a factor of 10.

DIGITAL LEVELS SCOPE DISPLAYS 2 LOGIC SIGNALS ON 1 SCOPE

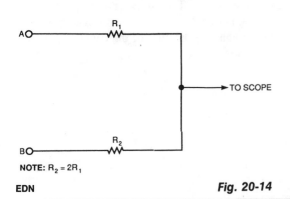

NOTE: $R_2 = 2R_1$

EDN

Fig. 20-14

Using this simple resistor circuit, you can trick your oscilloscope into displaying two logic signals on one channel. If you select R_2 to be twice R_1, the scope trace will show one of four distinct analog levels for each possible combination of the states of inputs A and B.

Of course, the voltage levels that your oscilloscope sees depends heavily on the current-sourcing capability of your digital logic. Because you must use high resistances, this technique has a limited frequency range.

DEGLITCHER

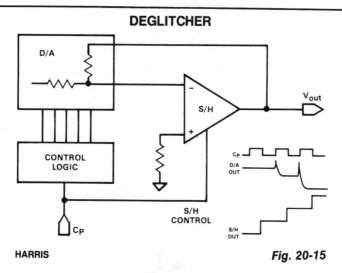

HARRIS

Fig. 20-15

Glitch has been a universal slang expression among electronics people for an unwanted transient condition. In D/A converters, the word has achieved semiofficial status for an output transient, which occurs when the digital input addressed is changed. The sample/hold amplifier does double duty, serving as a buffer amplifier as well as a glitch remover, delaying the output by $1/2$-clock cycle.

The sample/hold can be used to remove many other types of glitches in a system. If a delayed sample pulse is required, it can be generated using a dual monostable multivibrator IC.

STALLED-OUTPUT DETECTOR

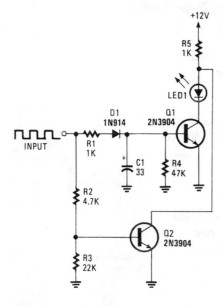

This circuit can be used to detect a stuck output or node in a circuit, or a loss of data or pulses. The pulse train charges C1 and biases Q1 on, which lights the LED. If the input remains high, Q2 extinguishes the LED.

POPULAR ELECTRONICS   *Fig. 20-16*

109

21

Converters

The sources of the following circuits are contained in the Sources section, which begins on page 664. The figure number in the box of each circuit correlates to the entry in the Sources section.

4-TO-20-mA CURRENT LOOP

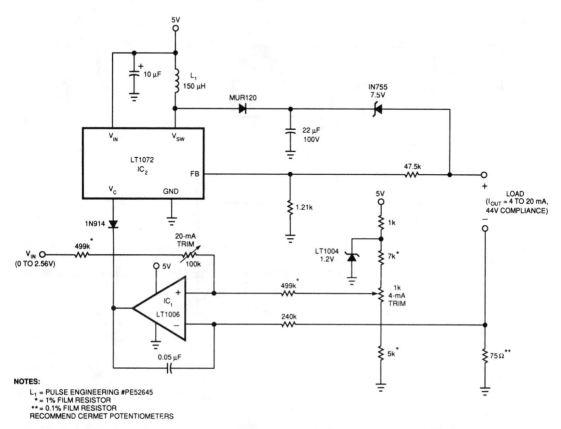

NOTES:
L₁ = PULSE ENGINEERING #PE52645
* = 1% FILM RESISTOR
** = 0.1% FILM RESISTOR
RECOMMEND CERMET POTENTIOMETERS

EDN

Fig. 21-1

This 5-V circuit utilizes a servo-controlled dc/dc converter to generate the compliance voltage necessary for loop-current requirements. This circuit will drive 4 to 20 mA into loads as high as 2 200 Ω with 44 V of compliance. It is inherently short-circuit protected. A current source by definition limits current regardless of the load.

The circuit's input voltage and the 4-mA trim network determine IC1's positive input voltage. IC1's output biases the LT1072 switching regulator's V_C pin. The resistors connected to the regulator's feedback pin, FB, prevent the circuit output from running away in the event that the load opens up.

Normally, IC1 controls the loop. However, if the load opens, IC1 receives no feedback. Under this condition, the FB pin becomes active when it equals 1.2 V and forces the loop to close locally around the regulator by activating IC2's internal amplifier. Thus, the circuit automatically changes from a current to a voltage regulator, thereby preventing excessive output voltages.

POWER-SAVING INTERMITTENT CONVERTER

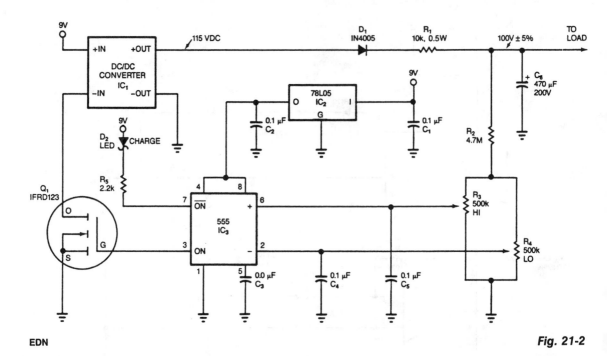

EDN

Fig. 21-2

This circuit switches its dc/dc converter, IC1, off whenever the large filter capacitor, C6, has sufficient charge to power the load. This particular circuit uses a dc/dc converter that produces 115 Vdc from a 9-Vdc input; you can tailor the circuit to suit other converters. The heart of the circuit is a 555 timer configured as a dual-limit comparator. Thus, the 555 turns the converter on or off, depending on the voltage across C6. The 555's complementary output lights the charge LED when the FET is on.

Initially, the voltage on C6 is zero, and the 555's output turns on the FET, Q1, in turn, enabling the converter to run, which charges C6. When the voltage on the capacitor reaches the value set by R3, the 555 turns the converter off. Then, C6 slowly discharges into the combined load of the voltage divider (R2, R3, and R4) and the reverse-biased blocking diode, D1.

When the voltage falls below $1/3$ V_{CC}, the 555 restarts the dc/dc converter. If this circuit powers a load that periodically goes into a zero-power, shutdown mode, the 555 switches the dc/dc converter on full time whenever the load kicks in.

When the supply voltage falls below 7.5 V, the output of the converter is no longer high enough to charge, the LED doesn't light. The circuit uses 205 mA when the converter is on and 10 mA when the converter is off. The duty cycle comprises a 5-s ON period, a 150-s OFF period, and it represents a 92% power reduction. You can further reduce power consumption by removing the charge LED and using a CMOS 555 and a CMOS 78L05 regulator.

CURRENT-TO-FREQUENCY CONVERTER

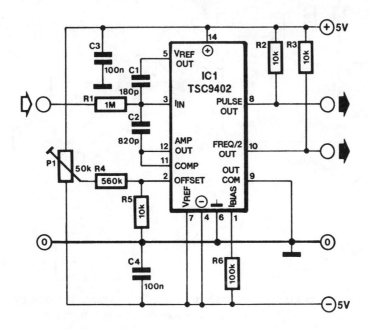

ELEKTOR ELECTRONICS   *Fig. 21-3*

Teledyne Semiconductor's Type TSC9402 IC is eminently suitable as an inexpensive current-to-frequency converter. The maximum input current of the design shown in the diagram is 10 μA (input voltage range is 10 mV to 10 V), while the output frequency range extends from 10 Hz to 10 kHz. The conversion factor is exactly 1 kHz/μA. The factor can be altered by changing the value of R1—as long as the maximum input current of 10 μA is not exceeded.

The circuit has two outputs. That at pin 8 is a short-duration pulse, whose rate is directly proportional to the input current; that at pin 10 is a square wave of half the frequency of the pulse at pin 8.

Calibrating the circuit is fairly simple. Connect a frequency meter to pin 8 (preferably one that can read tenths of a hertz) and connect a voltage of exactly 10 mV to the input (check with an accurate millivoltmeter). Adjust P1 to obtain an output of exactly 10 Hz. Next, connect a signal of exactly 10 V to the input and check that the output signal has a frequency of 10 kHz. If this frequency cannot be attained, shunt C1 with a small trimmer or replace R1 by a resistor of 820 kΩ and a preset of 250 kΩ.

The circuit may be adapted to individual requirements with the aid of:

$$f_{out} = I_{in} U_r (C_1 + 12 \text{ pF}) \qquad\qquad \text{[Hz]}$$

The reference voltage, U_r, here is -5 V.

SAWTOOTH CONVERTER

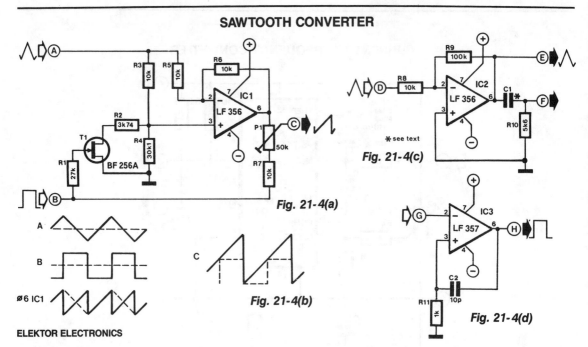

Fig. 21-4(a)

Fig. 21-4(b)

Fig. 21-4(c)

Fig. 21-4(d)

ELEKTOR ELECTRONICS

Simple function generators normally provide sinusoidal, rectangular, and triangular waveforms, but seldom a sawtooth. The circuit in Fig. 21-4(a) derives a sawtooth signal from a rectangular and triangular signal. Its quality depends on the linearity of the triangular signal, the slope of the edges of the rectangular signal and the phase relationship between the rectangular and triangular signals.

The conversion is carried out in IC1. Whether the triangular signal at input A is converted or not by IC1 depends on the state of T1. This FET is controlled by the rectangular signal at input B.

The signal at the output of the op amp is a sawtooth (see Fig. 21-4(b)) whose trailing edge is inverted. The frequency of this signal is double that of the input signals.

If in this state, the dc level of each inverted edge is raised sufficiently to make the lower level of that edge coincide with the higher level of the preceding edge, a sawtooth signal of the same frequency (but double the peak value of the input signals) is obtained. The dc level is raised by adding input B to the output of IC1 via R7 and P1. The preset should preferably be a multiturn type. Resistors R2 and R4 are 1% types.

If a rectangular signal is not available, or if its peak value is too small, the auxiliary circuits (shown in Figs. 21-4(c) and 21-4(d)) will be found useful. Figure 21-4(c) amplifies the triangular input at A by 10. Differentiating network C1/R10 derives rectangular pulses from the amplified triangular signal and these are available at F.

The pulses at F are shaped by the circuit in Fig. 21-4(d) to rectangular signals that have the same peak value as the supply voltage. Capacitor C2 increases the slope of the edge; it can be omitted for low-frequency signals.

The converter provides sawtooth signals over the frequency range of 15 Hz to 15 kHz. If the auxiliary circuits are used, capacitor C1 must be compatible with the frequency of the sawtooth signal (its value lies between 2 nF and 100 pF). The supply for all circuits can be between ± 10 V and ± 15 V. Each op amp draws a current of 4 to 6 mA.

PERIOD-TO-VOLTAGE CONVERTER

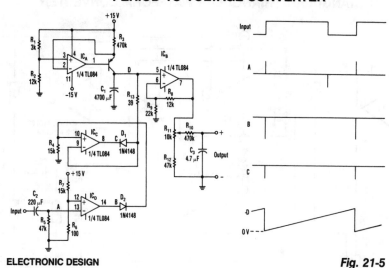

ELECTRONIC DESIGN

Fig. 21-5

ICA, R1, R2, R3, and Q1 form a current source. The current that charges C1 is given by:

$$I = \frac{V_{Dn} \times R_1}{(R_1 + R_2) \times R_3}$$

$$= \frac{15 \times 3\ k\Omega}{(3\ k\Omega + 12\ k\Omega) \times 470\ k\Omega}$$

$$= 6.4\ \mu A$$

The input signal drives ICD. Because ICD's positive input $(V+)$ is slightly offset to $+0.1$ V, its steady state output will be around $+13$ V. This voltage is sent to ICC through D2, setting ICC's output to $+13$ V. Therefore, point D is cut off by D1, and C1 is charged by the current source. Assuming the initial voltage on C1 is zero, the maximum voltage (VC_{max}) is given by:

$$VC_{max} = \frac{I \times T}{C_1}$$

$$= \frac{6.4 \times t}{0.004\ 7}$$

$$= 1362t$$

If $t = 1$ ms, then $VC_{max} = 1.362$ V.

When the input goes from low to high, a narrow positive pulse is generated at point A. This pulse becomes -13 V at point B, which cuts off D2. ICC's $V+$ voltage becomes zero. The charge on C1 will be absorbed by ICC on in a short time. The time constant of C2 and R5 determines the discharge period— about 10 μs. ICB is a buffer whose gain is equal to $(R_8 + R_9) \div R_9 = 1.545$. ICD's average voltage will be $(1362t \times 1.545) \div 2 = 1\ 052t$. R10 and C3 smooth the sawtooth waveform to a dc output.

RECTANGULAR/TRIANGULAR WAVEFORM CONVERTER

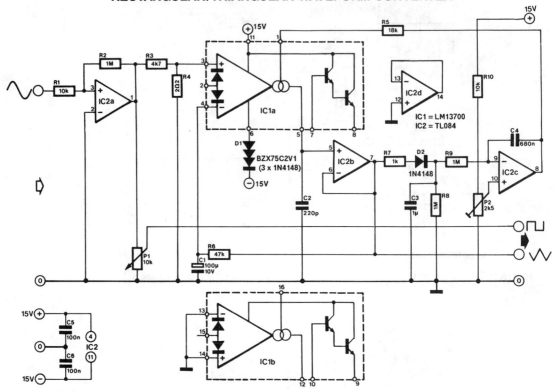

Fig. 21-6

Many function generators are based on a rectangular waveform generator that consists of a Schmitt trigger and integrator. The triangular signal produced by the integrator is then used to form a sinusoidal signal with the aid of a diode network. The converter presented here works the other way around. It converts the output of a good-quality sine-wave oscillator into a rectangular and a triangular signal.

The sinusoidal signal is converted into a rectangular signal by IC2A. Because the output of this gate varies between −15 V and +15 V, it is reduced to a value that is suitable for integration by potential divider R3/R4. It is then integrated by transconductance amplifier IC1A and C2. The amplifier has a current output that is controlled by the current through pin 1. The output therefore behaves as a resistance, with which it is possible to influence the integration time.

The voltage across C2 is available in buffered form at the output of impedance inverter IC2B; this is the triangular signal. The amplitude of this signal is compared with a voltage set by P2 and the difference between these voltages, which is the output of IC2C, is applied to the current source at the output of IC1 via R5. This arrangement ensures that the level of the output voltage is virtually independent of the frequency of the rectangular signal or of the sinusoidal input.

One problem with a precision integration is its being affected by offset voltages and bias currents. Feedback loop R6/C1 ensures that the output follows the potential across R4 accurately. However, tiny deviations might be caused by the bias current in circuit IC1, which is not greater than 8 μA at 70°C.

116

RECTANGULAR/TRIANGULAR WAVEFORM CONVERTER (Cont.)

The time constant R6/C1 is large for a purpose: to ensure that the triangular signal, even at low frequencies, cannot affect the waveform of the signal to be integrated—the rectangular shape must be retained. The converter can process signals at frequencies from 6 Hz (where the amplitude is not affected) to 60 kHz (where the amplitude is reached by 10%).

Because of the long time constants, the time taken for the recovery of the amplitude of the triangular signal at frequencies above 1 kHz is rather long. The peak value of this signal should be set to 1 V.

Diode D1 is a so-called *stabistor*—three diodes in one package. It might be replaced by three discrete type 1N4148 diodes. The current drawn by the converter is of the order of 9 mA.

μP-CONTROLLED NEGATIVE VOLTAGE CONVERTER

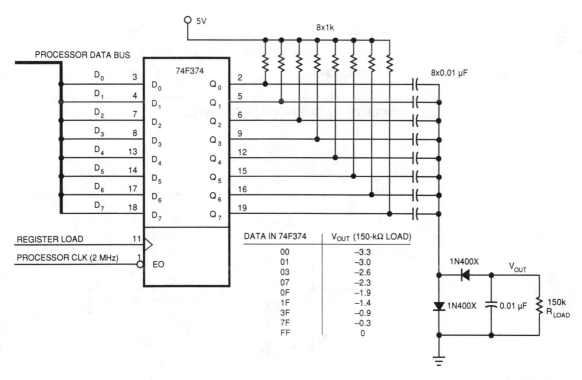

EDN

Fig. 21-7

This circuit was used to produce a variable negative voltage for contrast control of an LCD display. A 74F374 generates a square wave that is ac coupled to a rectifier and load. By using the μP clock and data from the processor bus, and properly timed load signal, the dc level generated can be controlled by the μP.

dc/dc CONVERTER

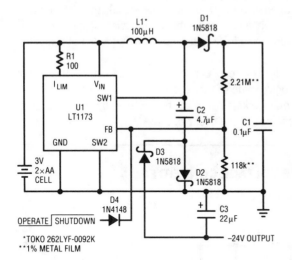

This circuit uses a Linear Technology LT1073 in a −24-V converter. The supply can be two AA cells (3 V) or 5 V. The circuit can deliver 7 mA.

LINEAR TECHNOLOGY **Fig. 21-8**

VOLTAGE-TO-CURRENT CONVERTER

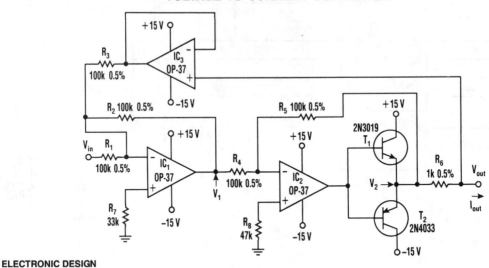

ELECTRONIC DESIGN **Fig. 21-9**

This voltage to current converter uses three op amps to drive a pair of power transistors. The current output is calculated as:

$$I_{OUT} = \frac{V_{in}}{R_6}$$

Output resistance is over 50 MΩ. I_{OUT} can range from 1 mA to the current ratings of T1 and T2.

1-TO-5 V dc/dc CONVERTER

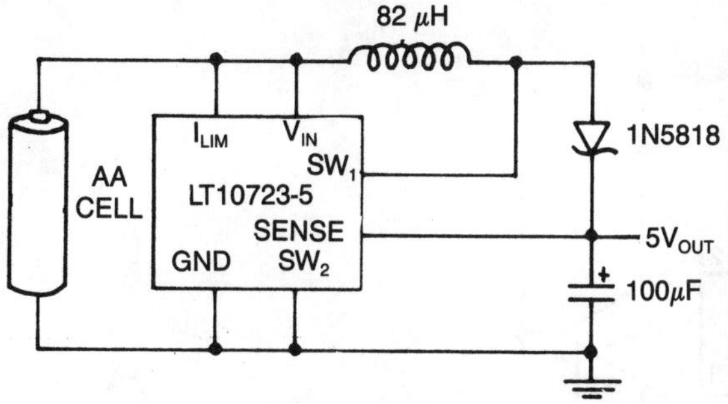

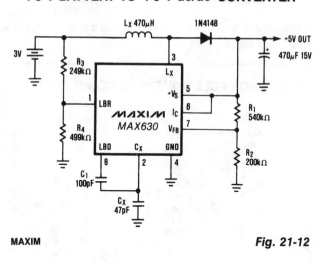

82 μH

1N5818

AA CELL

I_LIM V_IN SW_1

LT10723-5

SENSE

GND SW_2

5V_OUT

100μF

LINEAR TECHNOLOGY

Fig. 21-10

9-TO-5-V CONVERTER

220Ω

9V BATTERY

I_LIM V_IN SW_1

SENSE

LT1173-5

GND SW_2

270μH

5V_OUT

1N5818

100μF

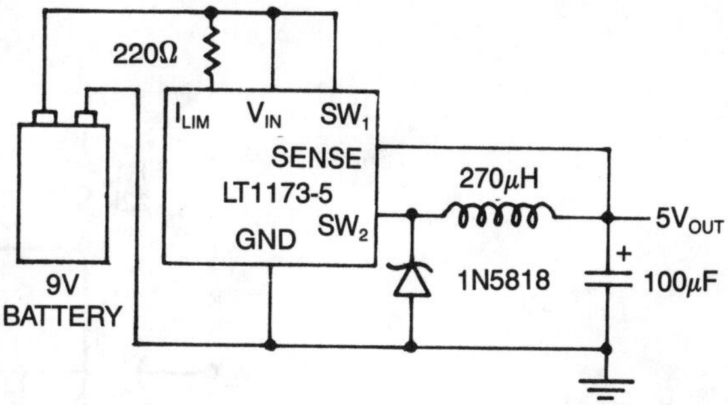

LINEAR TECHNOLOGY

Fig. 21-11

This circuit, using the Linear Technology LT1173, produces 5 V at 40 mA from a 1.5-V AA cell.

Using a Linear Technology LT1173-5, this converter produces regulated 5 V from a 9-V battery.

+3-V BATTERY TO +5-V dc/dc CONVERTER

L_X 470μH   1N4148   +5V OUT

3V

R_3 249kΩ

R_4 499kΩ

1 LBR

MAXIM

MAX630

I_C

V_FB

LBD C_X GND

8   2   4

C_1 100pF

C_X 47pF

L_X

+V_S

3

5

6

7

R_1 540kΩ

R_2 200kΩ

470μF 15V

MAXIM

Fig. 21-12

A common power-supply requirement involves converting a 2.4- or 3-V battery voltage to a 5-V logic supply. This circuit converts 3 V to 5 V at 40 mA with 85% efficiency. When I_C (pin 6) is driven low, the output voltage will be the battery voltage minus the drop across diode D1. The optional circuitry that uses C1, R3, and R4 lowers the oscillator frequency when the battery voltage falls to 2.0 V. This lower frequency maintains the output-power capability of the circuit by increasing the peak inductor current, which compensates for the reduced battery voltage.

SINE-WAVE/SQUARE-WAVE CONVERTER

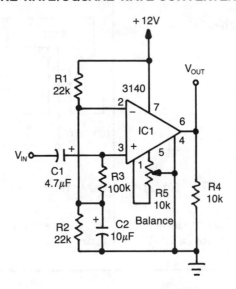

RADIO-ELECTRONICS   *Fig. 21-13*

An op amp used as a comparator produces a 10-V p-p square-wave output with 100-mV input, to 15 kHz. Adjust R5 for symmetry of square wave at low input levels.

PRECISION FULL-WAVE ac/dc CONVERTER

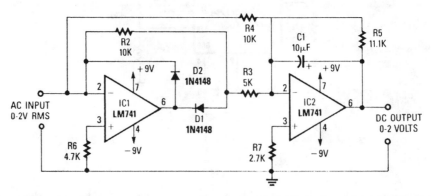

RADIO-ELECTRONICS   *Fig. 21-14*

A dc level is produced that corresponds to the ac input rms value (if sine wave). $\frac{R_1}{C_5}$ set the gain of IC2 to 1.11. This factor is the average-to-rms conversion factor. IC1 and IC2 act as a full-wave rectifier circuit, with D1 and D2.

22

Crystal Oscillators

The sources of the following circuits are contained in the Sources section, which begins on page 664. The figure number in the box of each circuit correlates to the entry in the Sources section.

MICROPOWER CLOCK

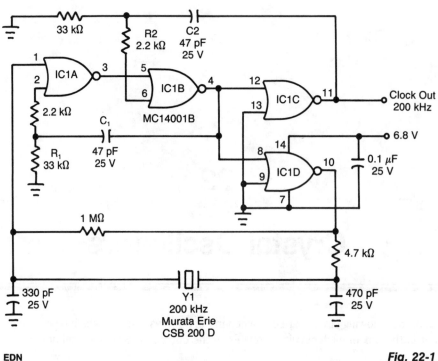

EDN

Fig. 22-1

Although ceramic resonators are a good choice for low-power, low-frequency clock sources (if you can stand their 30-ppm temperature coefficient), they have troublesome, spurious-resonance modes. This circuit rejects all but the resonator's fundamental mode. This clock circuit works from -40 to $+80°C$ and consumes only 2.8 mW.

The rising edge of resonator Y1 toggles IC1A low. ac-coupled positive feedback from IC1D via C1 and R1 immediately confirms this state change at IC1B so that Miller loading, harmonic components, or below-minimum rise times at IC1A cannot force IC1C to relapse to its previous state. This tactic also applies to resonator Y1's falling edge because IC1C, via C2 and R2, holds IC1B high.

Choose time constants R_1C_1, and R_2C_2 to be equal and ranging from 60 to 75% of one-half of the clock's period. Ceramic capacitors (10% tolerance) with X7R dielectric work well. With these time constants, the logic will be locked and unavailable to the ceramic resonator until just before it executes a legitimate transition. IC1D and IC1C are in parallel to isolate the resonator from external loads and, more importantly, from C2.

SIMPLE FUNDAMENTAL CRYSTAL OSCILLATOR

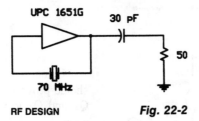

RF DESIGN   Fig. 22-2

This simple fundamental oscillator uses a μPC1651G IC and two components. The crystal is fundamental.

SIMPLE THIRD-OVERTONE OSCILLATOR

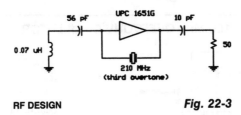

RF DESIGN   Fig. 22-3

Using a 210-MHz third overtone crystal, this circuit operates directly at the crystal frequency, with 210-MHz output and no multiplier stages.

COLPITTS 1-to 20-MHz CRYSTAL OSCILLATOR

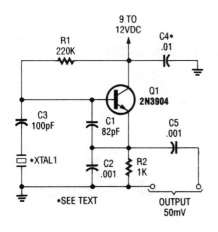

This is a simple Colpitts crystal oscillator for 1 to 20 MHz, can be easily made from junk-box parts (provided that a crystal is handy).

POPULAR ELECTRONICS   Fig. 22-4

100-kHz CRYSTAL CALIBRATOR

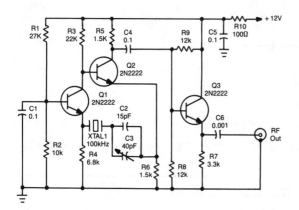

Using a 12-V supply, this crystal calibrator should prove a useful accessory for a SW receiver. Q1 and Q2 form an oscillator and Q3 is a buffer amp.

POPULAR ELECTRONICS

Fig. 22-5

100-MHz OVERTONE OSCILLATOR

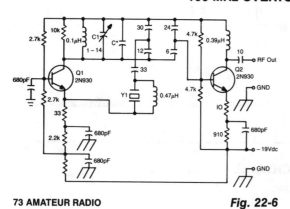

This oscillator circuit uses a 5th overtone crystal in the 85-to-106 MHz range. Y1 is the crystal. The circuit was originally used to frequency control a microwave oscillator.

73 AMATEUR RADIO

Fig. 22-6

VOLTAGE-CONTROLLED CRYSTAL OSCILLATOR

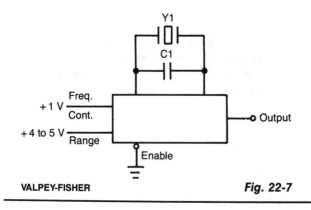

Notes:

1. For frequencies ≤ 1 MHz, C1 = 5 to 15 pF.
   For frequencies ≥ 1 MHz, C1 can be eliminated.

2. IC is SN74S124 for f_{max} of 60 MHz.
   IC is SN74LS124 for f_{max} of 35 MHz.

VALPEY-FISHER

Fig. 22-7

330-MHz CRYSTAL OSCILLATOR

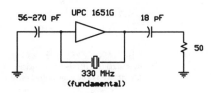

RF DESIGN

Fig. 22-8

A μPC 1651G IC operates in the fundamental mode with an experimental crystal at 330 MHz. The 56-to-270 pF capacitor is not critical; about +1-dBm RF output is available.

10-to 150-kHz OSCILLATOR

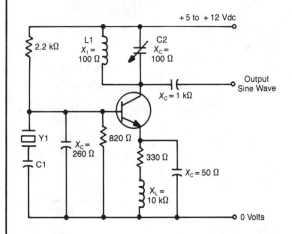

VALPEY-FISHER

Fig. 22-10

Note:

Y1 is "H", "NT," or "E" cut

10-to 80-MHz OSCILLATOR

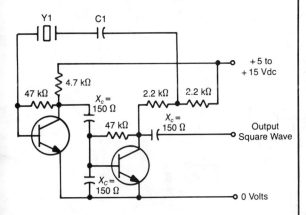

VALPEY-FISHER

Fig. 22-9

Notes:

1. Y1 is "AT" cut, fundamental, or overtone crystal.
2. Tune L1 and C2 to operating frequency.

1-to 4-MHz CMOS OSCILLATOR

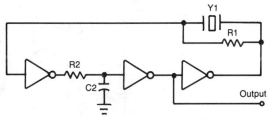

VALPEY-FISHER

Fig. 22-11

Notes:

1. $1\ M\Omega < R_1 < 5\ M\Omega$
2. Select R_2 and C_2 to prevent spurious frequency.
3. ICs are 74C04 or equivalent.

150-to 30 000-kHz OSCILLATOR

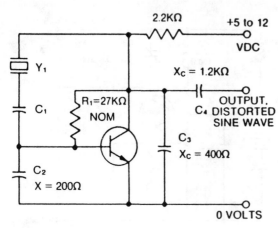

Notes:

1. Y1 is "AT", "CT", "DT", "NT", "SL", or "E" cut.
2. C1, C2, and C3 in series should equal the load capacity of crystal.
3. Adjust R1 for 1/2 supply voltage at collector of transistor.

VALPEY-FISHER

Fig. 22-12

C1 capacitor in series with the crystal may be used to adjust the output frequency of the oscillator. The value can range between 20 pF and 0.01 μF, or it can be a trimmer capacitor.

X values are approximate and can vary for most circuits and frequencies; this is also true for resistance values. Adequate power supply decoupling is required; local decoupling capacitors near the oscillator are recommended.

50-to 150-MHz OSCILLATOR

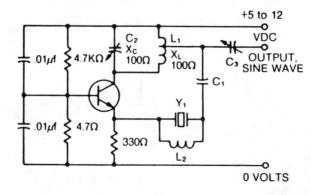

VALPEY-FISHER

Fig. 22-13

Notes:

1. Y1 is "AT" cut, overtone crystal.
2. Tune L1 and C2 to operating frequency.
3. L2 and shunt capacitance (C0) of crystal (approximately 6 pF) should resonate to the output frequency of the oscillator (L2 = 0.5μH at 90 MHz). This is necessary in order to tune out effect of C0 of the crystal.
4. C3 is varied to match output.

126

TWO-FREQUENCY COLPITTS OSCILLATOR

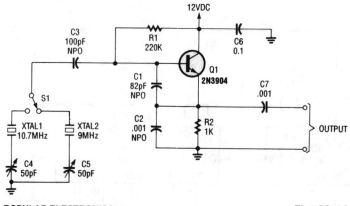

POPULAR ELECTRONICS

Fig. 22-14

Using switched crystals, this oscillator is intended for receiver alignment purposes.

1-to 20-MHz TTL OSCILLATOR

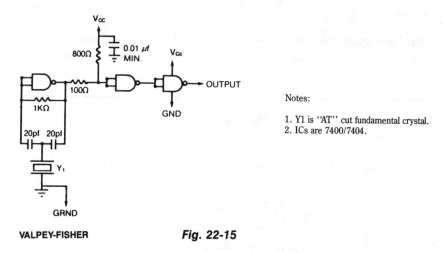

Notes:

1. Y1 is "AT" cut fundamental crystal.
2. ICs are 7400/7404.

VALPEY-FISHER

Fig. 22-15

CRYSTAL-CONTROLLED BRIDGE OSCILLATOR

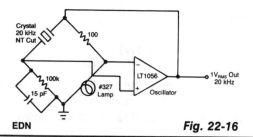

This crystal-controlled oscillator uses the current variations in a small lamp to stabilize amplitude variations.

EDN

Fig. 22-16

127

ECONOMICAL CRYSTAL TIME BASE

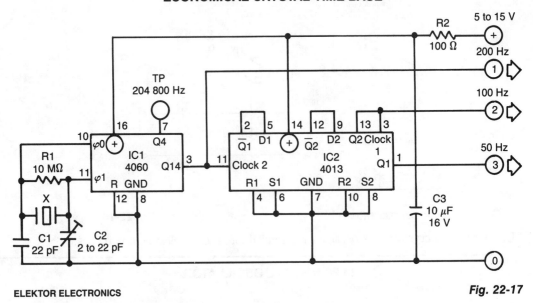

ELEKTOR ELECTRONICS

Fig. 22-17

The above time base circuit will provide 50-, 100-, or 200-Hz signals from an inexpensive crystal cut for 3.276 8 MHz, a common crystal used for microprocesser works. It requires a power supply of 5 to 15 V at 0.05 to 2.5 mA.

CRYSTAL OSCILLATOR

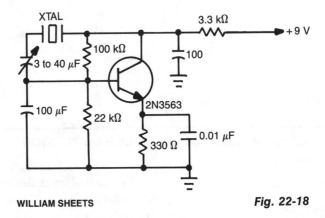

WILLIAM SHEETS

Fig. 22-18

This simple circuit will oscillate with a wide range of crystals. Connect several different types of crystal holders in parallel to improve versatility. The 3-to 40-pF capacitor adjusts crystal frequency over a small range for setting to standard-frequency transmissions when the unit is used as a crystal calibrator.

23

Data Circuits

The sources of the following circuits are contained in the Sources section, which begins on page 664. The figure number in the box of each circuit correlates to the entry in the Sources section.

3-Wire Receiver/Message Demuxer
Data Acquisition System I
Data Acquisition System II
Low-Frequency Prescaler
Analog Data-Signal Isolater

3-WIRE RECEIVER/MESSAGE DEMUXER

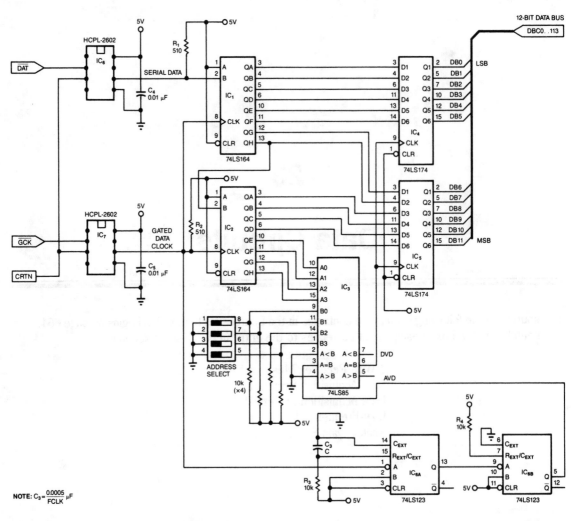

NOTE: $C_3 = \frac{0.0005}{FCLK} \mu F$

Fig. 23-1

This 3-wire receiver checks the first four data bits of 16 received bits against a preset address. If the two match, the remaining 12 bits are latched into two 6-bit flip-flop registers. Either CMOS or TTL logic families can be used in the design.

DATA ACQUISITION SYSTEM I

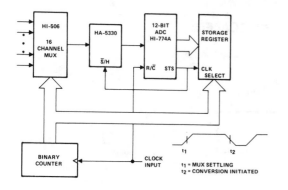

HARRIS

Fig. 23-2

The HI-506 multiplexer is used as an analog input selector, controlled by a binary counter to address the appropriate channel. The HA-5330 is a high-speed sample and hold. The sample/hold control is tied to the status (STS) output of the HI-774A; whenever a conversion is in process, the S/H is in the hold mode. A conversion is initiated when the clock input becomes low; when the clock becomes high, the mux address changes. The mux will be acquiring the next channel while the ADC is converting the present input, held by the S/H. The clock low time should be between 225 ns and 6.5 μs, with the period greater than 8.5 μs.

With this timing, T/C will be high at the end of a conversion, so the output data will be valid $\sim$ 100 ns before STS goes low. This allows STS to clock the data into the storage register. The register address will be offset by one; if this is a problem, a 4-bit latch can be added to the input of the storage register. With a 100-kHz clock rate, each channel will be read every 160 μs.

DATA ACQUISITION SYSTEM II

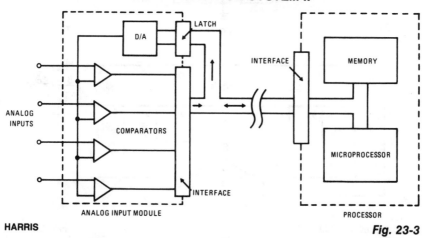

HARRIS

Fig. 23-3

In this circuit, an HA-4900 series comparator is used in conjunction with a D/A converter to form a simple, versatile, multichannel analog input for a data acquisition system. The processor first sends an address to the D/A, then the processor reads the digital word generated by the comparator outputs. To perform a simple comparison, the processor sets the D/A to a given reference level, then examines one or more comparator outputs to determine if the inputs are above or below the reference. A window comparison consists of two such cycles with two reference levels set by the D/A. One way to digitize the inputs would be for the processor to increment the D/A in steps. The D/A address, as each comparator switches, is the digitized level of the input. While stairstepping, the D/A is slower than successive approximation; all channels are digitized during one staircase ramp.

LOW-FREQUENCY PRESCALER

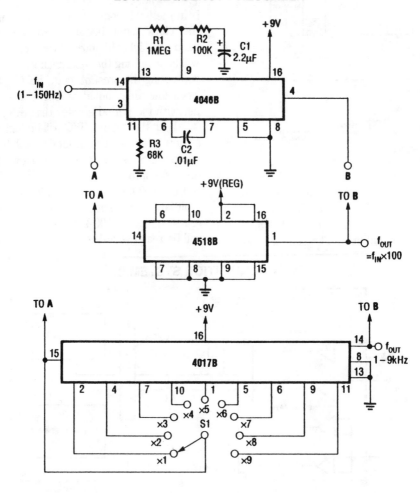

Fig. 23-4

For multiplying frequencies in the 1-to 150-Hz range, this circuit uses a 4046B and a ÷100 prescaler. The VCO output is phaselocked to the low-frequency input. This simplifies use of a frequency counter to measure LF signal frequencies.

By using a 4017B and a 1-kHz f_{IN}, the circuit can be used as a 1-to 9-kHz frequency synthesizer or as a ×10 frequency multiplier.

ANALOG DATA-SIGNALS ISOLATER

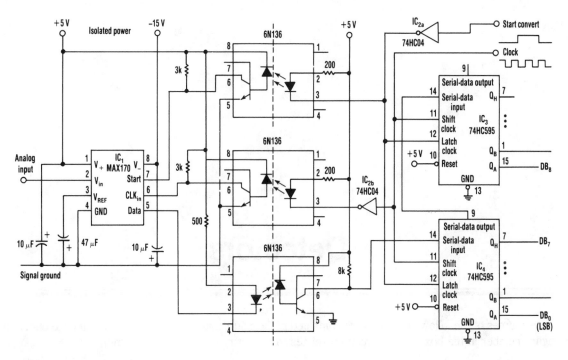

Fig. 23-5

By converting analog data to digital and using optocouplers, this circuit can be used to transmit analog signals across barriers, such as voltage levels, different ground systems, etc.

24

Detectors

The sources of the following circuits are contained in the Sources section, which begins on page 664. The figure number in the box of each circuit correlates to the entry in the Sources section.

Thermally Operated Direction Finder
Two-Sheets Detector
Metal Detector
Metal Detector
Peak Detector
Undervoltage Detector
SSB/CW Product Detectors
RF-Field Detector
Line-Operated Smoke Detector

NOAA Weather-Alert Decoder
Low-Level Diode Envelope Detector
Trip-Point Detector and Controller
AM Envelope Detector
IC Product Detectors
Peak Detector
Air-Pressure Change Detector
Duty-Cycle Detector
Op Amp Peak Detector

THERMALLY OPERATED DIRECTION DETECTOR

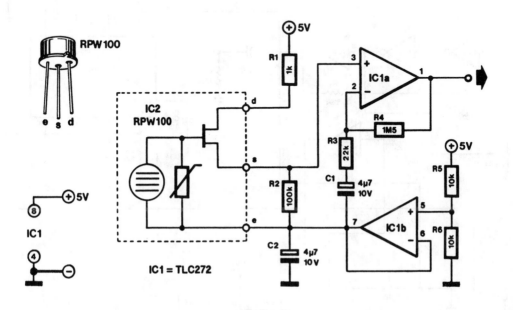

ELEKTOR ELECTRONICS

Fig. 24-1

A heat-sensitive sensor can be used to construct a direction detector. Such a sensor reacts to all animal heat. The one used in this design has a sensitive surface that has been divided into two. It, therefore, makes a difference, whether the heat approaches from the left or the right. The indication for cold objects is, of course, exactly the opposite.

Circuit IC1B forms a symmetric supply. Terminal s of the sensor is its output. The signal at s is amplified in IC1A by a factor of about 70 before it is available at the output of the detector.

To obtain good directivity, it is best to place the sensor behind a single narrow slit, rather than behind the usual raster of a multifacetted mirror. The circuit draws a current of only a few mA from a 5-V supply.

TWO-SHEETS DETECTOR

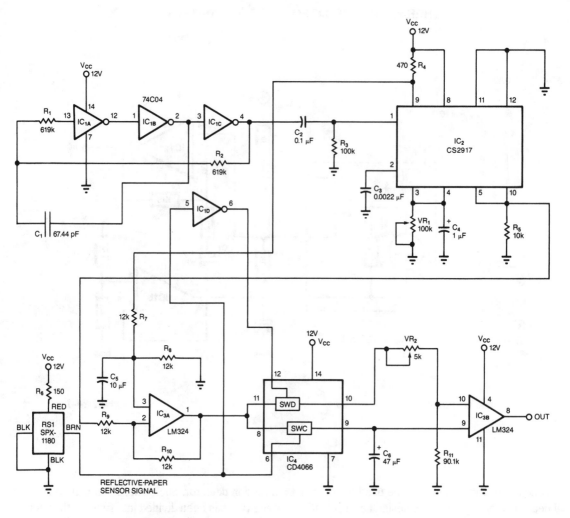

Fig. 24-2

Using the principle of capacitance between two plates this circuit senses when more than one sheet of paper goes between the sensing electrodes at a time. C1 is the sensing capacitor formed of two plates. It consists of two plates 2″ × 15″ with 0.1″ spacing. A change of capacitance causes a change in oscillator frequency of the IC1 circuit, which is detected by IC2 and IC3.

METAL DETECTOR

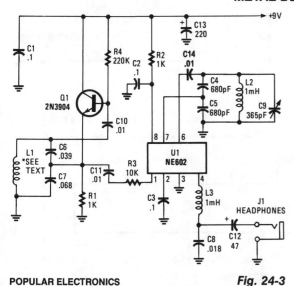

An NE602 acts as a heterodyne detector and Q1 as a sense oscillator. When L1 is brought near metal, it causes a charge in loop inductance, shifting the resonant frequency of L1 C6/C7. L1 is 5 turns #20 wire on a 9″ diameter wood or plastic form.

POPULAR ELECTRONICS

Fig. 24-3

METAL DETECTOR

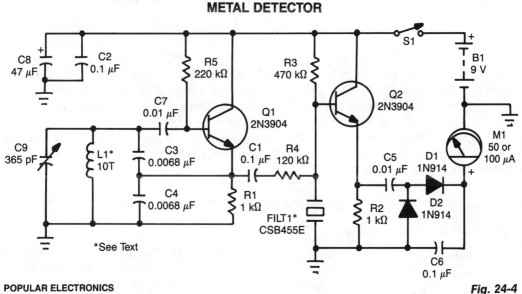

POPULAR ELECTRONICS

Fig. 24-4

Using an oscillator running at 455 kHz, the metal-detector circuit produces an indication on the meter M1. When the oscillator frequency changes because of metal in the field of L1, the change will show as an increase or decrease in frequency, which produces a charge in the meter reading. The ceramic filter FILT1 produces a selective bandpass that yields this effect. L1 can be a 4″ diameter coil wound on a suitable plastic form. About 10 turns of #26 wire are required. Use a frequency counter to adjust L1 and verify that Q1 is operating on or near 455 kHz.

PEAK DETECTOR

0 TO 5V INPUT

LM392

5V

2.2k

STATUS OUTPUT
HIGH=READY

PEAK OUT

15V

LM392

12M 6.2M 3.3M 1.6M 810k 402k 200k 100k 49.8k 24.9k

0.01µF

5V

Q4 Q5 Q6 Q7 Q8 Q9 Q10 Q12 Q13 Q14

V_{DD}

V_{SS}

CD4060BE IC_1

RESET CLOCK OUT1 OUT2

RESET

56k 0.01µF

120k

2N3904

Q_1

10k

EDN

Fig. 24-5

A 0-to-5 V input drives the negative input of LM392 comparator if reset (pin 12) if DC4060BE is pulled high then low, all outputs of ICF1 are forced low, forcing + input of comparator to go low. Q1 is cut off and IC1's clock oscillator, running at about 775 Hz, starts counting. The Q4 through Q14 outputs connect to a ladder. When the counter reaches a count so that the voltage on pin 3 of the LM392 equals the peak input voltage, the counter stops. This voltage is available at the output of the voltage follower LM392 (pin 7). The maximum time to acquire a peak is 22 seconds. This circuit is slow and was originally intended for battery-charging applications.

UNDERVOLTAGE DETECTOR

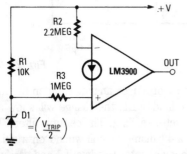

R2 2.2MEG

+V

R1 10K

R3 1MEG

LM3900

OUT

D1 $=\left(\dfrac{V_{TRIP}}{2}\right)$

The output goes high when the supply falls below a value determined by zener diode D1. If D1 is a 5.6-V zener, the op amp will switch high when the supply voltage falls below approximately 11 V. The precise trip point can be varied by replacing R3 with an 820-kΩ resistor in series with a 470-kΩ potentiometer.

RADIO-ELECTRONICS   *Fig. 24-6*

SSB/CW PRODUCT DETECTORS

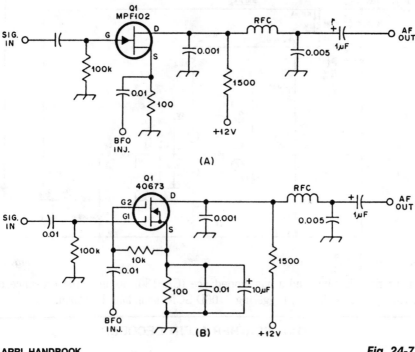

(A)

(B)

ARRL HANDBOOK

Fig. 24-7

These circuits are used for product detection of single-sidebound (SSB) and CW signals. BFO injection is typically 0.5 to 1 V rms for both circuits. Frequencies can be up to 25 MHz or so.

RF-FIELD DETECTOR

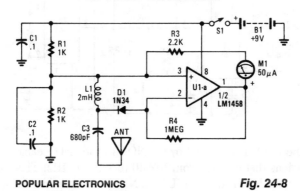

POPULAR ELECTRONICS

Fig. 24-8

This detector is a half-wave rectifier for RF, which then feeds an op amp. U1A acts as an amplifier, driving meter M1. This circuit can detect mW RF levels from below the AM broadcast band to well above the FM broadcast band.

LINE-OPERATED SMOKE DETECTOR

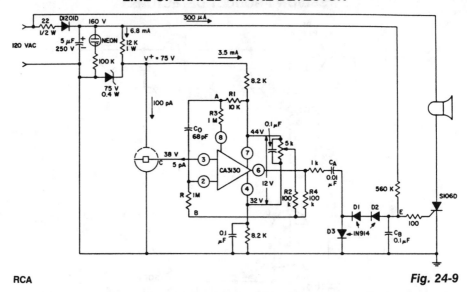

RCA

Fig. 24-9

Using an ionization chamber and a high-impedance (CA3130) op amp, the presence of smoke will cause the CA3130 to stop oscillating, triggering S106D SCR, sounding the alarm.

NOAA WEATHER ALERT DECODER

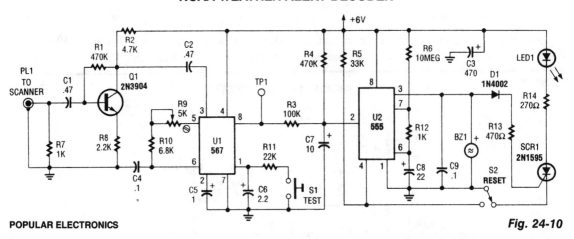

POPULAR ELECTRONICS

Fig. 24-10

This circuit detects the 1050-Hz tone sent by the NOAA (National Oceanic and Atmospheric Administration) Weather radio stations that operate from 162.40 to 162.55 MHz. This tone lasts for several seconds. Q1 is an amplifier that feeds tone detector U1, an NE567 detects this tone and produces a low on pin 8. This is coupled to a 555 timer (U3), which produces a high on its pin 3, sounds BZ1, triggers SCR1, and lights the LED. S2 is used to rest the circuit.

Audio is taken from the receiver that is used with the device. S1 is used to test the device and it will sound the alarm in two seconds if all is OK.

LOW-LEVEL DIODE ENVELOPE DETECTOR

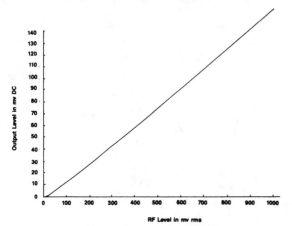

Envelope detector response.

Fig. 24-11(a)

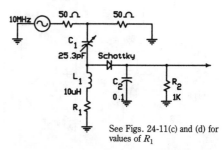

See Figs. 24-11(c) and (d) for values of R_1

Low-level envelope detector with lower L/C ratio to illustrate the effect of R.

Fig. 24-11(c)

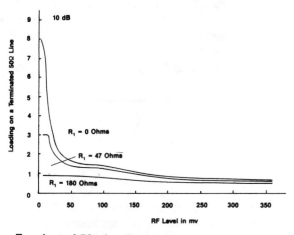

Terminated 50-ohm line.

Fig. 24-11(b)

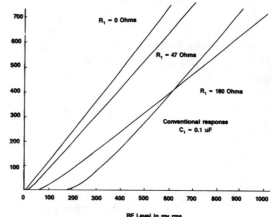

Envelope detector response for 3 values of R.

Fig. 24-11(d)

An approach to low-level RF detection and performance curves is shown here. This design is for 10 MHz, but values can be scaled to other frequencies, if needed.

DETECTOR AND CONTROLLER

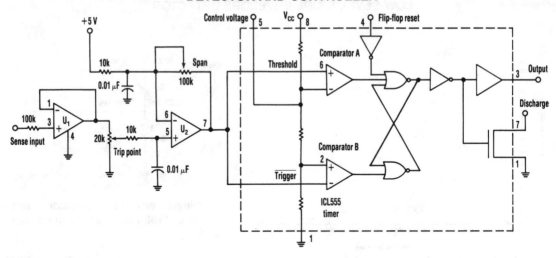

ELECTRONIC DESIGN

Fig. 24-12

Many applications require analog signals to be sensed and digital signals to be controlled. A way to detect these points is by using a 555 timer in an unconventional configuration. This method will also add hysteresis to the circuit and guard against oscillation. The 555 supplies two comparators and a flip-flop eliminates the oscillation. Using this classic timer in the new configuration also reduces the component count.

The circuit shows the 555's trigger and threshold pins tied together. This enables the comparators to set and reset the flip-flop. Op amp U2 supplies both the trip-point setting and a way to adjust the hysteresis for ON and OFF points. One application where this circuit would be useful is in a Nicad battery-charge controller.

AM ENVELOPE DETECTOR

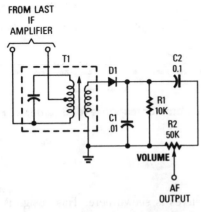

This general-purpose detector for AM envelope detection can be used in many receiver applications. T1 matches the IF amplifier impedance (typically 1 to 10 kΩ) to the 1 kΩ (approximately) detector impedance. D1 can be an IN60, IN82AG, IN270, or a similar type.

POPULAR ELECTRONICS

Fig. 24-13

IC PRODUCT DETECTORS

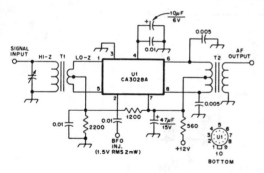

ARRL HANDBOOK　　　　*Fig. 24-14(a)*

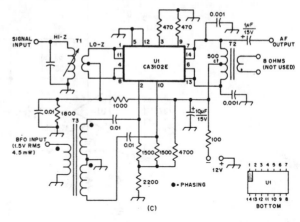

(c)

ARRL HANDBOOK　　　　*Fig. 24-14(c)*

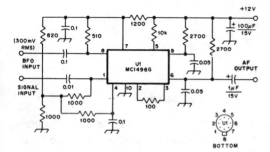

ARRL HANDBOOK　　　　*Fig. 24-14(b)*

These product detectors use IC devices. SSB and CW signals can be detected with them. The circuits should be useful up to 20 or 30 MHz. T3 in (c) is a 1:1:1 toroidal type, depending on the BFO frequency.

PEAK DETECTOR

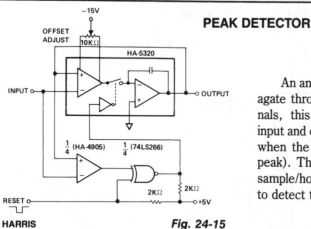

HARRIS　　　　*Fig. 24-15*

An analog signal requires about 100 ns to propagate through the HA-5320. For time-varying signals, this assures a voltage difference between input and output. Also, the voltage changes polarity when the signal slope changes polarity (passes a peak). This behavior makes the circuit a possible sample/hold peak detector, by adding a comparator to detect the polarity changes.

The exclusive NOR gate allows a reset function which forces the HA-5320 to the sample mode. The connections shown detect positive peaks; the comparator inputs can be reversed to detect negative peaks. Also, the offset must be introduced to provide enough step in voltage to trip the comparator after passing a peak. This circuit works well from below 100 Hz up to the frequency at which slew-rate limiting occurs. It captures the amplitude of voltage pulses, provided that the pulse duration is sufficient to slew to the top of the pulse.

AIR-PRESSURE CHANGE DETECTOR

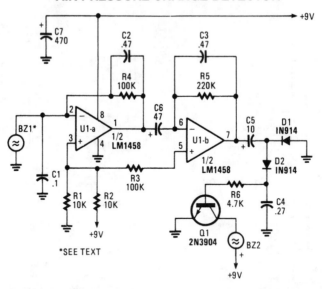

POPULAR ELECTRONICS

Fig. 24-16

A piezoelectric detector (BZ1) is used in this circuit to detect a change in air pressure. BZ1 produces a voltage that is amplified by U1A and U1B. Frequency response is limited to low frequencies. The signal is rectified by D1 and D2 and drives Q1, which activates BZ2, a piezoelectric buzzer.

DUTY-CYCLE DETECTOR

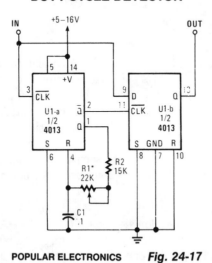

POPULAR ELECTRONICS   *Fig. 24-17*

This circuit looks at the time an incoming pulse is high. If the incoming pulse is shorter than the adjusted (VAR1) pulse, the output of U1B is high. Values are shown for a 1-to 2-μs pulse.

OP AMP PEAK DETECTOR

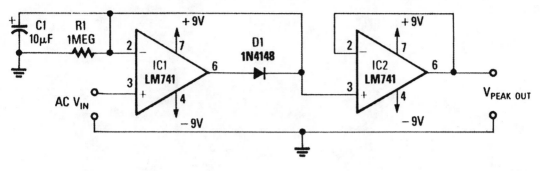

RADIO-ELECTRONICS

Fig. 24-18

The output of this circuit will be a voltage that is equal to the peak of the input. D1 and C1 detect the peak voltage and this is read by the IC2 voltage follower.

25

Direction Finders

The sources of the following circuits are contained in the Sources section, which begins on page 665. The figure number in the box of each circuit correlates to the entry in the Sources section.

Digital Compass
Radio Direction Finder

DIGITAL COMPASS

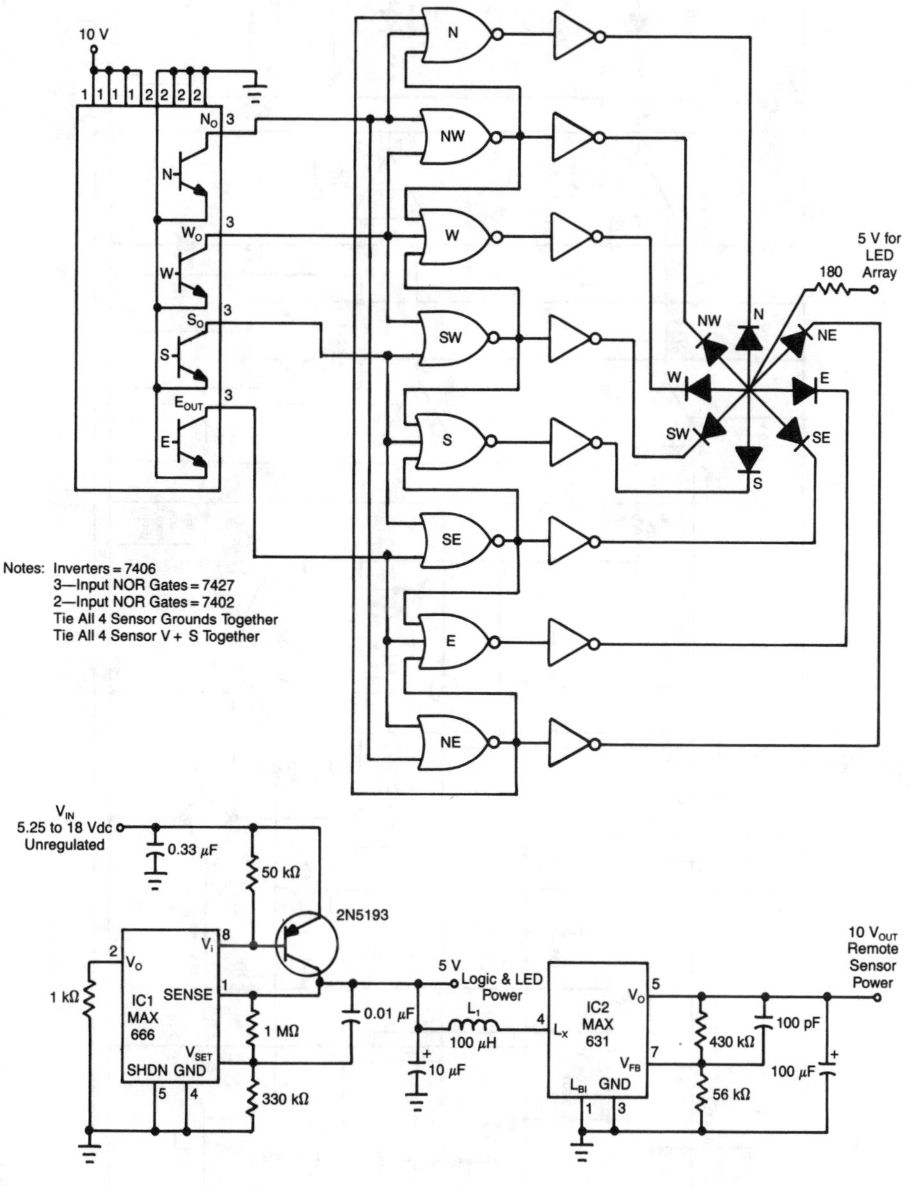

Notes: Inverters = 7406
3—Input NOR Gates = 7427
2—Input NOR Gates = 7402
Tie All 4 Sensor Grounds Together
Tie All 4 Sensor V + S Together

EDN

Fig. 25-1

A four output Hall sensor combined with a few logic gates produce this digital compass. The NOR gates resolve the four Hall outputs into eight distinct compass directions. LEDs to indicate direction are driven by eight inverters. A power supply for 5.25- to 18-Vdc operation is shown in the figure.

RADIO DIRECTION FINDER

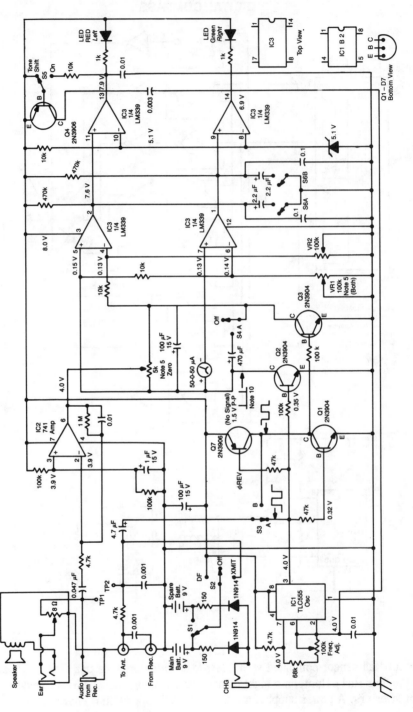

Fig. 25-2

73 AMATEUR RADIO

RADIO DIRECTION FINDER (Cont.)

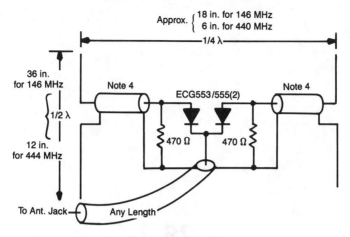

Antenna construction.

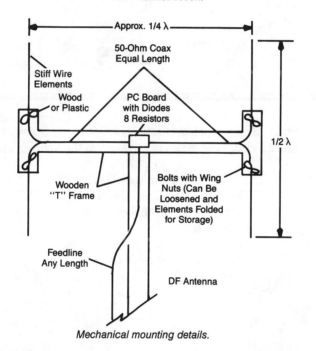

Mechanical mounting details.

73 AMATEUR RADIO

Fig. 25-2

This RDF circuit consists of a square-wave oscillator (IC1), which switches two antennas alternately at an audio rate. A phase detector (Q1, 2, 3, 7) is used to compare receiver output amplified by IC2 with the reference phase from IC2 with the reference phase from IC1. A 50-μA meter is used as a left-right indicator. IC3 is a comparator used to drive indicator LEDs.

26

Dividers

The sources of the following circuits are contained in the Sources section, which begins on page 665. The figure number in the box of each circuit correlates to the entry in the Sources section.

Clock Input Frequency Divider
Programmable Frequency Divider
Divide-by-Odd-Number Counter
$7490 \div N$ Circuits
Divide-by-2-or-3 Circuit
$1 +$-GHz Divide-by-N Counter
Divide-by-$N + 1/2$ Circuit

CLOCK INPUT FREQUENCY DIVIDER

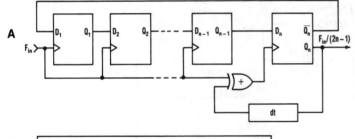

1. THE INPUT CLOCK frequency fed into this circuit is divided by $2n-1$. The circuit consists of n clocked flip-flops and one exclusive-OR gate. The dt delay is zero in most cases.

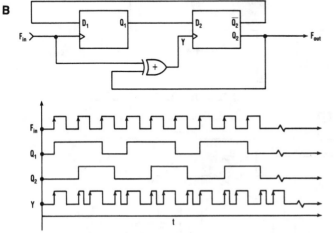

2. THIS CIRCUIT CONFIGURATION divides the input frequency by three (a). The circuit's timing diagram verifies the division (b).

ELECTRONIC DESIGN

Fig. 26-1

ICA, R1 through R3, and Q1 form a current source. The current that charges C1 is given by:

$$I = \frac{(V_D \times R_1)}{(R_1 + R_2) \times R_3}$$

$$= \frac{(15 \times 3 \text{ k}\Omega)}{(3 \text{ k}\Omega + 12 \text{ k}\Omega) \times 470 \text{ k}\Omega}$$

$$= 6.4 \ \mu\text{A}$$

The input signal drives ICD. Because ICD's positive input (V+) is slightly offset to +0.1 V, its steady-state output will be near +13 V. This voltage is sent to ICC through D2, setting ICC's output to +13 V. Therefore, point D is cut off by D1, and C1 is charged by the current source. Assuming the initial voltage on C1 is zero, the maximum voltage (VC_{max}) is given by:

$$tw_{clk} > tp_{ff} + dt + tp_{xr} + tw_{ff}$$

The right side of the inequality should be the minimum pulse width (either up time or down time) of the input clock. The circuit, when constructed with standard 74F-type parts, operates without any added delay in the exclusive-OR feedback path and with an input frequency of up to 22.5 MHz. The circuit's output signal will have the same duty cycle as the input clock.

151

PROGRAMMABLE FREQUENCY DIVIDER

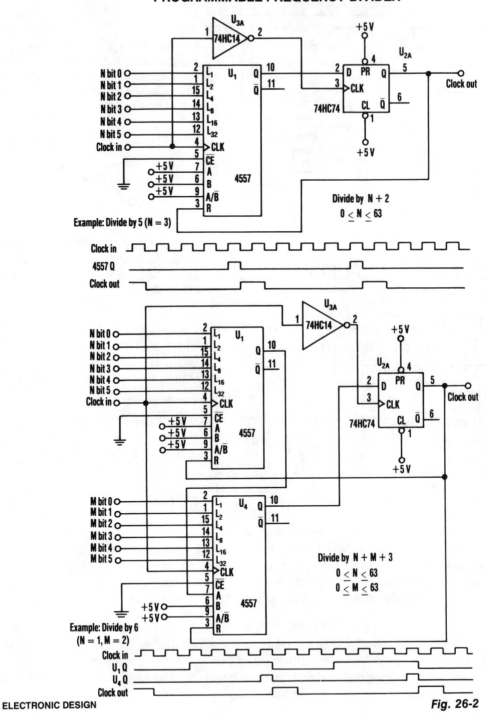

Example: Divide by 5 (N = 3)

Divide by N + 2
0 ≤ N ≤ 63

Example: Divide by 6
(N = 1, M = 2)

Divide by N + M + 3
0 ≤ N ≤ 63
0 ≤ M ≤ 63

Fig. 26-2

PROGRAMMABLE FREQUENCY DIVIDER (*Cont.*)

This divider uses a variable-length shift register, a type-D flip-flop, and an inverter. The clock feeds the flip-flop clock input and the output of the shift register feeds the D input of the flip-flop. The FF output is tied back to the reset input of the shift register so that each clock pulse shifts a "1" into the 4557. $N+1$ cycles after the reset pulse is removed. The first "1" will propagate through the register output. The "1" is latched into the FF on the clock's next falling edge and fed back to the 4557 reset pin, which resets the shift register to zero. When a zero is clocked into the flip-flop on the next falling clock edge, the reset is removed, restarting the process. The divide ratio is $(N+2)$, where N = the binary number that is programmed into 4557.

DIVIDE-BY-ODD-NUMBER COUNTER

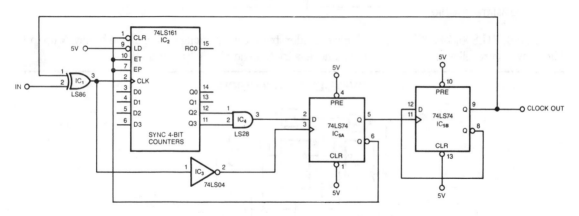

EDN

Fig. 26-3

This circuit symmetrically divides an input by virtually any odd number. The circuit contains $n+1/2$ clocks twice to achieve the desired divisor. By selecting the proper n, which is the decoded output of the 74LS161 counter, you can obtain divisors from 3 to 31. This circuit divides by 25; you can obtain higher divisors by cascading additional LS161 counters.

The counter and IC5A form the $n+1/2$ counter. Once the counter reaches the decoded counts, n, IC5A ticks off an additional $1/2$ clock, which clears the counter and puts it in hold. Additionally, IC5A clocks IC5B, which changes the clock phasing through the XOR gate, IC1. The next edge of the input clocks IC5A, which reenables the counter to start counting for an additional $n+1/2$ cycles.

Although the circuit has been tested at 16 MHz, a worst-case timing analysis reveals that the maximum input frequency is between 7 and 8 MHz.

7490 ÷ N CIRCUITS

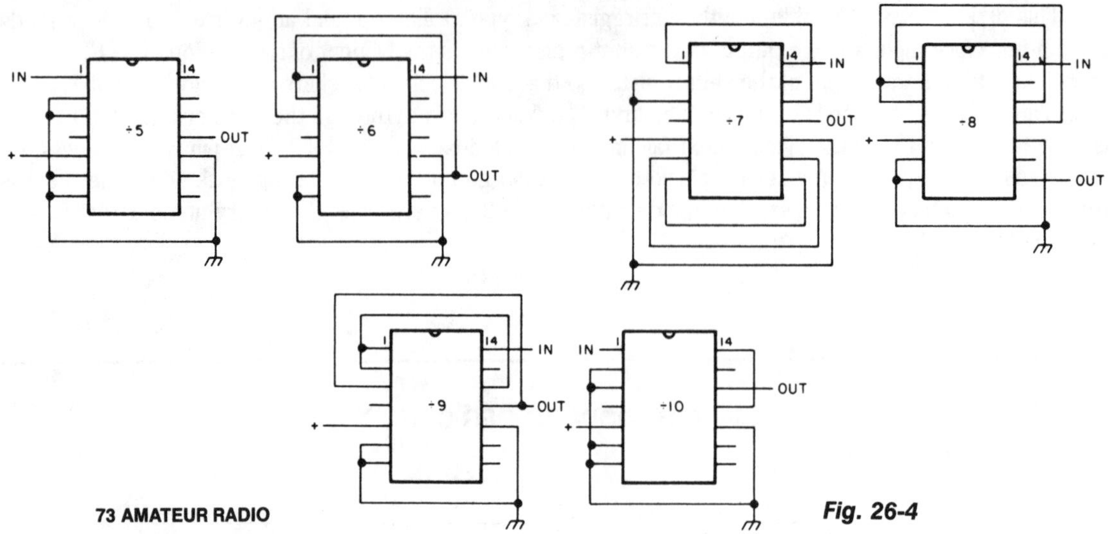

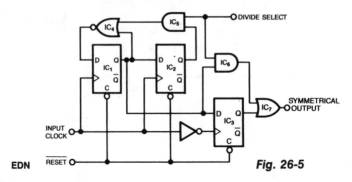

73 AMATEUR RADIO

Fig. 26-4

A 7490, 74LS90, 74C90, etc., is a decode divider, but it can be configured to divide by any N up to 10. The above figures illustrate the connections necessary to divide by N from 5 to 10.

DIVIDE-BY-2-OR-3 CIRCUIT

EDN

Fig. 26-5

This circuit produces a symmetrical waveform when dividing by either 2 or 3. The Divide Select input controls the division factor. When Divide Select is high, flip-flops IC1 and IC2, along with associated gates, form the classical divide-by-3 circuit.

When divide select is low, however, the output of the AND gate, IC5, goes low. Consequently, the NOR gate, IC4, inverts the feedback signal and passes it to the D input of the flip-flop, IC1. Now, IC1 acts like a toggle flip-flop and produces a divide-by-2 output.

IC3, which is, in effect, a negative-edge-triggered flip-flop, provides symmetrical output signals. When you select division by 2 (Divide Select is low), the output and AND gate IC6 is low, and IC3 simply clocks out the divider's output, delayed by one clock period. When you set Divide Select high, the path to the output through the AND and OR gates, IC6 and IC7, is enabled. This path means that the output goes high on the leading edge of IC3's input (not its output) and produces a symmetrical divide-by-3 output.

---

1 + -GHz DIVIDE-BY-N COUNTER

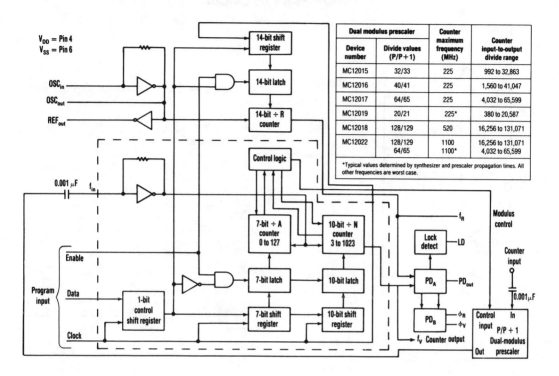

V_{DD} = Pin 4
V_{SS} = Pin 6

Dual modulus prescaler		Counter maximum frequency (MHz)	Counter input-to-output divide range
Device number	Divide values (P/P + 1)		
MC12015	32/33	225	992 to 32,863
MC12016	40/41	225	1,560 to 41,047
MC12017	64/65	225	4,032 to 65,599
MC12019	20/21	225*	380 to 20,587
MC12018	128/129	520	16,256 to 131,071
MC12022	128/129	1100	16,256 to 131,071
	64/65	1100*	4,032 to 65,599

*Typical values determined by synthesizer and prescaler propagation times. All other frequencies are worst case.

ELECTRONIC DESIGN

Fig. 26-6

Counter speeds for CMOS- and TTL-programmable counters are limited to under 100 MHz. ECL-type devices can approach a few hundred MHz, but with significant current requirements. However, coupling the dual-modulus-prescaling technique with the available phase-locked-loop synthesizer chips that control the prescaler circumvents these frequency and power-drain constraints.

With this approach, designers can also choose various counter-programming schemes (serial, parallel, or data bus), in addition to achieving higher frequency capabilities. Low-power drain (less than 75 mW) and low-cost devices can also be selected. Moreover, only two ICs are necessary to achieve divide values above 131 000.

Maximum input frequency and dividing range for the counter are controlled by choosing an appropriate 8-pin dual-modulus prescaler. The counter's output appears at synthesizer pin F_V (see the figure). The total input-to-output divide value is governed by the equation:

$$N_{\text{TOTAL}} = N \times P + A$$

N and A represent the value programmed through the serial port into the divide-by-N and divide-by-A counters. P is the lower dual-modulus value that is established by the synthesizer's modulus-control signal.

Typically, A varies from zero to $P-1$ to achieve steps within the system's divide range. N must be equal to or greater than A. $N > A$ then sets a lower limit on N_{TOTAL}, which is dictated by $A_{\text{MAX}} = P-1$.

DIVIDE-BY-$N + 1/2$ CIRCUIT

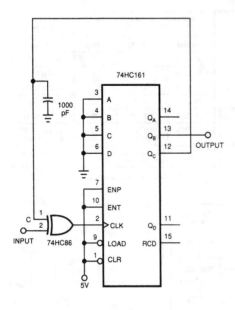

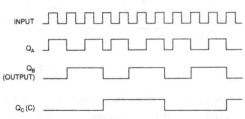

INPUT

Q_A

Q_B
(OUTPUT)

Q_C (C)

Fig. 26-7

Table 1—XOR feedback signals for $N + 1/2$ divider

Divide number	Feedback signal(s)			
N = 1.5	Q_1			
N = 2.5	Q_0	Q_2		
N = 3.5	Q_2			
N = 4.5	Q_0	Q_3		
N = 5.5	Q_0	Q_1	Q_3	
N = 6.5	Q_1	Q_3		
N = 7.5	Q_3			
N = 8.5	Q_0	Q_4		
N = 9.5	Q_0	Q_2	Q_4	
N = 10.5	Q_0	Q_1	Q_2	Q_4
N = 11.5	Q_0	Q_1	Q_4	
N = 12.5	Q_1	Q_4		
N = 13.5	Q_1	Q_2	Q_4	
N = 14.5	Q_2	Q_4		
N = 15.5	Q_4			
N = 16.5	Q_0	Q_5		
N = 17.5	Q_0	Q_3	Q_5	
N = 18.5	Q_0	Q_2	Q_3	Q_5
N = 19.5	Q_0	Q_2	Q_5	
N = 20.5	Q_0	Q_1	Q_2	Q_5

This circuit, instead of dividing by an integer, divides the input signal by $N + 1/2$. With the feedback connections exactly as the figure shows, the circuit divides by 3.5. Point C ultimately controls when the input clocks the 74HC161 4-bit counter. When $C = 0$, the positive edge of the input triggers the counter. If $C = 1$, the negative edge of the input triggers the counter. Each time that point C changes level, the circuit shortens the output pulse width of the counter by half of an input cycle. Thus, the counter's divisor depends on how many changes occur at point C during one output period.

Although the figure divides by 3.5, feeding back different counter outputs produces different divisors. Generally, an m-bit binary counter with pure exclusive-OR (XOR) feedback can form an $N + 1/2$ counter, where N ranges from $2^{m-2} + 1/2$ to $2^{m-1} - 1/2$. The divided output is available at the m-1 bit of the counter.

27

Driver Circuits

The sources of the following circuits are contained in the Sources section, which begins on page 665. The figure number in the box of each circuit correlates to the entry in the Sources section.

Op Amp Power Driver
Emitter/Follower LED Driver
Flip-Flop Independent Lamp Driver

OP AMP POWER DRIVER

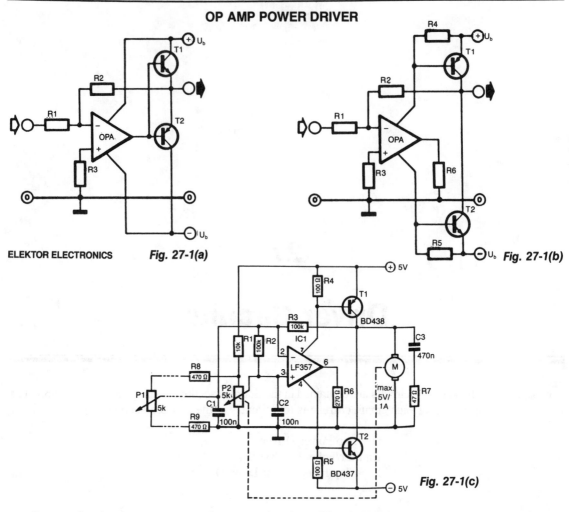

ELEKTOR ELECTRONICS *Fig. 27-1(a)*

Fig. 27-1(b)

Fig. 27-1(c)

Frequently, the output current of an operational amplifier is inadequate for the application as, for instance, when a small motor or loudspeaker has to be driven. Normally, this is resolved by adding an emitter follower to the circuit as shown in Fig. 27-1(a). Unfortunately, that circuit does not allow the full supply voltage, U_b, to be used, because the output voltage of the op amp must always be 1 to 2 V smaller than $\pm U_b$. To that must be added the drop across the base-emitter junction of transistors T1 and T2.

The circuit shown in Fig. 27-1(b) (principle) and Fig. 27-1(c) (practical) is a more appropriate solution: it was designed specifically for driving small motors. Since the output current of the op amp flows through its supply lines, the driver transistors may also be controlled over these lines.

The value of base-emitter resistors R4 and R5 has been chosen to ensure that in spite of the quiescent current through the op amp, T1 and T2 are switched off. Resistor R6 limits the output current of the op amp. If the op amp is a type with guaranteed short-circuit protection, R6 may be replaced by a jump lead.

The output voltage is only 50 to 100 mV (collector-emitter saturation voltage of the driver transistors) smaller than the supply voltage. When choosing these transistors, it is therefore essential to take into account the saturation voltage (in addition to the maximum current amplification and power rating).

OP AMP POWER DRIVER *(Cont.)*

The value of the resistors in an inverting circuit is calculated from:

$$\alpha = \frac{R_2}{R_1}$$

and:

$$R_3 \approx \frac{R_2}{R_1}$$

where α is the amplification.

In a noninverting circuit (R1 between the $-$input and ground and the input signal connected to the $+$input of the op amp), the amplification is:

$$\alpha = \frac{R_2}{(R1+1)}$$

and:

$$R_3 << R_e$$

$$R_4 < \frac{+\alpha}{= U_b}$$

$$R_5 < \frac{-0.5\alpha}{U_b}$$

$$R_6 \approx \frac{U_b}{I_{max}}$$

where R_e is the input impedance of the op amps.

The circuit can be used with discrete (single) op amps only, because double or quadruple types in one package share the supply voltage pins. The setting accuracy of the circuit in Fig. 27-1(c) is better than 1%.

EMITTER/FOLLOWER LED DRIVER

TYPICAL NPN LED DRIVER

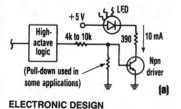

(a)

PNP EF LED DRIVER

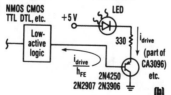

(b)

NPN EF DRIVER

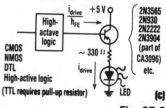

(c)

ELECTRONIC DESIGN

Fig. 27-2

Using emitter/followers saves parts and simplifies LED driver circuits and generally produces less loading on logic circuitry.

FLIP-FLOP INDEPENDENT LAMP DRIVER

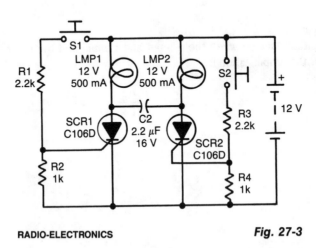

RADIO-ELECTRONICS

Fig. 27-3

Assume first that SCR1 is on and SCR2 is off so that C1 is fully charged, with its LMP2 end positive. The state of the circuit can be changed by pressing S2. As SCR2 turns on, it turns SCR1 off capacitively via its anode. Capacitor C1 then recharges in the opposite manner (i.e., the left end is now positive). The state of the circuit can be changed again by pressing S1, thus driving SCR1 on by way of its gate, and driving SCR2 off capacitively via its anode.

28

Electronic Locks

The sources of the following circuits are contained in the Sources section, which begins on page 665. The figure number in the box of each circuit correlates to the entry in the Sources section.

Digital Entry Lock
Keyless Lock

DIGITAL ENTRY LOCK

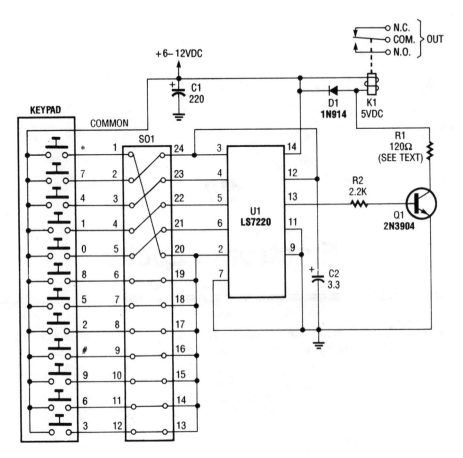

POPULAR ELECTRONICS

Fig. 28-1

A keypad enters a four-digit access code, which is programmed via jumpers on a 24-pin plug-in header and socket. U1 is an LST220, which detects a four-digit sequential data input. When the correct data is entered into the keyboard, pin 13 of U1 goes high, which activates Q1 and K1. K1 drives an external electric lock solenoid, etc.

KEYLESS LOCK

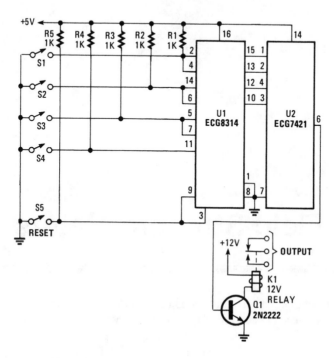

POPULAR ELECTRONICS

Fig. 28-2

The circuit uses a four-bit latch (U1). What makes the circuit sequential is that the set input of the first bit latch is tied to the reset of the second bit latch, and so forth. That ensures that any bit latched will be reset by the previous bit latch. The ECG8314 also has a master reset (pin 9) that is tied to the first bit-latch reset (pin 3), which provides an added measure of security for the lock.

The outputs of U1 are fed to a four-input AND gate (U2), then to Q1 (used as switching transistor), which is used to drive relay K1. The EGC8314 has an enable low (pin) that can be used as a timing circuit, if that is desired.

29

Field-Strength Meters

The sources of the following circuits are contained in the Sources section, which begins on page 665. The figure number in the box of each circuit correlates to the entry in the Sources section.

UHF Field-Strength Meter
Field-Strength Meter
Signal-Strength Meter

UHF FIELD-STRENGTH METER

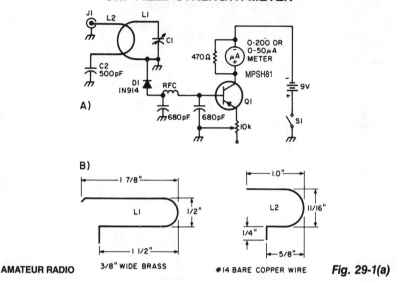

AMATEUR RADIO

Fig. 29-1(a)

Field strength meter schematic.

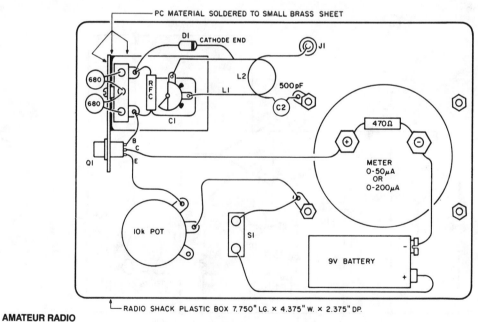

AMATEUR RADIO

Fig. 29-1(b)

Parts layout for the field strength meter.

Useful for transmitter or antenna alignment, this meter covers 400 to 500 MHz. An amplifier stage is included for improved sensitivity. Follow the layout in Fig. 29-1(b).

FIELD-STRENGTH METER

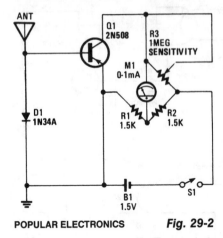

POPULAR ELECTRONICS   *Fig. 29-2*

This field-strength meter is basically a bridge circuit that is equipped with a 0-to-1-mA meter as a readout.

SIGNAL-STRENGTH METER

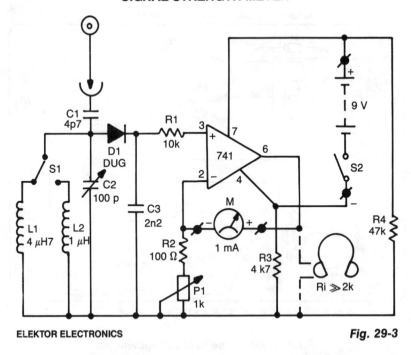

ELEKTOR ELECTRONICS   *Fig. 29-3*

This field-strength meter is useful for antenna testing. It covers 6 to 60 MHz and uses a rugged 0-to-1 mA meter. A 9-V battery supplies power. The unit can be mounted in a small plastic or metal case.

30

Filter Circuits

The sources of the following circuits are contained in the Sources section, which begins on page 665. The figure number in the box of each circuit correlates to the entry in the Sources section.

Fast-Response (settling) Low-Pass Filter
Tunable Audio Filter
Turntable Rumble Filter
Tunable Bandpass Filter
Low-Cost Crystal Filters
Antialiasing and Sync-Compensation Filter
Two-Section 300-to-3 000 Hz Speech Filter
300-to-3 400 Hz Second-Order Speech Filter
Fourth-Order 100-Hz High-Pass Filter
Simple Ripple Suppressor
Second-Order 100-Hz High-Pass Filter
Scratch Filter
Simple Rumble Filter
1 000:1 Tuning Voltage-Controlled Filter
Two Sallen-Key Low-Pass Active Filter

FAST-RESPONSE (SETTLING) LOW-PASS FILTER

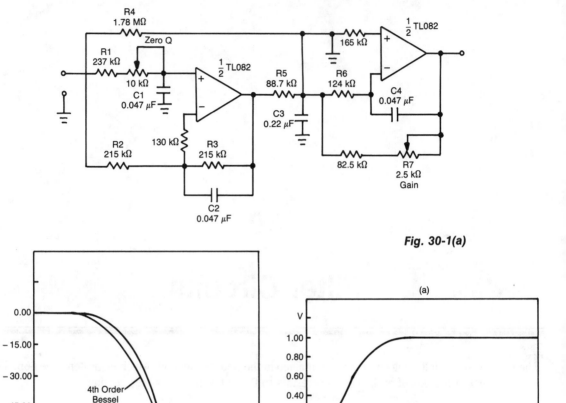

Fig. 30-1(a)

Fig. 30-1(b)

Fig. 30-1(c)

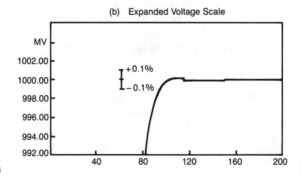

Fig. 30-1(d)

FAST-RESPONSE (SETTLING) LOW-PASS FILTER (*Cont.*)

By introducing an extra transmission zero to the stopband of a low-pass filter, a sharp roll-off characteristic can be obtained. The filter design example of Fig. 30-1(a) shows that the time-domain performance of the low-pass section can also be improved. Figure 30-1(b) shows the attenuation characteristic of the proposed circuit. Position of the transmission zero is determined by the passive components around the first op amp. It was chosen to obtain 60 dB of rejection at 60 Hz.

A suitable fourth-order Bessel filter has the frequency response, as shown by the dashed line. Its response to a step input is characterized by settling time to 0.1% of $1.8 \div F_C = 180$ ms.

Figure 30-1(c) and 30-1(d) represent the step response for the filter of Fig. 30-1(a) in both normal and expanded voltage scales. As you can see, settling time to 0.1% is below 100 ms; overshoot and ringing stay below 0.03%.

This quite significant speed and accuracy improvement can be a major factor, particularly for low-frequency applications. Averaging filter for low-frequency linear or true rms ac-to-dc converters is an example. Some anti-aliasing applications can also be considered.

For best results, resistance ratios $R_4 \div R_5 = 20$, $R_6 \div R_5 = 1.4$, and capacitance ratios $C_3 \div C_2 = C_3 \div C_4 = 4.7$ should be kept up for any selected F_C.

TUNABLE AUDIO FILTER

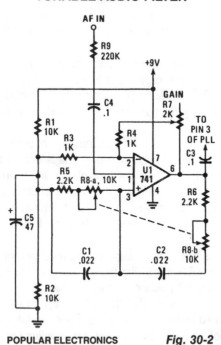

POPULAR ELECTRONICS   *Fig. 30-2*

This circuit uses a Wien Bridge and variable negative feedback. R7 controls the gain and R8A and R8B controls the tuned frequency.

TURNTABLE RUMBLE FILTER

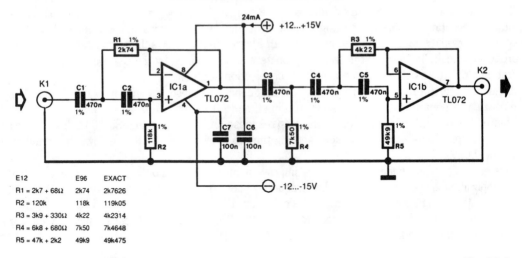

E12	E96	EXACT
R1 = 2k7 + 68Ω	2k74	2k7626
R2 = 120k	118k	119k05
R3 = 3k9 + 330Ω	4k22	4k2314
R4 = 6k8 + 680Ω	7k50	7k4648
R5 = 47k + 2k2	49k9	49k475

ELEKTOR ELECTRONICS

Fig. 30-3

Many record players unfortunately exhibit two undesired side effects: rumble (noise caused by the motor and the turntable) and other low-frequency spurious signals. The active high-pass Chebyshev filter presented here was designed to suppress those noises. The filter has a 0.1-dB ripple characteristic and a cut-off point of 18 Hz.

The choice of a Chebyshev filter might not seem optimum for audio purposes, but because of its 0.1-dB ripple in the pass band it behaves very much like a Butterworth type. Its advantage is that the response has steeper skirts (which are calculated curves). Frequencies below 10 Hz are attenuated by more than 35 dB. The phase behavior in the pass band shows a gradual shift so that its effect on the reproduced sound is inaudible.

If the filter is used in a stereo installation, the characteristics of both filters must be identical or nearly so. Phase differences between channels can be heard—perhaps not so much at lower frequencies, but certainly in the mid ranges. To ensure identity and also to obtain the desired characteristics, capacitors C1 through C5 must be selected carefully. It does not matter much whether their value is 467 or 473 pF; this difference only causes a slight shift of the cut-off point. However, they must be identical within that 1% tolerance. For symmetry of channels, the capacitors can be paired and then used in either channel at the corresponding position.

The diagram shows theoretical values for the resistors: their practical values are given in the table. The prototype was constructed with 5% metal-film types from the E12 series and these were used without sorting. Their tolerance was perfectly acceptable in practice.

The current drawn by the circuit is purely that through the op amp and it amounts to about 4 mA. The high cut-off point is also determined by the op amp and it lies at about 3 MHz.

The only problem that cannot be foreseen is a possible coupling capacitor in the signal source. That component will be in series with C1 and this might adversely affect the frequency response. However, if its value is greater than 47 μF, it will have little if any effect; if it is below that value, it is best removed; C1 will assume its function.

TUNABLE BANDPASS FILTER

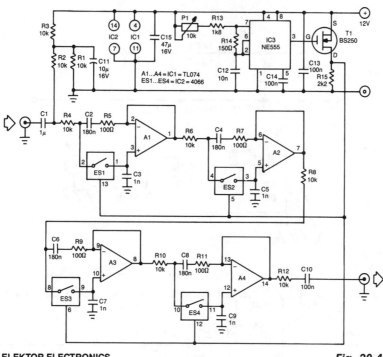

Fig. 30-4

One of the difficulties in the design of higher-order tunable bandpass filters is achieving correct tracking of the variable resistors in the RC networks. The use of switched capacitor networks can obviate that difficulty, as is shown in this filter.

The filter can be divided roughly into two stages: an oscillator that controls the electronic switches and the four phase-shift networks that provide the filtering proper. The oscillator, based on a 555, generates a pulsating signal whose frequency is adjustable over a wide range: the duty factor varies from 1:10 to 100:1.

Electronic switches ES1 through ES4 form the variable resistors whose value is dependent on the frequency of the digital signal. The operation of these switches is fairly simple. When they are closed, their resistance is about 60 Ω; when they are open, it is virtually infinitely high. If a switch is closed for, say, 25% of the time, its average resistance is therefore 240 Ω. Varying the open:closed ratio of each switch varies the equivalent average resistance. The switching rate of the switches must be much greater than the highest audio frequency to prevent audible interference between the audio and the clock signals.

The input signal causes a given direct voltage across C1, so the op amp can be operated in a quasisymmetric manner, in spite of the single supply voltage. The direct voltage is removed from the output signal by capacitor C10.

The fourth-order filter in the diagram can be used over the entire audio range and it has an amplification of about 40, although this depends to some extent on the clock frequency. The bandwidth depends mainly on the set frequency. The circuit draws a current of not more than 15 mA.

LOW-COST CRYSTAL FILTERS

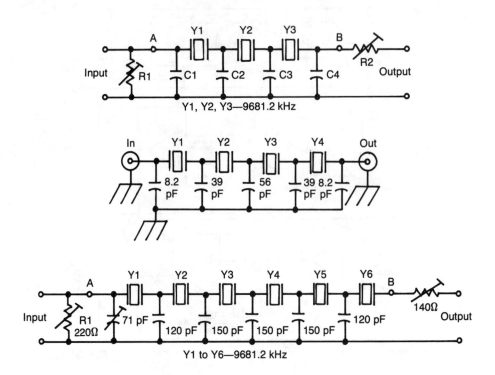

Y1, Y2, Y3—9681.2 kHz

Y1 to Y6—9681.2 kHz

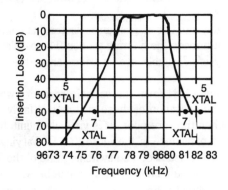

Fig. 30-5

Low-cost CB crystals can be used for these 9-MHz crystal ladder filters. Notice that the 27-MHz crystals (3rd overtone) are used on their fundamental frequencies.

ANTIALIASING AND SYNC-COMPENSATION FILTER

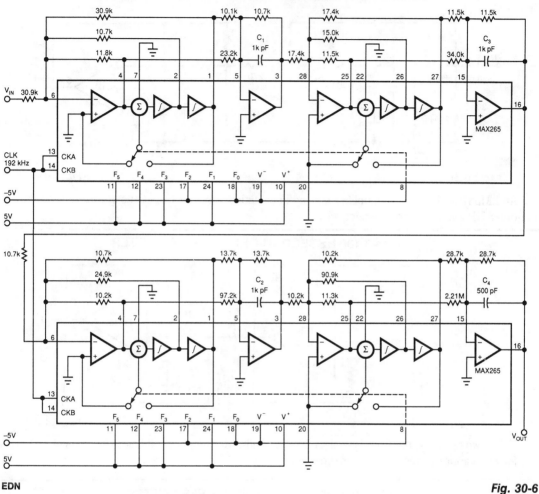

EDN

Fig. 30-6

Two dual-biquad filter chips and some external components form a multipurpose filter to reconstruct D/A converter signals. Connected to a converter's output, the filter provides antialiasing, reduces the D/A converter's quantization noise, and compensates for $\sin(\pi x) \div (\pi x)$—the "sync" function (attenuation).

The circuit incorporates an inverse-sync function that operates to one-third of the converter's sample rate. Beyond one-third, the filter's response shifts to a stopband filter, which provides -70 dB attenuation. This attenuation conforms to the converter's inherent signal-to-noise ratio and quantization error.

To prevent aliasing, the stopband edge must be no higher than the Nyquist frequency ($f_{sn} \div 2$). To achieve 70-dB stopband rejection with this eighth-order filter requires a transition ratio ($f_{STOPBAND} \div f_{PASSBAND}$) of 1.5, which sets the passband's upper limit at $f_s \div 3$.

Notice also that you can apply a simple divide-by-64 circuit to the 192-kHz clock frequency to set the necessary $3\times$ ratio between the converter's sample rate and the filter's 1-kHz corner frequency. The V^+, V^-, and the F_0 through F_5 connections program each filter chip for an f_{CLK}/f_0 ratio of 191.64.

TWO-SECTION 300-3 000 Hz SPEECH FILTER

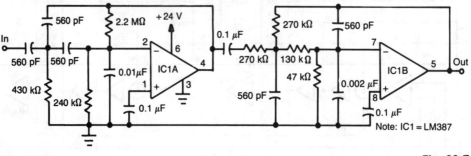

RADIO-ELECTRONICS

Fig. 30-7

An LM387 dual low-noise amplifier is used in an active filter. Both sections are used to produce second-order HP and LP filters, respectively.

300-to-3 400 Hz SECOND-ORDER SPEECH FILTER

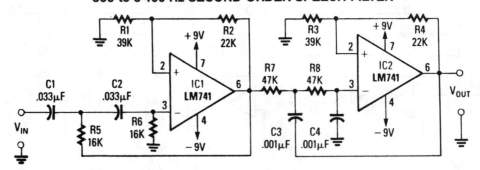

RADIO-ELECTRONICS

Fig. 30-8

Using two op amps, this filter is designed for second-order response. It has a bandpass of 300 to 3 400 Hz, for applications in speech or telephone work.

FOURTH-ORDER 100-Hz HIGH-PASS FILTER

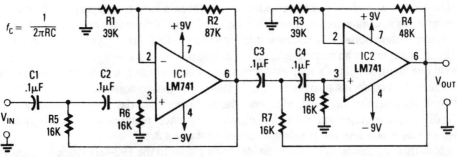

RADIO-ELECTRONICS

Fig. 30-9

This filter, using two sections of LM741, can be scaled in frequency, if desired.

SIMPLE RIPPLE SUPPRESSOR

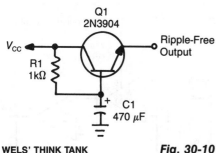

WELS' THINK TANK **Fig. 30-10**

This circuit, at times called a *capacitance multiplier*, is useful for suppression of power-supply ripple. C1 provides filtering that is equal to a capacitor of $(B+1)$ C_1, where B = dc current gain of Q1 (typically > 50).

SECOND-ORDER 100-Hz HIGH-PASS FILTER

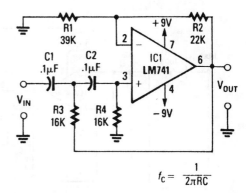

$$f_c = \frac{1}{2\pi RC}$$

RADIO-ELECTRONICS **Fig. 30-11**

This second-order filter can be scaled to change the cutoff frequencies.

SCRATCH FILTER

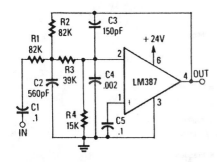

RADIO-ELECTRONICS **Fig. 30-12**

Designed to produce 12-dB/octave roll-off above the 10-kHz cutoff frequency, this LP active filter will help reduce needle scratch on records. It uses an LM387 low-noise amplifier IC.

SIMPLE RUMBLE FILTER

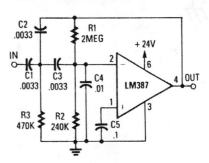

RADIO-ELECTRONICS **Fig. 30-13**

This circuit is a two-section active HP filter using an LM387, with a cutoff below 50 Hz at 12-dB per octave. It will help reduce rumble as a result of turntable defects in record systems.

175

1 000:1 TUNING VOLTAGE-CONTROLLED FILTER

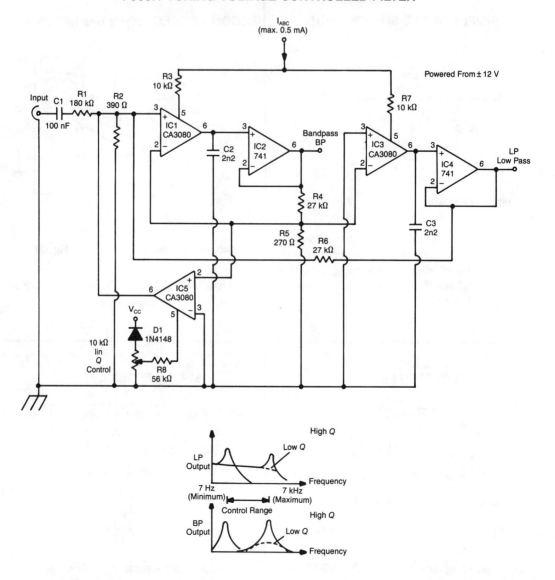

ELECTRONICS TODAY INTERNATIONAL

Fig. 30-14

A standard dual integrator filter can be constructed using a few CA3080s. By varying IABC, the resonant frequency can be swept over a 1 000:1 range. At IC1, three are current-controlled integrators. At IC2, four are voltage followers that serve to buffer the high-impedance outputs of the integrators. A third CA3080 (IC5) is used to control the Q factor of the filter. The resonant frequency of the filter is linearly proportional to I_{ABC}. Hence, this unit is very useful in producing electronic music. Two outputs are produced: a low-pass and a bandpass response.

TWO SALLEN-KEY LOW-PASS ACTIVE FILTERS

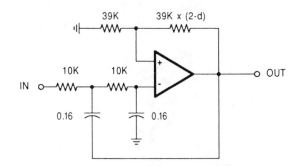

(A) SIMPLE SECOND ORDER SALLEN-KEY SECTION

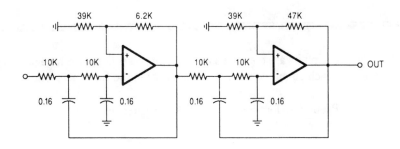

(B) FOURTH ORDER BUTTERWORTH LOW PASS AUDIO FILTER

RADIO-ELECTRONICS

Fig. 30-15

These filters are designed for 10-kΩ impedance level and 1-kHz cutoff frequency, but the components can be scaled as required for other impedances and cutoff frequencies.

31

Flashers and Blinkers

The sources of the following circuits are contained in the Sources section, which begins on page 666. The figure number in the box of each circuit correlates to the entry in the Sources section.

Pseudorandom Simulated Flicker Sequencer
Xenon Flasher
Strobe Alarm
Sequential Flasher
LED Flasher
Sequential LED Flasher with Reversible Direction
Multivibrator with LEDs
Flashing LED Controller
Flicker Light

PSEUDORANDOM SIMULATED FLICKER SEQUENCER

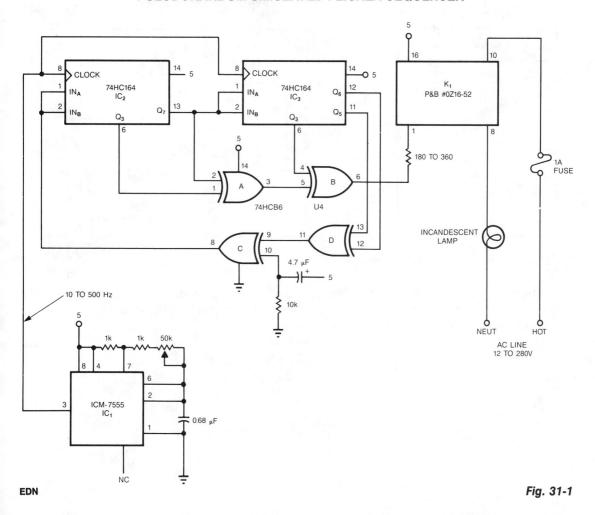

Fig. 31-1

The pseudorandom sequencer drives a solid-state relay. If you power a low-wattage lamp from the relay, the lamp will appear to flicker like a candle's flame in the wind; using higher-wattage lamps allows you to simulate the blaze of a fireplace or campfire. You can enhance the effect by using three or more such circuits to power an array of lamps.

The circuit comprises an oscillator, IC1, and a 15-stage, pseudorandom sequencer, IC2–4. The sequencer produces a serial bit stream that repeats only every 32 767 bits. Feedback from the sequencer's stages 14 and 15 go through IC4D and back to the serial input of IC2. Notice the RC network that feeds IC4C; the network feeds a positive pulse into the sequencer to ensure that it won't get stuck with all zeros at power-up. The leftover XOR gates IC4A and IC4B further scramble the pattern. The serial stream from IC4B drives a solid-state relay that features zero-voltage switching and can handle loads as high as 1 A at 12 to 280 Vac.

XENON FLASHER

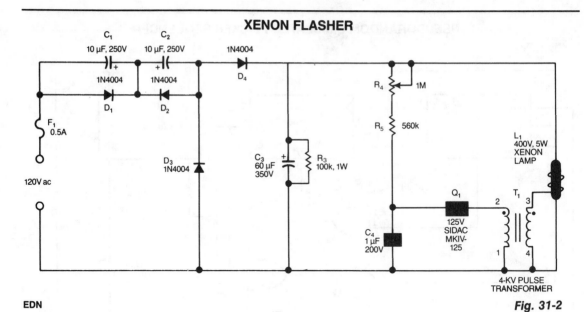

EDN

Fig. 31-2

Using a voltage-doubler supply, this circuit charges a 60-μF capacitor and discharges it through a Xenon lamp. The SIDAC device is manufactured by Motorola. It is a two-terminal device that breaks over at a specified voltage. R4, R5, and C4 determine the flash rate.

STROBE ALARM

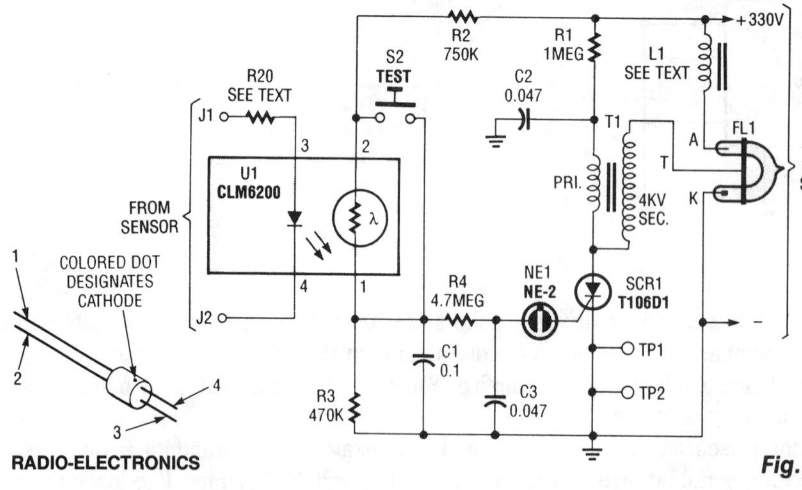

Fig. 31-3

RADIO-ELECTRONICS

This strobe gives a visual indication of a sensor input. The input signal causes U1, a light dependent resistor, to charge C1 and C3 through R4. When NE1 fires, C3 discharges into SCR1, which triggers it and causes C2 to discharge through trigger transformer T1, which triggers Flashlamp FL1. The 330-V supply should have about 50 to 100 μF output capacitance. L1 supplies about 25-mH inductance to prolong the flash and the life of FL1.

SEQUENTIAL FLASHER

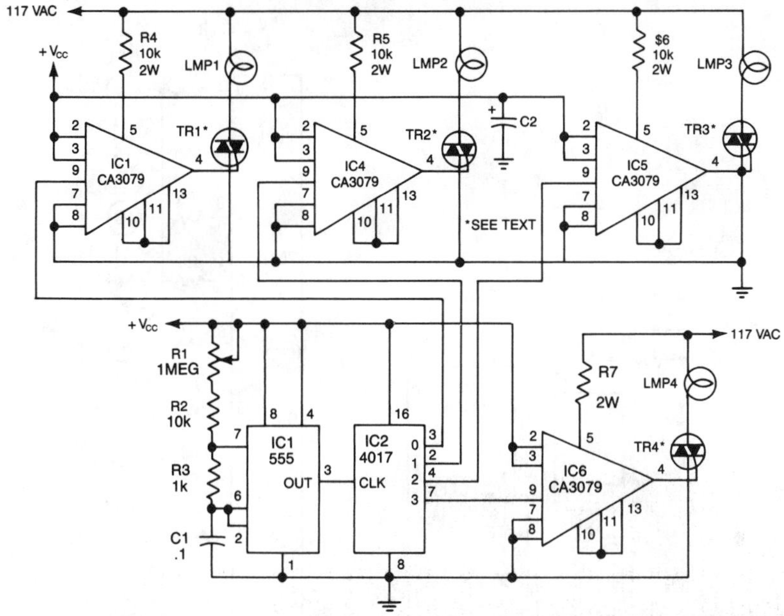

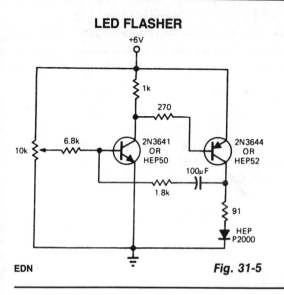

RADIO-ELECTRONICS

Fig. 31-4

Using a 555 timer to drive a CMOS counter, this device uses RCA CA3079 zero-voltage switch to control triacs TR1 through TR4. This circuit can be used to sequence lamp displays, etc.

Caution: The CA3079s are connected to the 117-V line, as is the clock and counter circuit and their power supplies. Use caution, good insulation, and safe construction practices.

LED FLASHER

This circuit is designed to flash an LED. The 100-μF capacitor can be changed to alter the flash rate as desired.

EDN

Fig. 31-5

181

SEQUENTIAL LED FLASHER WITH REVERSIBLE DIRECTION[1]

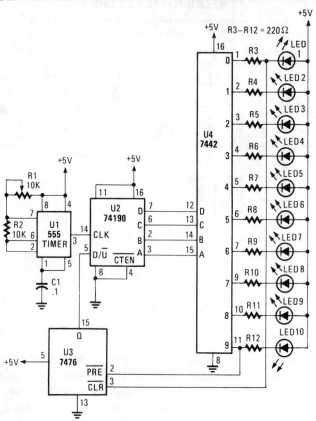

POPULAR ELECTRONICS

Fig.31-6

A 555 timer clocks a 74190 up/down counter. The 74190 drives BCD decoder driver 7442. The 7476 is used to reverse the count on 0 and 9, which results in an up-down-up-down count sequence.

MULTIVIBRATOR WITH LEDs

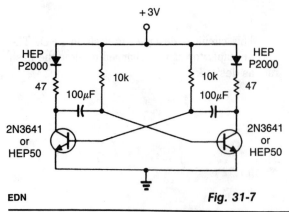

A simple astable multivibrator is used to alternately flash two LEDs. The approximate time constant is 0.69.

$$(R_1C_1 + R_2C_2) \quad R_1 = R_2 = 10 \text{ k}\Omega$$
$$C_1 = C_2 = 100 \ \mu\text{F}$$

EDN

Fig. 31-7

FLASHING LED CONTROLLER

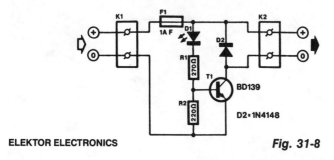

ELEKTOR ELECTRONICS

Fig. 31-8

The LED with integrated flasher is connected in series with the base-emitter junction of transistor T1. Thus, the load connected to K2 is switched on and off in rhythm with the flash rate. This load can be a relay or a lamp.

The maximum collector current of the transistor (of the BD139 = 750 mA) must not be exceeded. If that is not sufficient, a power Darlington can be used, which will give some amperes. The current drawn by the circuit under no-load conditions amounts to 20 mA.

FLICKER LIGHT

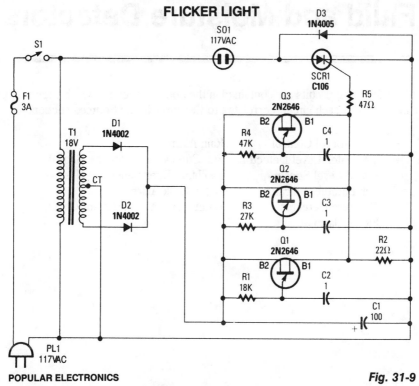

POPULAR ELECTRONICS

Fig. 31-9

This circuit will produce a flicker light effect with an ordinary incandescent lamp. Three UJT relaxation oscillators fire the SCR in a pattern.

32

Fluid and Moisture Detectors

The sources of the following circuits are contained in the Sources section, which begins on page 666. The figure number in the box of each circuit correlates to the entry in the Sources section.

Water-Level Control
3-V Water-Level Detector
Liquid-Level Sensor
Full Bathtub Indicator
Moisture Detector
Flood Alarm

Rain Alarm
Full-Cup Detector for the Blind
Latching Water Sensor
Water-Leak Alarm
Water-Level Measurement Circuit

WATER-LEVEL CONTROL

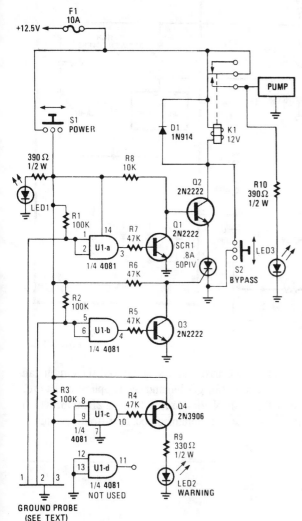

POPULAR ELECTRONICS

Fig. 32-1

This circuit will power up a water pump when the water reaches a predetermined level. Then it turns itself off when the water recedes to another predetermined point.

Gates U1A through U1C each have their two inputs tied together, and serve as probes. The probes are then placed at various levels to trigger a particular function at a predetermined time. The ground side of the circuit is placed below the minimum water level. The inputs to each gate are tied high through a 100-kΩ resistor connected to the +12.5-V bus.

As the water level slowly rises to probe 1, the input to U1A is pulled low by the conduction of current through the water to the ground probe. That turns Q1 off and Q2 on. With Q2 turned on, the circuit is placed in the standby mode, ready to activate the pump when conditions are right.

Probe 2 is placed at the maximum water level. If the water level reaches probe 2, the input of U1B is brought low, turning Q3 on, which, in turn, causes current to be applied to the gate of SCR1, turning it on. The circuit through K1, Q2, and SCR1 is now complete to ground, and the water pump is now turned on, which causes the water level to recede. When the water level falls below probe 2, U1B goes back to logic high.

However, because of the latching nature of SCR1, the pump continues to run until the water level falls below probe 1. At that point, the ground circuit opens and de-energizes K1, which turns the pump off. The pump will not turn on again until the water level again rises above probe 2.

Probe 3 was added as a warning. If the water level reaches probe 3, LED2 indicates that the pump is not working. Switch S2 is a manual override and S1 powers the sensing circuit. LED3 indicates that power has been applied to the pump. LED1 indicates that power has been applied to the sensor.

3-V WATER-LEVEL DETECTOR

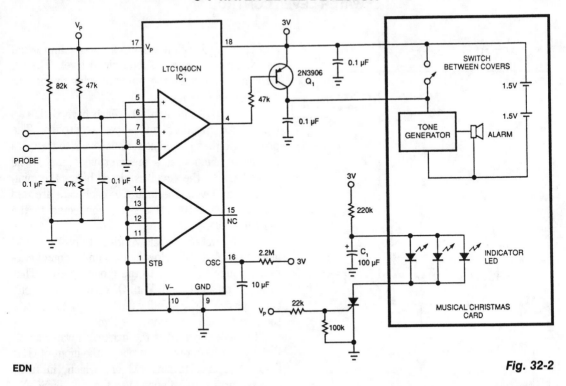

EDN

Fig. 32-2

Originally, this circuit was used to sense a low-water level in a Christmas tree stand, but the circuit can be used as a water-level detector for pump controls, water sensors (for garden and lawn applications), etc. A comparator and probe setup with a Linear Technology LTC1040CN comparator drives a 2N3906, which switches a tone generator. Sampling occurs every 20 seconds, which minimizes current drain. A pair of dry cells will power the circuit for several months.

LIQUID-LEVEL SENSOR

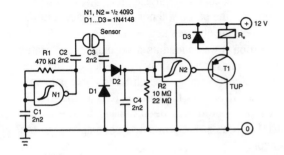

This circuit uses an ac-sensing signal to eliminate electrolytic corrosion. The ac signal is rectified and used to drive a transistor that controls a relay.

ELEKTOR ELECTRONICS

Fig. 32-3

FULL BATHTUB INDICATOR

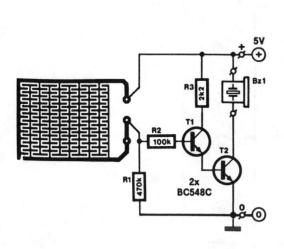
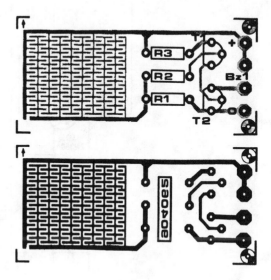

ELEKTOR ELECTRONICS   *Fig. 32-4*

Running a bath can end in a minor domestic disaster if you forget to turn off the taps in time. This indicator activates an active buzzer to provide an audible warning when a given water level is reached.

Because the water sensor and the driver circuit for the buzzer are contained on one PC board, the indicator, together with the 9-V battery and the buzzer, can be built into a compact case. Obviously, the sensor, which is etched on the PC-board, must not be fitted in case-iron or steel bath, the indicator is secured to it with the aid of a magnet glued onto the case. To prevent scratching the bath, the magnet can be covered in plastic or rubber. If you have a polypropylene bath, the indicator can be stuck to it with blue tack or double-sided adhesive tape.

When the water reaches the sensor, the base of T1 is connected to the positive supply line. As a result, T1 and T2 are switched on so that the buzzer BZ1, a self-oscillating type, is activated. The current drawn by the circuit in that condition is about 25 mA.

In case the circuit is actuated by steam, its sensitivity can be reduced by increasing the value of R2. It is best to tin the PC board tracks to prevent corrosion.

T1, T2 = BC548C
BZ1 = active piezo-ceramic resonator

MOISTURE DETECTOR

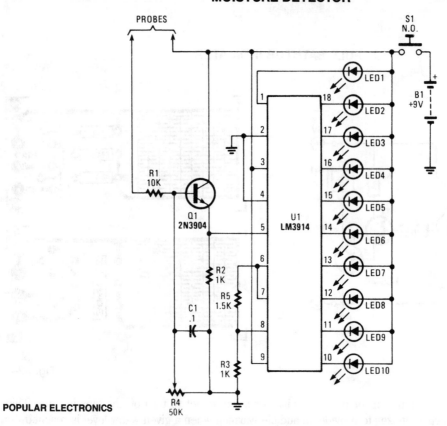

POPULAR ELECTRONICS

Fig. 32-5

A bar-graph LED driver is used to drive 10 LEDs to give a relative indication of moisture. The moisture probes are connected so that electrical conductivity due to moisture tends to forward bias Q1, providing a dc voltage at pin 5 of U1 that is proportional to leakage current. Ideally, the probes should be made of stainless steel.

FLOOD ALARM

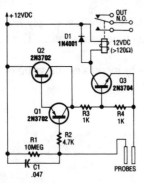

RADIO-ELECTRONICS   **Fig. 32-6**

Using a few bipolar transistors, this circuit acts as a flood alarm. When liquid touches the probes, leakage current biases Q1, Q2, and Q3 (a dc-coupled amplifier) into conduction, which activates the relay. The contacts can be hooked into the alarm system.

RAIN ALARM

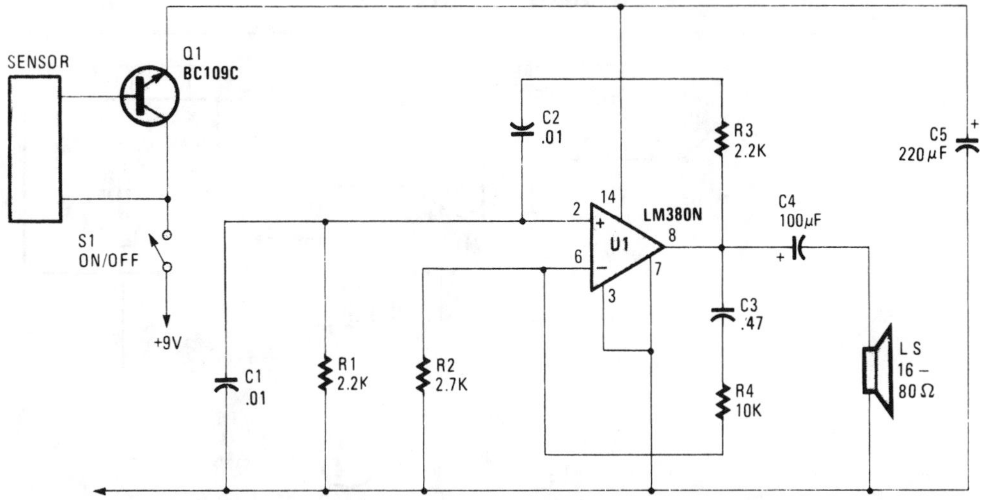

POPULAR ELECTRONICS

Fig. 32-7

This rain sensor causes Q1 to conduct when conductive liquid (rainwater, etc.) applies bias to its base. This bias triggers LM380N oscillator and causes LS to emit a tone.

FULL-CUP DETECTOR FOR THE BLIND

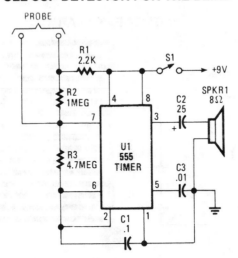

POPULAR ELECTRONICS

Fig. 32-8

At the heart of the Full-Cup Detector is a 555-oscillator/timer configured to produce a 15-Hz click, until its probe contacts are bridged, at which time its output frequency goes to about 500 Hz.

This circuit can be used by the visually handicapped to determine when a cup or bowl is full of liquid (coffee, soup, etc). U1, an NE555, produces ticks at 15 Hz. A set of probes (wire, etc.) is placed in the container at the desired level. When the liquid level contacts the probes, the frequency of clicks increases to several hundred hertz, depending on its conductivity.

LATCHING WATER SENSOR

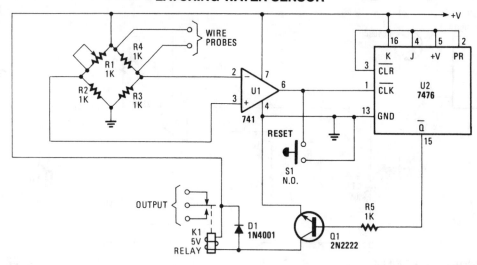

POPULAR ELECTRONICS

Fig. 32-9

A balanced Wheatstone bridge controls a JK flip-flop that uses an op amp as an interface. This in turn drives a relay circuit. R1 through R4 can be made larger for increased sensitivity.

WATER-LEAK ALARM

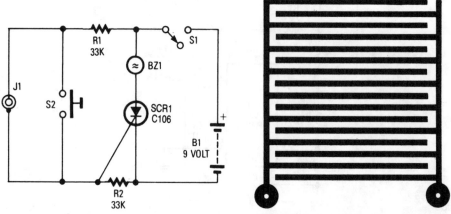

POPULAR ELECTRONICS

If you choose to make your own moisture sensor, this foil pattern should come in handy.

Fig. 32-10

A sensor connected to J1 causes SCR1 to conduct, which sounds buzzer BZ1. The sensor is a PC-board foil pattern grid. Several sensors can be wired in parallel for increased coverage or to monitor several places simultaneously.

WATER-LEVEL MEASUREMENT CIRCUIT

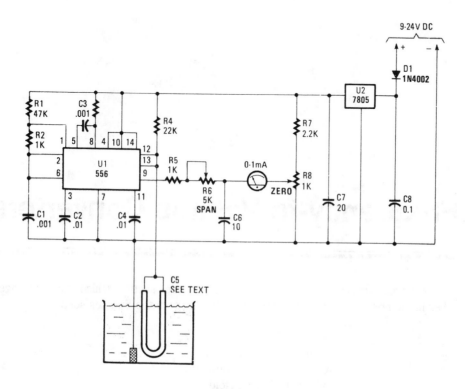

POPULAR ELECTRONICS

Fig. 32-11

Using a capacitor sensor to detect a water level is a simple method of sensing. This circuit uses C5, which is 10″ to 20″ of #22 enamelled wire as one electrode. This shifts the oscillator, an NE556 timer, in frequency. The frequency shift depends on the capacitance charge, which in turn varies with water level. A meter connected to pin 9 of the 556 is used as an indicator. C5 can be made larger or smaller to suit the intended application.

33

Frequency-to-Voltage Converters

The sources of the following circuits are contained in the Sources section, which begins on page 666. The figure number in the box of each circuit correlates to the entry in the Sources section.

FREQUENCY/VOLTAGE CONVERTER WITH OPTOCOUPLER INPUT

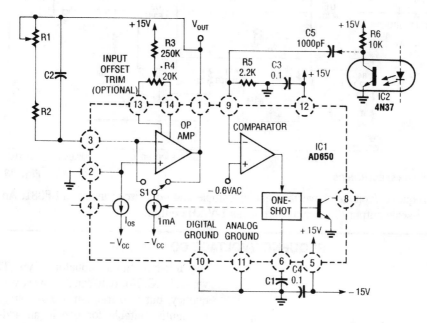

Full-Scale Frequency	Full-Scale Output	C1	C2	R1 (ohms)	R2 (ohms)
10 kHz	1 V	3300 pF	3.3 μF	1K	3.8K
10 kHz	10 V	3300 pF	0.33 μF	10K	38.3K
100 kHz	1 V	680 pF	0.33 μF	500	1.82K
100 kHz	10 V	680 pF	3300 pF	5K	18.2K
1 MHz	1 V	47 pF	3300 pF	500	1.33K
1 MHz	10 V	47 pF	1000 pF	5K	13.3K

CIRCUIT VALUES

RADIO-ELECTRONICS

Fig. 33-1

In this circuit, the input from IC2 optocoupler is fed to the comparator input of the AD650 (Analog Devices or Maxim Electronics) V/F converter. This internally generates a pulse that is fed to the op amp, which outputs a dc voltage that is proportional to frequency. Component values are shown in the figure.

FREQUENCY/VOLTAGE CONVERTER WITH SAMPLE AND HOLD

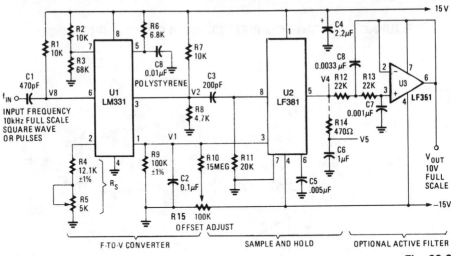

POPULAR ELECTRONICS

Fig. 33-2

U1 is a frequency/voltage converter, feeding sample-and-hold circuit using an LF381. An LF351 provides 10-V full-scale output. The circuit produces 1-V/kHz output.

FREQUENCY/VOLTAGE CONVERTER

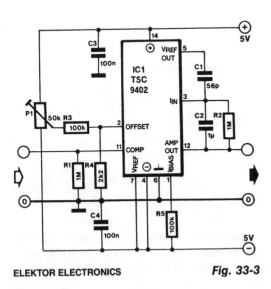

ELEKTOR ELECTRONICS

Fig. 33-3

Teledyne Semiconductor's Type TSC9402 is a versatile IC. Not only can it convert voltage into frequency, but also frequency into voltage. It is thus eminently suitable for use in an add-on unit for measuring frequencies with a multimeter. Only a few additional components are required for this.

Just one calibration point sets the center of the measuring range (or of that part of the range that is used most frequently). The frequency-proportional direct voltage at the output (pin 12—AMP OUT) contains interference pulses at levels up to 0.7 V. If these have an adverse effect on the multimeter, they can be suppressed with the aid of a simple RC network. The output voltage, U_o, is calculated by:

$$U_o = U_{ref}(C_1 + 12 \text{ pF}) \, R_2 f_{in}$$

Because the internal capacitance often has a greater value than the 12 pF taken here, the formula does not yield an absolute value. The circuit has a frequency range of dc to 10 kHz. At 10 kHz, the formula gives a value of 3.4 V. The circuit draws a current of not more than 1 mA.

FREQUENCY/VOLTAGE CONVERTER

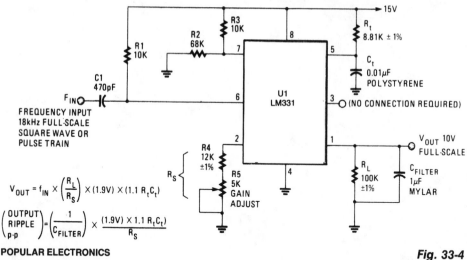

$$V_{OUT} = f_{IN} \times \left(\frac{R_L}{R_S}\right) \times (1.9V) \times (1.1\, R_tC_t)$$

$$\left(\begin{array}{c}\text{OUTPUT} \\ \text{RIPPLE} \\ \text{p-p}\end{array}\right) = \left(\frac{1}{C_{FILTER}}\right) \times \frac{(1.9V) \times 1.1\, R_tC_t)}{R_S}$$

POPULAR ELECTRONICS

Fig. 33-4

A dc output that is proportional to frequency can be derived with this circuit. It is useful for analog frequency meter or tachometer applications.

SINGLE-SUPPLY FREQUENCY/VOLTAGE CONVERTER

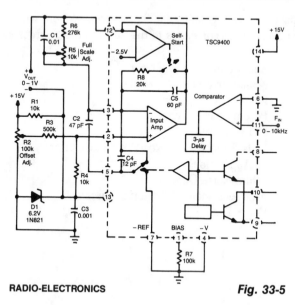

RADIO-ELECTRONICS

Fig. 33-5

A Teledyne TSC9400 provides 0-to-1-V output from a 0-to-10-kHz input. A single +15-V supply is used. Linearity is 0.25% to 10 kHz.

34

Function Generators

The sources of the following circuits are contained in the Sources section, which begins on page 667. The figure number in the box of each circuit correlates to the entry in the Sources section.

AUDIO FUNCTION GENERATOR

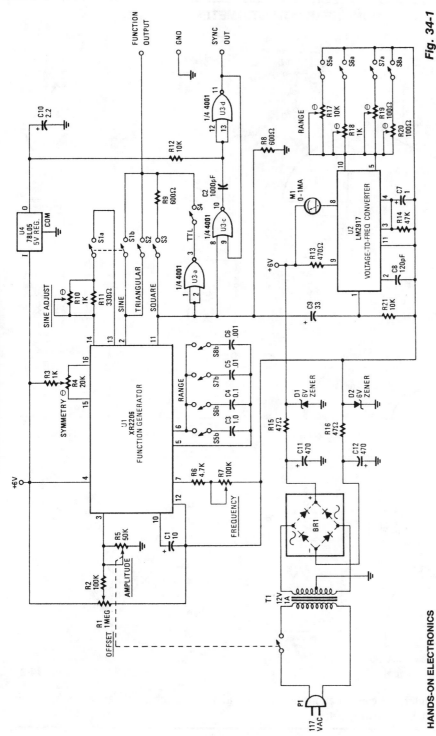

Fig. 34-1

HANDS-ON ELECTRONICS

Using an EXAR XR2206, this generator will produce sine, square, and triangular waves from 10 Hz to 100 kHz. U1 is the XR2206 chip, R7 controls frequency, and S5 through S8 select the frequency range. U3 produces a TTL-compatible square-wave output, while U3C and D produce a sync signal for scope use. U2 is a frequency/voltage converter that is used to drive analog meter M1, which reads the generator frequency.

NONLINEAR POTENTIOMETER OUTPUTS

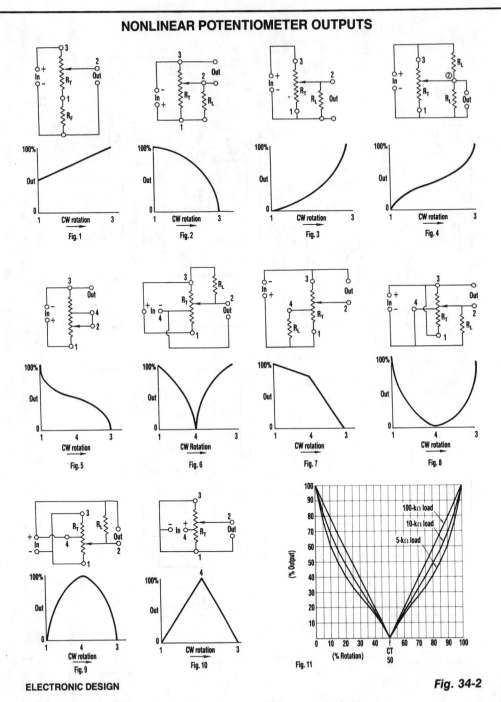

Fig. 1

Fig. 2

Fig. 3

Fig. 4

Fig. 5

Fig. 6

Fig. 7

Fig. 8

Fig. 9

Fig. 10

Fig. 11

ELECTRONIC DESIGN

Fig. 34-2

Using these illustrated configurations, various rotation output characteristics can be obtained from potentiometers and one or two resistors.

FUNCTION GENERATOR

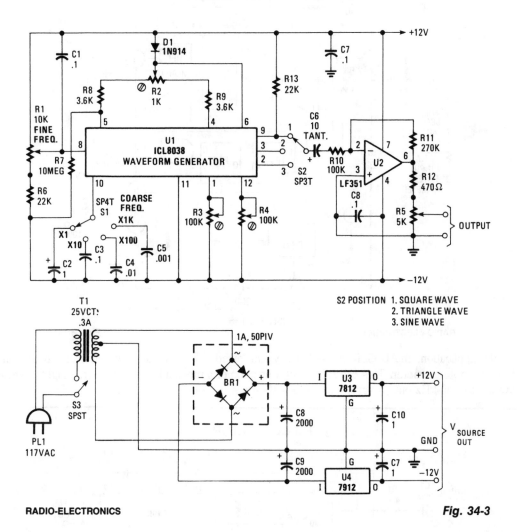

S2 POSITION 1. SQUARE WAVE
2. TRIANGLE WAVE
3. SINE WAVE

RADIO-ELECTRONICS

Fig. 34-3

Using an Intersil ICL8038, this function generator generates frequencies from 1 Hz to over 80 kHz. R1 is the fine frequency control and S1 is the coarse frequency control range switch. S2 selects square-, triangle-, or sine-wave output. U2 is a buffer amplifier and R5 sets output level. R2 is adjusted for a symmetrical triangle wave. R3 and R4 are adjusted for minimum sine-wave distortion. Power supply is ± 12 V at less than 100 mA.

POTENTIOMETER-POSITION V/F CONVERTER

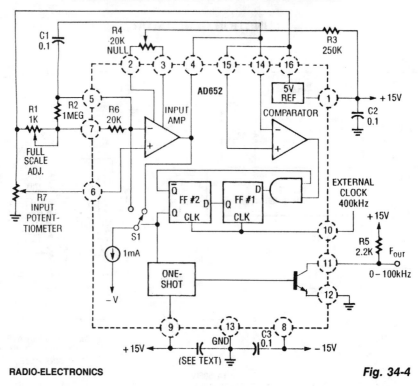

Fig. 34-4

In this application, an AD652IC is used in a synchronized V/F converter that derives its input from the position of a potentiometer. This can represent a position of a mechanical component, weight, size, etc., to give a 0-to-100-kHz output versus the 0-to-5-V output from the potentiometer.

FM GENERATOR

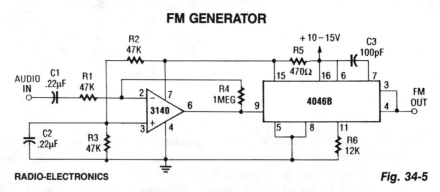

Fig. 34-5

The internal zener on pin 15 of the 4046B supplies a stable voltage to the 3140IC op amp. This amplifier modulates the 4046B VCO. The amplifier gain is about $20\times$ (26 dB voltage). The VCO produces a 220-kHz carrier that is FM modulated. C3 can be changed to vary this frequency.

1-Hz TIMEBASE FOR READOUT AND COUNTER APPLICATIONS

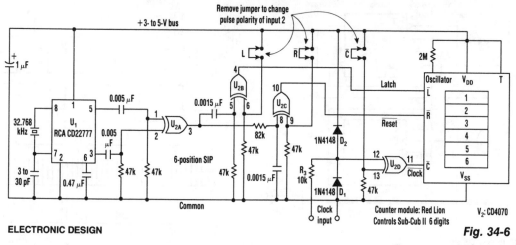

ELECTRONIC DESIGN

Fig. 34-6

This counter makes direct readout of frequency-generating equipment very easy when a 1-Hz timebase is added to latch, reset, and the count signal is conditioned. This design has the flexibility to select either polarity.

By differentiating, inverting, and ORing the clock pulses in XOR gate U2A, a stream of 1-Hz, positive, 200-μs pulses is generated. For latching, the 1-Hz stream is again differentiated in U2B, input 1 to supply a 50-μs pulse. Though U2B's output goes from high to low, it can be reversed, by making input 2 low.

Because the reset pulse must occur after the latch signal, the 1-Hz stream from U2A is delayed 100 μs at U2C input 1. The output-pulse polarity is determined by making U2C's input 2 either high or low.

WHITE NOISE GENERATOR

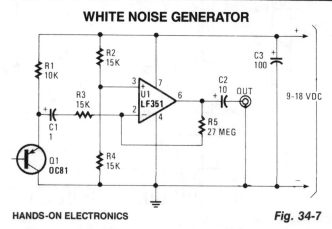

HANDS-ON ELECTRONICS

Fig. 34-7

Germanium transistor Q1 is used as a noise generator in the audio range. U1 acts as a high-gain amplifier. Q1 is not critical; most germanium transistors appear to be satisfactory. A germanium diode can also be substituted. This circuit is mainly used for sound effects and noise experiments. It is not flat over the audio range because of unpredictable effects in Q1, but it should be useful where high precision is not necessary.

FREQUENCY-RATIO MONITORING CIRCUIT

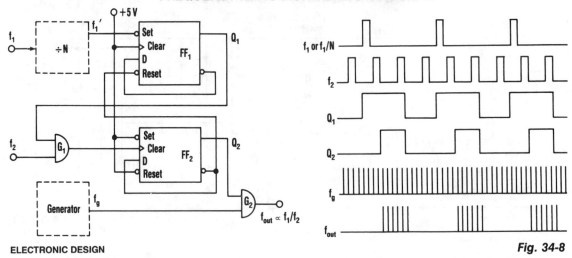

ELECTRONIC DESIGN

Fig. 34-8

This circuit produces an output frequency that is linearly proportional to the ratio of two input frequencies f_1/f_2. Each pulse of the bias f_1 (or f) will open G1 for a period $T = 1/f_2$ so that f_g/f_2 pulses pass to the output.

PULSE TRAIN

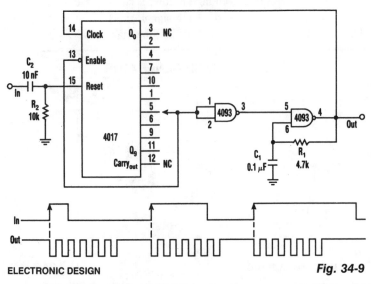

ELECTRONIC DESIGN

Fig. 34-9

This circuit has a rate multiplier using a 4093 Schmitt trigger as an oscillator, driving a 4017 decade counter. When a pulse present at the input (to C2) 4017 is reset, output zero goes high, and outputs 1 to 9 go low. The oscillator (4093) starts running and the 4017 counts the pulses until the 4017 output (1 to 9) connected to pin 1 and 2 of the 4093 goes high. The oscillator is inhibited and the output remains high until the next input pulse.

35

Games

The sources of the following circuits are contained in the Sources section, which begins on page 667. The figure number in the box of each circuit correlates to the entry in the Sources section.

Reaction Timer
Electronic Roulette Game
Run-Down Clock/Sound Generator
Wheel of Fortune
Simple Lie Detector
Electronic Dice

REACTION TIMER

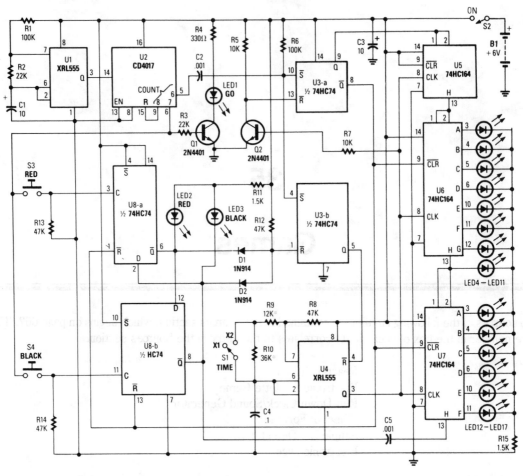

Fig. 35-1

This circuit uses a timer to generate pulses at a 5-ms clock rate. The pulses are shifted into the shift register, one at a time, lighting an LED. An auxiliary timer that generates one pulse per second is used to generate timing to activate the "go" LED and start the 5 ms pulses clocking into the registers. At the GO signal each player presses his buttons (S3 or S4). The delay (reaction time) is read out on LED 4 to LED 17; after six seconds, the sequence repeats.

ELECTRONIC ROULETTE GAME

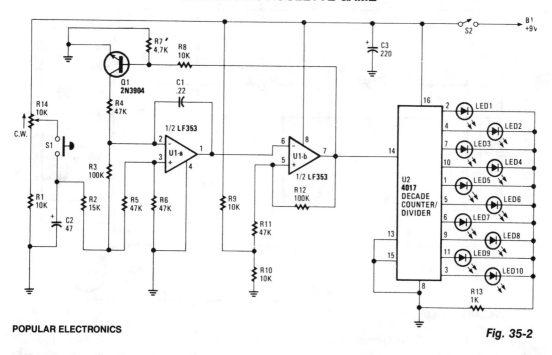

POPULAR ELECTRONICS

Fig. 35-2

R14 is set for an initial "starting" speed of the oscillator U1A and U1B. As C2 charges, oscillation begins slowing down as C2 discharges, giving a roulette-wheel effect on LED S1 through 10. The LED that remains on is the winning number.

RUN-DOWN CLOCK/SOUND GENERATOR

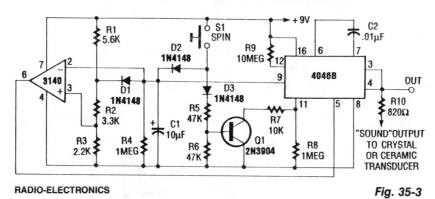

RADIO-ELECTRONICS

Fig. 35-3

Used in electronic roulette or dice games, this circuit produces a clock signal that initially is several tens of kHz (depending on C2) and gradually decreases to zero in about 15 seconds, as C1 discharges through R4.

WHEEL OF FORTUNE

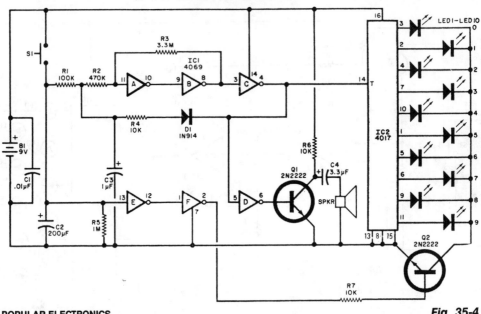

POPULAR ELECTRONICS

Fig. 35-4

This circuit is a 10-LED spinning wheel that "clicks" as the wheel passes each point. The rotation starts fast, then gradually slows down to a random stop (with a click at each position). After the rotation ceases, the selected LED stays lit for about 10 seconds, then goes out. The cycle restarts by depressing the pushbutton switch.

SIMPLE LIE DETECTOR

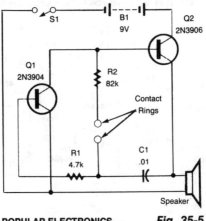

POPULAR ELECTRONICS

Fig. 35-5

The variation in skin resistance of the subject is used to vary the frequency of a tone oscillator. The contact rings are two brass rings, about 3/4" ID.

ELECTRONIC DICE

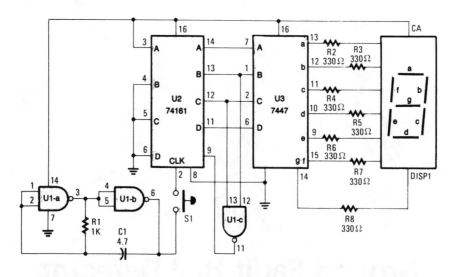

POPULAR ELECTRONICS

Fig. 35-6

When S1 is pressed, counter U2 is driven by oscillator U1A/U1B and the count (0 through 6) is read on DISP1. R1 and C1 determine the count rate, which should be fast enough to ensure a "random" count.

36

Ground-Fault Hall Detector

The source of the following circuit is contained in the Sources section, which begins on page 667. The figure number in the box correlates to the entry in the Sources section.

Ground-Fault Hall Sensor

GROUND-FAULT HALL SENSOR

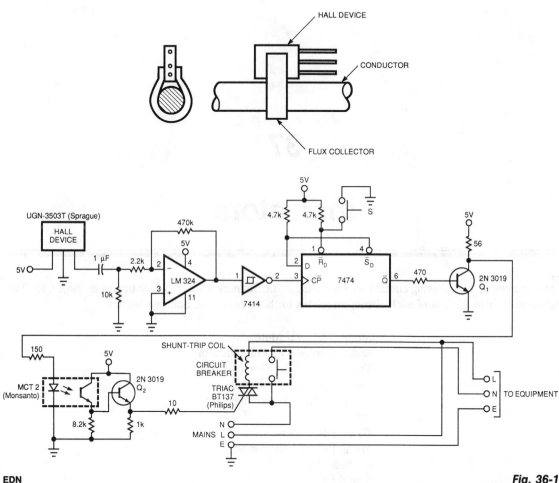

EDN

Fig. 36-1

No electrical contact exists between the circuit and the conductor. The 7474 flip-flop is triggered by the output from the Hall sensor, op amp, and Schmitt trigger. This triggering activates the optocoupler, turns on the triac, and trips the circuit breaker.

37

Indicators

The sources of the following circuits are contained in the Sources section, which begins on page 667. The figure number in the box of each circuit correlates to the entry in the Sources section.

Stereo LED VU Meter
Audio Amplifier Volume Indicator
Transistorized Bar-Graph Driver
Visual CW Offset Indicator
ac-Circuit LED Power Indicator
ac/dc Indicator
Balance Indicator
Mains Failure Indicator
On Indicator
Sound Sensor
Transmitter Output Indicator

STEREO LED VU METER

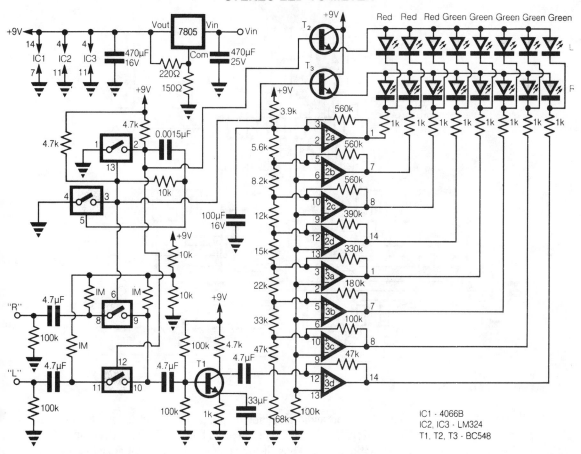

ELECTRONIC ENGINEERING

Fig. 37-1

This circuit provides a cheap alternative to the LM3915 series LED displays. The meter relies on a square-wave oscillator built around two CMOS analog switches, which alternatively selects the right and left channels for monitoring and display. The selected signal is amplified by the common-emitter stage T1, and the output is fed into the string comparators which control the display.

These eight comparators are from two LM324 quad op amps, each is connected to a resistor network, which has a 3dB step between each comparator. Each comparator has a positive feedback resistor to increase the hysteresis to provide a longer display, which is switched alternatively at about 10 kHz.

The two CMOS switches in line are biased at half the supply voltage by 1-MΩ resistors from a 100-kΩ divider, which allows them to handle analog signals up to 9 V peak to peak. As the voltage increases above the setpoint of each comparator, the output goes low and the corresponding LED lights, which produces a bar of light in response to the input voltage. For a linear response the resistor-network can be replaced by nine 10-kΩ resistors, giving an equal voltage gap before each LED comes on.

AUDIO AMPLIFIER VOLUME INDICATOR

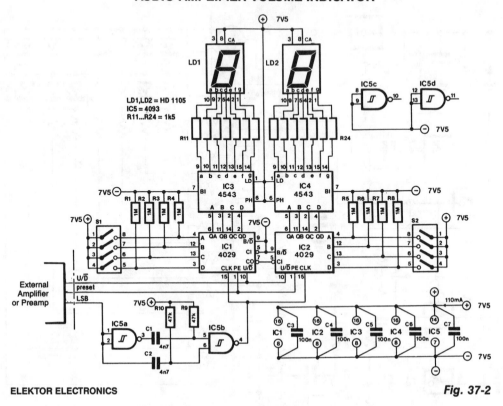

ELEKTOR ELECTRONICS

Fig. 37-2

The indicator is intended for use with an audio amplifier or preamplifier, but it can also be used in other applications where a number of steps or changes must be counted rapidly. To prevent interference with the audio signal, the circuit is a static design. Thus, if the volume control is not adjusted, the circuit does nothing.

The circuit does not need an external clock signal, because this is derived from any changes in the least significant bit (LSB). This is done by two differentiating networks: R9/C1 and R10/C2, which double the frequency of an available LSB signal.

Moreover, to ensure that the counters of the indicator remain in step with the volume control, signals "up/down" and "preset" from the preamplifier are used. It might seem rather extravagant to couple the state of the counters in the preamplifier with that of the present counters, but it is a good way to keep the connections between the two units to a minimum. Furthermore, the present counters operate in 8-bit BCD, instead of 6-bit binary as used by those in the volume control (in the preamplifier). All that is required to display the state of the volume control are a couple of BCD-to-7-segment decoders and 7-segment displays.

The preset in the indicator must be set in BCD code. Leading zeros are not suppressed so numbers up to and including 9 are displayed, starting with a 0. The DIP switches and resistors R1 through R8 in the diagram can be omitted if only one fixed preset is likely to be used. The resistors should be replaced by jump leads.

TRANSISTORIZED BAR-GRAPH DRIVER

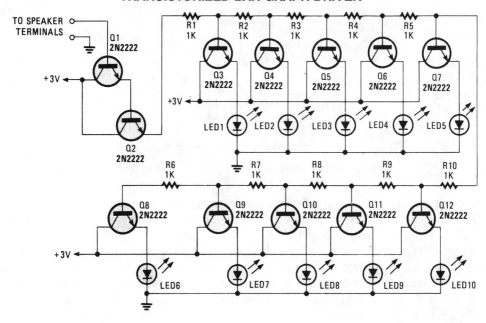

POPULAR ELECTRONICS

Fig. 37-3

A resistor network (R1 through R10) with emitter followers (Q1 and Q2) drives LED drivers (Q3 through Q7). This circuit was used as a "light organ" to provide visual volume indication. It can be hooked to a speaker, to another audio source, etc.

VISUAL CW OFFSET INDICATOR

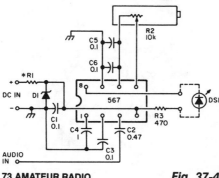

73 AMATEUR RADIO

Fig. 37-4

An NE567 tone decoder, tuned to the transceiver's CW offset frequency, ensures that the transceiver will be transmitting on the same frequency as the received CW signal. Simply tune the transceiver so that the LED lights. Eight to 13 Vdc is required; this can be taken from the transceiver supply or an extra battery. Audio is taken from the speaker or headphone output.

ac CIRCUIT LED POWER INDICATOR

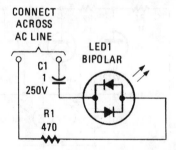

CONNECT
ACROSS
AC LINE

C1
1
250V

LED1
BIPOLAR

R1
470

POPULAR ELECTRONICS

Fig. 37-5

Many electronic circuits need an indication that they are under power; for most ac circuits, a neon lamp is the device of choice. A bidirectional tricolor LED can be used if a capacitor is connected in series with the LED to limit the current through the LED. A 1-μF, 250-WVdc capacitor, which has a reactance of 2 650 ohms at 60 Hz, is used in series with an LED to limit the current through the unit to 43 mA. The impedance of the LED is low compared to the reactance of the capacitor, so nearly all the impedance will be caused by the capacitor with the added advantage of no energy loss caused by the capacitor.

The power of the LED is 1.175 V × 0.043 A = 50 mW compared to an NE-2H at 250 mW. For 230 V, use a 0.47 − μF, 400-WVdc capacitor.

ac/dc INDICATOR

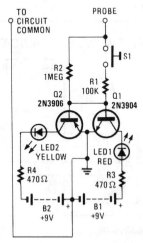

TO
CIRCUIT
COMMON

PROBE

S1

R2
1MEG

R1
100K

Q2
2N3906

Q1
2N3904

LED2
YELLOW

LED1
RED

R4
470 Ω

R3
470 Ω

B2
+9V

B1
+9V

RADIO-ELECTRONICS

Fig. 37-6

By using two switching transistors and two LEDs, this circuit can distinguish low-level ac and dc signals. If the red LED illuminates, the signal is positive dc. If the yellow LED lights, the signal is negative dc. If the signal is ac, both LEDs will light.

BALANCE INDICATOR

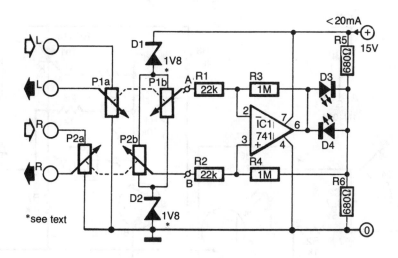

ELEKTOR ELECTRONICS

Fig. 37-7

If your amplifier is fitted with two level controls, it actually offers you a balance control and a level control. A drawback of this is that it is quite difficult to set the balance properly. This can be obviated, however, by replacing the two monopotentiometers with stereo versions P1 and P2 in the diagram.

One half of the pair, P1A and P2A, assumes the tasks of the removed components. The other half is connected in a bridge circuit. The voltage between wipers of the potentiometers is then a measure of the balance between the two channels. The lower the potential, the better the balance. If you are interested in knowing the degree of unbalance, connect a center-zero moving coil meter with a bias resistor between A and B. With this arrangement, zener diodes D1 and D2 can be omitted: they are necessary only with the LED indicator shown in the diagram to prevent the input voltage of the op amp from getting too close to the level of the supply voltage.

The circuit around IC1 is a classical differential amplifier. Resistors R5 and R6 provide a virtual earth for the LEDs, which is necessary to ensure that, in spite of the asymmetrical supply voltage, a positive and a negative output is obtained.

Because the LEDs have been included in the feedback loop of the indicator, the circuit is pretty sensitive. At only 40 mV, that is, just 1/400 of the supply voltage, one of the LEDs begins to light. The maximum current drawn by the LEDs is determined by the values of R5 and R6.

MAINS FAILURE INDICATOR

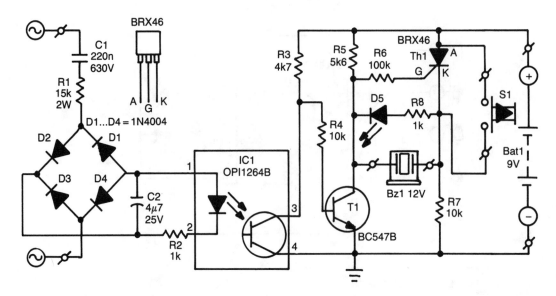

Fig. 37-8

When the mains voltage is present at the input terminals, the transistor in the optocoupler is on, T1 is off, and silicon-controlled rectifier Th1 is in the conducting state. Because both terminals of the piezoelectric buzzer are then at the same potential, the buzzer is inactive. If the mains voltage drops out, transistor T1 conducts and causes one of the terminals of the buzzer to be connected to earth; the thyristor remains in the conducting state. In this situation, a large enough potential difference is across both the buzzer and D5 to cause these elements to indicate the mains failure—both audibly and visibly.

When the mains is restored, the circuit returns to its original state. A touch on the reset button then interrupts the current through the SCR so that the thyristor goes into the blocking state, and the other terminal of the buzzer is connected to ground.

The unit is powered by a 9-V battery and draws a quiescent current of 1.7-2.5 mA. It is important for the enclosure to be well-insulated.

If by accident the circuit to the optocoupler and R2 is broken, electrolytic capacitor C2 might be damaged because it will be charged well above its 25-V rating. Secondly, where a plug is used for the mains connection, it is advisable to solder a 1-MΩ resistor across C1 so that this capacitor does not retain its charge after the plug is removed from the mains socket.

ON INDICATOR

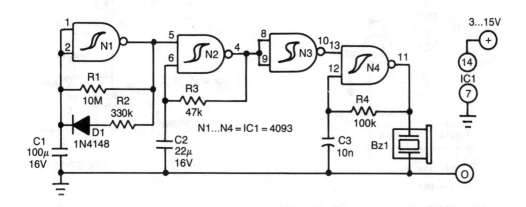

ELEKTOR ELECTRONICS

Fig. 37-9

Battery-operated equipment can work on one set of batteries for a long time nowadays. However, if it is left on inadvertently, that "long time" is over very quickly. Moreover, flat (dead) batteries are always found at the wrong moment. The circuit proposed here is a sort of *aide-memoire*. Every two minutes, it emits 5 to 10 pips to indicate that the equipment is still switched on.

Basically, the circuit consists of three rectangle-wave generators and an inverter. The first of the generators is formed by N1 and provides a signal with a period of about two minutes and a pulse duration of around 10 seconds. During those 10 seconds, the second generator starts operating in a one-second rhythm. Thus, N2 outputs 10 pulses every 2 minutes. That output is inverted so that N4, like N2, can only be enabled during the 10-second pulse train from N1. The difference is that during those 10 seconds, N4 is enabled and inhibited 10 times; this is what causes the pips.

Do not take the times and number of pulses too literally, because wide variances are between ICs from different manufacturers. On the other hand, component values are not critical, so it is fairly easy to adapt the circuit to personal taste or requirements. The buzzer can be a standard Toko type or equivalent. The current drawn by the circuit is negligible.

SOUND SENSOR

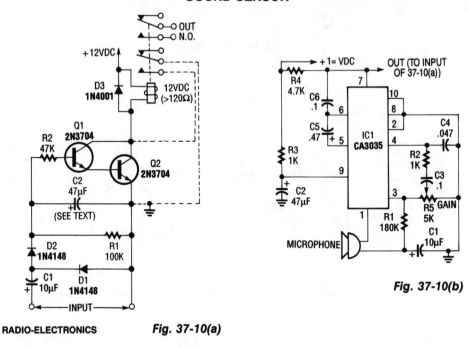

RADIO-ELECTRONICS *Fig. 37-10(a)*

Fig. 37-10(b)

By using a microphone, high-gain amplifier (Fig. 37-10(b)), and detector-relay driver (Fig. 37-10(a)) a sound-detecting alarm system can be constructed. If you want a latching setup, make the dotted connections to the relay shown in Fig. 37-10(a).

TRANSMITTER OUTPUT INDICATOR

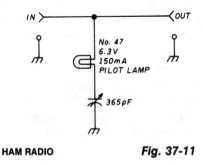

HAM RADIO *Fig. 37-11*

Relative power can be indicated with this simple circuit. Adjust the 365-pF variable capacitor for desired lamp brightness.

38

Infrared Circuits

The sources of the following circuits are contained in the Sources section, which begins on page 667. The figure number in the box of each circuit correlates to the entry in the Sources section.

IR RECEIVER I

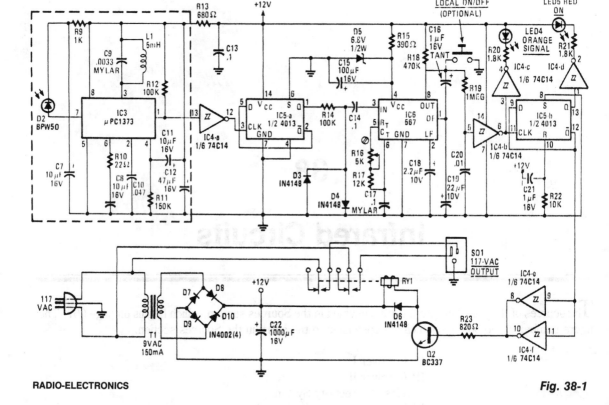

Fig. 38-1

This receiver is built around a uPC1373 IR remote-control preamplifier, a sensitive 30-to-40 kHz tuned detector, an automatic gain control, a peak detector, and an output waveshaping buffer. The demodulated signal from the preamp stage is sent to IC4A, a 74C14 Schmitt trigger. The squared-up 1 500-Hz signal is then sent to the clock input of IC5A, half of a 4013 dual "D" flip-flop. That 750-Hz signal is clipped to approximately 0.7-V p-p by diodes D3 and D4. The clipped signal is then fed to IC6, a 567 tone decoder. The output of that IC goes low whenever the frequency of the signal fed to it is within the lock range of its internal VCO.

When IC6 detects a signal of the proper frequency, pin 8 goes low. The output signal is fed through another Schmitt trigger (IC4B), which drives another "D" flip-flop, IC5B. Schmitt trigger IC4B also drives IC4C, which in turn drives LED4, SIGNAL, which lights up whenever a signal is received. The Q output of IC5B drives two parallel-connected inverters. IC4C and IC4F turn transistor Q2 on when Q goes low. That transistor energizes the relay; its contacts switch the controlled device on and off.

IR RECEIVER II

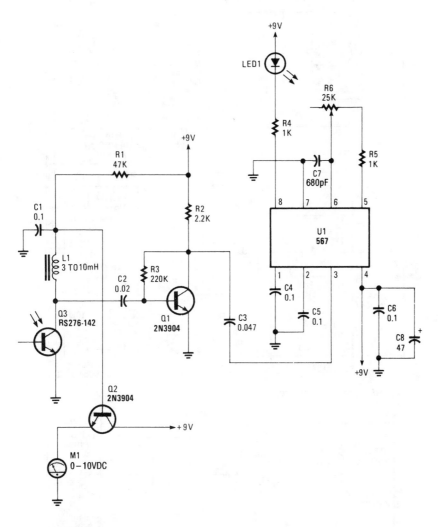

Fig. 38-2

Q3 is an IR phototransistor that responds to a modulated IR beam. Q1 amplifies the ac component of the IR beam. Q2 drives a meter as a relative indication of the strength of the light beam. A strong beam gives a lower meter reading. U1 is a tone decoder that produces a low output on pin 1 during reception for an IR beam that is modulated with the correct tone frequency, determined by R6.

WIRELESS IR SECURITY SYSTEM

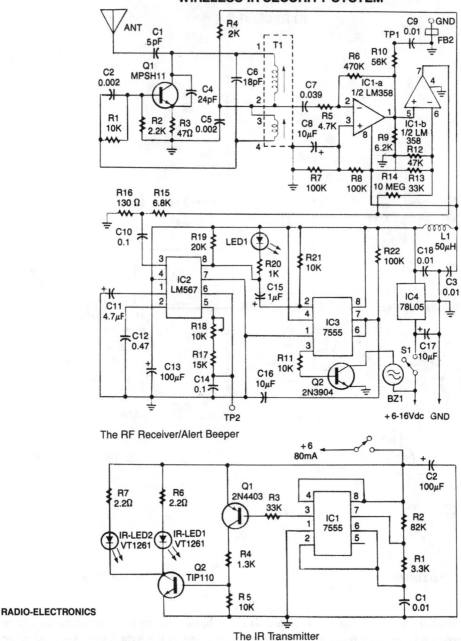

The RF Receiver/Alert Beeper

The IR Transmitter

Fig. 38-3

This system contains an IR transmitter, an IR receiver/RF transmitter, and an RF receiver/alert beeper. Two IR LEDs in the transmitter transmit a pulsed beam of invisible infrared light to the receiver, which contains an IR phototransistor. The phototransistor detects and amplifies the pulse-modulated IR beam. If the receiver section senses that the IR beam is momentarily interrupted by an object blocking the

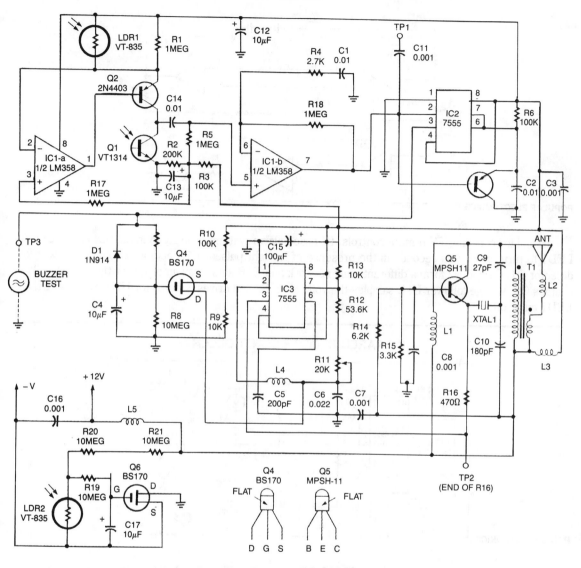

The IR Receiver/RF Transmitter

beam's path, it triggers the transmitter, which outputs a 49.890-MHz carrier that is amplitude-modulated by a 490-Hz tone.

Upon receiving the 490-Hz amplitude-modulated carrier, the RF receiver/beeper unit sounds an alarm that alerts the user to the intrusion. The system is not limited to just one RF transmitter. A single RF receiver/beeper can be used to monitor any number of RF transmitters (or locations). However, the receiver/beeper unit cannot discriminate between different transmitter sites in multiple-transmitter systems.

IR DETECTOR

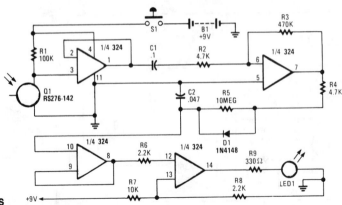

Fig. 38-4

Useful for checking TV remote controls, IR-based alarm systems, and IR sources, this circuit causes LED1 to turn on for two seconds in the presence of IR light pulses. U1A acts as a voltage follower for detector Q1. C1 and R2 form a differentiating network and U1B acts as an amplifier for the pulses, which charges C2. Voltage follower U1C samples the voltage on C2 and drives comparator U1D, which switches LED1 on or off.

INFRARED REMOTE CONTROLLER

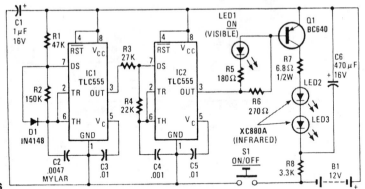

Fig. 38-5

The transmitter is built around two CMOS 555 timer ICs (TLC 555s). The transmitter generates a modulated 35-kHz IR signal. The 35-kHz carrier frequency is generated by IC2, and the 1 500-Hz modulating signal is generated by IC1. The output of IC2 drives LED1 through resistor R5; that LED provides visual indication that the transmitter is working. In addition, IC2 drives transistor Q1, which drives the two infrared LEDs (LED 2 and LED3).

To provide the high current needed to drive the two IR LEDs, capacitor C6 is precharged, the charge it contains is dumped when S1 is pressed. When S1 is not pressed, the power to the ICs is cut off. However, C6 is kept charged via R8. Then, when S1 is pressed, the current stored in C6 can be used to drive the LEDs for as much as 1/2 second. That's plenty of time for the receiver to pick up a signal.

IR-CONTROLLED SOLDERING STATION

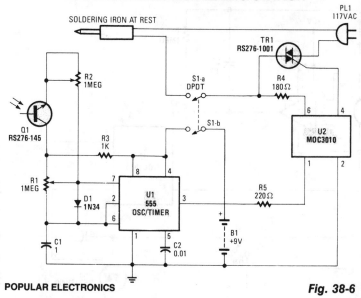

POPULAR ELECTRONICS

Fig. 38-6

An IR-sensitive phototransistor is used to sense soldering-iron temperature. The phototransistor must see the tip through an opaque tube to avoid stray light, and the phototransistor should preferably be fitted with an IR filter. An old photo negative, dark red plastic, or red or black glass can be used. The iron sits on a holder.

When the iron is removed from the holder, the iron is not being viewed by the detector. The heat will increase, but the circuit has a lag time; if the iron is returned to its holder after each use, overheating should not be a problem.

INFRARED "PEOPLE" DETECTOR

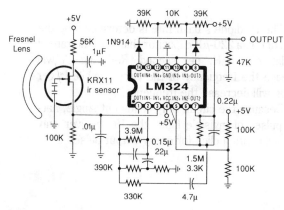

This circuit uses an Amperex pyroelectric IR sensor, an LM324 op amp (configured as a high-gain amplifier in the 0.3-to 5-Hz range), and a window detector. The output will go high on any motion, which will change the infrared signature seen by the sensor.

RADIO-ELECTRONICS

Fig. 38-7

IR HEAT-CONTROLLED KITCHEN FAN

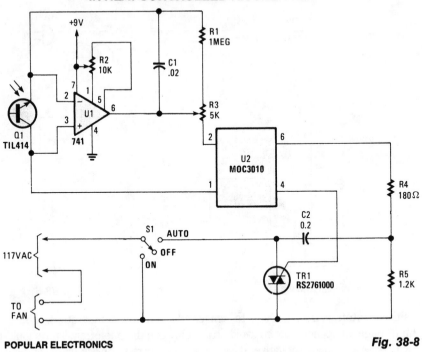

POPULAR ELECTRONICS

Fig. 38-8

Q1 senses IR from heat sources, causes U1 to switch, activates optocopuler U1, and triggers TR1. This controls a fan. The Triac is from Radio Shack, or else a 200-V, 6-A unit (C106B) can be used.

SIMPLE IR TRANSMITTER

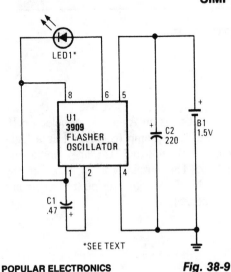

The IR diode's flash rate is determined by the value of C1, a 220-μF capacitor that sets the rate of oscillation at 1 Hz per second. Reducing C_1 will increase the frequency of the circuit, while larger values will decrease the frequency.

Because the circuit only sends out single, narrow pulses of invisible light, the IR receiver only responds with a click for every output pulse.

POPULAR ELECTRONICS   *Fig. 38-9*

IR TRANSMITTER

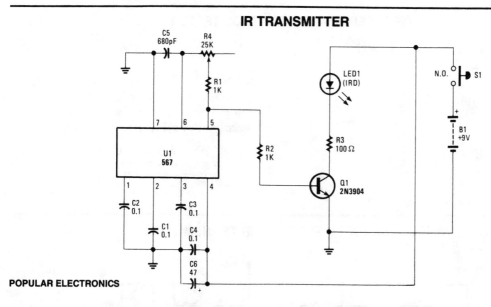

POPULAR ELECTRONICS

Fig. 38-10

Using an NE567 as a tone oscillator, this circuit produces an IR signal from the LED, which is modulated with a square wave. LED1 is an IR-emitting LED. The modulation helps improve performance under high ambient light conditions.

IR REMOTE EXTENDER

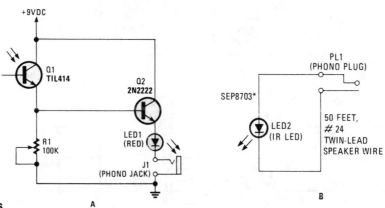

POPULAR ELECTRONICS

A

B

Fig. 38-11

This circuit can be used to operate a VCR or CD player from another room. It's really an infrared signal repeater. The signal from the remote is received and then retransmitted over wires to an infrared LED. The beam from the LED is then picked up by the receiving window on the VCR or CD player.

The visible light LED (LED1) in series with the IR unit (LED2) is used to indicate that the transmitted signal has been detected. The 100-kΩ trimmer potentiometer (R1) adjusts the repeater's sensitivity. The resistor that is usually found in series with the LEDs is omitted, because the voltage reading is about 1.0 Vdc as a result of the voltage drop across the lines.

INFRARED REMOTE CONTROL TESTER

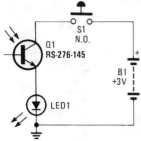

Using a battery, a phototransistor and a visible-light LED, this simple circuit is a "go/no go" tester for IR remote control devices. The illumination of the LED indicates that Q1 is being modulated by IR energy.

POPULAR ELECTRONICS   *Fig. 38-12*

VOICE-MODULATED PULSE FM IR TRANSMITTER

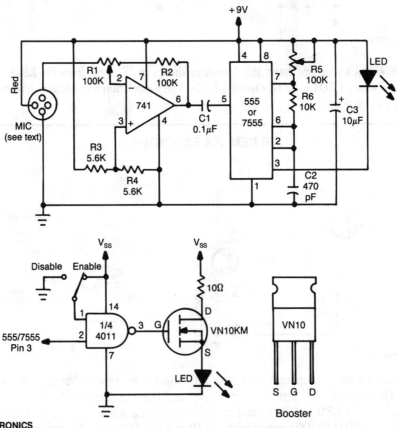

POPULAR ELECTRONICS   *Fig. 38-13*

This circuit has a 741 audio amplifier, which is fed by a microphone (use an amplified type), an FM modulator, and a CMOS timer that acts as a VCO. The LED is pulsed with the timer output (the booster circuit can be used for increased range). This yields an FM-modulated, pulsed IR beam.

39

Instrumentation Amplifiers

The sources of the following circuits are contained in the Sources section, which begins on page 668. The figure number in the box of each circuit correlates to the entry in the Sources section.

Oscilloscope Preamp
Low-Drift/Low-Noise dc Amplifier
Instrumentation Amplifier
Extended Common-Mode Instrument Amplifier

OSCILLOSCOPE PREAMP

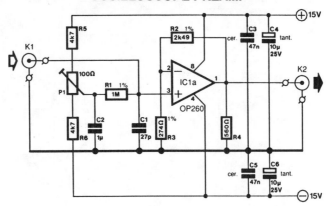

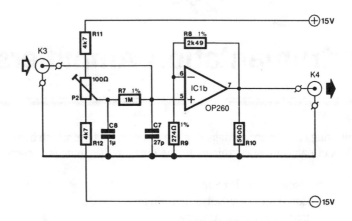

Fig. 39-1(a)

In many oscilloscopes, the most sensitive range is 2 to 5 mV, although it is often possible to improve this to 1 to 2 mV by a variable gain control. To obtain even better sensitivity, the present preamplifier, which has an amplification of about 10 (20 dB), might be useful.

Because most oscilloscopes have a bandwidth of 20 MHz or more, the amplifier must, of course, have a slightly wider bandwidth and that is achieved with a Type OP260 op amp. This has a slew rate of 550 V/μs (at an amplification of 10) and a bandwidth of 40 MHz that is virtually independent of the amplification. The gain vs. frequency response is not so good, however: as can be seen from Fig. 39-1(b), where the characteristics are given for a number of loads. The hump in the curves depends on the value of the feedback resistor, whose optimum value appears to be 2.5 kΩ.

The curves in Fig. 39-1(c) accord with different values of R_2/R_8 for an amplification factor of 10. Some experimentation with the value of R_2/R_8 for different amplification factors can be instructive. Remember, however, that the output impedance increases from 20 to 225 Ω over the frequency range of 10 MHz to 60 – 70 MHz. It is therefore important to keep all connections on the prototyping board as short as possi-

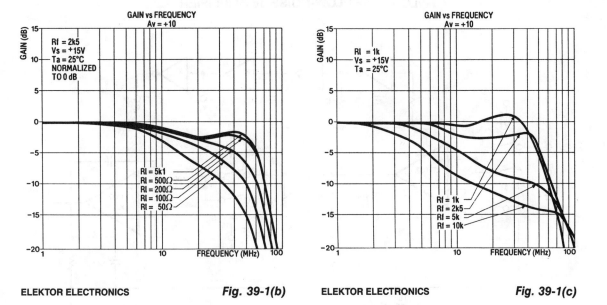

ELEKTOR ELECTRONICS · *Fig. 39-1(b)* · ELEKTOR ELECTRONICS · *Fig. 39-1(c)*

ble and to connect all earth points to a common ground via a separate, heavy track. Also, do not use an IC socket.

An input impedance of 1 MΩ was chosen, which results in a fairly high level of noise at the output (with open-circuit input). This value can be reduced, because otherwise the use of a 1:10 probe will be inhibited; it would give constant problems with the noise. However, when the amplifier is connected to a suitable source, the noise reduction is normally more than ample to obtain a good trace on the screen. Presets P1 and P2 provide compensation for the dc offset and input offset, caused by R1 and R7 respectively.

The input bias current for the noninverting input is about 10 times lower than that for the inverting input, which makes the OP260 more suitable for noninverting circuits. The inverting circuit can also give problems because of the low values of R_2 (R_8) and R_3 (R_9). The input bias current is typically 0.2 μA, and the input offset is about 3 mV (max. 7 mV).

In this type of circuit, it is important to use a well-regulated power supply. The power-supply suppression up to 10 kHz is roughly 70 dB, and this reduces with increasing frequency. Any noise or tiny ripple on the supply lines would make the application of the circuit as a small-signal amplifier impossible.

The circuit draws a current of about 14 mA. The slew rate, as with most op amps, is asymmetric and might lead to visible distortion of the signal when the drive to the 560-Ω resistor is high at the higher frequencies.

LOW-DRIFT/LOW-NOISE dc AMPLIFIER

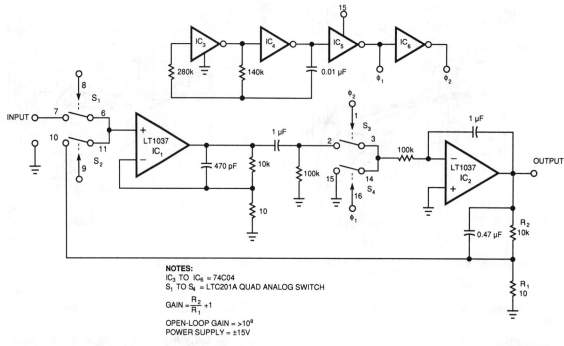

NOTES:
IC3 TO IC6 = 74C04
S1 TO S4 = LTC201A QUAD ANALOG SWITCH

$$\text{GAIN} = \frac{R_2}{R_1} + 1$$

OPEN-LOOP GAIN = >10^8
POWER SUPPLY = ±15V

Fig. 39-2

Figure 39-2's circuit combines a low-noise op amp, IC1, with a chopper-based carrier-modulation scheme to achieve a low-noise, low-drift dc amplifier whose performance exceeds any currently available monolithic amplifier. The amplifier's offset is less than 1 μV, and its drift is less than 0.05 μV/°C. This circuit has noise within a 10-Hz bandwidth less than 40 nV. The amplifier's bias current, which is set by the bipolar input of IC1, is about 25 nA.

The 74C04 inverters (IC3 to IC6) form a simple 2-phase square-wave clock that runs at about 350 Hz. The complementary oscillator signals (O_1 and O_2) provide drive to S1 and S2, respectively, causing a chopped version of the input to appear at IC1's input. IC1 amplifies this ac signal. S3 and S4 synchronously demodulate IC1's square-wave output. Because S3 and S4 switch synchronously with S1 and S2, the circuit presents proper amplitude and polarity information to IC2, the dc output amplifier. This output stage integrates the square wave to provide a dc voltage output. R1 and R2 divide the output and feed it back to the input chopper where the divided output serves as a zero signal reference. The ratio of R1 and R2 sets the gain, in this case to 1 000.

INSTRUMENTATION AMPLIFIER

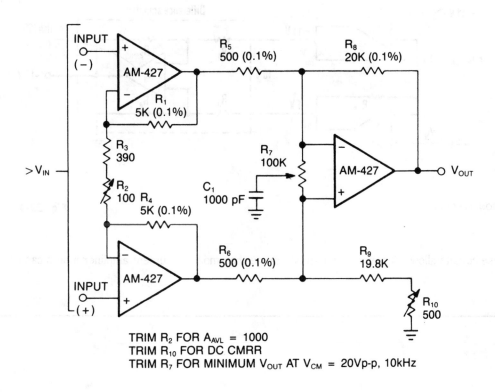

TRIM R_2 FOR A_{AVL} = 1000
TRIM R_{10} FOR DC CMRR
TRIM R_7 FOR MINIMUM V_{OUT} AT V_{CM} = 20Vp-p, 10kHz

DATEL

Fig. 39-3

In a single difference amplifier configuration, the AM-427 exhibits excellent common-mode rejection and spot noise voltage so low it is dominated by the resistor Johnson noise. The three-amplifier configuration shown avoids the low input-impedance characteristics of difference amplifiers. Because of the additional amplifiers used, the spectral noise voltage will increase from a typical of 3 nV/Hz to approximately 4.9 nV/Hz. The overall gain of the circuit is set at 1 000; with balanced source resistors, a CMRR of 100 dB is achieved.

233

EXTENDED COMMON-MODE INSTRUMENT

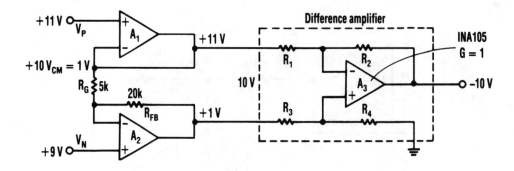

ELECTRONIC DESIGN

Fig. 39-4

These circuits allow a larger common-mode range than most instrument amplifier inputs can allow.

40

Intercom

The source of the following circuit is contained in the Sources section, which begins on page 668. The figure number in the box correlates to the entry in the Sources section.

Two-Wire Intercom

TWO-WIRE INTERCOM

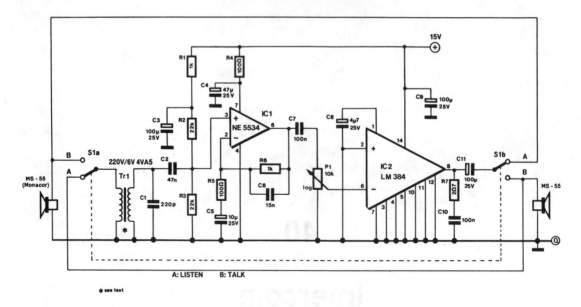

ELEKTOR ELECTRONICS

Fig. 40-1

The design consists of an amplifier, a double-pole changeover switch and two loudspeakers: one for the master station and one for the slave. More than one slave unit can be used, but each requires an additional changeover switch.

The power amplifier is a Type LM384, which can provide almost 2 W output at a supply voltage of 15 V. Pins 3, 4, 5, 10, 11, and 12 are connected to ground and at the same time afford some cooling of the device. Because of that, the IC should not be fitted in a socket, but be soldered direct to the circuit board.

The LM384 processes signals with respect to earth so that an asymmetric supply is sufficient. The amplification has been set internally to ×50 (34 dB). The IC's supply line is decoupled by C9.

To ensure adequate input sensitivity, a preamplifier, IC1, is provided, which has an amplification of 11 (21 dB). Because this stage is intended for speech only, its bandwidth is limited to 160 Hz to 10 kHz. Divider R2/R3 at the input of the op amp is decoupled by C3.

Special loudspeakers that can also serve as microphones are readily available: in the prototype, MS-55 units from Monacor were used, but a number of other makes will do just as well. The bandwidth of the MS-55 (used as loudspeaker) extends from 150 Hz to 20 kHz and (used as a microphone) from 20 Hz to 20 kHz. The MS-55 can handle up to 5-W output. To ensure satisfactory operation, particularly as a microphone, the loudspeaker must be fitted in a closed box.

Although it is advantageous for the "microphone" to have a low internal resistance, it is necessary for a transformer to be used at the input of the circuit. This has, however, the advantage that long cables can be used. The present circuit uses a standard mains transformer instead of a special microphone transformer. For this purpose, the secondary (6 V) winding is connected to the "microphone." The microphone impedance is thereby magnified from about 8 Ω to around 10 kΩ. The power handling of the

TWO-WIRE INTERCOM *(Cont.)*

transformer is quite high to ensure that signal losses in the primary winding are kept at a minimum. Capacitor C1 suppresses HF interference.

If the mains transformer and the ''microphone transformer'' are housed in the same enclosure, some trial and error and screening are necessary to eliminate hum. The ''microphone transformer'' itself might cause hum in the remainder of the circuit. In that case, the preamp stage must also be screened.

In the prototype, the speech bandwidth was limited from 400 Hz to 4 kHz and this proved perfectly acceptable for good speech transfer. Most of the current drawn by the circuit flows through the power amplifier. At worst, this is 210 mA (680 mA peak), when the amplifier delivers 1.8-W output. The LM384 can deliver a power of up to 5 W. The supply voltage should then be raised to 22 V and a heatsink for the device will be necessary.

41

Interface Circuits

The sources of the following circuits are contained in the Sources section, which begins on page 668. The figure number in the box of each circuit correlates to the entry in the Sources section.

Versatile One-Shot CPU Interface
Keyboard Matrix Interface
Low-Level Power FET Driver Method
Logic-Level Translators

VERSATILE ONE-SHOT CPU INTERFACE

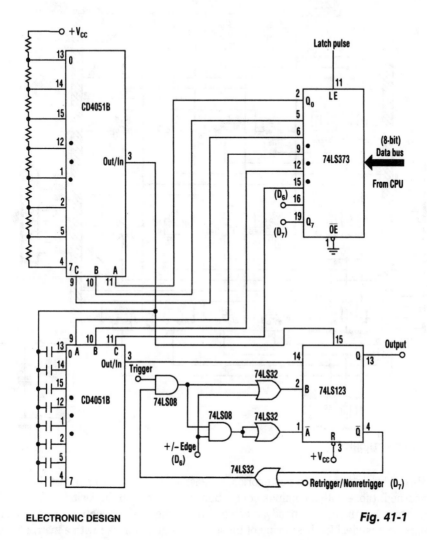

ELECTRONIC DESIGN

Fig. 41-1

Process-control applications often require a monostable multivibrator (one-shot) with a pulse width that can be selected on-the-fly.

This circuit uses two CD4051B analog multiplexers to select the required timing components for the multivibrator, and hence, the pulse width. The multiplexers' address input comes from an 8-bit latch. Bit D6 tells the multivibrator whether to trigger on the leading or trailing edge of the trigger input. Bit D7 determines whether the multivibrator should be in a retriggerable or nonretriggerable mode.

KEYBOARD MATRIX INTERFACE

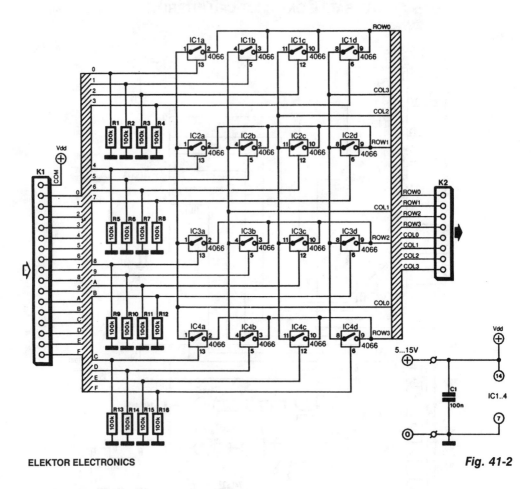

ELEKTOR ELECTRONICS

Fig. 41-2

Keyboards can be slotted into two categories, at least as far as the manner that the switches are connected is concerned: those with a common connection and those with the switches arranged in a matrix.

The matrix type has the important advantage that the number of connections is an absolute minimum. Such an arrangement is ideal for ICs; many of these are designed for use with a matrix keyboard.

However, many keyboards are available in job lots, for instance, that apart from a common connection also have a connector for each key. Such keyboards can be connected to ICs that require a matrix type with the aid of a number of electronic switches.

The principle is straightforward: each key of the keyboard controls an electronic switch that is included in a matrix. As an example, the diagram shows a hexadecimal keyboard that is arranged in a 4-×-4 matrix. Each of the electronic switches is held in the open position by a pull-down resistor.

The current drawn by the circuit is very small and is determined mainly by the value of the pull-down resistors and the number of keys being pressed. The CMOS switches draw virtually no current.

LOW-LEVEL POWER FET DRIVER METHOD

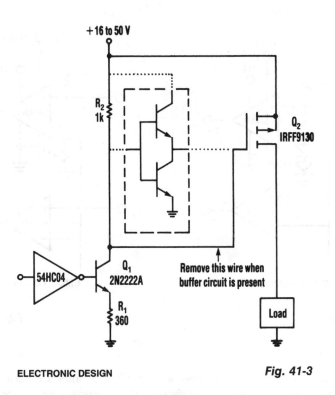

+16 to 50 V

R₂
1k

Q₂
IRFF9130

54HC04

Q₁
2N2222A

**Remove this wire when
buffer circuit is present**

R₁
360

Load

ELECTRONIC DESIGN

Fig. 41-3

This circuit operates from a 16- to 50-V supply. Adding the buffer circuit (within the dashed lines) offers 100-ns switching times. Otherwise, the circuit switches in 1 μs.

Q1 and R1 form a switched current source of about 12 mA. The current flows through R2, which supplies 12 V to the FET. The circuit works well over a wide range of supply voltages. Furthermore, it switches smoothly in the presence of large ripple and noise on the supply. The switching time (about 1 μs) can be reduced considerably by lowering the values of R_1 and R_2 at the expense of higher power dissipations in the resistors and Q1. Alternatively, a buffer circuit can be added to produce switching times of 100 ns without generating significant power dissipation.

LOGIC-LEVEL TRANSLATORS

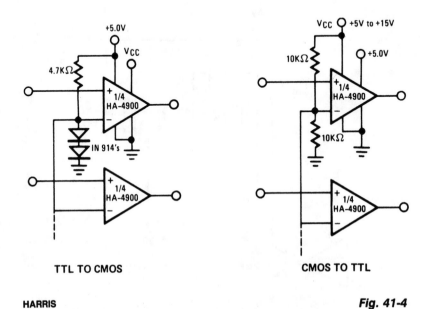

TTL TO CMOS

CMOS TO TTL

HARRIS

Fig. 41-4

The HA-4900 series comparators can be used as versatile logic interface devices, as shown in these circuits. Negative logic devices can also be interfaced with appropriate supply connections. If separate supplies are used for $V-$ and $V_{LOGIC}-$, these logic-level translators will tolerate several volts of ground-line differential noise.

42

Keying Circuits

The sources of the following circuits are contained in the Sources section, which begins on page 668. The figure number in the box of each circuit correlates to the entry in the Sources section.

CW Keyer
Transmitter Negative Key Line Keyer
Frequency Shift Keyer

CW KEYER

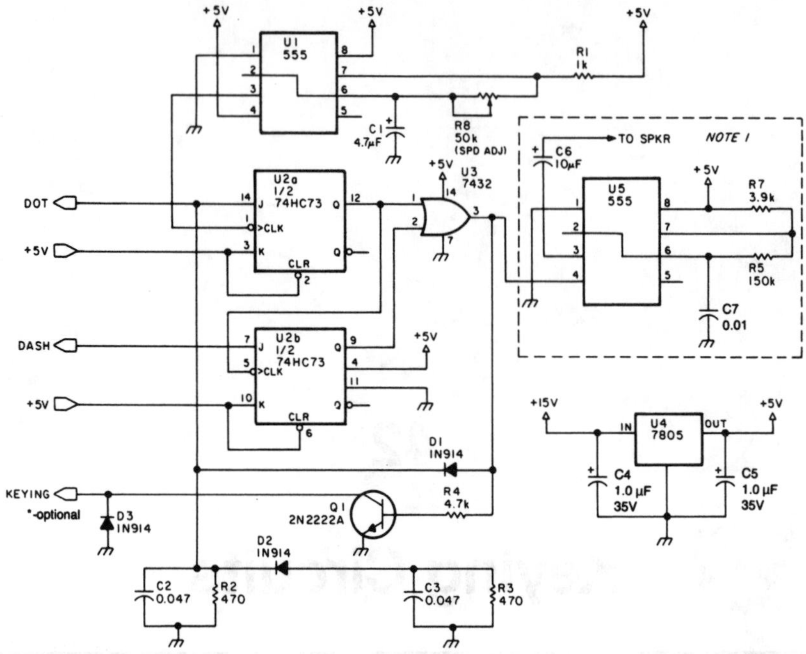

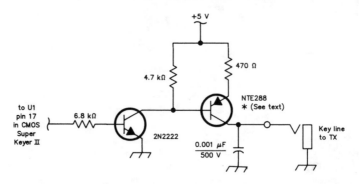

73 AMATEUR RADIO

Fig. 42-1

This electronic keyer uses four ICs (five, if the optional sidetone oscillator is desired) and operates from dc sources of 9 to 15 V. A 2N2222 is used as a keying transistor. If isolators or more power handling ability is desired, a 6-V relay can be keyed with the 2N2222 and the relay in turn can be used to key the transmitter.

TRANSMITTER NEGATIVE KEY LINE KEYER

QST

Fig. 42-2

Using an NTE288 (or ECG288, GE223, or SK3434), this circuit can key a negative line up to -300 V maximum. Do not use this circuit to key a vacuum-tube amplifier that draws grid current because the keying transistor might be damaged under these conditions.

FREQUENCY SHIFT KEYER

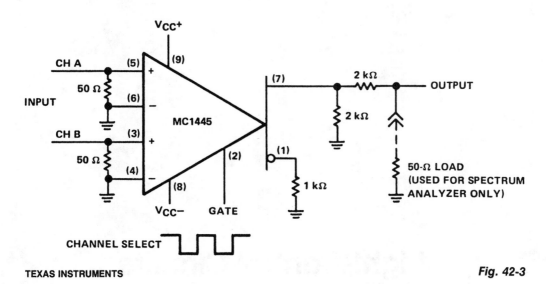

TEXAS INSTRUMENTS

Fig. 42-3

   Apply a signal to each differential amplifier input pair. When the gate voltage is changed from one extreme to the other, the output can be switched alternately between the two input signals. When the gate level is high (1.5 V), a signal applied between pins 5 and 6 (channel A) will be passed and a signal applied between pins 3 and 4 (channel B) will be suppressed. In this manner, a binary-to-frequency conversion is obtained that is directly related to the binary sequence, which is driving the gate input (pin 2).

43

Light-Control Circuits

The sources of the following circuits are contained in the Sources section, which begins on page 668. The figure number in the box of each circuit correlates to the entry in the Sources section.

THREE-WAY TOUCH LAMP

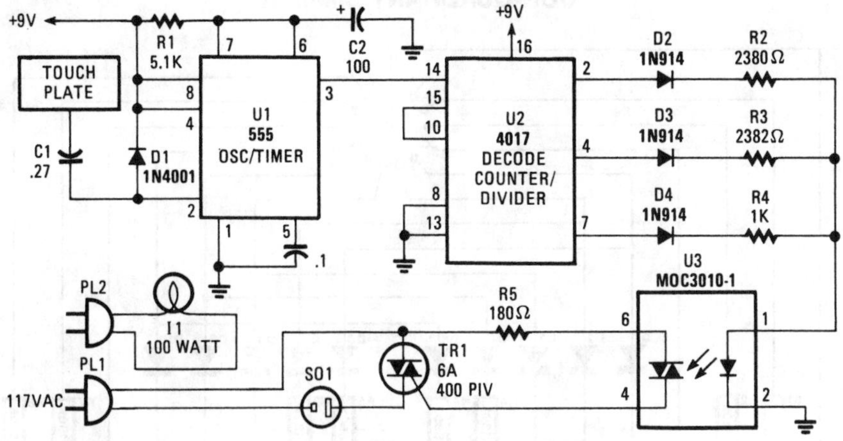

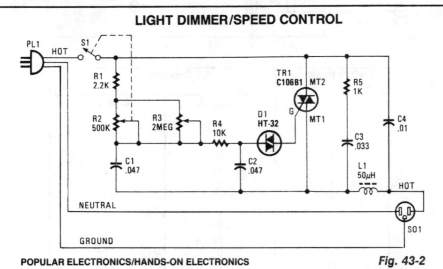

POPULAR ELECTRONICS

Fig. 43-1

A three-way switch to control a lamp (off-dim-bright, etc.) uses an NE555 timer to generate a one-second pulse, triggered by ambient ac fields that are picked up by the human body. C1 and D1 form an input network. U2 is a decode counter/divider and drives one of 10 outputs (three are used). The logic outputs drive various resistors in series with the LED in the optocoupler. The optocoupler controls a triac that is in series with a load (lamp, etc.).

By reconfiguring the outputs of U2, more than three brightness levels can be obtained, up to 10. An IN914 and resistor will be required for each output.

LIGHT DIMMER/SPEED CONTROL

POPULAR ELECTRONICS/HANDS-ON ELECTRONICS

Fig. 43-2

A phase-controlled triac (HT-32) circuit provides control of effective voltage at load. Do not omit L1 and C4 because they are for RFI suppression. The maximum load is about 500 W. WARNING: 120 Vac is present on this circuit—provide adequate insulation and construction techniques.

FOUR-QUADRANT DIMMER

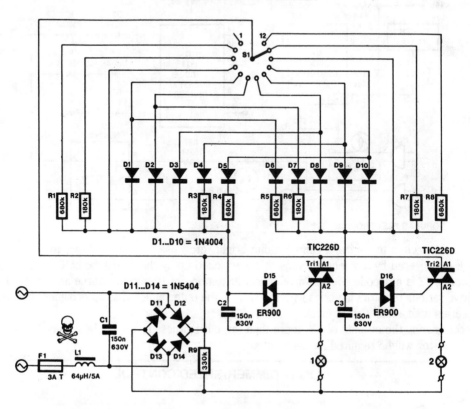

D1...D10 = 1N4004

TIC226D

TIC226D

D11...D14 = 1N5404

Switch position	Brightness	
	Group A	Group B
1	0	0
2	1/3	0
3	2/3	0
4	1	0
5	1	1/3
6	1	2/3
7	1	1
8	2/3	1
9	1/3	1
10	0	1
11	0	2/3
12	0	1/3

ELEKTOR ELECTRONICS

Fig. 34-3

FOUR-QUADRANT DIMMER (Cont.)

This very special mains-operated dimmer for domestic or industrial lights is not available in proprietary form: it enables brightness control of two groups of lights in one operation. The possible combinations of brightness are shown in the table. It will be clear that it is not possible to obtain continuous brightness control in the two groups. Instead, the circuit affords the setting of four states of brightness in either group: full on, fully dimmed, $1/3$ on, and $2/3$ on.

Both sections of the circuit operate on the well-known principle of the triac being switched from the blocking state to the conducting state with the aid of an RC network and a diode. The RC network provides the necessary phase shift and determines when the triac is switched. The rotary switch selects the resistor in a given network, and thus the brightness of the relevant group of lights. No resistor means that the group is off; a short-circuit gives maximum brightness, and resistors of 10 kΩ and 18 kΩ produce intermediate brightness. The diodes prevent the groups from affecting one another.

The 64-μH choke (L1) and the 150 nF capacitor across the bridge rectifier prevent the dimmer causing interference in other equipment connected to the mains.

If the triacs are fitted on a heatsink that is rated at 12° K/W, up to 500 W per group can be controlled. It is, of course, essential that the enclosure in which the dimmer is fitted provides ample cooling. A fair number of slots or holes in it are, therefore, essential; these should not permit the circuit elements to be touched.

The switch should have a nonmetallic spindle: this is not only safer than a metallic one, but it also enables the easy removal of the end-notch so that the switch can be rotated continuously, instead of having to be returned to the first stop every time it is operated.

The mains on/off switch S2 should be fitted with a built-in ON indicator bulb, which shows at a glance whether the circuit is on, even though S1 might be in the OFF position. Finally, remember that this circuit carries mains voltage in many places: good workmanship and insulation are, therefore, of the utmost importance.

LIGHT DIMMER

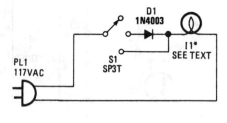

POPULAR ELECTRONICS   *Fig. 43-4*

Lamp I1 is a household lamp. When the switch is in the center position, the lamp is operated on half-wave rectified ac; the effective voltage the lamp sees is less, which dims it. I1 can be a lamp up to 200 W or 50 rated at 120 or 240 V, and D1 should be a 200-V PIV or better diode (400 PIV for 240-V operation).

AUTOMATIC EMERGENCY LIGHTING UNIT

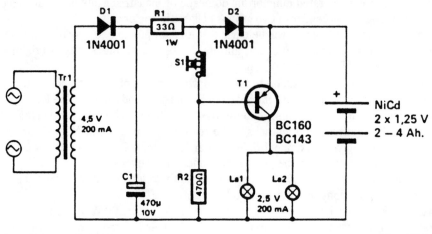

ELEKTOR ELECTRONICS

Fig. 43-5

This unit uses a Nicad battery to provide power to an emergency lighting setup. When power fails, T1 becomes forward biased, which lights L1 and L2. The batteries are normally kept charged. When power is on, T1 is cut off and it keeps the lamps extinguished.

LIGHTS-ON SENSOR

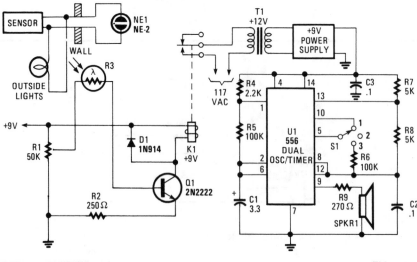

POPULAR ELECTRONICS

Fig. 43-6

Remote monitoring of a light source is possible with this circuit. Photocell R3 activates Q1 and relay K1. U1 is a tone generator that drives a small speaker.

LIGHT CHASER I

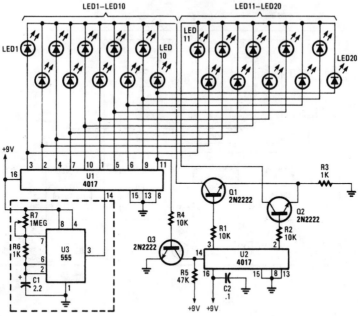

POPULAR ELECTRONICS

Fig. 43-7

Up to 100 lights, LEDs, or optocoupler triac circuits can be sequentially activated by this circuit. One (U1) 4017 decode counter sequences 10 LEDs whose common anode is returned through a second (U2) CD4017, which counts at one-tenth of the rate. The flash rate is controlled by U3, a clock circuit, with a 555 timer.

3-WAY LIGHT CONTROL

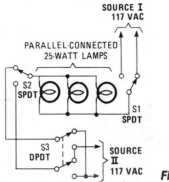

POPULAR ELECTRONICS

Fig. 43-8

This hookup is useful in some house wiring situations, where only two wires are available between switches, rather than the usual 3-way setup where 3 wires are required. S1 and S2 are ordinary three-way switches and S3, a DPDT switch, is commonly available as a four-way switch at hardware stores.

LIGHT CHASER II

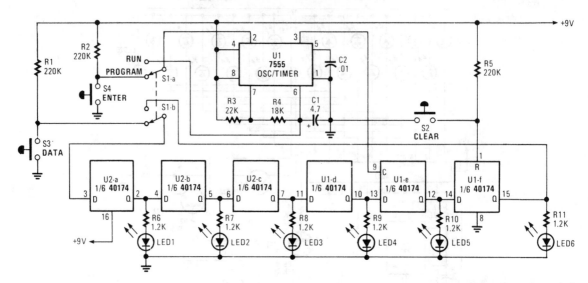

POPULAR ELECTRONICS

Fig. 43-9

Up to six lights can be sequentially flashed using this circuit. LED1 through LED6 can be replaced by optocouplers (MOC3010, etc.) to control 120-Vac loads via triacs. U1 generates pulses that clock the shift register mode up of the six D flip-flops in the CD 40174. By S1A – B, the register can be programmed either ON or OFF (low or high) and then switched to run in the programmed sequence. S2 clears the program.

LIGHT CONTROLLER

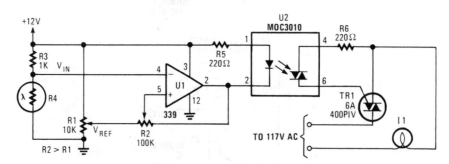

POPULAR ELECTRONICS

Fig. 43-10

A photocell drives U1, a comparator, which controls optocoupler U2. A 6A Triac is used to switch an ac load, such as a lamp, etc.

"AUTOMATIC" LIGHT BULB CHANGER

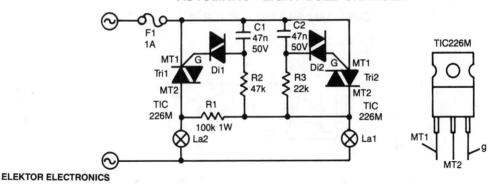

ELEKTOR ELECTRONICS

Fig. 43-11

The circuit presented here guarantees that if bulb La1 "gives up the ghost," bulb La2 will take over its task. In series with La1 is triac Tri2. Resistor R3 and C2 form a delay network. As soon as the voltage across C2 rises above about 30 V, diac (gateless triac) D2 is switched on, which causes Tri2 to conduct so that La1 lights.

The control circuit of La2 is parallel to that of La1, but because R2/C1 has twice the delay of R3/C2, Tri1 will not be triggered when Tri2 conducts; C1 discharges so that Tri1 cannot be triggered.

When, however, La1 is open-circuited, a voltage is across both RC networks via La2 and R1. Again, Tri2 will be triggered first, but because the current is smaller than its holding current, it will cease to conduct almost immediately. Capacitor C1 will then continue to charge and after a little while Tri1 is switched on.

Because the time constant for La2 is somewhat longer than that for La1, La2 will always be slightly less bright than La1. It is, of course, possible to give La2 a slightly higher wattage than La1 to ensure equal brightness.

Without heatsinks, the triacs can handle up to 100 W each; with heatsinks, powers of up to 1 000 W can be accommodated. It is not recommended to use bulbs with a wattage below 25 W, because these can flicker.

The triacs can be any type that can handle at least 400 V at no less than 5 A. The M types used in the prototype can handle 600 V at 5 A.

INDUCTIVE LOAD TRIAC SWITCH

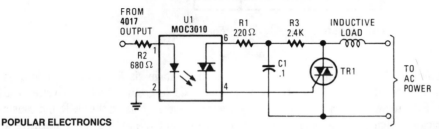

POPULAR ELECTRONICS

Fig. 43-12

An additional resistor and capacitor enable control of an inductive load, such as a small blower motor, fluorescent lamp, etc.

CHRISTMAS LIGHT DRIVER

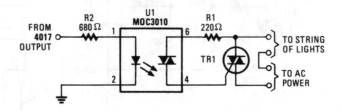

POPULAR ELECTRONICS

Fig. 43-13

This circuit will enable a CMOS logic chip, such as a 4017 decode driver, to control a string of Christmas lights or other lighting. The triac should be rated at 200 V and 3 A or higher. The 4017 should be powered from at least 10 V to ensure adequate drive to the optoisolator.

SCR CAPACITOR TURN-OFF CIRCUIT

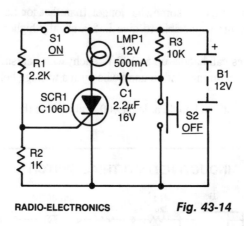

RADIO-ELECTRONICS

Fig. 43-14

After the SCR turns on, C1 charges up to almost the full supply voltage via R3 and the anode of the SCR. When S2 is subsequently closed, it clamps the positive end of C1 to ground, and the charge on C1 forces the anode of the SCR to swing negative momentarily, thereby reverse-biasing the SCR and causing it to turn off. The capacitor's charge bleeds away rapidly, but it has to hold the SCR's anode negative for only a few μs to ensure turn-off. C1 must be a nonpolarized type.

44

Limiter

The sources of the following circuits are contained in the Sources section, which begins on page 668. The figure number in the box of each circuit correlates to the entry in the Sources section.

Adjustable Voltage Limiter
One-Zener Precise Limiter

ADJUSTABLE VOLTAGE LIMITER

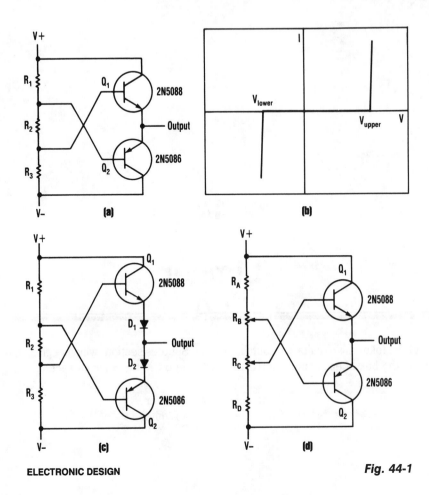

ELECTRONIC DESIGN

Fig. 44-1

This bipolar voltage clipper can be built with two transistors, two resistors, and two potentiometers. Notice that the maximum p-p range must be $\leq BV_{ebo}$ of the transistors used. The design equations are:

$$V_{\text{upper}} = V_- + \frac{(V_+ - V_-)\,(R_2 + R_3)}{(R_1 + R_2 + R_3\quad)} + V_{be}(Q_2)$$

$$V_{\text{lower}} = V_- + \frac{(V_+ - V_-)\,(R_3)}{(R_1 + R_2 + R_3)} - V_{be}(Q_1)$$

ONE-ZENER PRECISE LIMITER

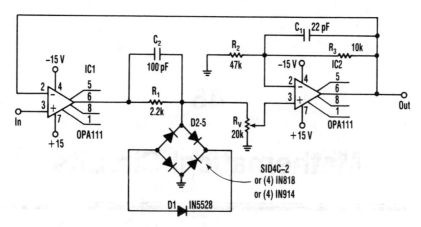

ELECTRONIC DESIGN

Fig. 44-2

A limiter circuit that requires matched zener diodes can instead use one zener with a full-wave diode bridge. The circuit's two limits are nearly equal when determined by the same zener—only two pairs of forward diodes need to be matched. For best results, an integrated quad of diodes can be used. But, after testing the circuit, four single controlled-drop diodes and four ordinary diodes gave about the same accuracy (better than 0.5%).

Because the limiting level can be adjusted, zener tolerance can be adjusted out. Gain stability can be optimized by connecting the inverting input to the first op amp to the output of the second to make the circuit inherently unity-gain.

The zener voltage must be increased to 8.2 V to compensate for the two diode drops. Placing small capacitors across the resistors in the loop stabilized the circuit adequately and response is orders of magnitude faster than conventional circuits. Moreover, it's limited primarily by the op amp's slew rate.

45

Mathematical Circuits

The sources of the following circuits are contained in the Sources section, which begins on page 668. The figure number in the box of each circuit correlates to the entry in the Sources section.

Fast Binary Adding Circuits
Programmable Slope Integrator
Multiplying Precise Commutating Amp

PROGRAMMABLE SLOPE INTEGRATOR

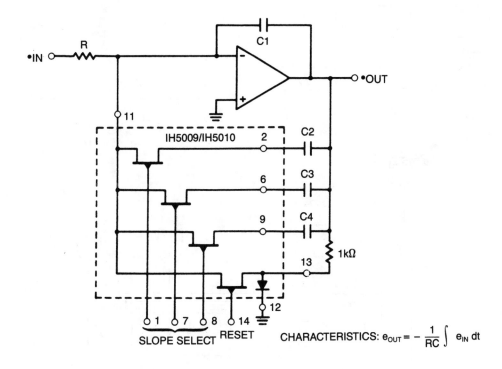

CHARACTERISTICS: $e_{OUT} = -\dfrac{1}{RC} \displaystyle\int e_{IN}\, dt$

INTERSIL

Fig. 45-1

By using analog switch IH5009/IH5010 to select various capacitors, a variable slope integrator can be had. If $C_3 = 2(C_2)$ and $C_4 = 4(C_2)$, seven different slopes can be obtained if binary information is fed to pins 1, 7, and 8 of the analog switch.

FAST BINARY ADDING CIRCUITS

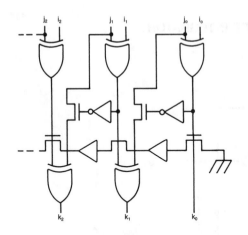

Fig. 45-2(a)

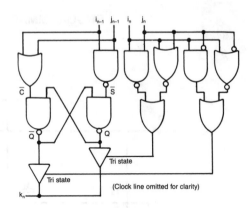

Fig. 45-2(b)

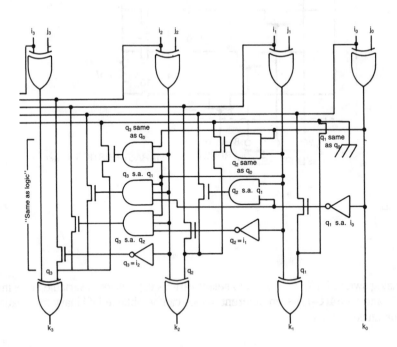

ELECTRONIC ENGINEERING

Fig. 45-2(c)

FAST BINARY ADDING CIRCUITS (*Cont.*)

Some circuits that add binary numbers have problems with time delay caused by carry propagation. This has been partially solved by the carry look-ahead adder. However, because of the complexity of this scheme, the carry look-ahead logic usually covers no more than 4 bits, and a ripple carry is implemented between the carry look-ahead blocks.

The Daniels Adder avoids these problems by presenting a scheme where carry bits are not used at all in the process of binary addition. It is based on recognition patterns, which exist with the binary addition truth table.

The addition is described by the following two sets of equations:

$$\text{if } i_{n-1} = j_{n-1} \qquad q_n = i_{n-1}$$

$$\text{if } i_{n-1} = j_{n-1} \qquad q_n = q_{n-1}$$

$$\text{if } i_n = j_n \qquad k_n = q_n$$

$$\text{if } i_n = j_n \qquad k_n = q_n$$

with the boundary condition that $q_{-1} = 0$, where i_n, j_n, and k_n are the bit of binary weight 2^n (n^{th} bit) of the addend, summand, and sum respectively, q_n is an intermediate variable and q_n is the inverse of q_n.

The value of the sum is (depending upon i_n and j_n) either the same as or the inverse of (depending upon i_{n-1} and j_{n-1} a0, a1, or the inverse of the $(n-1)^{\text{th}}$ bit of the sum. Figure 45-1(a) shows the logic diagram of the ripple through implementation of the adder.

Because each stage calculates whether its value of the intermediate variable q_n is the same as the previous stage's value (q_{n-1}) in parallel, it is possible to devise simple "same as" logic that does not have the complexity drawback of carry look-ahead logic and can be carried over any number of bits (Fig. 45-1(b)). A 32-bit adder built in this way will result in 11-gate delays (no gate having more than 4 inputs).

Especially compact and efficient is the pipelined implementation (Fig. 45-1(c)), which can produce the sum at a rate of 3-gate delays/bit.

The high-speed adder circuits can be used on gate arrays or full-custom ICs to implement fast calculation of addresses or data values. Because of their compact nature, they also use less space on the silicon than conventional adders do.

MULTIPLYING PRECISE COMMUTATING AMP

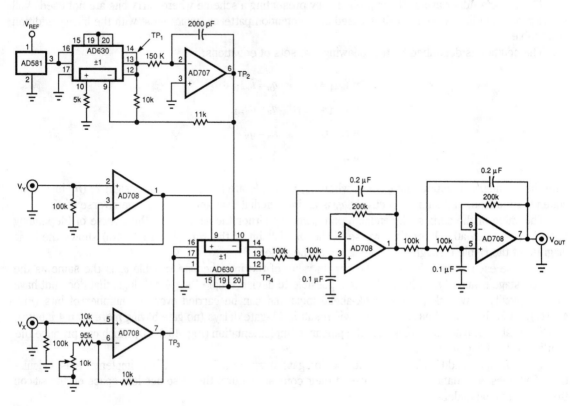

EDN

Fig. 45-3

By using a pulse-width-height modulation technique, this circuit implements a 0.015%-accurate multiplier. The circuit's output equals $V_X V_Y/10$. An AD581 voltage reference, an AD630 commutating amplifier, and an integrator comprising an AD707 op amp, 2000-pF capacitor, and 150-kΩ resistor first generate a precision triangle wave. For a given state of the AD630's output ($+ V_{REF}$ at TP$_1$, for example) the integrator ramps until its output reaches -11 V. Then, TP$_1$ changes state and the integrator begins ramping toward $+11$ V. The triangle wave's period is $4.4RC$ or 1.32 ms, where R and C are the values of the integrator components.

MULTIPLYING PRECISE COMMUTATING AMP (*Cont.*)

The circuit uses a second AD630, driven by the variable V_X to compare the triangle waveform at TP_2 to the signal at V_Y. The duty cycle, $T_1 + T_2$, at the output of this second commutating amplifier is:

$$T_1 = \frac{2RC(11 - V_Y)}{10}$$

and

$$T_2 = \frac{2RC(11 + V_Y)}{10}$$

During T_1, the voltage at TP_4 equals $-1.1V_X$. During the remaining period, T_2, the pulse height will equal $+1.1V_X$. V_{OUT} is the average, obtained by low-pass filtering, of this T_1 and T_2 combined waveform and equals:

$$V_o = \frac{-1.1V_X T_1 + 1.1V_X T_2}{T_1 + T_2} = \frac{V_X V_Y}{10}$$

You can use a higher bandwidth filter and a higher carrier frequency to build a faster multiplier.

46

Measuring and Test Circuits

The sources of the following circuits are contained in the Sources section, which begins on page 669. The figure number in the box of each circuit correlates to the entry in the Sources section.

Duty-Cycle Measurer
Simple Capacitor Tester
Magnetic Field Meter
Four-Trace Oscilloscope Adapter
Digital Tachometer Circuitry
Sensitive SWR Meter
Meter Tester
Broadband ac Active Rectifier
Bike Speedometer
B-Field Measurer
Low-Cost Barometer
VOR Signal Simulator
Simple Diode Curve Tracer
Simple Absolute Value Circuit

Microfarad Counter
Simple Duty-Cycle Meter
Static Detector
Electrolytic Capacitor Reforming Circuit
Digital VOM Phase Meter
Simple Electrometer
Digital Tachometer Counter
Capacitor ESR Measurer
Analog Tachometer Readout
Diode Matching Circuit
Permanent Magnet Detector
Transistor Tester
Bike Speedometer
Frequency Meter

DUTY-CYCLE MEASURER

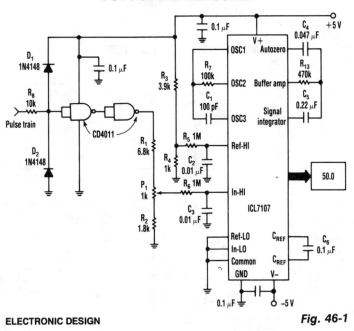

ELECTRONIC DESIGN

Fig. 46-1

An Intersil 7101 $3^1/_2$-digit A/D converter is used to display the duty cycle of a pulse train as a percentage. The frequency range of this circuit is 100 Hz to 250 kHz. The CMOS gates convert the pulse train to constant amplitude. This amplitude is then compared to a reference of 1 V, derived from R3 and R4. P1 is for calibration.

SIMPLE CAPACITOR TESTER

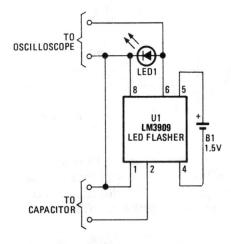

POPULAR ELECTRONICS   **Fig. 46-2**

An LM3909 LED flasher is used as an oscillator, and the capacitor connected to the terminals determines frequency.

C	f approx
4 700 μF	0.04 Hz
470 μF	0.4 Hz
4.7 μF	4 Hz
0.047 μF	4 kHz
0.004 7 μF	25 kHz
47 μF	660 kHz

The LED can be used to count frequency visually using a stopwatch for large capacitors ($C > 500$ μF).

MAGNETIC FIELD METER

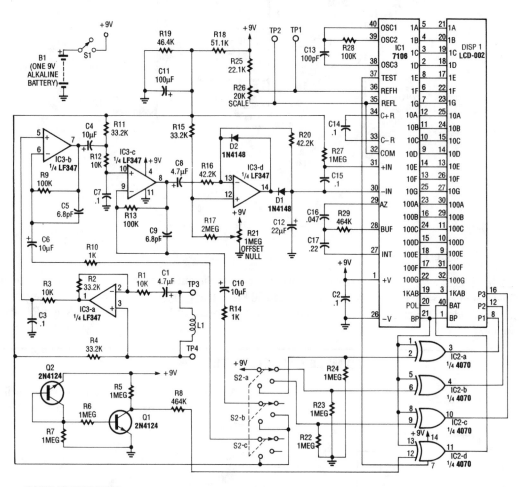

RADIO-ELECTRONICS

Fig. 46-3

Using a pickup coil to drive an amplifier (IC3A-B-C-D), this meter circuit can be directly calibrated in field-intensity units. R3/C3 and R12/C7 establish a frequency roll off that compensates for the pickup-coil sensitivity, and set a 20-kHz cut-off point. S2 is the range-select switch. L1 is an 18-turn 3″ diameter coil. The frequency range is 50 Hz to 20 kHz and the range of measurement is 0.1 to 20 000 microTesiers (μT).

FOUR-TRACE OSCILLOSCOPE ADAPTER

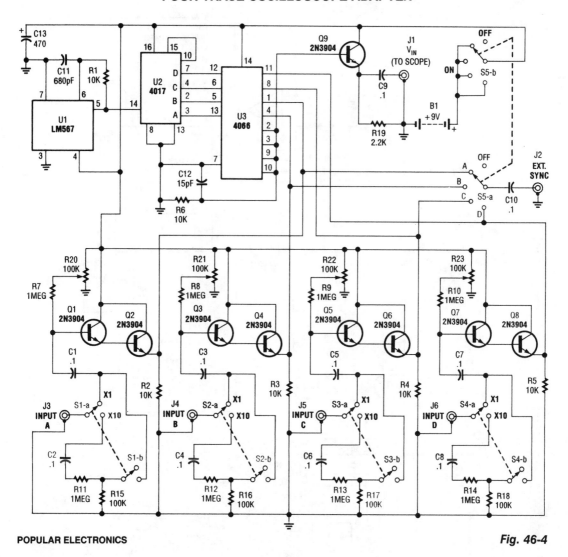

POPULAR ELECTRONICS

Fig. 46-4

This simple adapter uses an oscillator (567) to drive a counter (U2) and switch (U3) that selects the output of one of four scope preamps (Q1/Q2 through Q7/Q8) and feeds it to buffer Q9 and output jack J1. J2 provides synch to the scope. R20 through R23 are posting controls for channels A through D (J3 through J6). S1A-B through S4A-B are switched attenuators, one for each channel. Switching rate is about 125 kHz. This circuit is useful for adding four-trace operation to inexpensive oscilloscopes. Signal levels of 0 to 20 V can be handled.

DIGITAL TACHOMETER CIRCUITRY

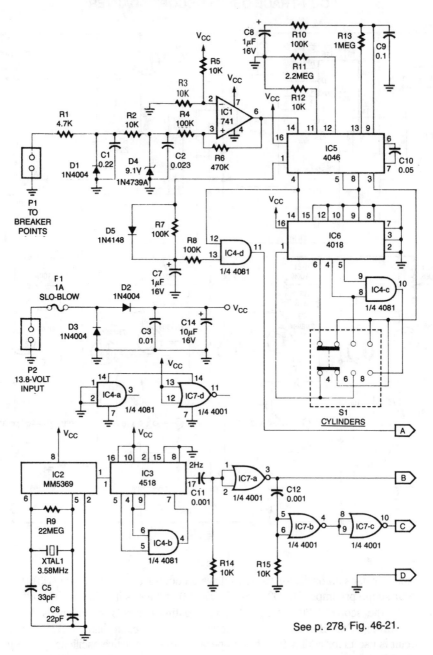

See p. 278, Fig. 46-21.

RADIO-ELECTRONICS

Fig. 46-5

DIGITAL TACHOMETER CIRCUITRY (*Cont.*)

This system can be used with 4-, 6-, or 8-cylinder automobiles. The timebase formed by IC5 is an oscillator that drives counter IC6, which divides by 6, 4, or 3 for 4-, 6-, or 8-cylinder engines, respectively. S1 selects this number. IC5 produces a signal that is phaselocked to this multiple of the ignition system frequency, which in turn depends on engine speed.

$$\text{freq} = \text{rpm} \times \frac{\text{\# cylinders (4, 6, or 8)}}{120} \text{ Hz}$$

IC1 conditions the ignition input at P1 to feed IC5. The output of IC4D, which is the same frequency as the VCO in IC5, is fed to the frequency display.

IC2 generates a 60-Hz signal using a 3.58-MHz reference. IC3 and IC4B divide this by 30 to produce 2 Hz. IC7B/IC7C and C12/R15 produce a delayed 2-Hz signal. These signals are fed to the counter circuit.

SENSITIVE SWR METER

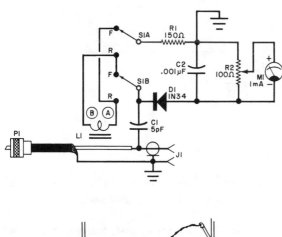

Reflected Meter Reading (% of full scale)	SWR
0	1:1
10	1.22:1
20	1.5:1
30	1.85:1
33.5	2:1
40	2.33:1
50	3:1
60	4:1
66.67	5:1
70	5.66:1
80	9:1
90	19:1
100	∞:1

POPULAR ELECTRONICS

Fig. 46-6

Using a toroidal pickup coil around the center conductor of a coaxial cable, this circuit can be used to measure the SWR of an antenna. L1 is two turns #26 enamelled wire on a Fair-Rite 5963000301 toroidal core.

METER TESTER

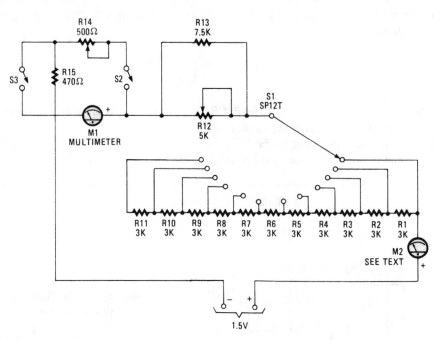

POPULAR ELECTRONICS

Fig. 46-7

This unit uses switches and resistors to provide a number of current ranges. It allows you to test most of the meters available at surplus outlets, and without damaging the sensitive movements when you have no idea of internal resistance or full-scale current of the unit.

M1 is a multimeter set to measure current, and M2 is the meter-under-test. Starting with S1 set at the maximum resistance and S2 open, decrease the resistance setting of S1, fine tuning with R12, until M2 reads full scale. Then, read M1. It will tell you the full-scale current for the unknown meter. As the meters are connected in series, the same current flows through both.

Now, close S2 and adjust R14 and R15 until M2 reads exactly mid-scale and M1 reads the same current as determined earlier to be the maximum current for M2. Half the current is flowing in M2 and half is going through R14 and R15. The voltage drop is the same across the meter and R14 and R15, because they're in parallel. Thus, the sum of the resistance of R_{14} and R_{15} is the same as the internal resistance of meter M2.

If the internal resistance of M2 is less than 470 Ω, set R14 at maximum resistance and close S3. Readjust R14. Both R14 and R12 should be linear-taper potentiometers.

BROADBAND AC ACTIVE RECTIFIER

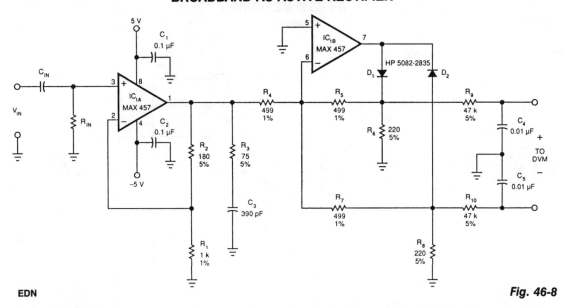

EDN

Fig. 46-8

This circuit converts sine waves of up to 1-V rms into an equivalent dc level. It should prove useful as an ac broadband voltmeter. IC1 is an input amplifier that converts level from rms to an equivalent dc level and feeds ICB1. R3 and C3 are stabilizing components. IC2 acts as a full-wave rectifier. R6, R5, and D1 rectify positive levels, R7, R8, and D2 negative-going signals. R9, R10, C4, and C5 are low-pass filters. The output can feed a DVM or another meter.

BIKE SPEEDOMETER

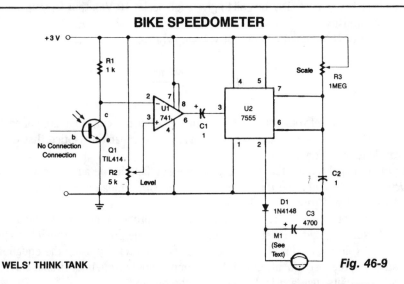

WELS' THINK TANK

Fig. 46-9

A TIL414 photo transistor senses reflection from a spoke-mounted reflector. This generates a pulse and sends it to U1 and U2, a monostable multivibrator, which drives meter M1.

B-FIELD MEASURER

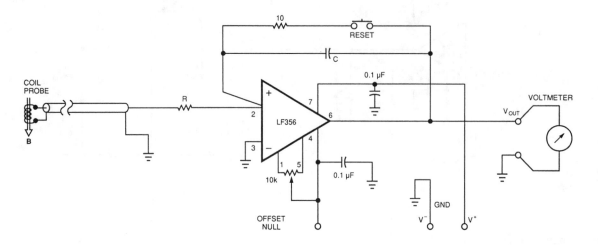

EDN

Fig. 46-10

This circuit develops an output voltage that is proportional to the magnetic induction, B, flowing through its probe's coil. You must size the coil to give a full-scale, 10-V output for your maximum expected magnetic-induction intensity.

For a given value of B (in tesla) and output voltage, V_{OUT}:

$$B = \frac{(R \times C \times V_{OUT})}{A}$$

where A is the effective area of your coil in m^2 (A = number of turns × average area of each turn), R is the resistance of the coil and the probe, and C is the value of the capacitor. Notice that C should be a low-leakage polypropylene or Teflon device.

For most practical applications that measure a magnetic field in the air, the coil will be either tiny or very thin. If R = 1 kΩ, C = 1 μF, and the coil is 100 turns with a mean area per turn of 1 cm^2, then the circuit's output will be 1 mV/gauss ($1T = 104G$).

To use the circuit, push the reset button and place the probe in an area that you know is devoid of magnetic fields. Be sure to avoid magnets and iron. Then, put the probe into the field to be measured and read the V_{OUT} with a voltmeter. Finally, calculate the B field's intensity using the equation.

When constructing the instrument, guard the op amp's inputs from undesirable currents at the minus input. For full-scale outputs, use a ±15-V supply for the op amp.

LOW-COST BAROMETER

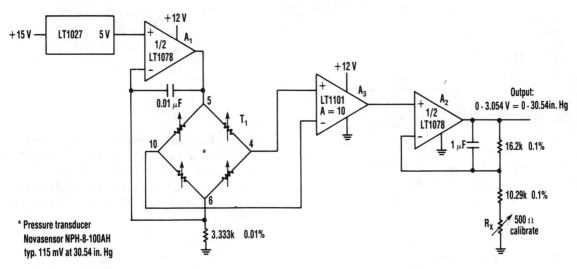

ELECTRONIC DESIGN

Fig. 46-11

Using Linear Technology LT1027 reference and LT1078 op amps, transducer T1 is fed with 1.5 mA. The pressure transducer feeds an amplifier with a gain of 10, then it feeds a voltage follower. Output can either drive an analog meter or a DVM circuit.

VOR SIGNAL SIMULATOR

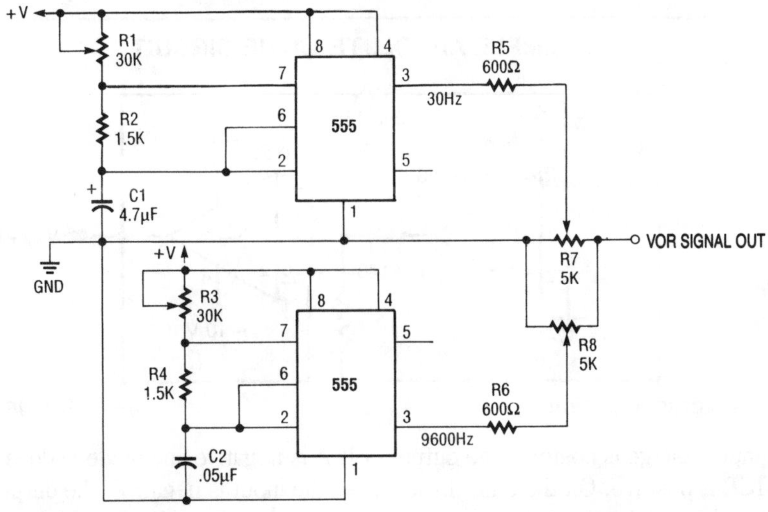

RADIO-ELECTRONICS

Fig. 46-12

An output of 9 960 and 30 Hz at 0.5 Vrms is produced by this VOR (VHF Omni Range) signal simulator, which uses a pair of 555 timers. Alternatively, a 556 (two timers in one package) could be used.

SIMPLE DIODE CURVE TRACER

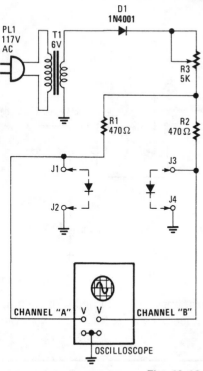

Suitable for matching diodes or examining V/I characteristics of two terminal devices (diodes, etc.), this circuit should be handy for lab use. R_1 and R_2 can be increased in value and a higher voltage transformer can be used for higher voltage test using this principle.

POPULAR ELECTRONICS   *Fig. 46-13*

SIMPLE ABSOLUTE VALUE CIRCUIT

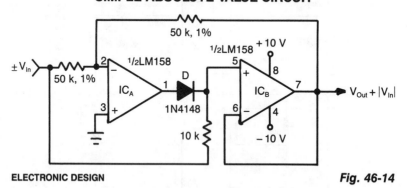

ELECTRONIC DESIGN   *Fig. 46-14*

When an input voltage is positive, the output of ICA is negative and diode D does not conduct; hence the output of ICB is positive. On the other hand, when the input is negative, the output of ICA is positive. D will conduct and cause the absolute value, expressed as a positive voltage, to appear on the noninverting input of ICB and on the circuit's output.

The circuit's dynamic range extends from zero to the point at which the op amp saturates. The bandwidth is determined by the characteristics of the diode and the high-frequency performance of the op amp.

MICROFARAD COUNTER

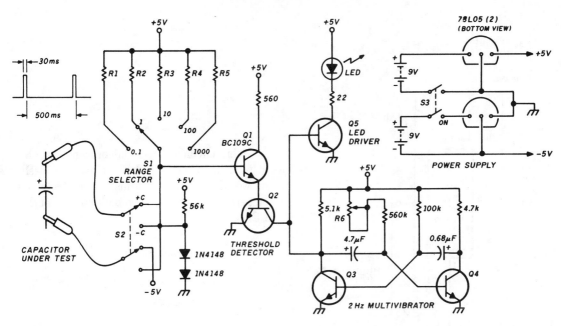

HAM RADIO

Fig. 46-15

This circuit measures capacitance by the time it takes an unknown capacitor to reach 6.32 V (10 V × 63% or 1 *RC* time constant) when charged through resistor R. The LED used as a timebase is pulsed by Q3, Q4, and Q5. By counting seconds (two flashes per second) until threshold detector Q1/Q2 stops the count, you can directly read the number of microfarads. R_1, R_2, R_3, R_4, and R_5 can be any convenient values, such as 1 kΩ, 10 kΩ, 100 kΩ, 1 MΩ, 10 MΩ.

SIMPLE DUTY-CYCLE METER

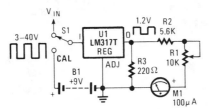

POPULAR ELECTRONICS

Fig. 46-16

Using an LM317T as a precision clipper, this displays the duty cycle of a pulse train from 0 to 100% on a standard 100-mA analog panel meter. This circuit will work up to about 50 kHz.

STATIC DETECTOR

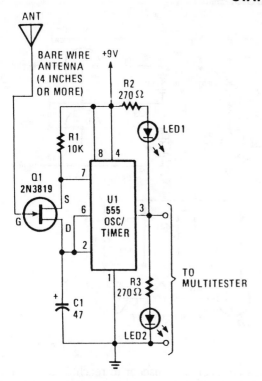

This circuit detects the presence of electrostatic fields by using an FET to alter the flash rate of two LEDs. The FET is installed in the timing circuit and causes a change in the flash rate of the two LEDs.

POPULAR ELECTRONICS   *Fig. 46-17*

ELECTROLYTIC CAPACITOR REFORMING CIRCUIT

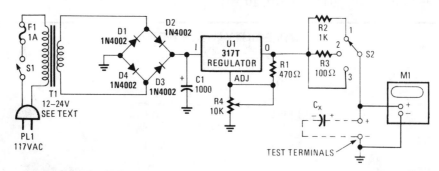

POPULAR ELECTRONICS   *Fig. 46-18*

Sometimes electrolytic capacitors that are stored for some time will exhibit high leakage currents. Before placing these in service, the capacitors might need to be reformed. This power supply can be used for reforming. Adjust R4 for the capacitor's rated voltage and set S2 in position 1. When M1 indicates the rated voltage, reforming is complete. For large capacitors, use position 2, and then position 3 after reforming starts. T1 can be any transformer with a rating of 12 to 24 Vac at about 1 A.

DIGITAL VOM PHASE METER

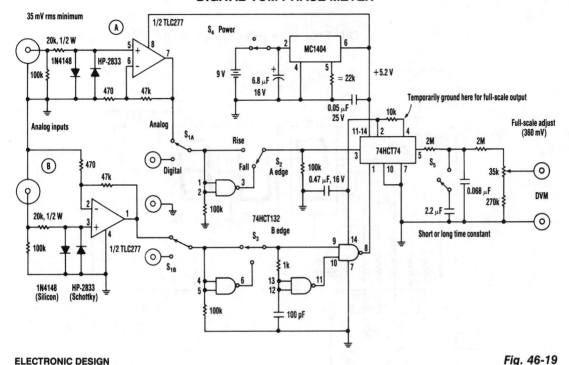

ELECTRONIC DESIGN

Fig. 46-19

The phase-angle meter will work with either analog or digital inputs. A DVM is used as a readout device. The output is 1 mV per degree (360 mV or degrees full-scale). The MC1404 precision regulator maintains calibration with a battery source (9 V).

SIMPLE ELECTROMETER

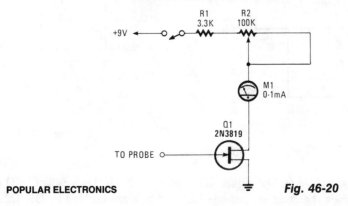

POPULAR ELECTRONICS

Fig. 46-20

This electrometer is useful as a relative indicator of static charges or as an electric field in a charge-free environment. An induced negative charge on the probe will reduce drain current toward zero.

DIGITAL TACHOMETER COUNTER

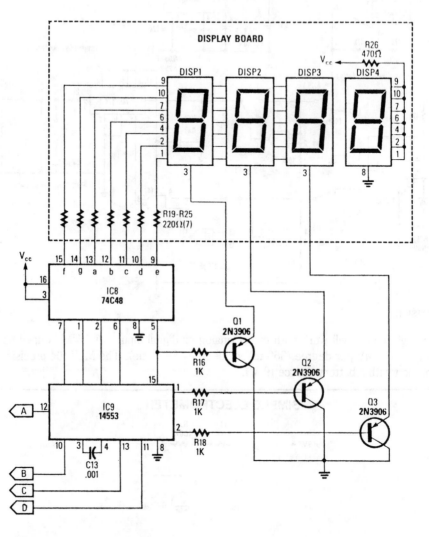

Fig. 46-21

This circuit produces a readout for the digital tachometer circuit. IC9 is a 3-digit LED display driver, counter, and latch. IC8 drives the common-cathode LEDs, which are enabled by Q1, Q2, and Q3. See page 268, Fig. 46-5 for the matching project.

CAPACITOR ESR MEASURER

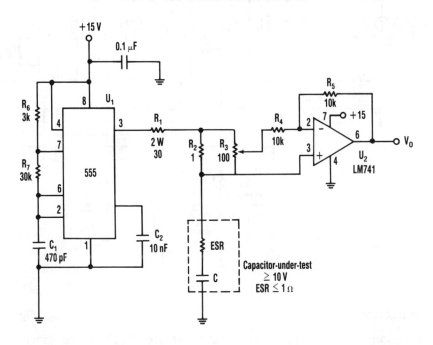

ELECTRONIC DESIGN

Fig. 46-22

The equivalent series resistance (ESR) of a capacitor can be measured using this circuit and an ac voltmeter. U1 functions as a 50-kHz square-wave generator. It drives a current waveform of about ± 180 mA in the capacitor-under-test, through R1 and R2. When R3 is adjusted to the proper value, the voltage drop across the equivalent series resistor is precisely nulled by the inverting amplifier (U2). Thus, V_0 is the pure capacitor voltage which is the minimum voltage that can be produced at V_0.

To make an ac voltage measurement, adjust R3 until V_0 is minimized. Then, note the position of the potentiometer and multiply it by the value of R2, 1 Ω in this case. That product equals the capacitor's ESR. The capacitor is biased at about 7.5 V. Lower-voltage capacitors won't work with this circuit. By changing the value of R2, other ranges of ESR can be measured. However, for small R2 values, the current level should be increased to keep a reasonable voltage across R2. This requires some sort of buffer. The circuit is intended for capacitors greater than 100 μF. The ripple voltage gets large for smaller values and accuracy decreases.

ANALOG TACHOMETER READOUT

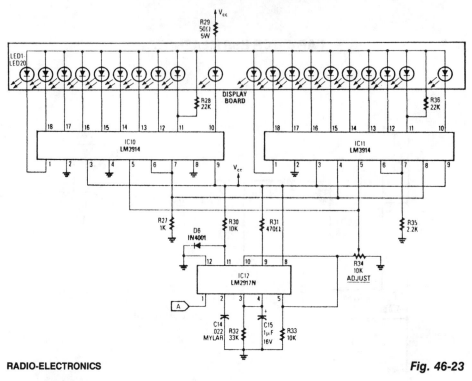

RADIO-ELECTRONICS

Fig. 46-23

The analog display consists of a frequency/voltage converter (IC12) and bar-graph segment drivers IC10 and IC11. R34 is the calibration adjustment and is set so that an engine rpm of 5 000 to 7 000 rpm lights the first LED (redline value).

DIODE MATCHING CIRCUIT

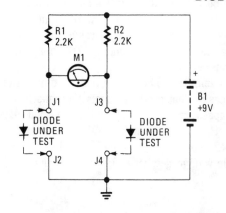

This circuit can be used to match diodes for use in circuits where such a balance is necessary (a balanced modulator, for instance). The diode matching circuit will indicate the forward-voltage drop of the two diodes in millivolts.

POPULAR ELECTRONICS   *Fig. 46-24*

PERMANENT MAGNET DETECTOR

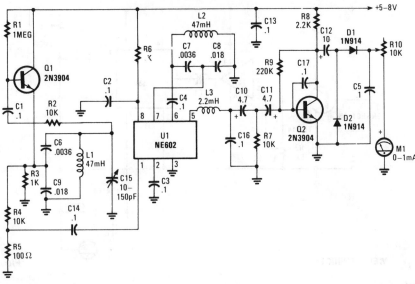

POPULAR ELECTRONICS

Fig. 46-25

In this circuit, oscillator Q1 runs at about 15 kHz and feeds mixer U1. U1 has an internal oscillator that runs at around 15 kHz. C15 is used to zero-beat both oscillators. When a magnet is brought near L1 or L2, the magnetic field shifts the permeability of their respective cases. This changes the oscillator frequency, and the audio note is passed through filter L3, C16, and C10/R7 to amplifier Q2. There the audio is amplified and drives meter M1 via rectifiers D1 and D2.

TRANSISTOR TESTER

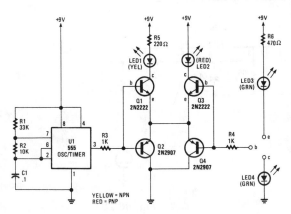

A 555 timer drives Q1 through Q4 with a square wave. LED1 and LED2 light when an npn or pnp transistor respectively, are connected to the text terminals. If LED3 and LED4 light equally as LED1 and LED2, the transistor is functional. If LED3 and LED4 are brighter than LED1 or LED2, the transistor is shorted.

POPULAR ELECTRONICS

Fig. 46-26

BIKE SPEEDOMETER

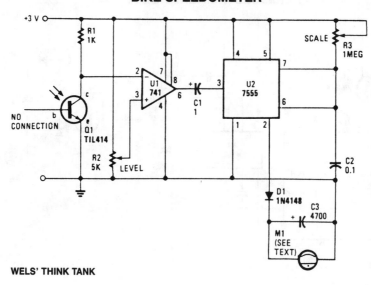

WELS' THINK TANK

Fig. 46-27

A TIL414 photo transistor senses reflection from a spoke-mounted reflector, generates a pulse, and sends it to U1 and U2, a monostable multivibrator, which drives meter M1.

FREQUENCY METER

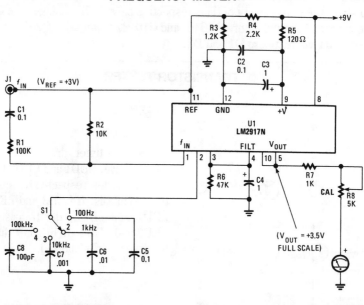

POPULAR ELECTRONICS

Fig. 46-28

Using an LM2917N, this F/V converter-based circuit indicates the frequency on a meter. S1 selects full-scale range (up to 100 kHz). R8 is recommended to obtain the battery source.

47

Measuring and Test Circuits
(*E*, *I*, and *R*)

The sources of the following circuits are contained in the Sources section, which begins on page 669. The figure number in the box of each circuit correlates to the entry in the Sources section.

Hall-Sensor Current Monitor
Synchronous System
Digital LED Voltmeter
Continuity Checker
Quick Multiconductor Cable Tester
Continuity Tester
4-Range ac Millivoltmeter
Low-Ohms Adapter
ac Current Indicator
5-Range Linear-Scale Ohmmeter

High-Resistance-Measuring DMM
High-Gain Current Sensor
Sensitive Continuity Checker
4-Range dc Microammeter
Multimeter Shunt
Continuity Tester
LED Millivoltmeter
Latching Continuity Tester
dc Millivoltmeter
Continuity Tester

HALL-SENSOR CURRENT MONITOR

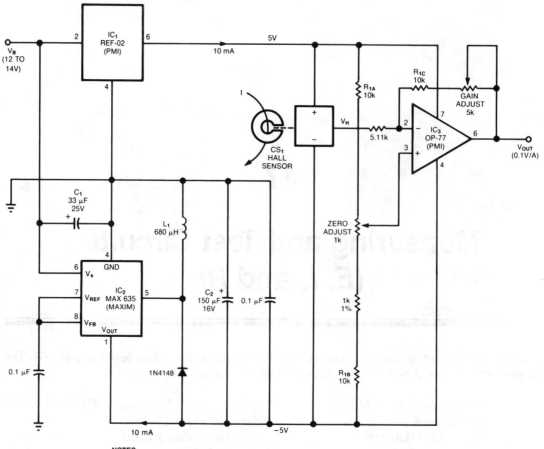

NOTES:
1. C_1 and C_2 ARE 199D TANTALEX CAPACITORS FROM SPRAGUE.
2. L_1 IS A 6860-23 INDUCTOR FROM CADDELL-BURNS.
3. R_{1A}, R_{1B}, AND R_{1C} ARE PART OF A THIN-FILM RESISTOR NETWORK SUCH AS THE CADDOCK T914-10K.
4. CS_1 IS A HALL-EFFECT CURRENT SENSOR (CSLA1CD) FROM MICROSWITCH.

EDN

Fig. 47-1

The circuit uses a Hall-effect sensor, which consists of an IC that resides in a small gap in a flux-collector toroid, to measure dc current from 1 to 40 A. Wrap the current-carrying wire through the toroid; the Hall voltage V_H is then linearly proportional to the current (I). The current drain from V_B is less than 30 mA.

To monitor an automobile alternator's output current, for example, connect the car's battery between the circuit's V_B terminal and ground, and wrap one turn of wire through the toroid (or, you could wrap 10 turns—if they'd fit—to measure 1 A full-scale). When $I=0$ V, the current sensor's (CS_1's) V_H output equals one-half of its 10-V bias voltage, V_H and V_{OUT} are zero when I is zero; you can then adjust the output gain and offset to scale V_{OUT} at 1 V per 10 A.

SYNCHRONOUS SYSTEM

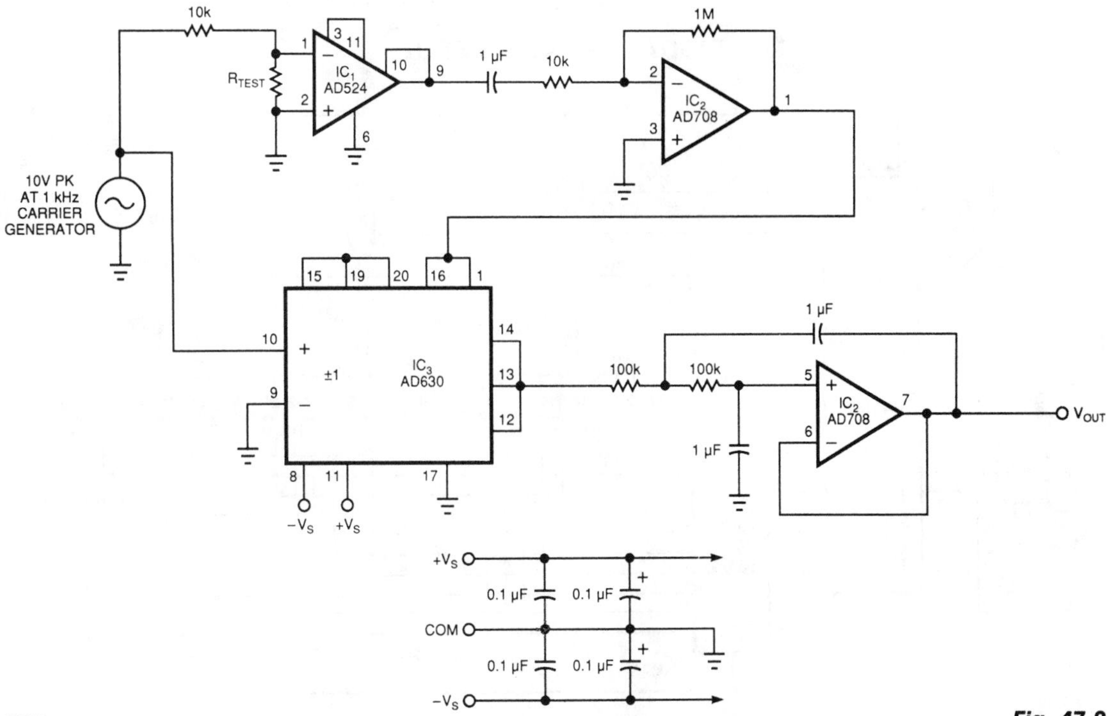

EDN

Fig. 47-2

The circuit uses a synchronous-detection scheme to measure low-level resistances. Other low-resistance-measuring circuits sometimes inject unacceptable large currents into the system-under-test. This circuit synchronously demodulates the voltage drop across the system-under-test and can hence use extremely low currents while measuring resistance.

The 10-V (pk), 1-kHz carrier generator injects a 1-mA reference current into unknown resistor, R_{TEST}. Instrument amplifier IC1 and precision op amp IC2A amplify the voltage across R_{TEST} by a gain of 100 000. Synchronous detector IC3 demodulates this voltage, then op amp IC2B acts a low-pass filter on the demodulated voltage. The low-pass filtering will attenuate all uncorrelated disturbances (such as noise, drift, or offsets), while passing a dc voltage that is proportional to the unknown resistance.

The relationship between the output voltage and the unknown resistance is:

$$V_{OUT} = 10 \times \left(\frac{2 \text{ V}}{\pi} \right) \times R_{TEST} \times \frac{10_5}{10 \text{ k}\Omega} \text{ ,or}$$

$$R = 0.0157 \times V_{OUT},$$

which is $\dfrac{15.7 \text{ m}\Omega}{\text{V}}$ at the circuit's output.

DIGITAL LED VOLTMETER

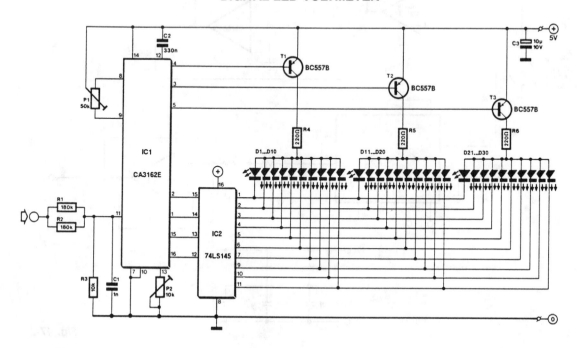

Fig. 47-3

The voltage to be measured is digitized in an analog-to-digital (A/D) converter and then displayed in three decimal digits. The display consists of three groups of 10 LEDs. The meter can only be used for measuring direct voltages.

The A/D converter is based on CA3162, which can process direct voltage up to 999 mV (1-V full-scale deflection—FSD). The FSD is extended to 10 V with the aid of potential divider R1/R2/R3. Other ranges are possible by altering the values of the resistors.

The measured value is read from three bars of LEDs: the first one of these, D1 through D10, shows units; the second, D11 through D20 tens; and the third, D21 through D30, hundreds.

The circuit is nulled with P1 when the input is open. Zero here means that diodes D1, D11, and D21, light. Diodes D10, D20, and D30, represent the figure 9. Next, a known voltage is applied to the input and P2 is adjusted until the LEDs read the correct value. When the input voltage is too high, the display goes out. When the input is negative, the unit LEDs do not light. Notice that variations in the supply voltage affect the measurement adversely.

CONTINUITY CHECKER

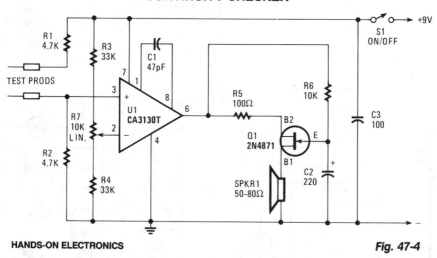

HANDS-ON ELECTRONICS

Fig. 47-4

U1 is an op amp used as a comparator. When the test probes are shorted together, resistors R1 and R2 bias the noninverting input to half the supply voltage. The inverting input is biased by a voltage divider that consists of R3, R7, and R4. Resistor R7 is adjusted so that the voltage to the inverting input is lower than that to the noninverting input when the probes are shorted together. With continuity across the probes, U1's output goes high and supplies power to Q1, which is configured as a relaxation oscillator. The output of Q1 is fed to a high-impedance loudspeaker for an audio tone. When the probe is open, the noninverting input goes to the negative supply rail via R2. This action forces U1's output low, which results in no output from the oscillator.

CONTINUITY TESTER

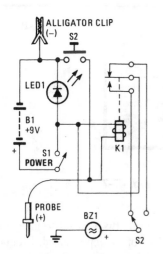

Using an LED and an audible indicator, relay K1 actuates buzzer BZ1. S2 is a buzzer/battery test switch for testing the battery in both NO and NC mode.

POPULAR ELECTRONICS   **Fig. 47-5**

287

QUICK MULTICONDUCTOR CABLE TESTER

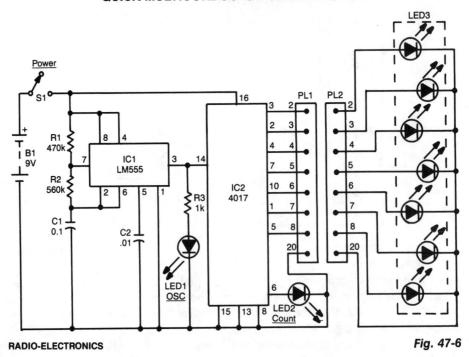

RADIO-ELECTRONICS

Fig. 47-6

This circuit can be used to check up to an eight-conductor cable. IC1, a 555 timer, drives decade counter IC2, a 4017. Each LED should light in sequence. The cable to be tested is connected between PL1 and PL2. If the cable is miswired, the LEDs will light out of sequence. If it is shorted or open, some LEDs will not light.

4-RANGE ac MILLIVOLTMETER

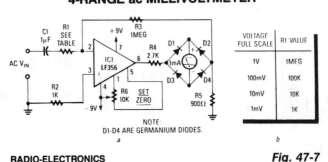

VOLTAGE FULL SCALE	R1 VALUE
1V	1MEG
100mV	100K
10mV	10K
1mV	1K

NOTE:
D1-D4 ARE GERMANIUM DIODES.

a

b

RADIO-ELECTRONICS

Fig. 47-7

By placing the rectifier in the op amp feedback path, nonlinearity is greatly reduced to insignificant levels. The meter will read the full-wave rectified average (absolute value) of the input signal. Frequency response is a few Hz to about 50 kHz.

LOW-OHMS ADAPTER

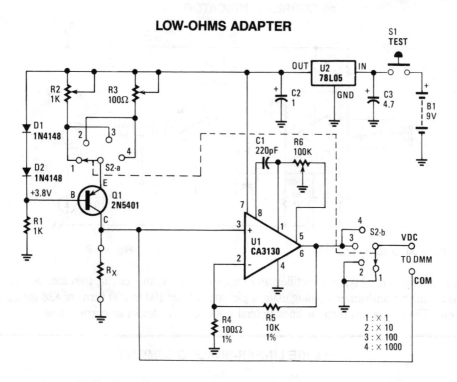

Fig. 47-8

The circuit consists of a 5-V regulator, constant-current source D1, D2, and Q1, and op amp gain stage U1. Power is provided by a 9-V battery whose output is regulated to +5 Vdc by the 3-terminal regulator. The emitter of Q1 is always 0.6 V below the +5-V line. Resistor R1 sets the current through both diodes D1 and D2 to 5 mA.

The resulting 0.6 Vdc across one of the multiturn trimmer potentiometers (R2 and R3), as selected by switch section S2A, sets the current through Q1 and the resistor-under-test.

When R2 is selected, the test current is 1 mA; when R3 is selected, the test current is 10 mA. On the lower two ranges, ×1 and ×10, the voltage across resistance-under-test is applied directly to the DMM terminals. On the upper two ranges, op amp gain stage U1 is switched into the circuit and the DMM measures the voltage between op amp output pin 6 and the test resistor.

When switch S2 is in position 3 (×100) the current set by the constant-current source is 1 mA; the multiplying factor is ×100. When S2 is in position 4, ×1 000, the current is 10 mA and the multiplying factor is 100×10 = 1 000. Multiturn trimmer-potentiometer R6 adjusts the offset of the op amp so that, with no voltage across the resistor-under-test (i.e., with the measurement terminals short-circuited), the output is zero.

ac CURRENT INDICATOR

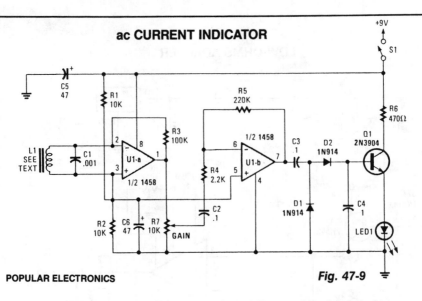

POPULAR ELECTRONICS

Fig. 47-9

Using a dual op amp driving a rectifier and emitter-follower, this circuit indicates ac current on an LED. L1 is an audio transformer winding using a pick-up coil, or 100 to 200 turns of #28 gauge wire 2″ in diameter, etc. The circuit can trace ac lines behind walls, etc. or detect ac current flow.

5-RANGE LINEAR-SCALE OHMMETER

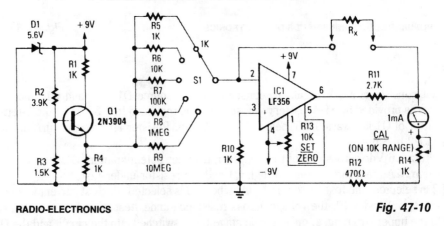

RADIO-ELECTRONICS

Fig. 47-10

R_x is inserted in the feedback path of IC1. A known reference current is selected from reference voltage generator Q1, D1, and R5 through R9. A meter reading will be produced:

$$I = \frac{R_x}{R_5} \times 1 \text{ mA}$$

where $R_5 = R_5$ through R_9, as selected. This corresponds linearly to the value of R_x.

HIGH-RESISTANCE-MEASURING DMM

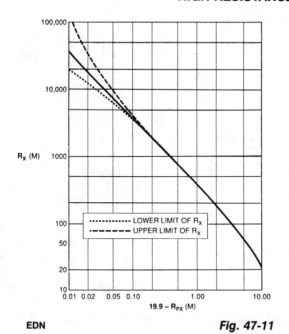

Using a simple technique, you can extend the resistance-measurement range of your $3^1/2$-digit DMM from the usual 19.99 MΩ to 40 GΩ. You could measure, for example, the leakage resistances of transformers, motor windings, and capacitors.

For a 19.99-MΩ DMM range, select a stable 20-MΩ resistor whose value is slightly below nominal, 19.99 MΩ for example. An unknown high resistance, R_X, is:

$$R_X = \frac{(R_P \times R_{PX})}{(R_P - R_{PX})}$$

where R_P is the high-value parallel resistor and R_{PX} is the measured value of R_P in parallel with R_X. An even easier way to determine the value of R_X is to use the graph.

EDN

Fig. 47-11

HIGH-GAIN CURRENT SENSOR

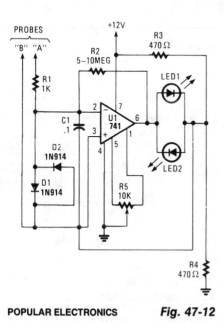

A high-gain amplifier using a UA741 is used to sense relative voltage drop in a conductor, and therefore current in the conductor. R_2 can be increased to 10 MΩ for increased sensitivity. LED1 and LED2 provide polarity indication. This circuit can be used to detect current flowing in a PC board trace, and also for locating shorts and opens.

POPULAR ELECTRONICS   **Fig. 47-12**

SENSITIVE CONTINUITY CHECKER

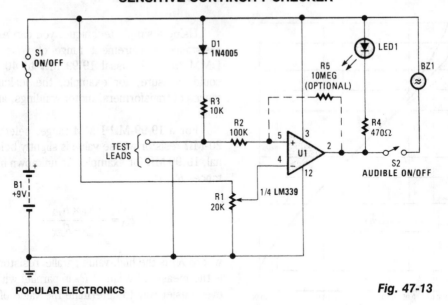

POPULAR ELECTRONICS

Fig. 47-13

This continuity checker (built around an LM339 quad comparator with open-collector outputs) eliminates false readings because of coils or low-resistance devices in a circuit.

U1 is a comparator that acts as a sensing amplifier for the bridge circuit (R1 and D1, R3 and the unknown resistance, R_x that is connected across the test leads. When R_x is less than this predetermined value (by the setting of R1), the LED lights and BZ1 sounds.

4-RANGE dc MICROAMMETER

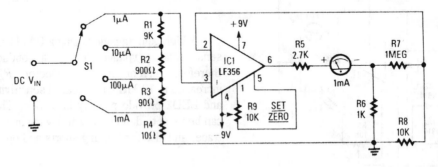

RADIO-ELECTRONICS

Fig. 47-14

IC1 produces a full-scale reading on M1 when pin 3 has a 10 mVdc level on it, as determined by R7 and R8. R1 through R4 are shunts that produce a 10-mV drop for the desired full-scale range. R9 zeros the meter.

MULTIMETER SHUNT

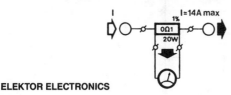

ELEKTOR ELECTRONICS

Fig. 47-15

The current range in multimeters, particularly the more inexpensive ones, is restricted by the load limits of the internal shunts to 1 to 2 A. The figure shows how easily a precision heavy-duty resistor from Dale or RCL (0.1 Ω; 20 W; 1%) can be used as an external shunt. These resistors were not designed for this purpose, but they are much cheaper than custom-made shunt resistors. The 20-W rating applies only, by the way, if a heatsink is used: without that its rating is only 8 W.

The maximum current through the device on a heatsink is about 14 A; the larger versions draw up to 17.5 A. When mounting the shunt, make sure that the test terminals (as well as the device terminals) are soldered properly, otherwise the resistance of the terminals is added to the shunt.

CONTINUITY TESTER

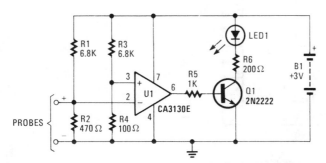

POPULAR ELECTRONICS

Fig. 47-16

Using a comparator, this continuity tester applies only about 300 mV to the circuit to be tested. This avoids false readings in semiconductor circuits.

LED MILLIVOLTMETER

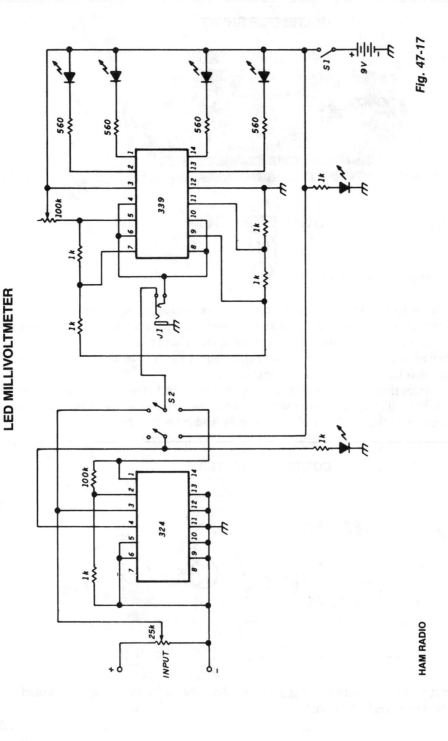

Fig. 47-17

HAM RADIO

This circuit uses a dc amplifier (324) in a bar-graph circuit that uses an LM339 Quad comparator IC to sense dc levels. The LEDs will light every 100 mV. The 324 op amp is configured to provide a gain of 100 × to increase this sensitivity to 1 mV. Two auxiliary LEDs indicate "power on" and "×100" gain setting.

LATCHING CONTINUITY CHECKER

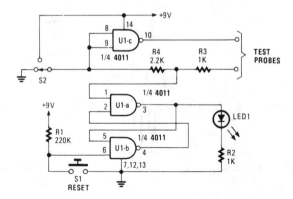

POPULAR ELECTRONICS

Fig. 47-18

This circuit detects brief shorts or opens. When S2 is in the up position, the circuit indicates if there is or was continuity by lighting LED1. U1A and U1B are connected as an R-S flip-flop. S1 resets the tester. When S2 is in the down position, a momentary interruption in continuity will light LED1. This tester is good for detecting intermittent shorts or opens.

dc MILLIVOLTMETER

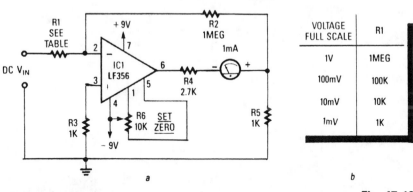

VOLTAGE FULL SCALE	R1
1V	1MEG
100mV	100K
10mV	10K
1mV	1K

a

b

RADIO-ELECTRONICS

Fig. 47-19

An LF356 op amp is used as a gain amplifier with the output taken across R5. When a full-scale current of 1 mA is flowing through the meter, exactly 1 V appears across R5 (should be 1% tolerance or better). This is fed back to R2 to the summing junction of IC1 (a full-scale produces $1\mu A$). This offsets the current through R1. R1 has a value of 1 MΩ/V which is used to zero the meter. R4 provides some overcurrent protection for the meter.

CONTINUITY TESTER

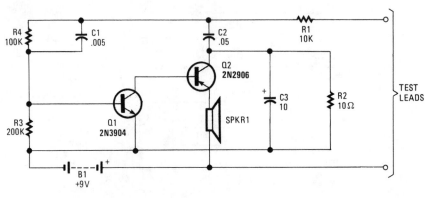

POPULAR ELECTRONICS

Fig. 47-20

A continuity tester that has an audible indicator can be more useful in some cases than a visual indicator, because you need not take your eyes from the job at hand. Q1 and Q2 form an audio oscillator. When the test leads are connected to a continuous circuit, the oscillator operates, and sounds a tone from the speaker.

48

Measuring and Test Circuits (High-Frequency and RF)

The sources of the following circuits are contained in the Sources section, which begins on page 670. The figure number in the box of each circuit correlates to the entry in the Sources section.

Bandswitched Grid-Dip Meter
UHF ''Source'' Dipper
Modulation Monitor
Frequency Counter
455-kHz AM IF Signal Generator
Precision Crystal Frequency Checker
Tuned RF Wavemeter
AM Broadcast Band Signal Generator
Wideband Test Amplifier
Deviation Meter

BANDSWITCHED GRID DIP METER

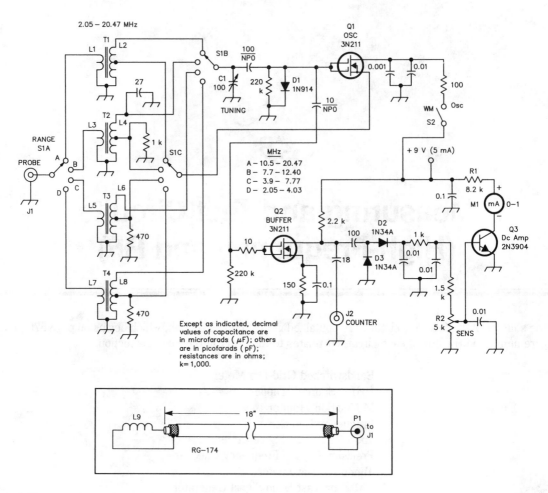

Fig. 48-1

For checking resonances, tuned circuits, antennas, etc., this circuit covers the 2-to 20-MHz range. Q1 serves as an oscillator tunable over this range via C1 and bandswitched coils L1 through L8. When the probe is coupled to a circuit resonant at the oscillation frequency, some RF power will be absorbed and the oscillator output will drop. Q2, D2, D3, and Q3 form an RF detector and dc amplifier to drive meter M1, which will show the drop in Rf level, indicating resonance. R2 is a sensitivity control.

UHF "SOURCE" DIPPER

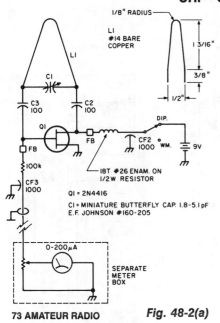

73 AMATEUR RADIO   *Fig. 48-2(a)*

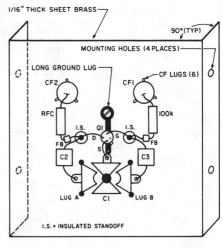

Fig. 48-2(b)

This dipper is useful for UHF experiments in the 420-to-450 MHz amateur band. Because layout can affect performance, follow Fig. 48-2(b).

MODULATION MONITOR

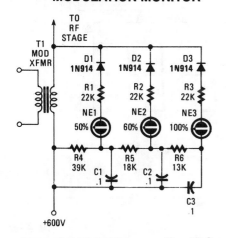

POPULAR ELECTRONICS   *Fig. 48-3*

Suitable for AM transmitters, this circuit uses neon lamps to indicate 50%, 85%, and 100% modulation on negative peaks.

RF OUTPUT INDICATOR

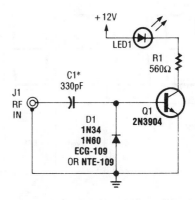

POPULAR ELECTRONICS   *Fig. 48-4*

A simple RF detector circuit using a visual indicator can be useful for an RF output indicator, etc. This circuit was used for a transmitter ON indicator.

FREQUENCY COUNTER

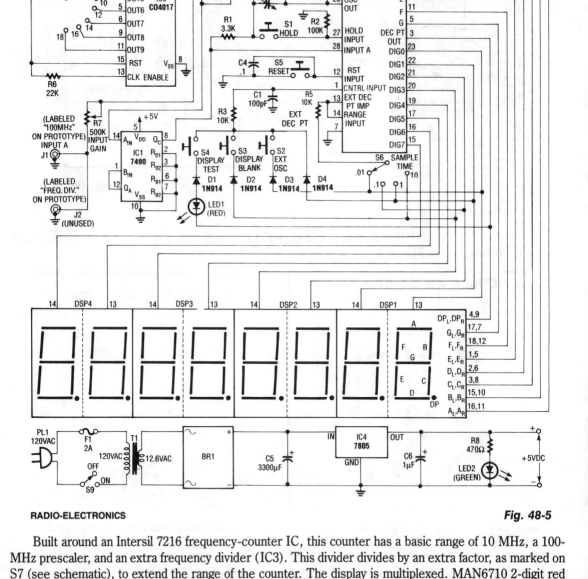

RADIO-ELECTRONICS

Fig. 48-5

Built around an Intersil 7216 frequency-counter IC, this counter has a basic range of 10 MHz, a 100-MHz prescaler, and an extra frequency divider (IC3). This divider divides by an extra factor, as marked on S7 (see schematic), to extend the range of the counter. The display is multiplexed. MAN6710 2-digit red common anode 7-segment LED displays were used on the prototype.

455-kHz AM IF SIGNAL GENERATOR

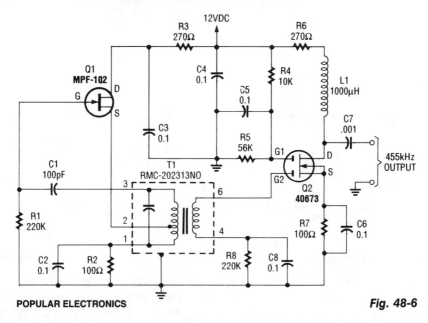

POPULAR ELECTRONICS

Fig. 48-6

An MPF102 FET oscillator drives a dual-gate MOSFET buffer. The MPF102 is configured as a Hartley oscillator. If desired, an audio voltage can be coupled to the junction of R4, R5, and C5 with an extra coupling capacitor ($\approx 1\mu$F) to AM modulate the signal. T1 = Toko P/N RMC—202313NO.

PRECISION CRYSTAL FREQUENCY CHECKER

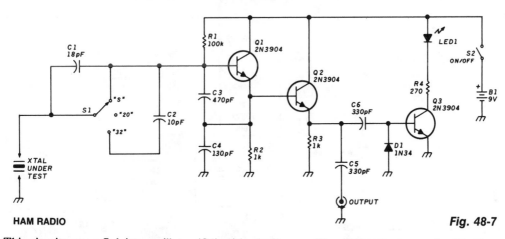

HAM RADIO

Fig. 48-7

This circuit uses a Colpitts oscillator (Q1) with a buffer amplifier (Q2) to test crystals. S1 selects three load conditions—series (S), 20 pF, and 32 pF. Leads to S1 and the crystal should be kept short. The circuit should be useful over 2-to 20-MHz.

TUNED RF WAVEMETER

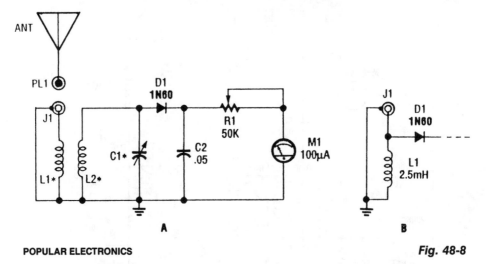

POPULAR ELECTRONICS

Fig. 48-8

L1 and L2 form a tuned transformer. About a 1:3 turns ratio is optimum. L2 and C1 tune to the desired frequency. The frequency range can be 10 kHz to over 200 MHz, depending on the value of C1. For HF use, C1 can be a 140-pF variable. For VHF, use about 25 pF. Use of a 2.5 μH RF choke will yield an untuned wavemeter.

AM BROADCAST-BAND SIGNAL GENERATOR

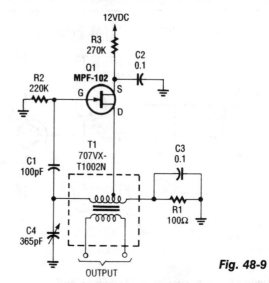

POPULAR ELECTRONICS

Fig. 48-9

A Hartley oscillator using an MPF102 covers the band from 530 to 1 600 kHz. T1 is a Toko P/N T1-707VXT1002N 217 μhy transformer.

WIDEBAND TEST AMPLIFIER

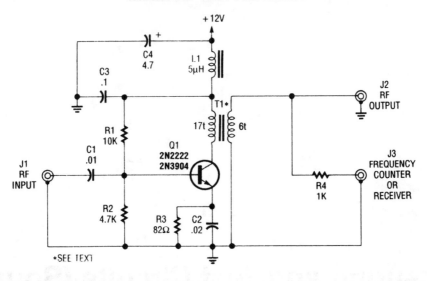

POPULAR ELECTRONICS

Fig. 48-10

This single-stage amplifier (using a 2N2222 or 2N3904 general-purpose transistor) is useful for interfacing test instruments. T1 is an Amidon Associates FT-23-43 core wound with 17 and 6 turns of #26 wire. J3 is a lower-level output for a monitoring device (such as a receiver frequency counter or spectrum analyzer).

DEVIATION METER

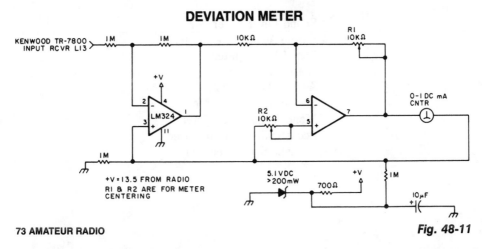

73 AMATEUR RADIO

Fig. 48-11

You can use this circuit in most FM VHF receivers; the hookup is off the FM discriminator. Because every signal transmitted has its own deviation signature, this can be a real plus in hunting jammers.

49

Measuring and Test Circuits (Sound)

The sources of the following circuits are contained in the Sources section, which begins on page 670. The figure number in the box of each circuit correlates to the entry in the Sources section.

SOUND-LEVEL METER

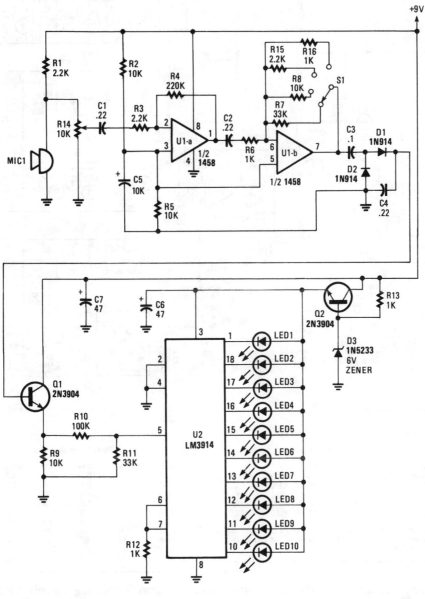

POPULAR ELECTRONICS

Fig. 49-1

An electret microphone feeds an audio amplifier/rectifier combination. The amplifier has switchable gain. The rectifier output drives an LM3914 bar-graph generator. R14 provides fine gain control.

STEREO AUDIO POWER METER

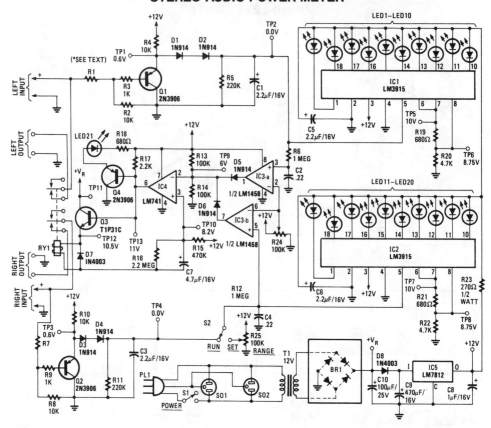

RADIO-ELECTRONICS

Fig. 49-2

Attenuation Resistor Values

Speaker Impedance	50 W	100 W	200 W	400 W	800 W
2 Ω	*	3.9K	10K	18K	30K
4 Ω	3.9K	10K	18K	30K	47K
8 Ω	10K	18K	30K	47K	68K
16 Ω	18K	30K	47K	68K	100K

(*To read a maximum power level of 50 W into 2 Ω, R1 and R7 should be replaced by a piece of wire between the appropriate printed circuit board pads.)

Peak Power Displayed

LED	50 W	100 W	200 W	400 W	800 W
1,11	0.1	0.2	0.4	0.8	1.5
2,12	0.2	0.4	0.8	1.5	3
3,13	0.4	0.8	1.5	3	6
4,14	0.8	1.5	3	6	13
5,15	1.5	3	6	13	25
6,16	3	6	13	25	50
7,17	6	13	25	50	100
8,18	13	25	50	100	200
9,19	25	50	100	200	400
10,20	50	100	200	400	800

This circuit is used to meter the audio power output of an amplifier feeding a speaker. RY1 is actuated if excess power is fed to the speaker. Two channels are included for stereo applications. R1 and R2 and R3 form an attenuator. When a signal level is reached that produces a voltage across C1, comparator IC3A goes high, and IC4 and Q4 produce enough drive to Q3 to trip relay 1, which cuts off the speakers. LED21 will light as well. In addition, IC1 reads the voltage across C1. IC1 is a bar-graph driver, which lights bar-graph display LED1 through LED10.

SOUND-LEVEL METER

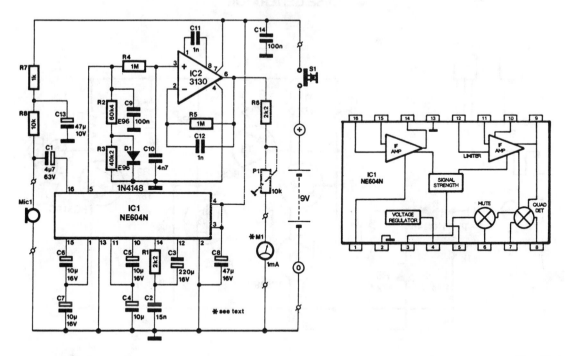

ELEKTOR ELECTRONICS

Fig. 49-3

The NE604's signal-strength indicator section is used, based on an internal logarithmic converter. This enables a linear decibel scale so that the moving-coil meter (shown in the diagram) can be replaced by a digital instrument.

The signal source is assumed to be an electret microphone that converts ambient noise into an electrical signal. Because this type of microphone normally contains a buffer stage, R7, R8, and C13 have been included to provide the supply voltage for this stage.

The NE604 delivers an output current (at pin 5) of 0 to 50 μA, which causes a potential difference across R2 + R3 of 0 to 5 V. The input and output signal range is equivalent to a sound range of 70 dB. To compensate for the effects of temperature changes, the required resistance of 100 kΩ is formed by two resistors (R2 and R3) and a diode (D1).

Any ripple remaining on the output voltage is removed by R4/C9/C10 before the output is buffered by IC2. The indicating instrument, here a moving-coil meter, is connected to the output (pin 6) of IC2 via a series resistance, $R_6 + P_1$. The preset is adjusted to give full-scale deflection (FSD) for an output voltage of 4 V.

Calibrating the meter is a little tricky, unless you have access to an already calibrated instrument. Otherwise, if you know the efficiency of your loudspeaker, that is, how many decibels for 1 W at 1 m, you can use that as reference. The scale of the meter can then be marked with the (approximate) value. In any case, the meter deflection must at all times be seen as an indication, not as an absolute value: it was not thought to be worthwhile to add a filter to the circuit to enable absolute measurements to be made.

NOISE GENERATOR

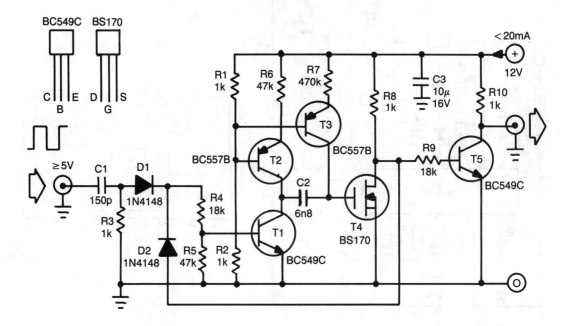

Fig. 49-4

This noise generator provides constant noise energy over its bandwidth, which results from the non-linear behavior of its switching components, more particularly T4. It is very useful for measurements where limited noise bands are required. Varying the ratio $R_6:R_7$ and the clock frequency enables the generated noise to be adapted to specific requirements.

Transistors T2 and T3 are current sources. The current through T2 is about 10 times the level of that through T3. Assuming that T4 is on and that the clock input is low, T1 is off, and C2 discharges. The capacitor is pulled to about half of the supply voltage by the two current sources. When that state is reached, stability ensues because the potential then present at the gate of T4 keeps the FET switched on.

When the clock goes high, T1 is switched on so that C2 is connected between the gate of T4 and the earth. Because C2 is only partly charged, the FET is switched off. Transistor T1 is kept switched on by OR gate D1/D2 so that the clock pulses are blocked. Capacitor C2 then charges via T3 until the potential across it becomes high enough to switch on T4. Transistor T1 is then switched off and the circuit is ready to receive another clock pulse (or rather a leading edge of one).

Because it is not known when the clock pulse arrives, it is not known to what potential C2 will be discharged by T2 (and countered by T3). It is therefore also not known when the next clock pulse will arrive. In other words, the pulse width of the output signal is varying constantly, which is characteristic of a noise signal.

AUDIO FILTER ANALYZER

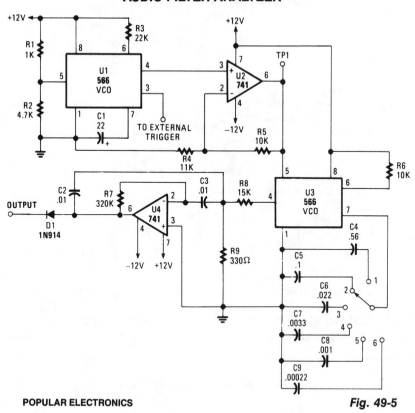

POPULAR ELECTRONICS

Fig. 49-5

When this circuit is connected to a filter and an oscilloscope, the scope displays the filter's frequency response. A frequency that sweeps from low to high is applied to a filter. An oscilloscope is triggered by the start of the sweep and ends its trace at the highest frequency of the sweep. The filter output goes to the vertical amplifier of the oscilloscope. Using bandpass filters as an example, as the bandpass frequency is approached, reached, and passed, the scope follows the peaking output and draws the response curve. A neat effect!

The 566 VCO (U1) produces a VLF triangle wave to frequency modulate the next stage. It also produces a square wave to externally trigger the scope. Op amp U2 (a 741 unit) optimizes the amplitude and the dc component. Another VCO (U3) produces the actual sweeping triangle wave. Its frequency is selectable via S1. Op amp U4 (another 741 op amp) is set up as a bandpass filter and has been included as an example filter. Finally, diode D1 chops off the bottom half of the output, and leaves a nice bell curve.

To set up and operate, power-up the circuit and scope. Set the scope's TIME/CM to 50 ms/cm. Set the VOLTS/CM control to 2 V. Attach a probe from the circuit's trigger to the scope's external trigger input. Set the triggering mode to normal, external. Attach a probe from the vertical amplifier to TP1. You'll see a diagonal line that runs across the CRT. Input coupling should be set to dc. Adjust the triggering level until the diagonal runs from the upper left to the lower right of the CRT to ensure a displayed sweep from low to high. Now, disconnect the probe from TP1 and attach it to the filter output past the diode.

STEREO AUDIO-LEVEL METER

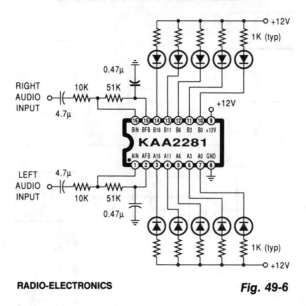

RADIO-ELECTRONICS

Fig. 49-6

A Samsung KAA2281 and a few LEDs make up a simple stereo indicator. Levels displayed are −16, −11, −6, −3, and 0 dB. Input sensitivity is 1 mV. LEDs can be any suitable types or a bar-graph display.

MONO AUDIO-LEVEL METER

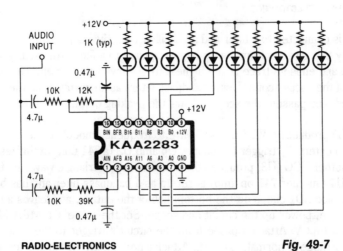

RADIO-ELECTRONICS

Fig. 49-7

This mono indicator uses both halves of a Samsung KAA2283. Levels displayed are −18 to 0 dB in 2-dB steps. Sensitivity is 0.1 to 0.9 mV.

ACOUSTIC SOUND TRANSMITTER

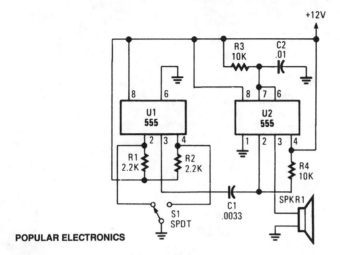

POPULAR ELECTRONICS

Fig. 49-8

Pulsed sound is produced by this circuit. U1 is used as a bistable multivibrator, which acts as a contact "debouncer" for S1. C1 feeds a trigger pulse to U2, which feeds a pulse to SPKR1, to piezo transducer. Values are shown for a pulse width of 110 μs.

ACOUSTIC SOUND RECEIVER

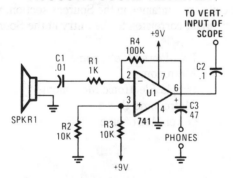

POPULAR ELECTRONICS   *Fig. 49-9*

The receiver is an audio amplifier fed by SPKR1, a piezo speaker that is used as a microphone. A scope or headphones can be used as a detector. The scope can be triggered horizontally by the transmitted acoustic pulse; the vertical display can be used to drive the delay time, and hence the distance.

$$\text{distance (feet)} = \frac{\text{delay time (ms)}}{1100}$$

(at 25°C air temperature)

311

50

Metronomes

The sources of the following circuits are contained in the Sources section, which begins on page 670. The figure number in the box of each circuit correlates to the entry in the Sources section.

LOW-POWER METRONOME

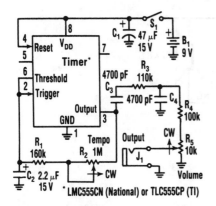

Using only 0.25 mA, this metronome is ideal for battery operation. The tempo range is 34 to 246 beats per minute. Use a CMOS timer, such as an LM555 CN or TLC55CP.

ELECTRONIC DESIGN   *Fig. 50-1*

METRONOME I

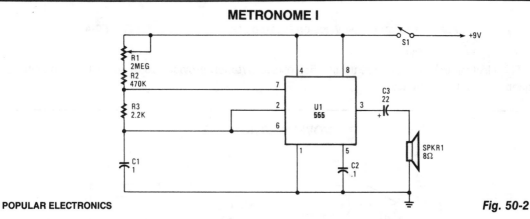

POPULAR ELECTRONICS   *Fig. 50-2*

This simple metronome circuit offers a range of speeds from *largo* to *prestissimo*! The parts values are set so that the repetition rate adjusted by R1 runs from nearly 45 to 184 per minute.

SIMPLE ELECTRONIC METRONOME

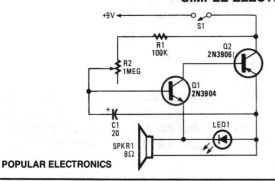

Two complementary transistors form a simple oscillator whose frequency range is from about 0.5 to several Hz. This circuit is useful as a metronome, timer, or pacer for exercise equipment.

POPULAR ELECTRONICS   *Fig. 50-3*

METRONOME II

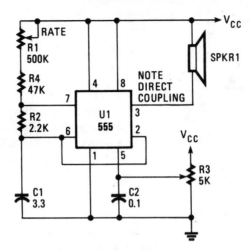

POPULAR ELECTRONICS/HANDS-ON ELECTRONICS

Fig. 50-4

This electronomic metronome, using a 555 oscillator/timer, provides 10 to 40 beats per minute. The frequency is controlled by R3.

NOVEL METRONOME

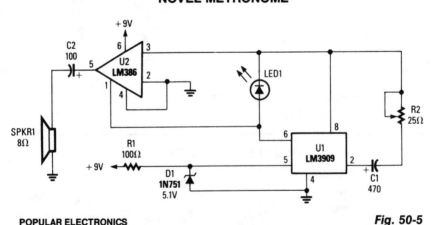

POPULAR ELECTRONICS

Fig. 50-5

The LM3909 is configured so that the frequency of oscillation is dependent on a single RC timing circuit, which consists of C1 and R2. LED1 discharges capacitor C1 and the resultant pulse is directed into pin 3 as well as pin 1 of the LM386 audio amplifier to externally control that unit, thereby providing adequate volume. The circuit, as it is configured, provides frequency ranges from 57 to 204 beats per minute, and plenty of volume.

51

Microwave Amplifier Circuits

The sources of the following circuits are contained in the Sources section, which begins on page 671. The figure number in the box of each circuit correlates to the entry in the Sources section.

2.3-GHz Microwave Preamp
3.4-GHz Microwave Preamp
5.7-GHz Microwave Preamp
10-GHz Single-Stage Preamp
Bias Supply for Microwave Preamps
10-GHz 2-Stage Preamp

2.3-GHz MICROWAVE PREAMP

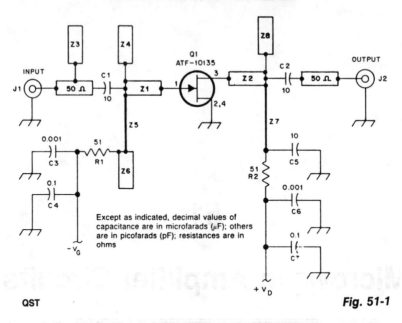

Except as indicated, decimal values of
capacitance are in microfarads (μF); others
are in picofarads (pF); resistances are in
ohms

QST

Fig. 51-1

This low-noise amplifier requires no tuning, has a gain of 13 dB, and a typical NF of 0.6 dB at 2.3 GHz.

3.4-GHz MICROWAVE PREAMP

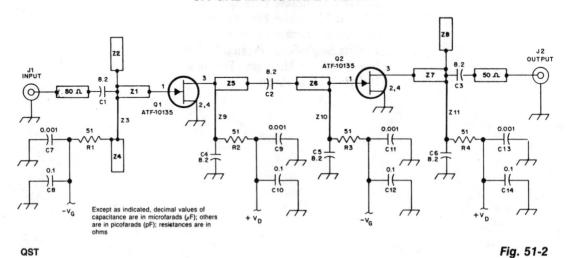

Except as indicated, decimal values of
capacitance are in microfarads (μF); others
are in picofarads (pF); resistances are in
ohms

QST

Fig. 51-2

At 3.45 GHz, this 2-stage preamp has a gain of 23 dB (typical) and less than 1 dB NF.

5.7-GHz MICROWAVE AMPLIFIER

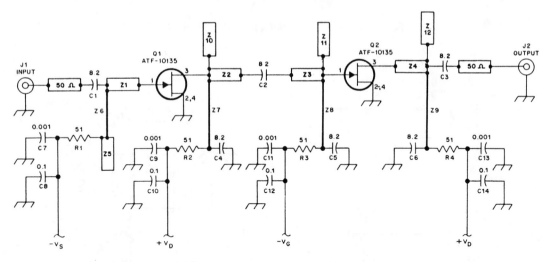

Except as indicated, decimal values of capacitance are in microfarads (μF); others are in picofarads (pF); resistances are in ohms.

QST

Fig. 51-3

This preamplifier has a typical gain of 17 dB and NF = 1.2 dB or better. If a 0.031" PC board (with a dielectric constant of 2.2) is used, the reverse side is unetched.

10-GHz SINGLE-STAGE PREAMP

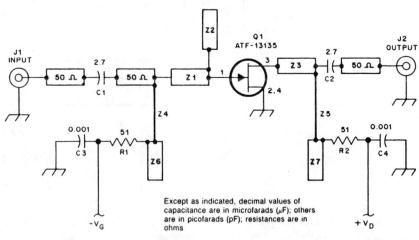

Except as indicated, decimal values of capacitance are in microfarads (μF); others are in picofarads (pF); resistances are in ohms

QST

Fig. 51-4

Using a single Avantek ATF 13135 GASFET, this preamplifier has 8-dB gain (typically) and 1.7-dB noise figure. The PC board is 0.031", doublesided, with $E = 2.2$.

BIAS SUPPLY FOR MICROWAVE PREAMPS

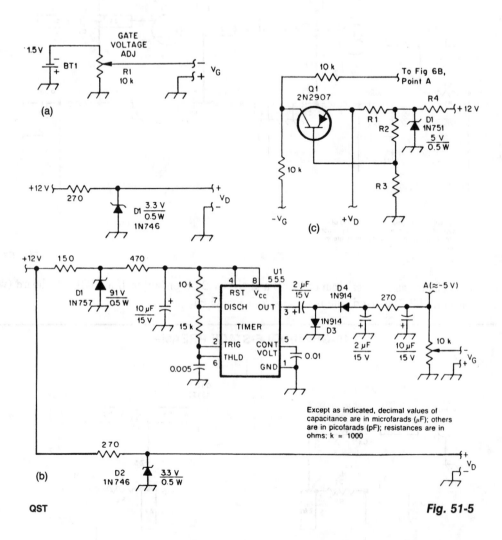

Except as indicated, decimal values of
capacitance are in microfarads (µF); others
are in picofarads (pF); resistances are in
ohms; k = 1000

QST

Fig. 51-5

These two circuits provide bias for the microwave preamps shown in this text. The circuit in Fig. 51-5(a) is a simple passive supply. Figures 51-5(b) and 51-5(c) are active supplies, with U1 generating a negative supply and Q1 setting the drain voltage and current, independent of GASFET characteristics.

10-GHz 2-STAGE PREAMP

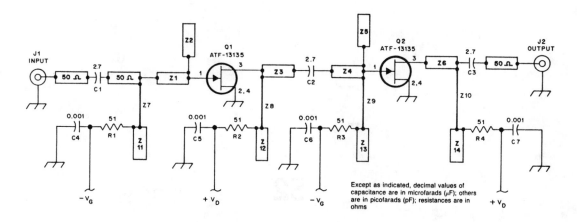

QST

Fig. 51-6

This preamp uses two ATF 13135 stages for typically 17-dB gain and less than 2 dB NF.

52

Miscellaneous Treasures

The sources of the following circuits are contained in the Sources section, which begins on page 671. The figure number in the box of each circuit correlates to the entry in the Sources section.

Diode Sensor for Lasers
RF Attenuator
SSB Generators
Sonic Defender
GASFET Frequency Doubler
Low-Frequency Multiplier
Precision Half-Wave Rectifier
Pulse-Width Modulator

Programmable Identifier
Variable-Voltage Reference Source
0-to 200-nA Current Source
Long-Line *IR* Drop-Voltage Recovery
Precision Full-Wave Rectifier
Fast Symmetrical Zener Clipper
Electronic Level

DIODE SENSOR FOR LASERS

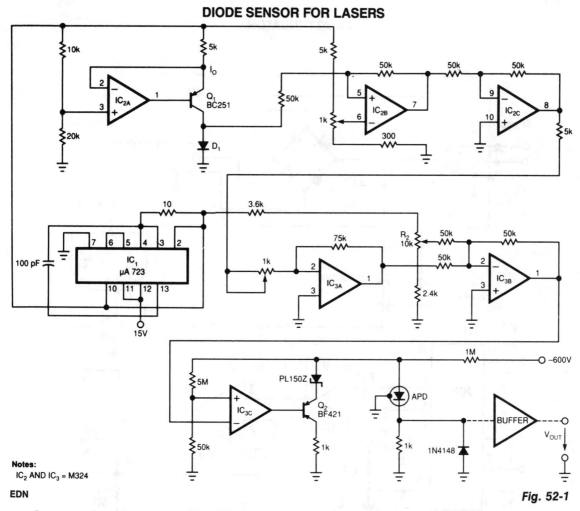

Notes:
IC₂ AND IC₃ = M324

EDN

Fig. 52-1

Laser-receiver circuits must bias their avalanche photo diodes (APD) to achieve optimal gain. Unfortunately, an APD's gain depends on the operating temperature. The circuit controls the operating voltage of an APD over a large temperature range to maintain the gain at the optimal value. The circuit uses D1 as a temperature sensor, thermally matched with the APD.

A voltage regulator, IC1, supplies the necessary reference voltage to the circuit. IC2A and Q1 bias D1 at a constant current. IC2B, IC2C, IC3A/IC3B, and IC3C amplify D1's varying voltage and set Q2 to the optimal-gain corresponding value. Potentiometer R1 controls the amplification over a range of 5 to 15. R2 controls the voltage level, which corresponds to the optimal gain of the APD at 22°C (the temperature is specific to the type of APD). The circuit shown was tested with an RCA C 30954E APD. The tests covered −40 to +70°C and used a semiconductor laser. The laser radiation was transmuted on the APD's active surface in the climatic room via fiberoptic cable. The gain varied by, at most, ±0.2 dB over the entire temperature range.

RF ATTENUATOR

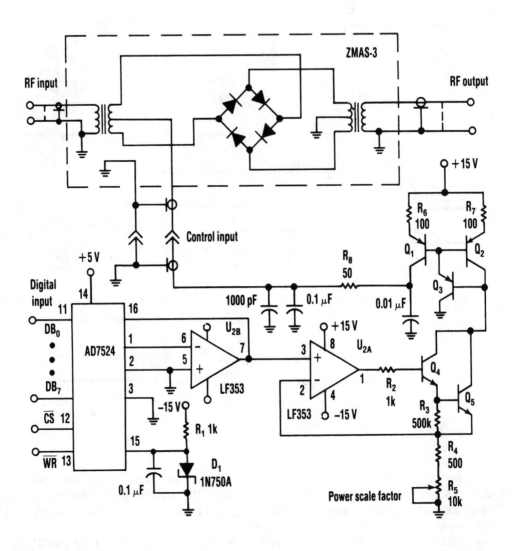

ELECTRONIC DESIGN

Fig. 52-2

A balanced mixer is used as a control element in this circuit. An Analog Devices AD7524 D/A converter drives a voltage-controlled current source using two LF353s and several transistors to control the balanced mixer, a Mini Circuits Lab Z MAS-3.

SSB GENERATORS

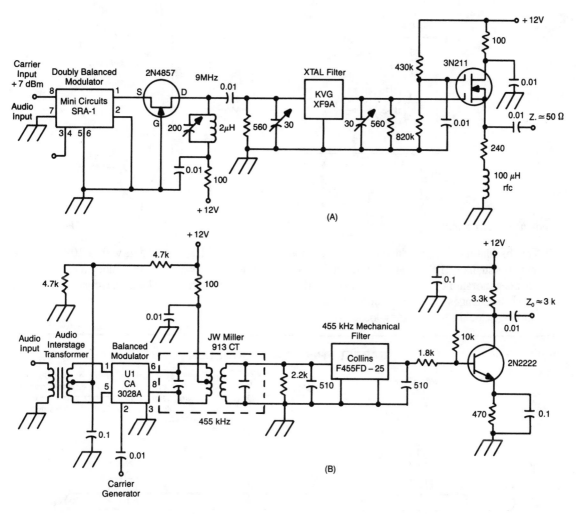

(A)

(B)

Fig. 52-3

These two circuits are SSB generators. One uses a crystal filter by KVG Electronics at 9 MHz, the other uses a 455-kHz mechanical filter. By feeding the outputs into a mixer, the frequency of the SSB generator can be converted to other frequencies. Keep signal levels low enough so that distortion does not occur.

SONIC DEFENDER

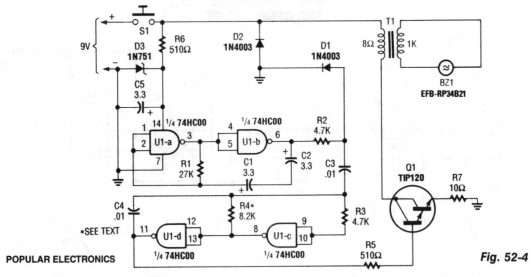

POPULAR ELECTRONICS

Fig. 52-4

This oscillator-driver produces a deafening sound of about 3 kHz, modulated with a 10-Hz warble. BZ1 is a Matsushita EFB-RP34BZ. U1A and B generate the 10-Hz waveform that modulates the 3-kHz tone that is generated by U1C and D. Q1 drives the transducer through a 8-Ω:1-kΩ transformer. R4 might require slight changes (±1 kΩ) to optimize sound intensity.

Warning: Could cause hearing loss if improperly used.

GASFET FREQUENCY DOUBLER

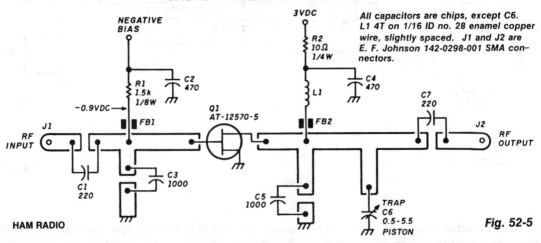

All capacitors are chips, except C6. L1 4T on 1/16 ID no. 28 enamel copper wire, slightly spaced. J1 and J2 are E. F. Johnson 142-0298-001 SMA connectors.

HAM RADIO

Fig. 52-5

This circuit will produce over +10 dBm in the 1 800-3 000-MHz range. Drive power is 7 dBm in the 900-to-1 500-MHz range. The PC board is G-10 Epoxy doublesided. Artwork is shown above, as well as parts placement connectors suitable for these frequencies (such as SMA) should be used. A negative bias supply of 0 to 3 V is required.

LOW-FREQUENCY MULTIPLIER

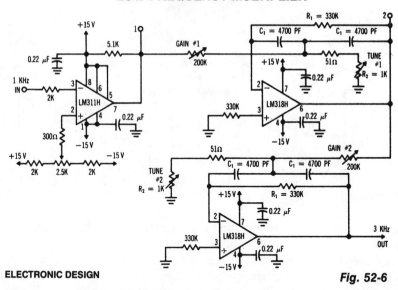

ELECTRONIC DESIGN

Fig. 52-6

This circuit uses a comparator as a Schmitt trigger (311H) and two active bandpass filters (LM318H). 3-kHz output is obtained. Higher harmonics (preferably odd) can be obtained by tuning the active filters to the desired frequency. N can be 1, 3, 5, 7, 9, etc. Even harmonics can be produced by substituting a full-wave rectifier or absolute-value circuit for the Schmitt-trigger comparator.

PRECISION HALF-WAVE RECTIFIER

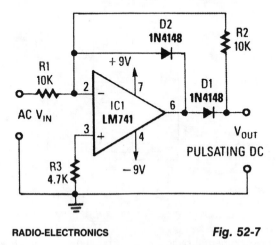

RADIO-ELECTRONICS

Fig. 52-7

An ac input voltage will cause an output that is a half-wave rectified version of the input voltage.

PULSE-WIDTH MODULATOR

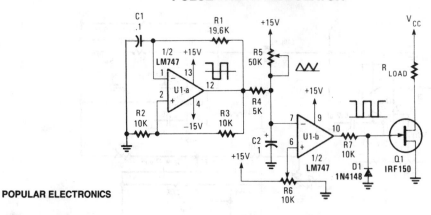

POPULAR ELECTRONICS

Fig. 52-8

This circuit allows the effective power in a load to be controlled by varying the duty cycle of the on/off ratio of load current. No power is dissipated in the switching circuit.

U1A generates a bipolar square wave that is integrated into a triangle by R4 and C2. Reference voltage from R6 is fed to a comparator. The triangle wave on C2 goes to the comparator as well. By varying the reference voltage (R6), the output waveform is a variable width pulse, that drives Q1. R6 controls on/off ratio and therefore load power. R5 sets the offset of the triangle wave across C2.

PROGRAMMABLE IDENTIFIER

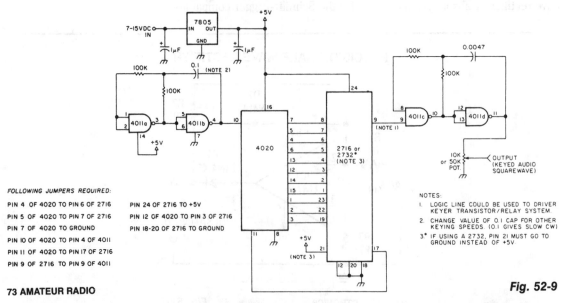

FOLLOWING JUMPERS REQUIRED:

PIN 4 OF 4020 TO PIN 6 OF 2716   PIN 24 OF 2716 TO +5V
PIN 5 OF 4020 TO PIN 7 OF 2716   PIN 12 OF 4020 TO PIN 3 OF 2716
PIN 7 OF 4020 TO GROUND   PIN 18-20 OF 2716 TO GROUND
PIN 10 OF 4020 TO PIN 4 OF 4011
PIN 11 OF 4020 TO PIN 17 OF 2716
PIN 9 OF 2716 TO PIN 9 OF 4011

NOTES:
1. LOGIC LINE COULD BE USED TO DRIVER KEYER TRANSISTOR/RELAY SYSTEM.
2. CHANGE VALUE OF 0.1 CAP FOR OTHER KEYING SPEEDS. (0.1 GIVES SLOW CW)
3.* IF USING A 2732, PIN 21 MUST GO TO GROUND INSTEAD OF +5V.

73 AMATEUR RADIO

Fig. 52-9

Used on an amateur or experimental radio beacon, the above ID circuit will generate any callsign or message programmed into the EPROM. A CD4020 CMOS counter is driven by an RC clock circuit. This addresses the EPROM, and serial data is available at pin 9.

VARIABLE-VOLTAGE REFERENCE SOURCE

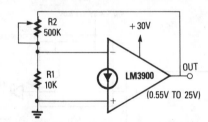

RADIO-ELECTRONICS

Fig. 52-10

The noninverting terminal of the op amp is grounded, and the circuit uses the voltage at the inverting terminal as a reference. Its voltage gain is determined by the R_2/R_1 ratio. When R2 is set at zero, the circuit has unity gain and a 0.55-V output. When R2 is set to the maximum value, the circuit has a gain of 50 and an output of about 25 V. The circuit provides good regulation and can supply output currents of several milliamps. The output voltage however, is not temperature compensated.

0-TO-200-nA CURRENT SOURCE

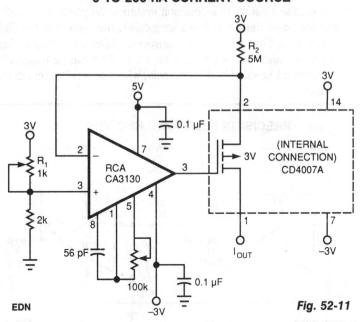

EDN

Fig. 52-11

This circuit uses readily available parts to implement a 0-to-200-nA current source. The circuit borrows a PMOS transistor from the input stage of a DC4007A, which is easier to obtain than a discrete PMOS transistor. The CA3130 op amp operates as a follower so that its positive input sets the current that flows through R2. The MOSFET input stage of this op amp exhibits low-input current. The op amp must be able to produce an output voltage high enough to turn the CD4007A's internal FET off. Thus, the op amp requires a positive supply voltage of 5 V. The circuit presents an output voltage from 0 to 3 V, and R1 controls the amount of output current.

LONG-LINE *IR* DROP-VOLTAGE RECOVERY

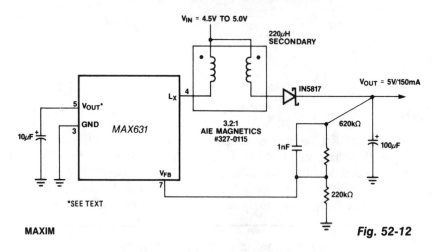

MAXIM

Fig. 52-12

This circuit provides a unique solution to a common system-level power distribution problem: When the supply voltage to a remote board must traverse a long cable, the voltage at the end of the line sometimes drops to unacceptable levels. This +5-V/+5-V converter addresses this by taking the reduced voltage at the end of the supply line and boosting it back to +5 V. This can be especially useful in remote display devices, such as some point-of-sale (POS) terminals, where several meters of cable could separate the terminal from the readout.

PRECISION FULL-WAVE RECTIFIER

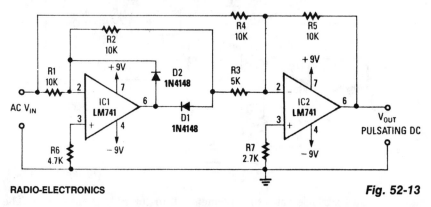

RADIO-ELECTRONICS

Fig. 52-13

Using two op amps, this circuit produces a full-wave rectified version of the input signal. Op amp IC1 inverts the negative-going signal, but because of D2, it stays near zero. IC2 produces a positive-going signal. For positive-going signals, IC1 produces a negative output through D1 to IC2, where it is combined with positive V_{IN} from R4/R5. At the summing junction of IC2, the negative output of IC1 is doubled and inverted via IC2, R3, and R5 to produce $+V_{OUT}$. This is summed with negative output of IC1 to produce $+V_{OUT}$.

FAST SYMMETRICAL ZENER CLIPPER

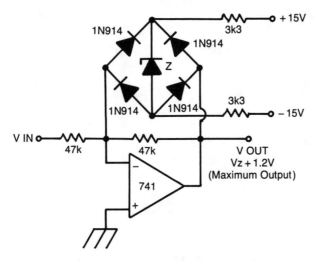

ELECTRONICS TODAY INTERNATIONAL   **Fig. 52-14**

The problem with using two zeners back to back in series to get symmetrical clamping is that the knee of the zener characteristics is rather sloppy. Also, charge storage in the zeners causes speed problems and the zeners will have slightly different knee voltages, so the symmetry will not be all that good. This circuit overcomes these problems.

By putting the zener inside a diode bridge, the same zener voltage is always experienced. The voltage errors caused by the diodes are much smaller than those caused by the zener. Also, the charge storage of the bridge is much less. By biasing the zener ON all the time, the knee appears to be much sharper.

ELECTRONIC LEVEL

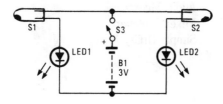

POPULAR ELECTRONICS   **Fig. 52-15**

An electronic level can be constructed using two mercury switches mounted on an absolutely flat board, along with the LEDs. If one LED lights, the surface is not level. If both light, the surface is level.

53

Mixers

The sources of the following circuits are contained in the Sources section, which begins on page 671. The figure number in the box of each circuit correlates to the entry in the Sources section.

Dynamic Audio Mixer
Stereo Mixer with Pan Controls
4-Channel Mixer
Digital Mixer
4-Input Unity-Gain Mixer
Audio Mixer
Mixer Diplexer
Simple Utility Mixer

DYNAMIC AUDIO MIXER

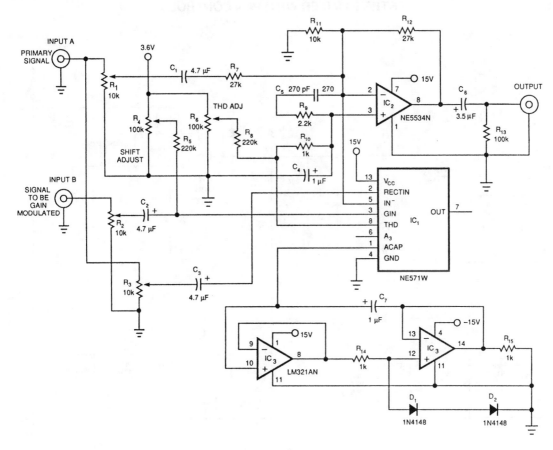

Fig. 53-1

The dynamic mixer combines two audio inputs by adding the primary signal, Input A, to a gain-controlled signal, Input B. The unusual aspect of this circuit is that the average voltage level of Input A controls the gain of Input B.

IC1 has the averaging function and many of the specialized gain blocks that the circuit requires. R1 sets the level of the primary input, Input A, to be passed to the output. R2 governs Input B's level to the modulator, while R3 sets the level of the modulating signal. IC1 can be either an NE571N or an NE570N. The average ac signal at pin 2 controls the amount of signal that shows up at IC1's output, pin 3.

The primary signal gets to IC2, an NE5534N low-noise op amp, via C1 and R7; the gain-modulated secondary signal arrives via pin 5 of IC1. IC2 sums the two signals. Potentiometers R4 and R6 make dc-offset and distortion adjustments, respectively. IC3, C7, R14, R15, D1, and D2 form a filter for IC1.

STEREO MIXER WITH PAN CONTROLS

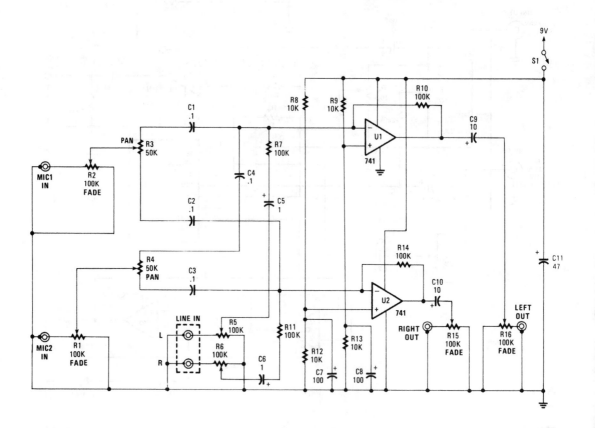

Fig. 53-2

This stereo mixer has two mono mixers and a modification to the microphone inputs. When a microphone is in use, the microphone's output is fed to the microphone input of the circuit. The signal is then run into R1 and R2 (which are used as faders). The signal is then split into two different paths by resistors R3 and R4, with which it is possible to change the place of the microphone inputs within the stereo panorama. The stereo line inputs are for that purpose. Joining the microphone inputs with the output of some other source (such as a tape deck, turntable, etc.), all the signals are fed to the inverting input of an op amp. The output reaches the master-fade potentiometers, which control output level.

4-CHANNEL MIXER

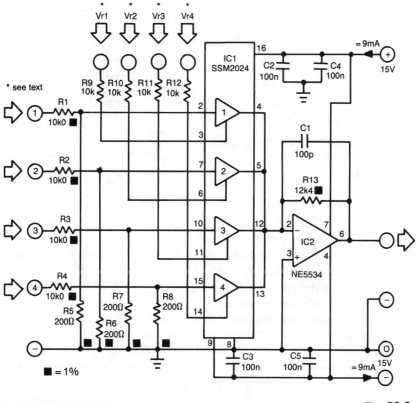

ELEKTOR ELECTRONICS   *Fig. 53-3*

The proposed mixer is designed around four current-driven transconductance amplifiers contained in an SSM2024 from Precision Monolithics. To obtain a low offset and high control rejection, the four inputs should have an impedance to earth of about 200 Ω. These impedances are obtained from resistors R5 through R8, which also form part of a potential divider at each input.

With the values in the diagram, the nominal input signal is 1 V (0 dBV). Distortion at that level is about 1%; at lower levels, it is not more than 0.3%.

The amplification of the current-driven amplifiers (CDAs) is determined by the current fed into the control inputs. These inputs form a virtual earth so that calculating the values of the bias resistors (to transform the inputs into voltage-driven inputs) is fairly simple.

The output currents of the amplifier are summed by simply linking the output pins. The current-to-voltage converter, IC2, translates the combined output currents into an output voltage. The value of R13 ensures that the amplification of IC2 is unity.

DIGITAL MIXER

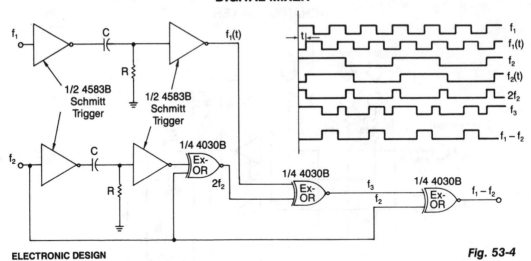

ELECTRONIC DESIGN

Fig. 53-4

A simple digital mixer, based on two dual-Schmitt triggers (4583B) and three exclusive-OR gates, uses an RC time-delay circuit to permit easy adjustment of the output-signal pulse width. The exclusive-OR gates can also be used separately as a symmetrical frequency doubler.

As shown, a signal passing through the Schmitt triggers is delayed by t, a value equal to RC In (V_{tp}/V_{tn}), where V_{tp} and V_{tn} are the positive and negative threshold voltages of the triggers.

To function properly, the same time delay must be introduced to signals f1 and f2. Also, the time delay must be less than 50% of the period of f1. Provided that f1 is more than twice the value of f2, the output of the circuit will equal the difference of the two signals (i.e., f1−f2).

4-INPUT UNITY-GAIN MIXER

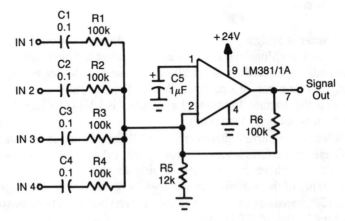

RADIO-ELECTRONICS

Fig. 53-5

An LM381/1A is used as a four-input unity-gain audio mixer. Gain can be increased by decreasing R1 through R4 or increasing R6.

AUDIO MIXER

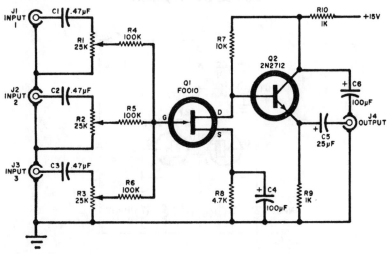

POPULAR ELECTRONICS

Fig. 53-6

Three audio circuits are combined in the circuit shown. Each input is coupled to its own level potentiometer (R1, R2, or R3) and they are combined at the gate of FET Q1. The output of Q1 is coupled to the external audio amplifier through emitter-follower Q2 and capacitor C6.

MIXER DIPLEXER

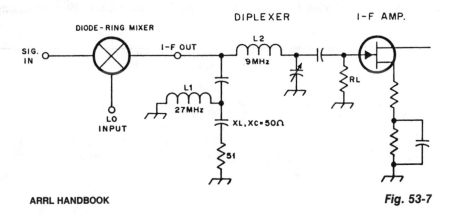

ARRL HANDBOOK

Fig. 53-7

By inserting a high-pass filter section in the IF lead, this mixer is terminated at all frequencies, besides the IF, for other mixer products, which results in improved IMD. In this example, high-pass filter section L1 and capacitors cut off below 28 MHz. Above this frequency, the mixer is terminated.

SIMPLE UTILITY MIXER

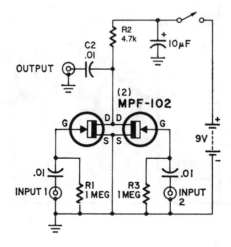

RADIO-ELECTRONICS   *Fig. 53-8*

Here's an interesting mixer circuit. With it you can effectively combine signals from audio to high-frequency RF. Also, as a special bonus, this circuit will provide some gain at a low noise figure. The inputs can be of almost any level or impedance, and the output (low-Z) will drive most tuned circuits or transistors.

Basically, the device consists of two similar FET amplifier stages with a common load resistor (R2). Each FET develops a signal across this resistor, a form of cancellation occurs, and a difference signal results.

If you want less gain, try reducing R_2 to 2 200-Ω. This modification will not affect the mixing ability.

54

Model and Hobby Circuits

The sources of the following circuits are contained in the Sources section, which begins on page 671. The figure number in the box of each circuit correlates to the entry in the Sources section.

Model Train and Slot-Car Controller
Model Train Throttle Control

MODEL TRAIN AND SLOT-CAR CONTROLLER

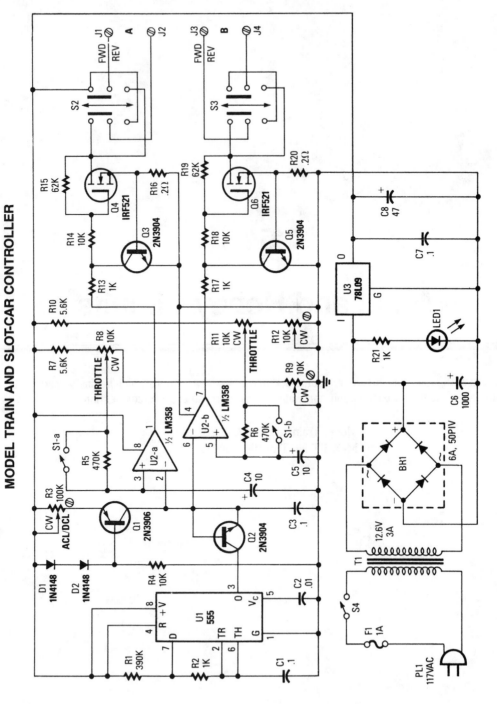

Fig. 54-1

POPULAR ELECTRONICS

As shown, a 555 timer (U1) is configured as an astable multivibrator (oscillator) with a 400:1 duty cycle and a frequency of 40 Hz.

When power is applied to the circuit, capacitor C1 (connected to pin 6 of U1) is discharged and the output of the 555 (which is used to sink current) is low. Capacitor C1 begins to charge via R1 and R2 toward the positive supply rail. When the charge on C1 reaches about 66% of +V, the output of U1 at pin 3 goes high.

At that point, C1 begins to discharge through R2. When the charge on C1 decreases to about 33% of the supply voltage, the output of U1 returns to the low state, and the cycle is repeated until power is removed from the circuit.

When the output of U1 is low, C3 is discharged into U1 via transistor Q2. When U1 pin 3 goes high, C3 charges through a current source that consists of D1, D2, R3, R4, and Q1. The charge/discharge cycling of C3 produces a stream of pulses that are fed to the inverting inputs of U2A and U2B (an LM358 dual op amp). Two voltage-divider networks (R7, R8, R9, and R10, R11, R12) set the reference voltage that is applied to the noninverting inputs of U1A and U1B at pins 3 and 5.

Potentiometers R9 and R12 set the low-level duty cycle (5 to 10%) of U1A and U1B. They are adjusted so that the train head-lights glow, but the motor hums only slightly. Potentiometer R3 adjusts the ramp rate of C3 for 100% duty cycle at the full throttle setting. A double-pole, single-throw switch (S1A and S1B) is used to place R3/C4 and R4/C5 in the circuit.

The R5/C4 and R6/C5 combinations cause the reference voltages presented to the noninverting inputs to U2A and U2B to change very slowly when the throttle is turned up and down. When the ACL/DCL switch is turned off, the resistance of the throttle-divider networks are much smaller than those of R5 and R6, so the reference voltages on C4/C5 change "instantly" to the new throttle setting.

The output drivers consist of resistors R13 and R15, and transistors Q3 and Q4 for output "A"; and resistors R17 to R20, and transistors Q5 and Q6 for output "B." Components R13/R16/Q3 and R17/R20/Q5 limit the output drive currents of Q4 and Q6 to about 3 A each. Resistors R14/R15 and R18/R19 turn on Q4 and Q5, respectively, before the breakover voltage is reached to prevent damage to the output drivers and dissipate the energy that is stored in an inductive field (such as in a motor).

The power supply delivers 18 V to the track. Voltage regulator U3 (a 78L09 9-V, 100-mA voltage regulator) supplies power to the control circuits.

MODEL TRAIN THROTTLE CONTROL

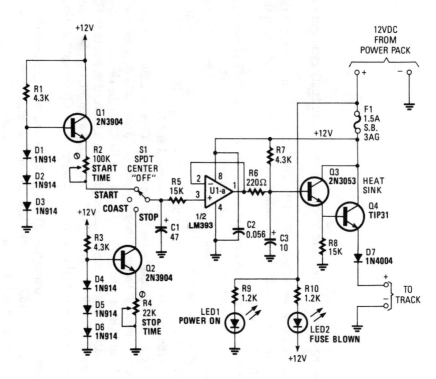

POPULAR ELECTRONICS

Fig. 54-2

What makes this control unique is its momentum feature, which adds a degree of realism. The circuit will operate well for trains that draw up to 1 A at 15 V. None of the components are critical.

In the start mode, current source Q1 charges capacitor C1. The charge current and start-up time are adjusted by resistor R2. In the stop mode, current-sink Q2 discharges capacitor C1. The discharge current and stop time are set by resistor R4. In the coast mode, op amp U1 draws very little current from C1, so the speed will remain nearly constant for some time, and then gradually decrease. Transistors Q3 and Q4 form a Darlington emitter-follower to amplify the output of U1. Diode D7 reduces the output by about 0.8 V. Another diode could be added in series to decrease the output to 0 in the stop mode.

55

Motion and Proximity Detectors

The sources of the following circuits are contained in the Sources section, which begins on page 671. The figure number in the box of each circuit correlates to the entry in the Sources section.

SENSITIVE LOW-CURRENT-DRAIN MOTION DETECTOR

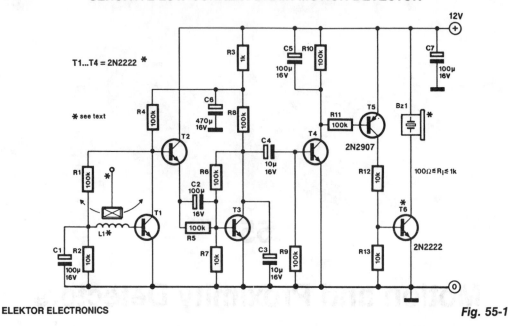

ELEKTOR ELECTRONICS

Fig. 55-1

This highly sensitive movement detector is designed from bipolar transistors and draws a current of only 0.3 mA during quiescent operation. It is intended primarily as a protection device, but it can also be used in certain games.

The principle is simple: a magnet is suspended by a thin thread 20 to 30 mm long, and a few millimeters above the coil of a relay (whose contacts are not used). Even a minute movement of the protected object will disturb the magnet. The resulting changes in the magnetic field above the relay coil will induce a tiny varying voltage across the coil.

The first stage consists of a common emitter design with automatic regulation. The collector resistors and the resistors in the regulation bridge have unusually high values.

Feedback from the bridge ensures stability of operation for T1. Each increase in collector voltage will be opposed by an increase in base-emitter current. Conversely, each reduction in collector voltage will be opposed by a decrease in base-emitter current. Consequently, the collector voltage will stabilize at a value that corresponds to a base voltage of about 0.6 V. Capacitor C1 delays the immediate effect of the feedback when the collector voltage changes rapidly.

The small varying voltage induced in the relay coil is magnified appreciably by T1 because C1 prevents automatic regulation. The output impedance of the first stage is very high, which is, of course, the price to be paid for low consumption. It would not make sense to follow this stage by one with a low output impedance, because this would adversely affect the overall amplification. Because of that, T1 is followed by an emitter-follower, T2, which provides the coupling between T1 and T3. Resistor R5 allows a partial discharge of C2 if T2 is switched off by a reduction in the output of T1. Because this resistor, as a result of the low-consumption requirement, has a high value, the circuit will attain its maximum sensitivity 10 seconds after the last movement detection. This is the time that is required for the charge on capacitor C2 to stabilize.

The detection proper is carried out by T4, which switches on when the voltage variations in the ampli-

SENSITIVE LOW-CURRENT-DRAIN MOTION DETECTOR (*Cont.*)

fier, passed on by C4, reach a level of 0.6 V. Saturating T4 leads to the instant charging of C5. This capacitor will discharge partly via R10 and R11 to the base of T5 when T4 switches off again. When C5 discharges, T5 is on, which will make T6 conduct. This in turn will actuate a load, for instance, a buzzer, in the collector circuit of T6.

The sensitivity of the detector depends to a large extent on the distance between the magnet and relay and the length of the "pendulum."

If the circuit is powered by a battery, there is a little problem: batteries have large internal resistances. Thus, a supply voltage can vary by some tenths of a volt if a sudden, large current is drawn. If the buzzer has stopped after a detection, such a situation can retrigger the circuit and cause undesired oscillations. To prevent this happening, the supply of the amplifier stage is decoupled by R3 and C6.

ACOUSTIC DOPPLER MOTION DETECTOR

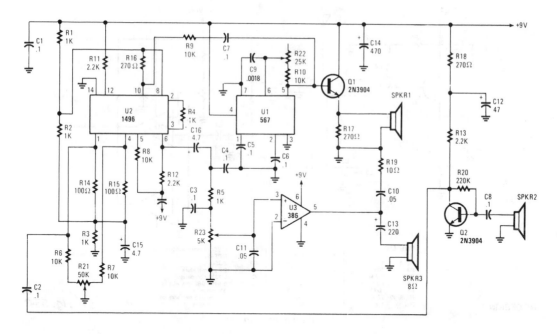

POPULAR ELECTRONICS

Fig. 55-2

A high-frequency audio signal (15 to 25 kHz) generated by U1 is fed to buffer Q1 and SPKR1. A portion is fed to balanced mixer U2. Received audio picked up by SPKR2 (used as a microphone) is amplified by Q2 and fed to U2. When sound is reflected from a moving object, the Doppler effect will cause an apparent shift in frequency. U2 produces a signal equal to the frequency difference. This is coupled via C16 and gain control R23 to amplifier U3, where the beat note is heard in SPKR3.

UHF MOTION DETECTOR

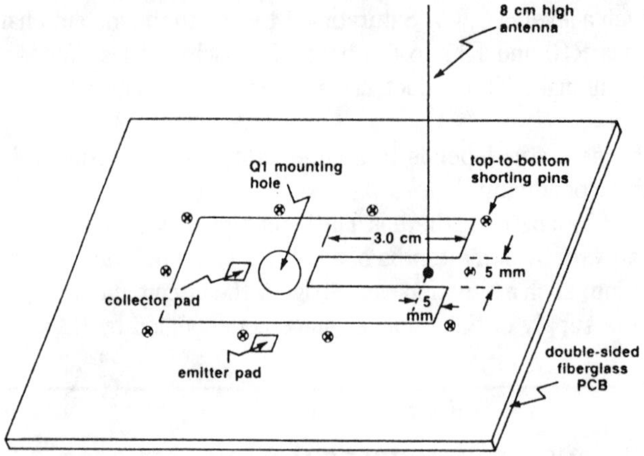

Construction layout.

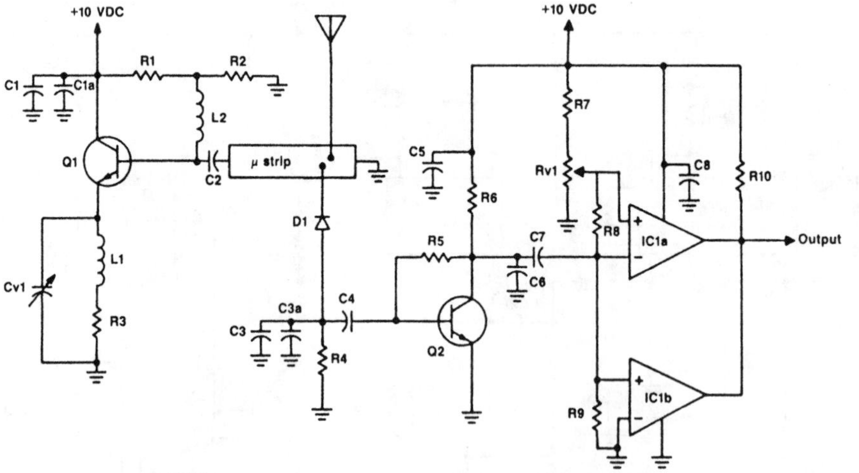

Parts List					
R1	3.9 k	C1	1 nF	D1	ISS97 or other Schottky type
R2, 4	1 k	C1a	470 uF	Q1	MRF961
R3	100	C2	47 nF	Q2	BC548
R5	2.2 M	C3a	1 uF	Cv1	2-7 pF miniature
R6	6.8 k	C4	22 uF	IC1	LM339 comparator
R7	100 k	C5, 8	100 uF	L1	5 turns 0.86 mm wire on
R8, 9, 10	22 k	C6	100 nF		3.5 mm core
Rv1	1 k	C7	10 uF	L2	4 turns 0.86 mm wire on
					3.5 mm core

RF DESIGN

Fig. 55-3

Circuit diagram.

The UHF motion detector operates on the Doppler radar principle. Q1 is an oscillator that creates a radiated signal. An object in the radiated field reflects some of this energy back to the detector. If the object is moving, the reflected signal will have a different frequency because of Doppler shift.

Q1 is an oscillator coupled to a small (8-cm) antenna. This antenna also receives the reflected signal. D1 acts as a mixer and produces a beat note of frequency that is equal to the difference in reflected and radiated signal. Q2 amplifies this signal and couples it to comparator/detector IC1A and IC1B. The output can continue on to an alarm, relay, lamp, etc.

IR REFLECTION PROXIMITY SWITCH

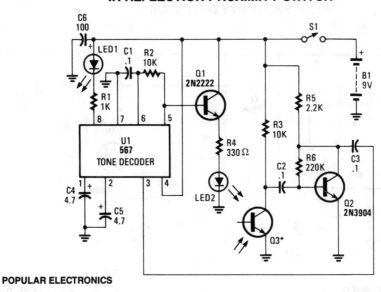

POPULAR ELECTRONICS

Fig. 55-4

IR radiation from LED2 (modulated by a 1-kHz wave) is keyed by U1, and Q1 is radiated. Reflected IR energy is picked up by Q3, and the audio signal from Q3 is amplified by Q2 and sent to the decoder. The LED1 lights to indicate presence of reflected IR. LED1 can be the input of an isolator so that a triac or SCR can be controlled.

RELAY OUTPUT PROXIMITY SENSOR

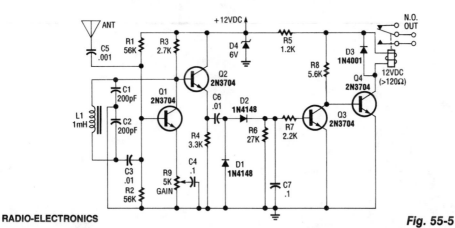

RADIO-ELECTRONICS

Fig. 55-5

Q1 is used as an oscillator around 300 kHz. R9 is set so that the oscillator just begins to run. An object near the antenna will load the circuit down, and stop the oscillations. This is detected by buffer Q2, diodes D1 and D2, and this activates relay driver Q4, which operates the relay.

PROXIMITY DETECTOR

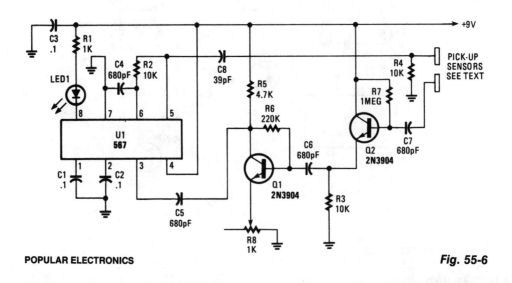

POPULAR ELECTRONICS

Fig. 55-6

In this proximity detector an NE567 tone decoder provides a signal of about 100 kHz that is fed to one sensor. The sensors are copper or aluminum wires or plates, or any other suitable conductive material. When another object is near the sensors, it causes an increase in capacitance between the sensors. Q1 and Q2 amplify the signal and feed it to U1. Notice that C8 and R4 phase shift the VCO signal from the NE567 so that U1 can detect its own output signal.

56

Motor Control Circuits

The sources of the following circuits are contained in the Sources section, which begins on page 672. The figure number in the box of each circuit correlates to the entry in the Sources section.

Mini-Drill Control
Tachometerless Motor Speed Control
Half-Step Drive Stepper Motor
Quarter-Step Stepper Motor Driver
Stepper Motor Speed and Direction Controller
Compressor Protector
dc Motor-Speed Control
Cassette Motor Speed Calibrator

MINI-DRILL CONTROL

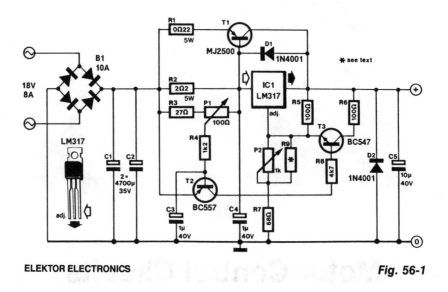

ELEKTOR ELECTRONICS

Fig. 56-1

This circuit is intended as a revolution control for small dc motors as fitted, for instance, in small electric drills (such as used for precision engineering and for drilling boards, among others). The behavior of these motors, which are normally permanent magnet types, is comparable to that of independently powered motors.

In theory, the rpm of these motors depends solely on the applied voltage. The motor adjusts its rpm until the counter emf generated in its coils is equal to the applied voltage. There is, unfortunately, a drop across the internal resistance of the motor, which causes the rpm to drop in relation to the load. In other words, the larger the load, the larger the drop across the internal resistance and the lower the rpm.

This circuit provides a kind of compensation for the internal resistance of the motor: when the current drawn by the motor rises, the supply voltage is increased automatically to counter the fall in rpm.

The circuit is based on an enhanced voltage regulator that consists of IC1 and T1, which provides a reasonably large output current (even small drills draw 2-to-5 A). The "onset" supply voltage, and thus the rpm, is set by P2. Because of emitter resistance R1, the currents through IC1 and T1 will be related to one another in the ratio that is determined by R1 and R2. Owing to this arrangement, the internal short-circuit protection of IC1 will also, indirectly, provide some protection to T1.

As soon as the current drawn exceeds a certain value, T2 will be switched on. This results in a base current for T3 so that R5 is in parallel (more or less) with R6. This arrangement automatically raises the output voltage to counter a threatened drop in rpm. The moment at which this action occurs is set by P1, so this circuit can be adapted pretty precisely to the motor used.

If only very small motors are likely to be used, the power supply (transformer and bridge rectifier) can be rated more conservatively. As a guide, the current in the transformer secondary should be about 1.5 times the maximum dc output current.

TACHOMETERLESS MOTOR SPEED CONTROL

This provides bidirectional speed regulation for small motors and requires no tachometer. The voltage that summing amplifier IC1A applies to the motor's windings equals:

$$\left(V_C + R_1 \left(\frac{R_3}{R_{2A} + R_{2B}} \right) \right) I_M \left(\frac{R_6}{R_5} \right)$$

where V_C is the command voltage and I_M is the motor current.

If you set the motor's winding resistance and brush resistance (R_M) equal to:

$$R_1 \left(\frac{R_3}{R_{2A} + R_{2B}} \right)$$

the command voltage will be proportional to the motor winding's counter emf. C1 provides compensation. Set R1's value so that it equals 5 to 10% of RM's value. You can generally find RM's value in a motor's spec sheet.

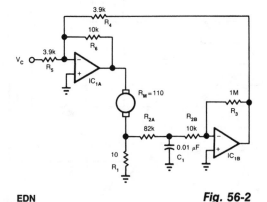

EDN

Fig. 56-2

HALF-STEP DRIVE STEPPER MOTOR

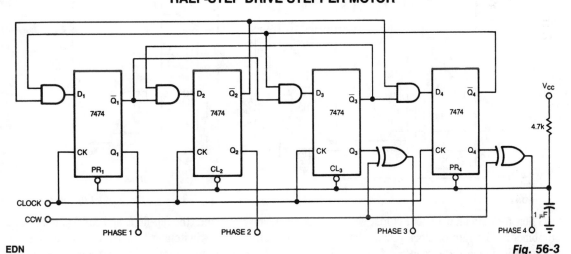

EDN

Fig. 56-3

Clock pulses are converted to the switching sequence necessary for a four-phase stepper through 400 half steps of 0.9° each.

QUARTER-STEP STEPPER MOTOR DRIVER

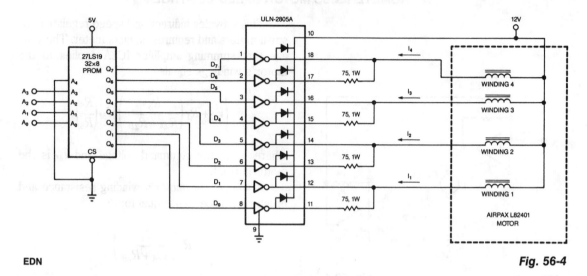

EDN

Fig. 56-4

Using a counter and a PROM, this circuit drives the stepper windings with two levels of current.

STEPPER MOTOR SPEED AND DIRECTION CONTROLLER

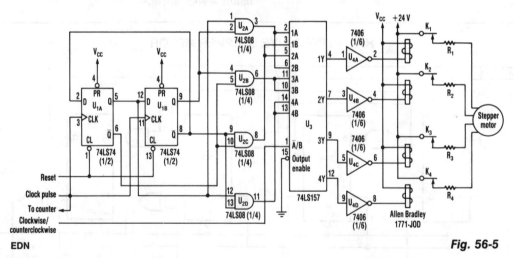

EDN

Fig. 56-5

This new circuit uses four chips, with an option of using just three (the flip-flops and AND gates can be combined). The rate of clock pulses determines the motor's rpm. Switching transistors can replace the relays to increase the circuit's efficiency. This circuit drives a stepper motor whose speed depends on clock rate. Standard SS1 LSTTL chips are used. Switching transistors can be used in place of the solid-state relays to improve the circuit's efficiency.

COMPRESSOR PROTECTOR

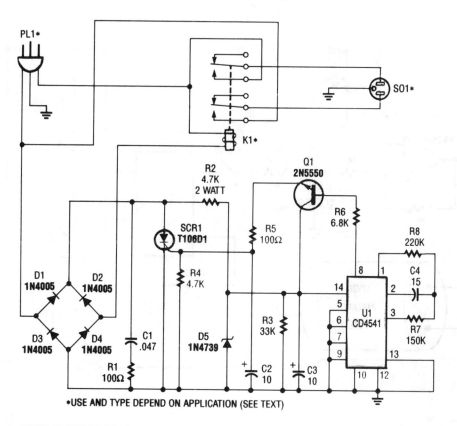

POPULAR ELECTRONICS

Fig. 56-6

This circuit monitors the power-line (ac) voltage. When a power failure occurs, on restoration of power, the circuit adds a five-minute delay before energizing K1, which protects the compressor against limited low voltage.

U1 is a 16-stage counter with an integral oscillator that is set to divide by 8192. R7, R8, and C4 set the oscillator frequency to about 25 Hz, which produces a total count interval of 300 seconds (5 minutes). After this time, pin 8 U1 goes high, which forward biases Q1, triggers SCR1, and activates K1. Up to 30 A can be switched.

dc MOTOR-SPEED CONTROL

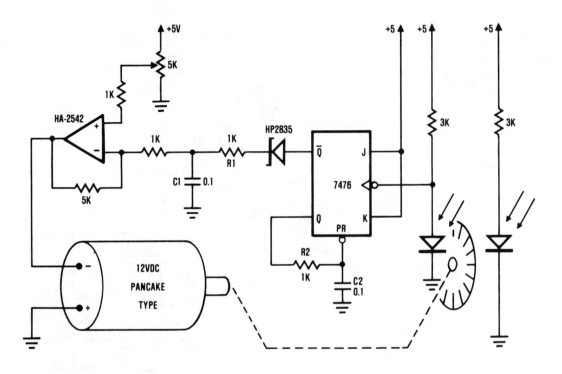

HARRIS

Fig. 56-7

The system shown consists of the HA-2542, a small 12-Vdc motor, and a position encoder. During operation, the encoder causes a series of ''constant-width'' pulses to charge C1. The integrated pulses develop a reference voltage, which is proportional to motor speed and is applied to the inverting input of HA-2542. The noninverting input is held at a constant voltage, which represents the desired motor speed. A difference between these two inputs will send a corrected drive signal to the motor, which completes the speed control system loop.

CASSETTE MOTOR SPEED CALIBRATOR

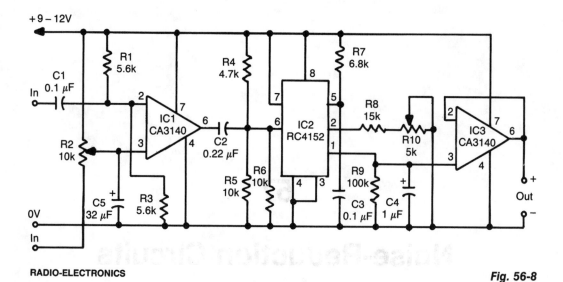

Fig. 56-8

This frequency/voltage converter enables calibrations of cassette-deck speed. It records a steady 1-kHz tone on the cassette deck, and monitors the frequency on the converter. Then, play the tone back and adjust the motor speed until you get the same frequency reading on the converter. The frequency/ voltage converter can drive an analog meter to indicate small variations in tape speed.

57

Noise-Reduction Circuits

The sources of the following circuits are contained in the Sources section, which begins on page 672. The figure number in the box of each circuit correlates to the entry in the Sources section.

Audio Shunt Noise Limiter
Simple Audio Clipper/Limiter
Noise Blanker

AUDIO SHUNT NOISE LIMITER

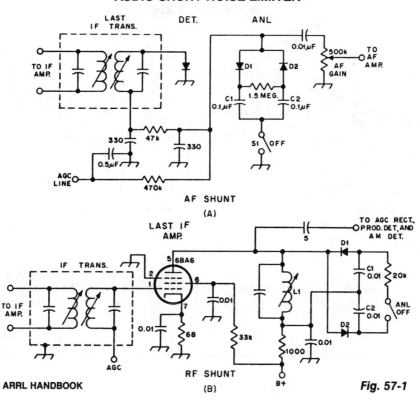

AF SHUNT

(A)

RF SHUNT

(B)

ARRL HANDBOOK

Fig. 57-1

Examples of RF and audio ANL circuits. Positive and negative clipping occurs in both circuits. The circuit at A is self-adjusting. This noise limiter operates at the IF output. It is self-adjusting. Adequate gain is needed at the IF frequency so that several volts p-p of audio is available.

SIMPLE AUDIO CLIPPER/LIMITER

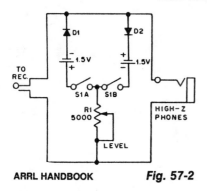

ARRL HANDBOOK

Fig. 57-2

For use with headphones, this circuit sets the audio clipping level via a 5-kΩ pot. This type of noise clipper works best for pulse-type noise of low duty cycle, such as ignition noise. R1 sets the bias on the diodes for the desired limiting level.

NOISE BLANKER

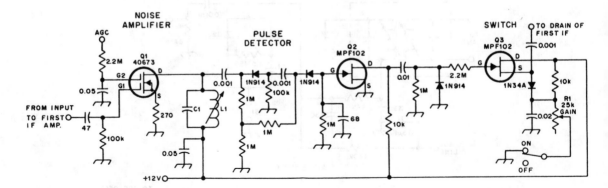

Fig. 57-3

This noise blanker takes a sample of IF input voltage, amplifies it, and drives a switch. The switch (when activated by a noise pulse) adds a heavy load to the first IF stage and kills the gain for the pulse duration.

58

Operational-Amplifier Circuits

The sources of the following circuits are contained in the Sources section, which begins on page 672. The figure number in the box of each circuit correlates to the entry in the Sources section.

Efficient Power Booster
Op Amp Regulator
Increased Feedback-Stabilized Amplifier
Gain-Controlled Op Amp
Bidirectional Compound Op Amp
Compound Op Amp VCO Driver
Make LM324 Op Amp Swing Rail-to-Rail
3-Input AND Gate Comparator
Programmable Inverter/Rectifier
Compound Op Amp

EFFICIENT POWER BOOSTER

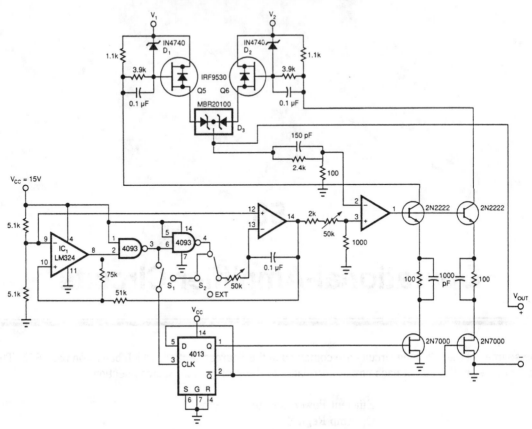

Fig. 58-1

This power booster functions as a high-efficiency "power multiplexer" or, if you supply an external signal-source, as a high-power linear amplifier.

If you want to drive a load with a high-power square wave, the circuit simply draws power from two external power sources, V1 and V2, alternately. In this mode, the circuit's power-handling devices function as switches, dissipating minimal power. The RC time constant of the integrator, IC1, determines the circuit's oscillation period.

If you supply an external drive waveform, the circuit functions as a linear amplifier, and, consequently, inherently dissipates varying portions of that power. The power amplifier is stable for gains ≥ 15.

Diodes D1 and D2 limit the FET's gate-voltage swing to less than 15 V. D3 is a dual Schottky diode that protects the FETs from short circuits between the two supplies, V1 and V2, through a FET's parasitic diode. With D3 in place, you can choose either power channel for the higher voltage input. To drive the FETs, Q5 and Q6, at switching frequencies greater than 1 kHz, you will have to use gate drivers for them.

OP AMP REGULATOR

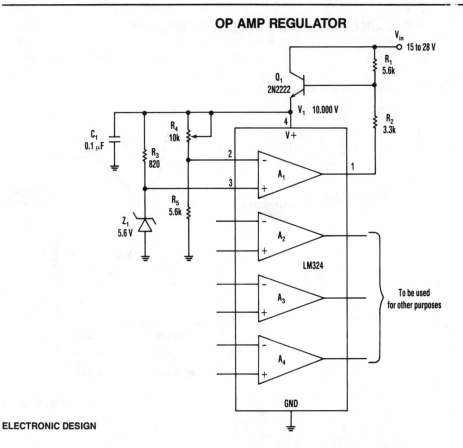

ELECTRONIC DESIGN

Fig. 58-2

This op amp offers a straightforward method of developing a single-polarity stable voltage source (see the figure). Transistor Q1 gets a base drive through resistor R1, and conducts to develop a voltage (V_1) across the IC's supply pins. Amp A1, R2, and Q1 form a positive-feedback closed loop, along with R3 and the zener diode. A1, R2, and Q1 also form a negative-feedback closed loop with R4 and R5.

The effect of positive feedback is predominant as the noninverting input receives V_1 while the inverting input receives only:

$$V_1 \times \frac{R_5}{R_4 + R_5}$$

This happens until the zener comes into play. When the voltage at the inverting input exceeds the voltage at the noninverting input, A1's output takes away Q1's base current through R2, which reduces V_1. Hence, an equilibrium condition is reached. Now:

$$V_1 = \frac{V_Z(R_4 + R_5)}{R_5}$$

This circuit can source more than 30 mA.

INCREASED FEEDBACK-STABILIZED AMPLIFIER

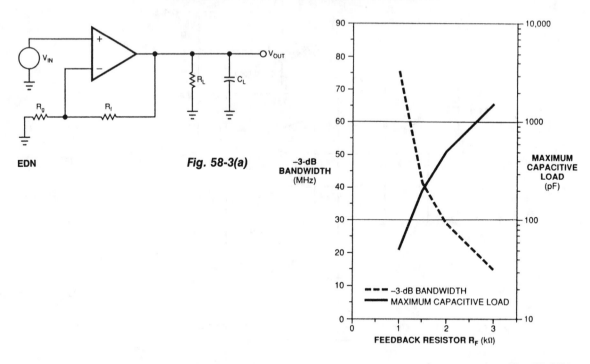

EDN

Fig. 58-3(a)

Fig. 58-3(b)

The usual method for using a current-feedback amplifier to drive a capacitive load isolates the load with a resistor in series with the amplifier's output.

A better solution involves only the amplifier's feedback resistors (Fig. 58-3(a)). Because the feedback resistors determine the amplifier's compensation, you can select the optimal value for these feedback resistors for almost any capacitive load.

Feedback resistance R_F sets the amplifier's bandwidth. Increasing R_F reduces the amplifier's bandwidth, which significantly improves the amplifier's ability to drive capacitive loads. Feedback resistor RG sets the amplifier's gain.

You cannot get the data necessary to calculate alternate values for R_F from most data sheets. However, a few minutes at the bench with a network analyzer will generate the data to make a graph of the value of the feedback resistor vs. the amount of capacitive load the amplifier can drive (Fig. 58-3(b)).

Start with the recommended data-sheet value for feedback resistor R_F and measure the amplifier's frequency response without any capacitive load. Note the bandwidth, then add capacitive loading until the response peaks by about 5 dB. Record this value of capacitance; it is the maximum amount for that feedback resistor. Then, increase the value of the feedback resistor and repeat the procedure until you develop a graph like the one in Fig. 58-3(b).

GAIN-CONTROLLED OP AMP

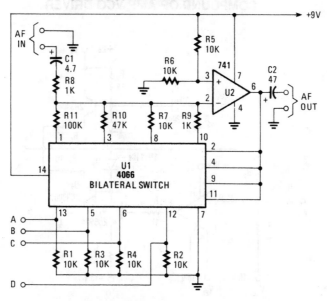

POPULAR ELECTRONICS

Fig. 58-4

The gain controller uses a 4066 quad bilateral switch to electronically select a feedback resistor for the 741 op amp. One or more switches can be turned on at the same time to produce a stepped, variable-gain range from less than 1 to 100.

BIDIRECTIONAL COMPOUND OP AMP

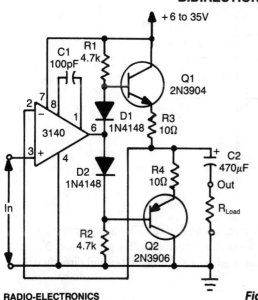

Using two transistors (Q1 and Q2), a bidirectional op amp can source or sink up to 50 mA. D1 and D2 provide bias for Q1 to eliminate "dead-zone" effects.

RADIO-ELECTRONICS

Fig. 58-5

COMPOUND OP AMP VCO DRIVER

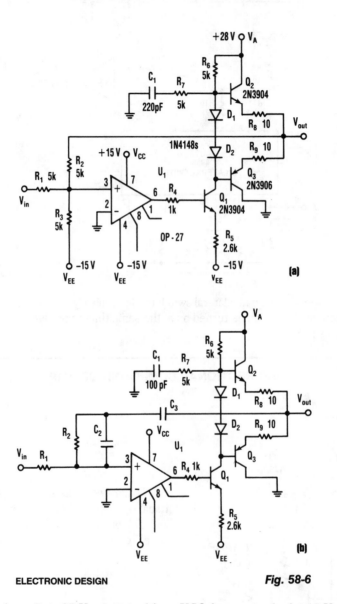

ELECTRONIC DESIGN

Fig. 58-6

This circuit produces 5- to 25-V output to drive a VCO from a standard ±15-V supply system. R7 and C1 supply frequency compensation. Q1 through Q3 form an inverting amplifier with a gain of two. Negative feedback through R2 closes the loop. This circuit can act as a an active load-log filter and directly drive a voltage-controlled oscillator.

MAKE LM324 OP AMP SWING RAIL-TO-RAIL

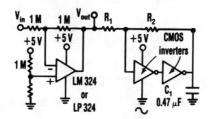

Op Amp		TYPICAL OUTPUT SWING				
	$R_1 =$	Open circuit	2.2 k	1 k	330 Ω	150 Ω
	$R_2 =$	Don't care	15 k	4.7 k	1.5 k	470 Ω
LM324	V_{out} Hi =	3.71 V	4.30 V	4.73 V	4.89 V	4.94 V
	V_{out} Lo =	0.04 V	0.023 V	0.015 V	0.010 V	0.003 V
LP324	V_{out} Hi =	4.16 V	4.91 V	4.965 V	4.982 V	4.987 V
	V_{out} Lo =	0.53 V	0.064 V	0.022 V	0.007 V	0.003 V

ELECTRONIC DESIGN   *Fig. 58-7*

By using two CMOS inverters, the output for an LM324 op amp can be increased from 3.5 Vpp to 4.9 Vpp. This circuit is only recommended for light loads (<30 mA) and for relatively slow op amps. Any CMOS inverter (74COO, 74CO2, 74C14, CD4001, CD4011, etc.) can be used.

3-INPUT AND GATE COMPARATOR

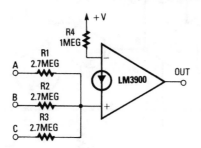

RADIO-ELECTRONICS   *Fig. 58-8*

This circuit has high output only when all three inputs are high. The noninverting-input current, when all three inputs are high, must exceed that of the inverting input, as determined by R4. The circuit can be converted to a NAND gate by transposing the two inputs of the op amp.

PROGRAMMABLE INVERTER/RECTIFIER

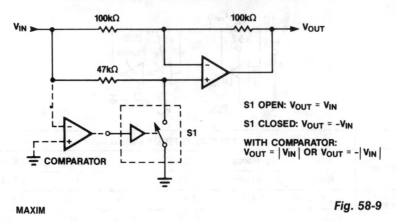

S1 OPEN: $V_{OUT} = V_{IN}$

S1 CLOSED: $V_{OUT} = -V_{IN}$

WITH COMPARATOR:
$V_{OUT} = |V_{IN}|$ OR $V_{OUT} = -|V_{IN}|$

MAXIM

Fig. 58-9

The op amp is alternately an inverter or buffer, under control of the switch polarity. As a buffer, the gain is always 1, but as an inverter, the gain is set by the ratio of the input and feedback resistors. By adding a comparator, the function can be synchronously switched as the input polarity changes, which effectively rectifies the output. The output polarity is determined by the switch logic (normally open or normally closed) and the comparator input polarity.

COMPOUND OP AMP

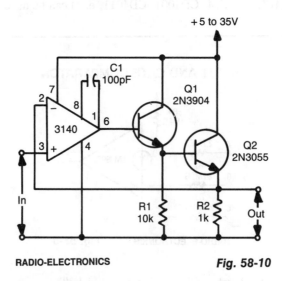

RADIO-ELECTRONICS

Fig. 58-10

By using an emitter-follower or a Darlington pair, a voltage-follower op amp configuration can source higher currents than the op amp otherwise could.

59

Optical Circuits

The sources of the following circuits are contained in the Sources section, which begins on page 672. The figure number in the box of each circuit correlates to the entry in the Sources section.

Optical Interruption Sensor
Light Receiver
Optical Receiver
Light Transmitter
Optical/Laser Receiver
Light Detector
Light Probe
Simple Photoelectric Light Controller

OPTICAL INTERRUPTION SENSOR

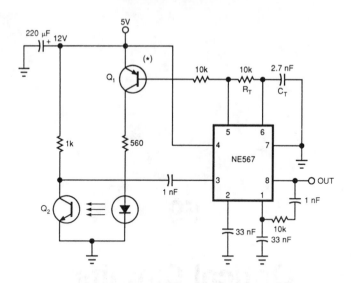

Source	Detector	Manufacturer	Notes
TIL 32	TIL 78	Texas Instruments	IR LED and phototransistor
TIL 38	TIL 414	Texas Instruments	IR LED and phototransistor
CQY58A	BPW22A	Philips	IR LED and phototransistor
CQY89A	BPW 50	Philips	IR LED and pin photodiode
TIL 139		Texas Instruments	Transmissive source and detector assembly
TIL 149		Texas Instruments	Reflective source and detector assembly

EDN   Fig. 59-1

Using only an 8-pin IC and a few discrete components, you can build the infrared optical interrupter. The NE567 tone decoder has all the necessary circuit elements: a local oscillator, a PLL decoder, and a 100-mA output-drive capability. The local oscillator, which is tuned to 40 kHz by RT and CT, drives Q1, a universal low-power silicon pnp transistor (such as a 2N3906, BC559, or ZTX500). Q1 drives the IR-emitting diode. The receiving part of the circuit surrounds the IC's internal PLL input at pin 3. When the photodetector, Q2, detects the oscillating IR light beam, the 40-kHz signal appears at pin 3 of the IC. Under this condition, the circuit locks and the IC's output is high. When something opaque comes between the LED and Q2, the 40-kHz signal doesn't reach the PLL input, and the IC's output goes low.

The feedback network between pins 1 and 8 prevents the output from chattering. If you connect this circuit to a high-inertia load (such as a mechanical relay), the output doesn't tend to oscillate and you can eliminate these feedback components. The circuit works with virtually any LED-photodetector pair, but matched pairs allow for longer distances between the emitter and receiver. The table lists some of the best choices.

LIGHT RECEIVER

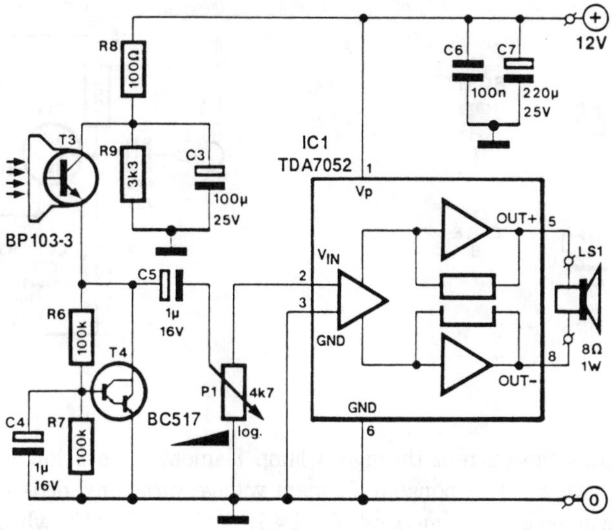

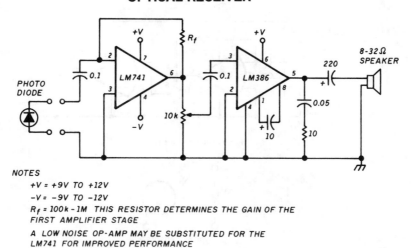

ELEKTOR ELECTRONICS

Fig. 59-2

T3 is a photocell or phototransistor. T4 controls the emitter voltage of T3. IC1 is an audio amplifier to provide amplification of the signal from the photocell.

OPTICAL RECEIVER

NOTES

+V = +9V TO +12V

−V = −9V TO −12V

R_f = 100k - 1M THIS RESISTOR DETERMINES THE GAIN OF THE FIRST AMPLIFIER STAGE

A LOW NOISE OP-AMP MAY BE SUBSTITUTED FOR THE LM741 FOR IMPROVED PERFORMANCE

HAM RADIO

Fig. 59-3

An optical receiver for light-wave communications, this circuit works with AM-type light signals.

LIGHT TRANSMITTER

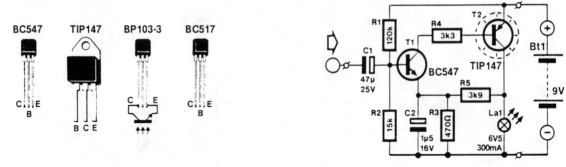

ELEKTOR ELECTRONICS

Fig. 59-4

This circuit modulates the current through a lamp filament. Use a low-voltage lamp with a thin, straight filament. They have a fast response to filament voltage variations. dc is applied to "bias" the filament that is on, and the audio is superimposed. A BC457 drives a TIP147, which modulates the filament current.

OPTICAL/LASER RECEIVER

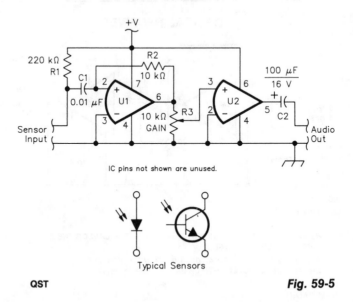

QST

Fig. 59-5

Using a single 741 op amp, a photodiode sensor, and an LM386, this simple receiver operates from a 9-V battery. The circuit will drive a pair of earphones or a small speaker.

LIGHT DETECTOR

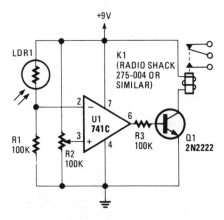

The circuit's threshold is set by resistor R2. When the intensity of the light falling on the LDR is lowered, resistance of that unit increases, and lowers the voltage applied to the inverting input of the 741. The reference voltage at the noninverting input of the 741 is set (via R2) so that the comparator switches from low to high when the light falling on the LDR is reduced. That high activates transistor Q1, which causes the relay contacts to close.

HANDS-ON ELECTRONICS   *Fig. 59-6*

LIGHT PROBE

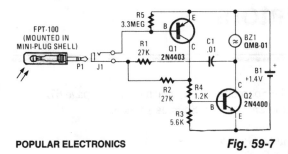

Originally designed as an aid for blind people, this probe was used as a light detector in order to tell if a device or room lights are on or off.

POPULAR ELECTRONICS   *Fig. 59-7*

SIMPLE PHOTOELECTRIC LIGHT CONTROLLER

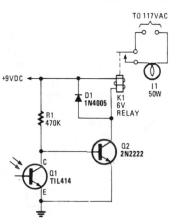

A phototransistor senses daylight. At dusk, it ceases to conduct and R1 biases Q2, activates K1, and switches on the light. At dawn, Q1 starts to conduct, and Q2 is cut off. K1 drops out and the light goes out.

POPULAR ELECTRONICS *Fig. 59-8*

60

Oscillators

The sources of the following circuits are contained in the Sources section, which begins on page 672. The figure number in the box of each circuit correlates to the entry in the Sources section.

Beat-Frequency Audio Generator
Simple Wien-Bridge Oscillator
Stable VFO
Code-Practice Oscillator I
Phase-Locked 20-MHz Oscillator
Audio Oscillator
Code-Practice Oscillator II
Audio Oscillator II
Code-Practice Oscillator III
Relaxation Oscillator
Wien-Bridge Oscillator

BEAT-FREQUENCY AUDIO GENERATOR

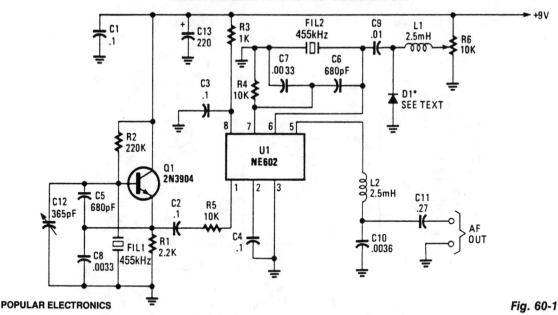

Fig. 60-1

Q1 is a fixed oscillator operating at 455 kHz. U1 is a mixer, with its own internal oscillator running at 455 kHz. FIL1 and FIL2 are Murata CSB455E filters or equivalent. D1 is a varactor diode (an IN4002 used as a varactor works well here). R6 controls the bias on D1. When R6 is varied, the oscillator frequency varies a few kHz. Audio beat note is taken, through RF filter L2 and C10, from pin 5 of U1.

SIMPLE WIEN-BRIDGE OSCILLATOR

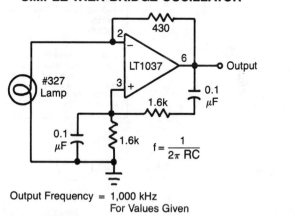

$$f = \frac{1}{2\pi RC}$$

Output Frequency = 1,000 kHz
For Values Given

Fig. 60-2

In this circuit, the Wien-bridge network provides phase shift, and the lamp regulates the amplitude of the oscillations. The smooth, limiting nature of the lamp's operation, in combination with its simplicity, gives good results. Harmonic distortion is below 0.3%.

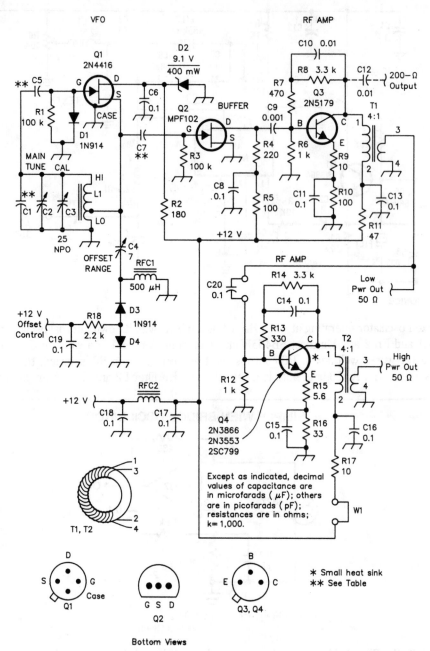

Fig. 60-3

STABLE VFO (Cont.)

This VFO circuit covers from 2.13 to 2.58 MHz and is intended for use with an external mixer to heterodyne the signal to desired frequencies. Coil data is shown in the parts list. Two power output levels are available—a few hundred mW (0 to +3 dbm) from Q3. Q4 is a class A amplifier for boosting the power to +22 dbm for driving a high-level mixer. The VFO can be operated on other frequencies, with suitable component charges (see the table).

Component Information

f(MHz)	C1(pF)	C2(pF)	C5, C7(pF)	L1
1.8-2	220	100	100	24 µH; 71 turns of no. 30 enamel on a T-68-6 toroid core. Tap at 18 turns from bottom end.
3.5-4	150	50	68	9.5 µH; 44 turns of no. 26 enamel on a T-68-6 toroid core. Tap at 11 turns from bottom end.
5-5.5	130	50	47	5 µH; 33 turns of no. 24 enamel on a T-68-6 toroid core. Tap at 8 turns from bottom end.
7-7.3	110	25	47	3.6 µH; 27 turns of no. 24 enamel on a T-68-6 toroid core. Tap at 7 turns from bottom end.

C3 is a 25-pF NPO ceramic trimmer. C4 is a 7-pF air trimmer. The total capacitance of C2 may be reduced to restrict the VFO tuning range. C1, C5 and C7 are NPO ceramic.

CODE-PRACTICE OSCILLATOR I

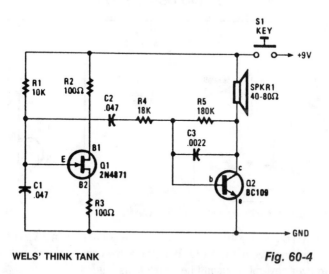

WELS' THINK TANK

Fig. 60-4

Q1, a unijunction transistor, generates a sawtooth of about 1.5 to 2 kHz, depending on C1 and R1. Q2 acts as a speaker driver. A 9-V battery is used, and the keying is done by keying the supply line.

PHASE-LOCKED 20-MHz OSCILLATOR

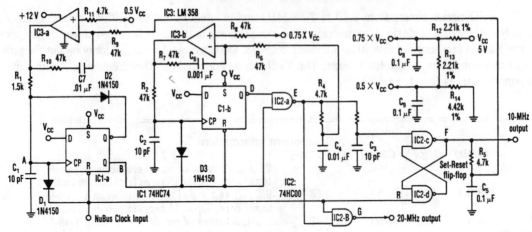

ELECTRONIC DESIGN

Fig. 60-5

This circuit produces a 20-MHz clock phase locked to a 10-MHz clock present in the Apple MAC II. To generate the 20-MHz signal, the circuit produces a 25 ns negative-going pulse delayed 50 ns from the falling edge of the 10-MHz Nubus clock input at point E. NORing that pulse with the Nubus clock produces the 20-MHz clock at point G. Applying the 25-ms pulse to the set input of an S/R flip-flop and the Nubus clock to the reset input results in a 10-MHz square wave at F.

AUDIO OSCILLATOR

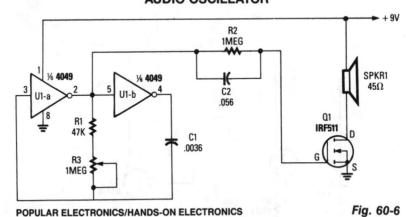

POPULAR ELECTRONICS/HANDS-ON ELECTRONICS

Fig. 60-6

Two gates, U1A and U1B (1/3 of a 4049 hex inveter), are connected in a VFO circuit. Components R1, R3, and C1 set the frequency range of the VFO. With the values given, the circuit's output can range from a few hundred hertz to over several thousand hertz by adjusting R3.

The simplest way to change the frequency range of the oscillator is to use different capacitance values for C1. A rotary switch, teamed up with a number of capacitors, can be used to select the desired frequency range.

CODE-PRACTICE OSCILLATOR II

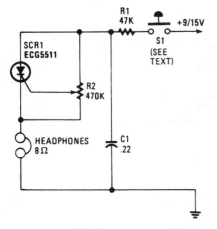

Capacitor C1 charges through resistor R1, and when the gate level established by potentiometer R2 is high enough, the SCR is triggered. Current flows through the SCR and earphones, discharging C1. The anode voltage and current drop to a low level, so the SCR stops conducting and the cycle is repeated. Resistor R2 lets the gate potential across C1 be adjusted, which charges the frequency or tone. Use a pair of 8-Ω headphones. The telegraph key goes right into the B+ line, 9-V battery.

HANDS-ON ELECTRONICS/POPULAR ELECTRONICS

Fig. 60-7

AUDIO OSCILLATOR II

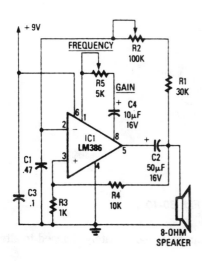

The circuit's frequency oscillation is $f = 2.8/[C_1 \times (R_1 + R_2)]$. Using the values shown, the output frequency can be varied from 60 Hz to 20 kHz by rotating potentiometer R2.

A portion of IC1's output voltage is fed to its noninverting input at pin 3. The voltage serves as a reference for capacitor C1, which is connected to the noninverting input at pin 2 of the IC. That capacitor continually charges and discharges around the reference voltage, and the result is a square-wave output. Capacitor C2 decouples the output.

WELS' THINK TANK

Fig. 60-8

CODE-PRACTICE OSCILLATOR III

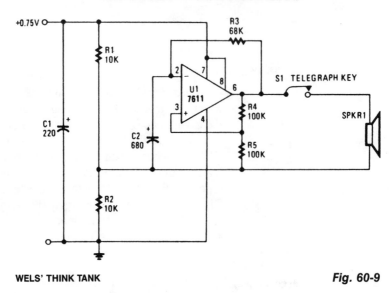

WELS' THINK TANK

Fig. 60-9

U1 is used as an oscillator; the frequency is determined by C2 and R3. Use an 80 Ω or similar high-impedance speaker, and a 1.5-V battery for a power source.

RELAXATION OSCILLATOR

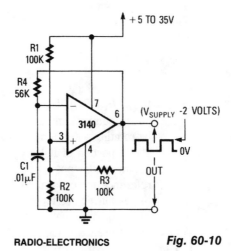

RADIO-ELECTRONICS

Fig. 60-10

This oscillator runs at about 150 Hz, but C1 and/or R4 can be proportionately changed to alter this frequency. Rise and fall times are 12 and 7 μs, respectively.

WIEN-BRIDGE OSCILLATOR

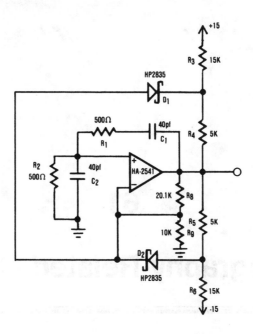

HARRIS

Fig. 60-11

The HA2541 is well-suited for use as the heart of an oscillator. In spite of the rudimentary diode limiting that is provided by R3 through R7 and D1 and D2, a good-quality sine wave of 40 MHz is readily attainable with an upper limit of 50 MHz, which exceeds the unity-gain bandwidth of HA-2541.

R1/C1 and R2/C2 provide the required regenerative feedback needed for adequate frequency stability. In theory, the feedback network requires a gain of three to sustain oscillation. However, the practical gain needed is just over three and is provided by R8 and R9.

61

Photography-Related Circuits

The sources of the following circuits are contained in the Sources section, which begins on page 673. The figure number in the box of each circuit correlates to the entry in the Sources section.

Photo-Event Timer
SCR Slave Flash
Time-Delay Photo-Flash Trigger
Camera Trip Circuit
Slide Projector Auto Advance
Sound-Trigger Flash
Slave Flash Trigger

PHOTO-EVENT TIMER

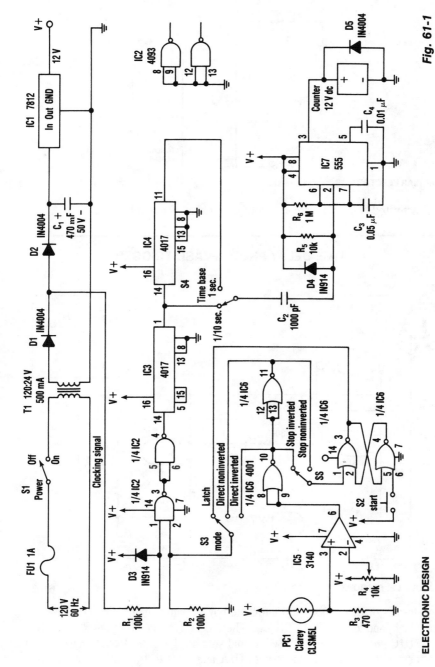

Fig. 61-1

ELECTRONIC DESIGN

S2 is used to initiate timing. A light-to-dark or dark-to-light transition stops this timer, depending on the setting of S5. S3 offers a direct operating mode, rather than through the latch. IC3 and IC4 supply 0.1- or 1-second timing pulses. IC7 drives a time display counter, a 12-Vdc unit that draws less than 200 mA.

379

SCR SLAVE FLASH

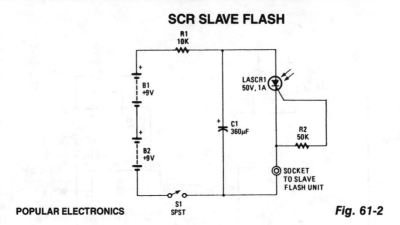

POPULAR ELECTRONICS

Fig. 61-2

Using a light-activated SCR, this circuit can trigger a slave-flash unit.

TIME-DELAY PHOTO-FLASH TRIGGER

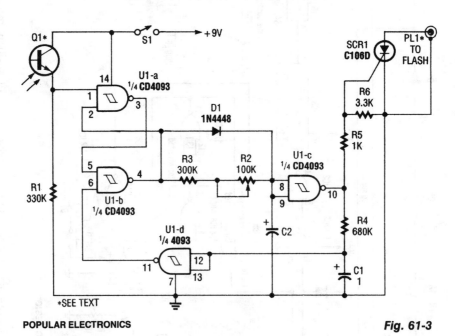

POPULAR ELECTRONICS

Fig. 61-3

Q1 is a phototransistor that is normally illuminated by a beam of light. When the beam is intercoupled, pin 1 of U1A goes high, and forces pin 4 U2B low. Then, C2 discharges through R2 and R3. After a certain time delay, pin 10 U1C goes high, triggers SCR1, and sets off the flash. R4/C1 charges, causes U1D output to go low after about 1/2 second, and resets U1A and U1B to the initial state. This delay prevents accidental double flash exposure.

CAMERA TRIP CIRCUIT

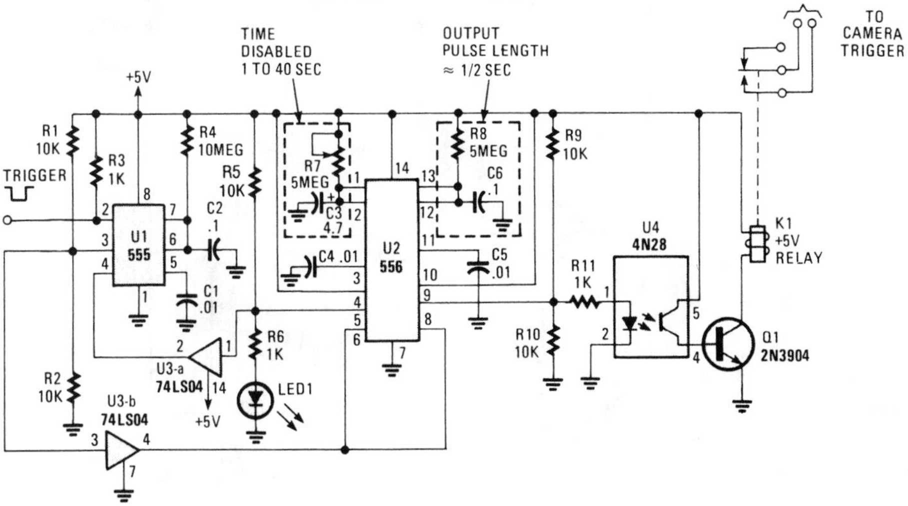

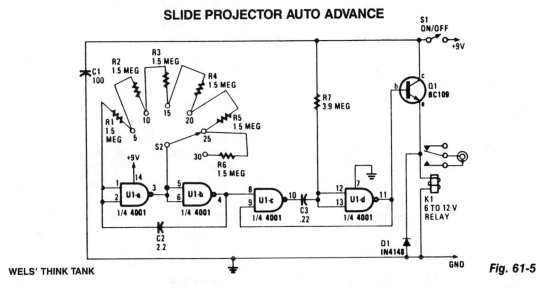

POPULAR ELECTRONICS

Fig. 61-4

This circuit was used to trip a camera shutter. Grounding pin 2 of U1 makes pin 4 of U1 go high. This triggers both timers of dual timer U1. One output holds reset (pin 4) of U1 low to keep U1 from accepting another trigger, depending on the time constant of R7 and C3. This prevents camera film waste. The other timer is used to generate a 1/2-second pulse to drive U4 and Q1, the relay driver. K1 triggers the camera.

SLIDE PROJECTOR AUTO ADVANCE

WELS' THINK TANK

Fig. 61-5

A 4001 CMOS Quad NORgate is set up as an astable multivibrator, which drives a simple differentiator and relay driver. Depending on the setting of S2, a delay of 5 to 30 seconds is generated. S2 and R1 through R6 can be replaced by a single 10-MΩ pot, if desired.

SOUND-TRIGGERED FLASH

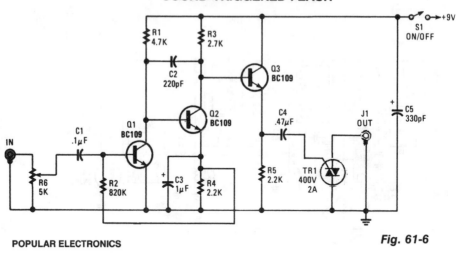

Fig. 61-6

POPULAR ELECTRONICS

Audio input from a microphone drives amplifier Q1/Q2/Q3 to produce an ac voltage across R5. C4 couples this to TR1, causing it to conduct, triggering photoflash or other device that is connected to J1.

SLAVE FLASH TRIGGER

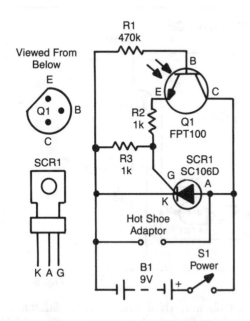

POPULAR ELECTRONICS/HANDS-ON ELECTRONICS

Fig. 61-7

The SCR is wired across the trigger circuit of the flash gun. Normally, the SCR is off, so the flash gun is able to charge to its trigger voltage. Photo transistor Q1 is used to monitor the light level. When a high-intensity flash occurs, Q1 briefly conducts and supplies gate current to the SCR. That causes the SCR to turn on, which then triggers the slave flash gun via the hot-shoe adapter terminals.

Once the flash gun has triggered, the SCR quickly turns off again. That happens because the current in circuit quickly falls below the SCR's holding current. The resistor at the base of Q1 (R1) determines the sensitivity of the circuit. If you wish, you can reduce the sensitivity, simply by reducing the value of the resistor from that shown. The 1-kΩ resistor between the gate and cathode of the SCR (R3) prevents the SCR from false triggering if high voltages are applied between the anode and the cathode. Q1 can also be a GEL14G2.

62

Power-Control Circuits

The sources of the following circuits are contained in the Sources section, which begins on page 672. The figure number in the box of each circuit correlates to the entry in the Sources section.

5-V SUPPLY HIGH-SIDE SWITCHER

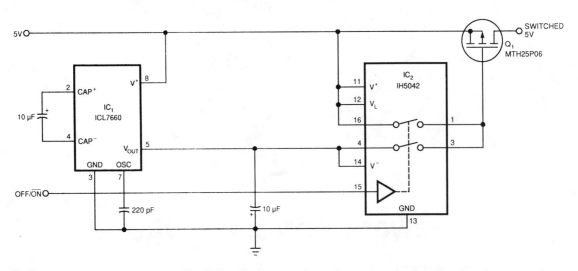

EDN

Fig. 62-1(a)

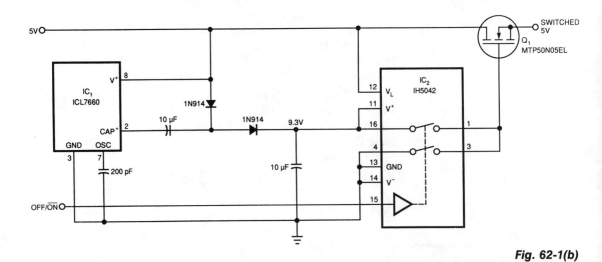

Fig. 62-1(b)

5-V SUPPLY HIGH-SIDE SWITCHER *(Cont.)*

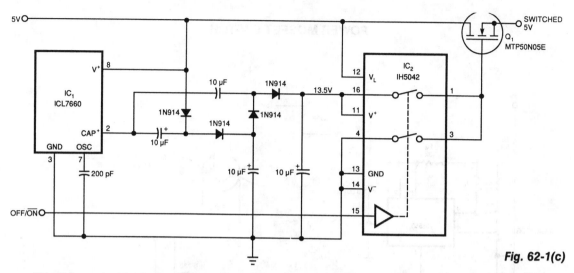

Fig. 62-1(c)

Requiring only 10 μA of quiescent current, the circuit of (Fig. 62-1(a)) produces only 0.1-Ω ON-resistance. IC1 is a charge pump voltage converter to produce a –5-V level, so analog switch IC2 can provide a 10-V swing to MOSFET Q1.

This circuit uses a voltage converter to enable the analog switch to apply a 4.3-V swing to logic level NMOS power transistor Q1. ON resistance is 0.03 Ω typical.

This circuit uses additional stages in the voltage-multiplying circuit to provide a higher gate voltage swing. This would enable the use of a converter for an NMOS switching transistor.

SCR POWER MONITOR

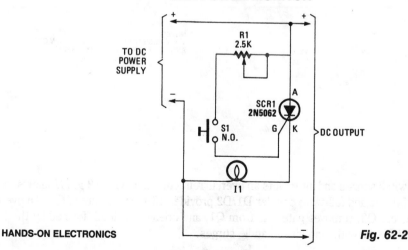

HANDS-ON ELECTRONICS

Fig. 62-2

Pressing R1 lights I1, which will extinguish if there has been a power failure.

POWER MOSFET SWITCH

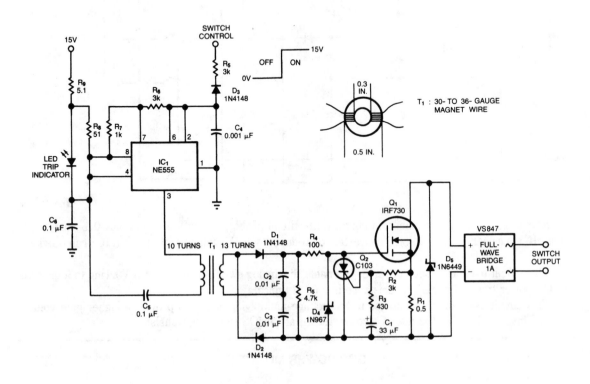

NOTES:
1. ALL RESISTORS EXCEPT R₁ ARE ¼W CARBON.
2. T₁ IS A FERROXCUBE 204XT250-3E2A.

Fig. 62-3

This solid-state switch senses and interrupts an overcurrent condition within 2 μs. I1 allows the circuit to float. IC1 runs at 150 kHz and full-wave doubler D1/D2 provides 15 V to the gate of Q1. An overcurrent sensed across R1 triggers Q3, removes gate bias from Q1, and opens the circuit formed by the full-wave bridge and Q1. C1 and R3 allow the circuit to handle surges.

CURRENT-LOOP SCR CONTROL

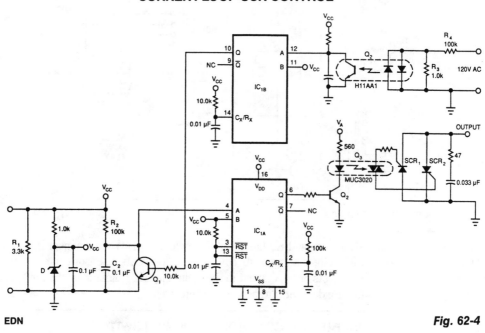

EDN

Fig. 62-4

This circuit allows a 4-to 20-mA current loop to control an isolated SCR drive. IC1A and B are one-shots. Q2 detects zero crossings of the 120 Vac line, which triggers one-shot IC1B. IC1A causes Q1 to discharge C2. When C2 recharges through R2, it triggers IC1A, and the optoisolator and SCR1/SCR2. Triggering of SCR1 and SCR2 is a function of input current, which can control motor speed, light intensity, etc.

BATTERY-TRIGGERED ac SWITCH

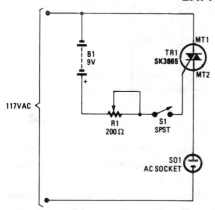

Using this method, a small switch (S1) can control a large ac load. R1 is adjusted for reliable triggering and should be as large as possible.

POPULAR ELECTRONICS   **Fig. 62-5**

UNIVERSAL POWER CONTROLLER

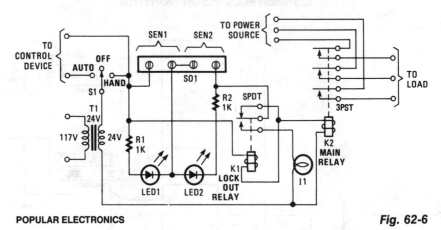

POPULAR ELECTRONICS

Fig. 62-6

Relay K1 has a low-impedance coil and K2 has a high-impedance coil. When a sensor opens, current is routed through the coil of K1. K1 activates, opens its contacts, and prevents a sensor contact reclosure from affecting the circuit. When K1 contacts open, current to the main relay K2 is limited by the impedance of K1. K2 controls power to a load (air conditioner, furnace blower, etc.).

PUSHBUTTON-CONTROLLED POWER SWITCH

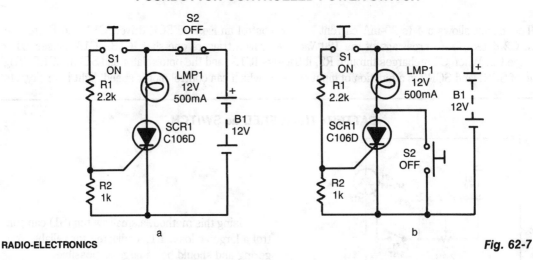

RADIO-ELECTRONICS

Fig. 62-7

In both circuits, the SCR (and thereby the lamp) can be latched on by momentarily closing S1, thereby feeding gate drive to the SC via R1. In both circuits, the gate is tied to the cathode via R2 to improve circuit stability.

Of course, after the SCR turns on, it can be turned off again only by momentarily reducing anode current below the device's I_H value. The SCR is turned off by momentarily opening S2, by using S2 to short the anode and cathode terminals of the SCR momentarily.

BANG-BANG CONTROLLERS

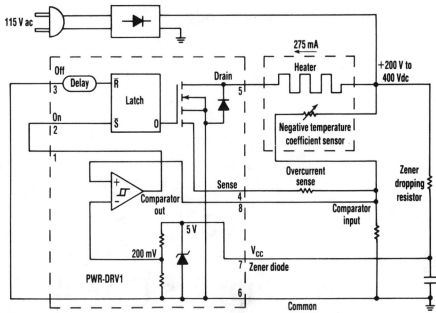

ELECTRONIC DESIGN

Fig. 62-8

Just one chip, the PWR-DRV1 from Power Integrations, builds a "bang-bang" controller that switches 275 mA and runs off the rectified 115-Vac mains. An on-chip zener diode powers the chip from high voltage through a dropping resistor.

SCR OVERVOLTAGE PROTECTOR

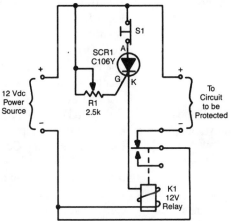

HANDS-ON ELECTRONICS

Fig. 62-9

Depending on the setting of R1, when the voltage exceeds a certain amount, SCR1 triggers, which activates K1 and opens the circuit. S1 resets the SCR.

63

Power Supplies (Fixed)

The sources of the following circuits are contained in the Sources section, which begins on page 673. The figure number in the box of each circuit correlates to the entry in the Sources section.

±15-V AND 5-V CAR BATTERY SUPPLY

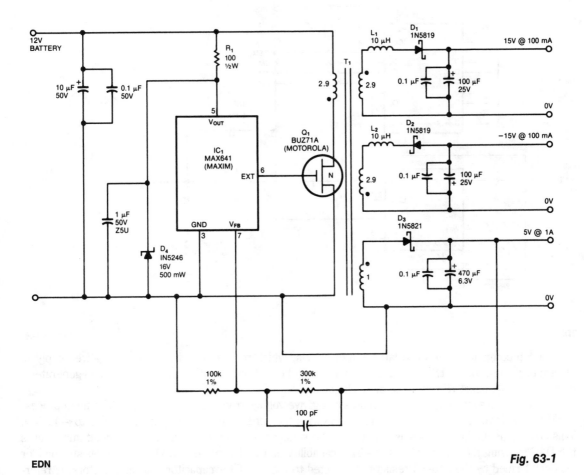

EDN

Fig. 63-1

IC1 is a switching regulator that generates a 45-kHz signal that drives the gate of MOSFET Q1. D1, D2, and D3 are Schottky diodes. The 5-V output is sensed as a reference; feedback to the chip turns off the gate signal to Q1 if the voltage rises above 5 V.

T1 has Trifilar windings that assume about 2% regulation for a 10-to 100-mA load change on the ±15-V supplies. R1/D4 provide overvoltage protection. T1 has a primary inductance of about 21 μH. Core size should allow 4-A peak currents. The turn ratios are $11\frac{1}{2}$ turns each for the 15-V supplies, $11\frac{1}{2}$ turns for the primary, and four turns for the 5-V secondary. The efficiency is about 75%.

SIMPLE LCD DISPLAY POWER SUPPLY

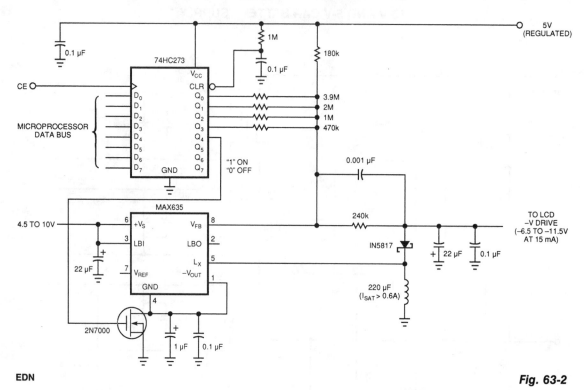

EDN

Fig. 63-2

Laptop computers often use large-screen LCDs, which require a variable and a negative supply to ensure maximum contrast. This circuit operates from the system's positive battery supply and generates a digitally variable negative voltage to drive the display.

This figure's switching regulator creates a negative voltage from the battery supply. The microprocessor data bus drives a 4-bit DAC, which in turn varies the actual regulator output from −6.5 to −11.5 V. This arrangement allows a staircase of 16 possible voltages between these limits. The circuit implements the DAC by using the rail-to-rail output-drive capability of a 74 HC-series CMOS gate. A resistor divider network formed by the 240-kΩ resistor, connected to the −V filter capacitor and the resistors, is referenced to the 5-V supply control (the MAX635 regulator).

When the voltage at the V_{FB} pin is greater than ground, the switching regulator turns on. The inductor dumps this energy into the −V filter capacitor. When the voltage at V_{FB} is less than ground, the regulator skips a cycle. The MAX635 regulates the voltage at the junction of the resistor divider to 0 V. Thus, any resistor that the DAC connects to ground (logic 0) will not contribute any current to the ladder. Only the resistors that are at 5 V (logic 1) will be part of the voltage-divider equation.

The entire switching-regulator supply draws less than 150 μA. You can place the circuit in an even lower power mode by interrupting the ground pin. The high-current path is from the battery input through the internal power PMOSFET to the external inductor. Disconnecting the ground connection simply disables the gate drive to the FET and turns off the internal oscillator.

OUTPUT STABILIZER

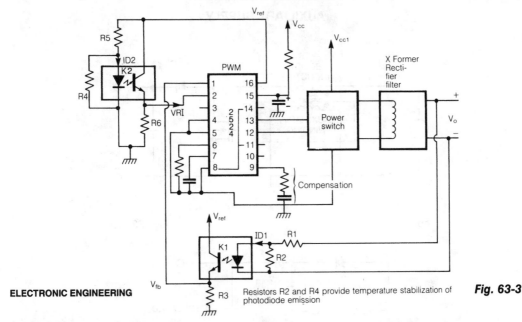

ELECTRONIC ENGINEERING

Resistors R2 and R4 provide temperature stabilization of photodiode emission

Fig. 63-3

Optically isolated SMPS and dc-to-dc converters face the variance of output voltage owing to the change of transmission characteristics of an optoisolator with (a) temperature and (b) aging. The photo diode emission decays with temperature and time, and causes the output voltage to change. The problem is solved using a homoeopathic principle. An additional optical isolator is used to derive the $+V_e$ input voltage, instead of a conventional potential divider from internally stabilized reference. A scheme is shown using IC2524 as PWM element:

$$1)\ V_{RI} = K_2 \times I_{D2} \times R_6 = K_2 \left(\frac{V_{Ref} - V_D}{R_5} - \frac{V_D}{R_4} \right) \times R_6$$

$$2)\ V_{fb} = K_1 \times I_{D1} \times R_3 = K_1 \left(\frac{V_0 - V_D}{R_1} - \frac{V_D}{R_2} \right) \times R_3$$

where V_D = forward drop of a photo diode. In equations 1 and 2, the terms:

$$\frac{V_D}{R_4} \text{ and } \frac{V_D}{R_2}$$

decrease with temperature. Thanks to $-V_e$ temperature coefficient of V_D any changes in K_1 and K_2 as a result of temperature and aging, track each other to maintain the output voltage constant. With proper selection of R_2 and R_4, the output voltages are found to vary by less than 0.01%/°C over a wide temperature range.

AUXILIARY SUPPLY

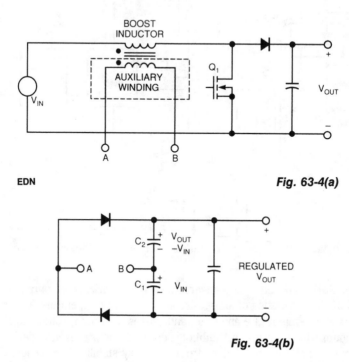

EDN

Fig. 63-4(a)

Fig. 63-4(b)

Many power-factor-correction circuits use a boost converter to generate a regulated dc output voltage from the ac line input while forcing the load to draw sinusoidal current, which maximizes the power factor.

This circuit's full-wave rectifier the auxiliary winding's output to completely cancel out line variations and provide a regulated output voltage. The circuit essentially sums the two phases of the boost inductor's voltage to eliminate the 120-Hz components. The regulated output tracks the power-factor-controlled pre-regulator output voltage and it can be used in the corrected output voltage's feedback loop.

An isolated auxiliary winding consists of the desired number of turns wound on the boost inductor. You can vary the exact value of the auxiliary supply's output voltage by adjusting or scaling the auxiliary winding's number of turns. Figure 63-4(b)'s rectifier develops two separate, but individually unregulated voltages, across capacitors C1 and C2. Each of these voltages varies in amplitude at twice the ac-line frequency. When switch Q1 is on, the boost inductor connects directly across the input supply, and a voltage proportional to the instantaneous input voltage develops across capacitor C1.

Once the switch turns off, the inductor voltage reverses and clamps to a voltage equal to $V_{OUT} - V_{IN}$. During this interval, a voltage proportional to $V_{OUT} - V_{IN}$ develops across C2. The sum of these two capacitor voltages produces a regulated auxiliary voltage that is proportional to V_{OUT}. The voltage across the output capacitor equals $V_{IN} + (V_{OUT} - V_{IN})$, which cancels the input-line variations.

ac-TO-dc CONVERTER

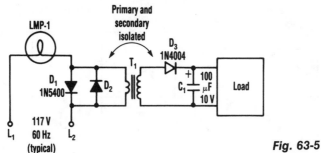

ELECTRONIC DESIGN

Fig. 63-5

By coupling two back-to-back diodes in series with an ac power circuit, a voltage of about 1.4 Vpp can be obtained. This voltage is useful for exciting the primary coil of a small transformer. The voltage induced in the secondary coil can then be rectified and used to power solid-state control circuits. The forward-voltage drop of the diodes is inherently constant and stable over a wide range of ac-circuit power variations. The resulting voltage developed across the transformer windings is also free from variation that might be caused by changes in the circuit's current or voltage.

In the circuit, a lamp (LMP-1) is connected to the primary ac input line (L1 and L2) through a pair of inverse-parallel-connected power diodes (D1 and D2). As power flows to the lamp, a drop of about 0.7 V is alternatively developed across each of the diodes. This voltage feeds the primary of a small transformer (T1). T1 can be a small 8-Ω to 500-Ω transistor radio output, etc. This will deliver about 11 Vpp across its secondary winding. LMP1 can be a small 120-V lamp of 5 to 25 W, etc.

12-V INPUT SIMPLE INVERTER

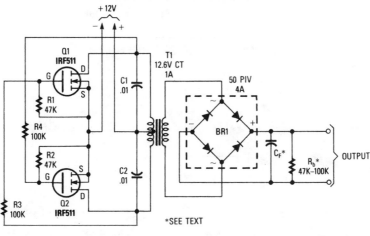

POPULAR ELECTRONICS

Fig. 63-6

Using two power MOSFETs, this inverter can deliver ac or dc up to several hundred volts. T1 is a 12.6-V CT to 120-, 240-, or 480-V transformer for 60-Hz application.

REGULATED CHARGE PUMP

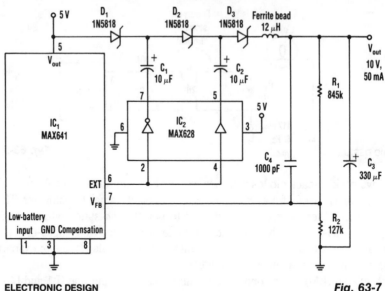

ELECTRONIC DESIGN

Fig. 63-7

The dc-dc converter substitutes a voltage tripler in place of the external inductor and the diode that's typically associated with the switching regulator, IC1. Inverting and noninverting amplifiers in the MOS-FET-driver (IC2) activate a diode-capacitor tripling network (D1 through D3, C1 through C3).

A 50-kHz oscillator residing within IC1 produces the EXT signal (pin 6). IC2 converts this signal into drive signals (180° out of phase) for the tripler. The resulting charge-discharge action in the capacitors recharges C3 toward 10 V every 20 μs. The ferrite bead limits output ripple to about 20-mVpp for a 50-mA load. Conversion efficiency is about 70% for the 5-V input, 10-V output configuration.

SIMPLE RIPPLE SUPPRESSOR

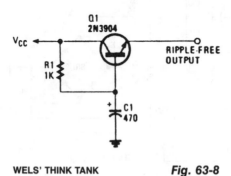

WELS' THINK TANK

Fig. 63-8

This circuit, at times called a *capacitance multiplier*, is useful for suppressing power-supply ripple. C1 provides filtering equal to a capacitor of $(B+1)$ C_1, where $B =$ dc current gain of Q1 (typically > 50).

3-V POWER SUPPLY FOR PORTABLE RADIOS

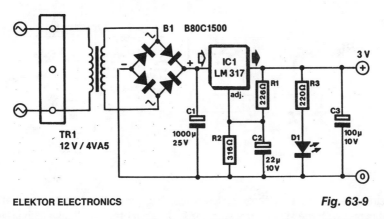

ELEKTOR ELECTRONICS

Fig. 63-9

Most small portable radios require a 3-V supply, which is normally provided by two AA or AAA batteries. Because rechargeable batteries are an option with many of these radios, most of them are fitted with a charger socket. When such radios are used in a stationary condition (e.g., in the kitchen or in the office), it is useful (and economical) to use the mains-operated supply described here.

The supply is small enough to be fitted inside the radio or in a mains adapter case (less than transformer). Voltage regulator IC1 is adjusted for an output of 3 V by resistors R1 and R2, which are decoupled by C2. Capacitor C3 provides additional filtering. Diode D1 indicates whether the unit has been connected to the mains. The diode also provides the load necessary for the regulator to function properly; in its absence, the secondary voltage of the transformer might become too high when the unit is not loaded.

The transformer should be a short-circuit-proof miniature type, which is rated at 12 V and 4.5 VA. The secondary voltage is slightly higher than needed for a radio, but this reserve is useful when the unit is used with a cassette or CD player. It is advisable to check the output voltage of the unit when it is switched on for the first time before connecting it to a radio or cassette player.

NEGATIVE VOLTAGE FROM A POSITIVE SUPPLY

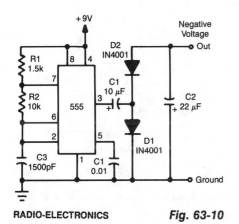

RADIO-ELECTRONICS

Fig. 63-10

By using a 555 timer to generate a square wave and voltage-doubling the output, a negative voltage that is almost equal to the positive supply can be obtained. The current available is up to 20 to 30 mA or so, depending on the regulation and voltage needed.

BRIDGE RECTIFIER

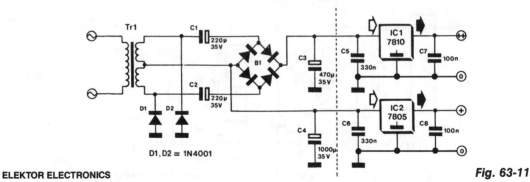

ELEKTOR ELECTRONICS

Fig. 63-11

This bridge circuit is intended for those cases where two unequal supply voltages are required. The lower voltage is obtained with the aid of a transformer with symmetric windings and half-wave rectification of the potential across one winding.

For the higher voltage, the potential across both windings is rectified. To that end, the output of the transformer is linked to the bridge rectifier via two electrolytic capacitors that provide isolation of the two direct voltages.

A bonus with this type of circuit is that although the two supplies can be loaded unequally, the currents through the two transformer windings are the same. Thus, the transformer is loaded symmetrically so that its full capacity can be used. Moreover, no unnecessary dissipation is in the voltage regulators.

The load on the lower voltage supply depends primarily on the rating of the transformer. The load on the higher voltage supply is limited by the reactance of C_1 and C_2 ($= 1/2 \pi 50 C$) and the required minimum output voltage.

±35-V SUPPLY FOR AUDIO AMPLIFIERS

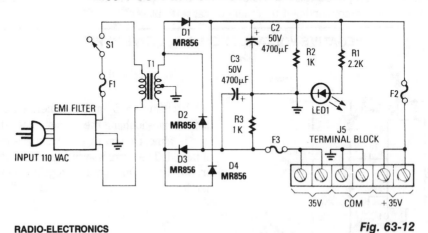

RADIO-ELECTRONICS

Fig. 63-12

This supply will be found useful for operating various transistor AF power amplifiers in the 50-to 100-W output range. T1 = 120 V: 70 V CT at 5 A.

PRECISE LOW-CURRENT SOURCE

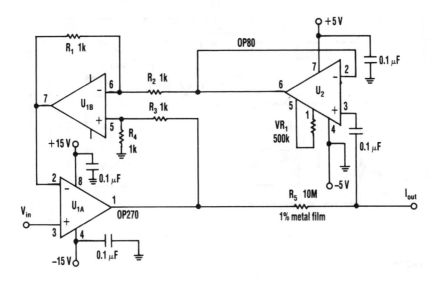

ELECTRONIC DESIGN

Fig. 63-13

A current source that attains a resolution as low as 10 pA is useful in applications where a precise, low-value current is needed. When the circuit forces current into ground, the output remains within 2% of the ideal current over the ±100-nA range. Over the −4- to +3.5-V compliance range, the error that appears is less than 5%. This accuracy results from using an OP80 op amp from Precision Monolithics in the feedback loop. The OP80 has an I_B of 200 mA typical.

For a given voltage (V_{in}), amp U1A generates an output voltage so that the current through R5 equals V_{in} divided by R_5 (10 MΩ). This current causes a voltage drop across R5, which is sensed by the unity-gain differential amp (U1B and U2). That amp's output is connected to the inverting input of U1A, completing the feedback loop.

The noise in the circuit is of particular concern, especially that produced by resistor R5. The circuit is limited to low-frequency and dc applications as a result of its 400-n V/kHz noise. For applications that don't require the circuit's 10-pA output, lower values of R_5 can be substituted. This will increase the bandwidth.

3- TO 15-V dc-dc CONVERTER

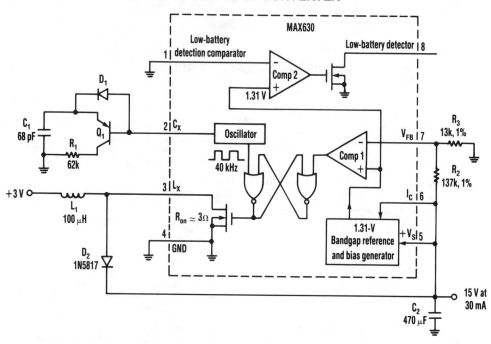

ELECTRONIC DESIGN

Fig. 63-14

This circuit supplies 15 V at 30 mA from a 3-V source. The MAX630 IC is dc-dc converter, Q1 and D1 modify the duty cycle from 50% to 80% to optimize the output power.

CURRENT SUPPLY FOR RTTY MACHINES

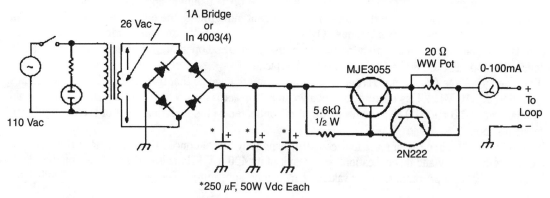

AMATEUR RADIO

Fig. 63-15

Suitable for powering an old Model 15 Teleprinter, this simple power supply uses few parts and is simple to construct. The 20-Ω pot adjusts loop current.

+24-V 1.5-A SUPPLY FROM A +12-V SOURCE

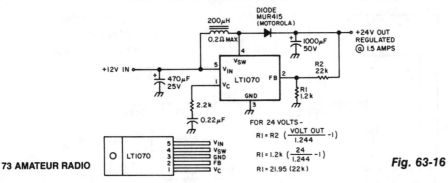

Fig. 63-16

This switching regulator will produce 1.5 A at 24 V from a 12-V auto battery. It operates as a boost-switching supply using an LT 1070. The diode should be a fast-switching type because this regulator operates above 10 kHz.

NEGATIVE SUPPLY FROM A +12-V SOURCE

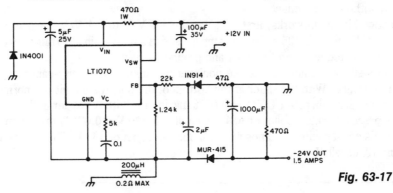

Fig. 63-17

This switching regulator will produce a −24-V/1.5-A source from a +12-V supply, for applications where a positive-grounded source is necessary.

1-A 12-V REGULATED SUPPLY

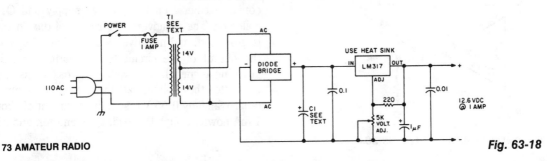

Fig. 63-18

Using a junked VCR power transformer, this circuit supplies 12 V at 1 A.

POSITIVE AND NEGATIVE VOLTAGE POWER SUPPLY

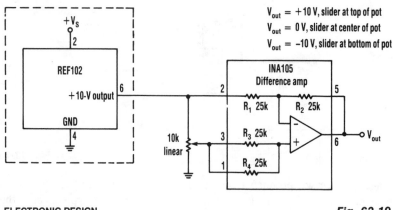

$V_{out} = +10\ V$, slider at top of pot
$V_{out} = 0\ V$, slider at center of pot
$V_{out} = -10\ V$, slider at bottom of pot

ELECTRONIC DESIGN   *Fig. 63-19*

This circuit provides a precision voltage source that can be adjusted through zero to positive and negative voltages, which eliminates reversing connections on the power supply. Also, it is possible to get exactly 0 V, without some offset.

As to how this circuit works, first consider the -1 V/V to $+1$ V/V linear gain-control amp (see the figure). A Burr-Brown INA105 difference amp is used in a unity-gain inverting amp configuration. A potentiometer is connected between the input and ground.

The pot's slider is connected to the noninverting input of the unity-gain amp; this input is typically connected to ground. With the slider at the bottom of the pot, the circuit is a normal-precision unity-gain inverting amp with a gain of -1.0 V/V $\pm 0.01\%$ maximum. With the slider at the top of the pot, the circuit is a normal-precision voltage follower with a gain of ± 1.0 V/V $\pm 0.001\%$ maximum. With the slider in the center, there's equal positive and negative gain for a net gain of 0 V/V. The accuracy between the top and the bottom will usually be limited by the accuracy of the pot.

1-mA CURRENT SINK

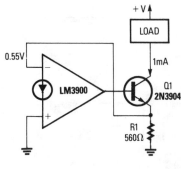

RADIO-ELECTRONICS   *Fig. 63-20*

A fixed current flows through any load that is connected between the positive supply and Q1's collector. The noninverting terminal of the op amp is grounded, and negative feedback flows between the output of the circuit (Q1's emitter) and the inverting terminal. The voltage across R1 is thus equal to the voltage at the inverting terminal (approximately 0.55 V), so a fixed current of about 1 mA flows through the load, Q1's emitter, and R1.

POSITIVE AND NEGATIVE VOLTAGE SWITCHING SUPPLY

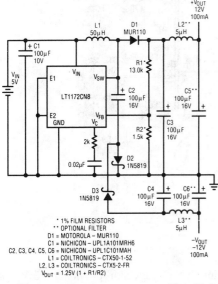

An LT1172 generates positive and negative voltages from a 5-V input. The LT1172 is configured as a step-up converter. To generate the negative output, a charge pump is used. C2 is charged by the inductor when D2 is forward-biased and discharges into C4 when LT1172's power switch pulls the positive side of C2 to ground.

LINEAR TECHNOLOGY   *Fig. 63-21*

LCD DISPLAY CONTRAST CONTROL POWER SUPPLY

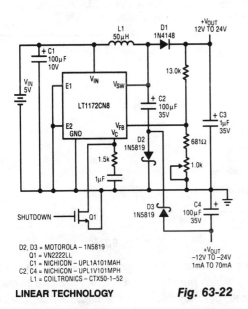

The LT1172 is configured as a step-up converter. C2 is charged by L1 and discharges into C4 when the LT1172's power switch goes to ground. Resistor R3 adjusts the output voltage between −12 and −24 V.

LINEAR TECHNOLOGY   *Fig. 63-22*

5- AND ±12-V ac-POWERED SWITCHING SUPPLY

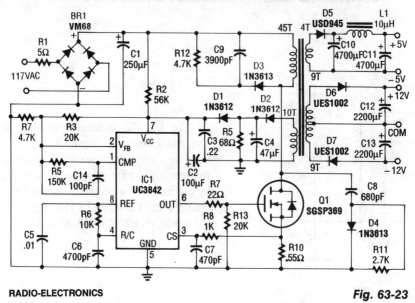

RADIO-ELECTRONICS

Fig. 63-23

This supply uses an SGS-Thomson UC3842 IC in an off-line flyback regulator, providing +5 V at 4 A and ±12 V at 300 mA. This enables a small high-frequency (50 kHz) transformer, to handle large amounts of power that are normally handled by a 60-Hz transformer. Q1 is a 5-A 500-V MOSFET, and the diodes are fast-recovery types. T1 has a 45-turn primary winding of #26 wire. The 12-V windings are each 9 turns of #30 wire, bifilar wound. The 5-V winding is 4 turns of four bifilar #26 wires. The control (feedback) winding is two bifilar, parallel 10-turn, #30 windings. The core is Ferroxcube EC35-3C8 with a 3/8" center leg.

AUXILIARY NEGATIVE dc SUPPLY FOR BIAS OR REFERENCE APPLICATIONS

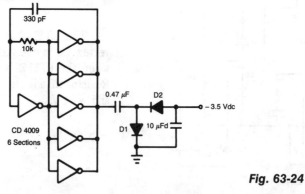

ELEKTOR ELECTRONICS

Fig. 63-24

In this circuit, IC1 (CD4009) is used as a square-wave oscillator at approximately 25 kHz. C1 and R1 set this frequency. C2, D1, D2, and C3 form a p-p rectifier, which outputs about −3.5 Vdc. This circuit should be useful where a small negative dc supply is required, but only positive dc voltages are available.

GASFET POWER SUPPLY

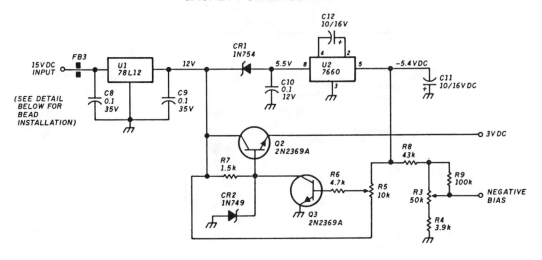

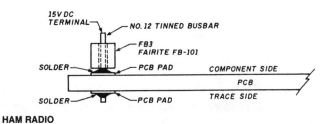

HAM RADIO   **Fig. 63-25**

Suitable for use with the GASFET doubler in this text, and other similar applications, this supply delivers +3 Vdc and 0 to −3 Vdc bias.

FAST DIFFERENTIAL INPUT CURRENT SOURCE

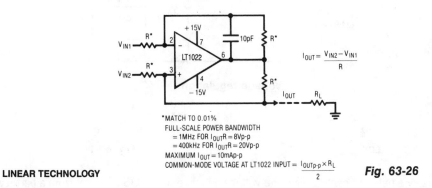

LINEAR TECHNOLOGY   **Fig. 63-26**

An LT1022 op amp used in this configuration can provide a rapidly switched current source. Use the equations in the figure to select component values.

BOOTSTRAPPED AMP CURRENT SOURCE

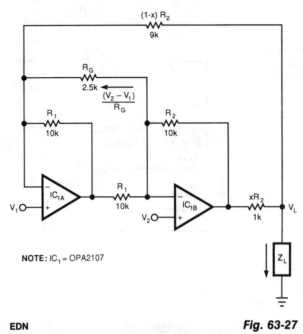

EDN

Fig. 63-27

This circuit responds to the difference between V_1 and V_2. R_G on sets gain. Resistors $XR2$ and $(1-X)$ R2 produce the bootstrap effect. These two resistors convert the circuit's output voltage to a current. IC1 and IC2 are Burr-Brown OPA2107 or equal.

DIODE CMOS STABILIZER

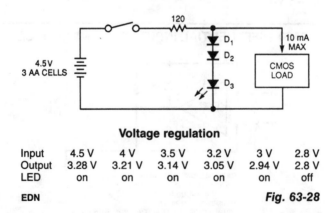

Voltage regulation

Input	4.5 V	4 V	3.5 V	3.2 V	3 V	2.8 V
Output	3.28 V	3.21 V	3.14 V	3.05 V	2.94 V	2.8 V
LED	on	on	on	on	on	off

EDN

Fig. 63-28

The simple diode network can stabilize the voltage supplied to CMOS circuitry from a battery. D1 and D2 must have a combined forward-voltage drop of about 1.5 V. And D3 is an LED with a forward-voltage drop of about 1.7 V. The table shows the network's output voltage as the battery's voltage declines.

MOBILE ±35-V 5-A AUDIO AMPLIFIER SUPPLY

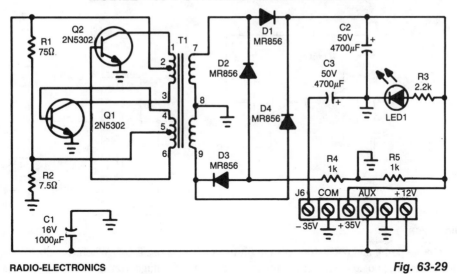

RADIO-ELECTRONICS

Fig. 63-29

This ±35-V supply uses a two-transistor multivibrator with a toroidal transformer. The transformer core is obtainable from Magnetics, Inc. Specifications for T1 are:

	1 mil tape wound	**$1.460'' \times 0.915'' \times 0.345''$**
Core	**Magnetics, Inc.**	**P/N 50029 ID**
Primary	14T CT	#12 AWG
Base Drive	7T CT	#18 AWG
Secondary	19T CT	#12 AWG

ISOLATED 15-V TO 2 500-V SUPPLY

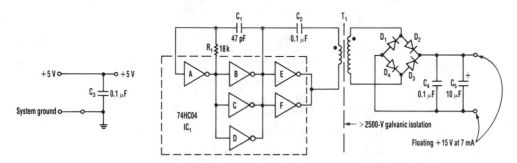

ELECTRONIC DESIGN

Fig. 63-30

A dc-dc converter using a 74HC04 drives T1. T1 is a ferrite-core transformer using a Fair-Rite, Inc. P/N 5975000201 ($\mu_o + 5\,000$) and has a 7-turn primary and a 25-turn secondary. Kynar, #30 wirewrap wire is used. With T1, the circuit isolation is good to 2 500 V.

GET NEGATIVE RAIL WITH CMOS GATES

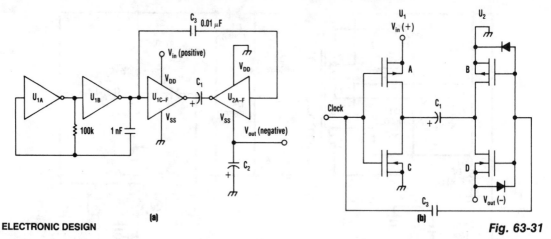

ELECTRONIC DESIGN

(a)

(b)

Fig. 63-31

Using a charge pump and oscillator, this circuit uses a 7-kHz oscillator. When the clock is high, C and D are on, grounding the positive side of C1 and making negative voltage available at the V_{SS} terminal of U2. With $C = C_1 + C_2$, the p-p output ripple is:

$$V_{pp} \text{ ripple} = \frac{V_{in}}{2RFC}$$

Converter output impedance is about 100 Ω and maximum current is 10 mA dc.

3-A SWITCHING REGULATOR

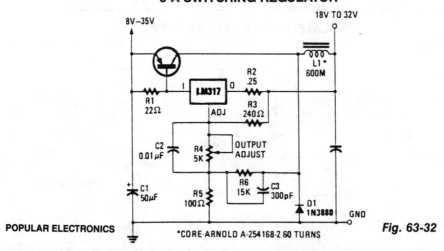

POPULAR ELECTRONICS

*CORE·ARNOLD A·254168·2 60 TURNS

Fig. 63-32

This switching regulator uses an LM317 and a pnp switching transistor of 3- to 5-A rating. L1 is wound on a commercially available core.

64

Power Supplies (High-Voltage)

The sources of the following circuits are contained in the Sources section, which begins on page 674. The figure number in the box of each circuit correlates to the entry in the Sources section.

40-W 120-Vac INVERTER

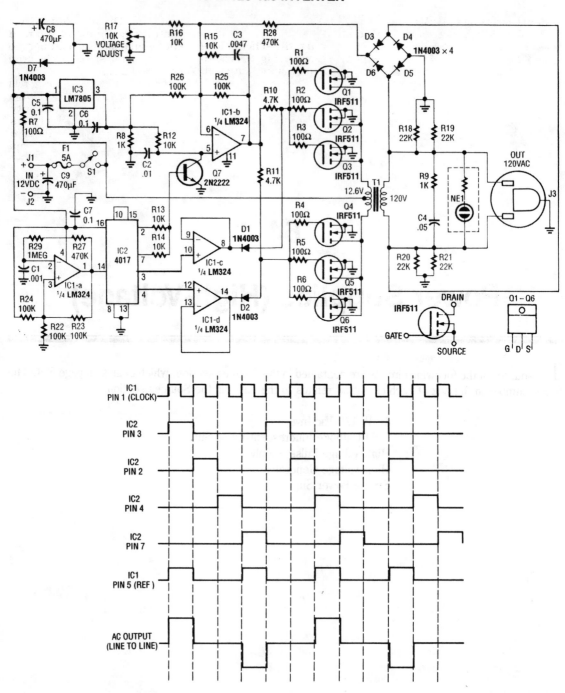

This inverter uses a 12.6-V to 120-V transformer to deliver a quasi-sine wave that has the same rms and peak voltage as a pure sine wave. Q1 to Q6 must be heatsinked. A 1.5″ × 4″ aluminum heatsink was used on the prototype. The transformer should be a 3-A unit. The circuit uses feedback to help regulate the output voltage to 120 Vac. Notice that the output frequency is 75 Hz to avoid saturating the core of T1.

COLD-CATHODE FLUORESCENT-LAMP SUPPLY

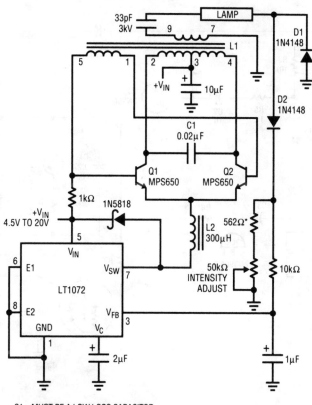

C1 = MUST BE A LOW LOSS CAPACITOR.
METALIZED POLYCARB
WIMA FPK 2 (GERMAN) RECOMMENDED.
L1 = SUMIDA 6345-020 OR COILTRONIX CTX110092-1.
PIN NUMBERS SHOWN FOR COILTRONIX UNIT
L2 = COILTRONIX CTX300-4
* = 1% FILM RESISTOR
DO NOT SUBSTITUTE COMPONENTS

LINEAR TECHNOLOGY

Fig. 64-2

For back-lit LCD displays, this supply will drive a lamp. LT1072 drives Q1 and Q2, and a sine wave appears across C1. L1 is a transformer that steps up this voltage to about 1 400 V. D1 and D2 detect lamp current and form a feedback loop to the LT1072 to control lamp brightness.

HIGH-VOLTAGE PULSE SUPPLY

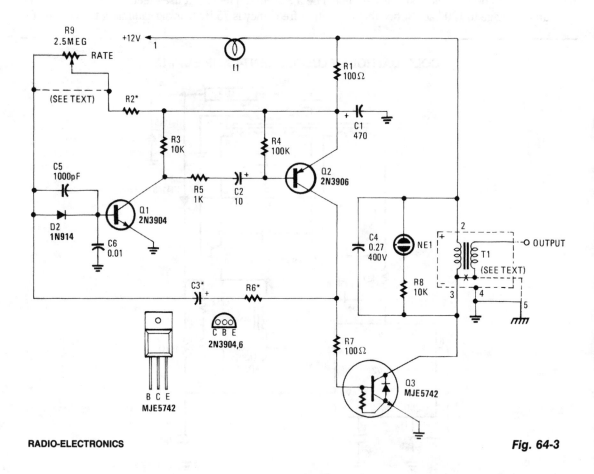

Fig. 64-3

This high-voltage pulse supply will generate pulses up to 30 kV. Q1 and Q2 form a multivibrator in conjunction with peripheral components R1 through R6 and C1, C2, C3, C5, C6, and D2. R9 adjusts the pulse repetition rate. R2 should be selected to limit the maximum repetition rate to 20 Hz. I1 is a type 1156 lamp used as a current limiter. R9 can be left out and R2 selected to produce a fixed rate, if desired. Try about 1 MΩ as a start.

Q3 serves as a power amplifier and switch to drive T1 (an automotive ignition coil). NE1 is used as a pulse indicator and indicates circuit operation. Because this circuit can develop up to 30 kV, suitable construction techniques and safety precautions should be observed.

HIGH-VOLTAGE GENERATOR

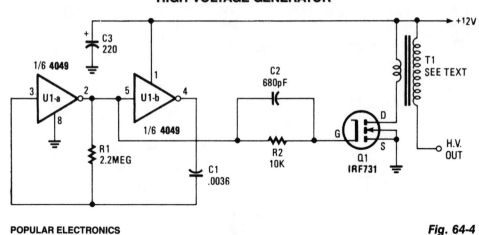

POPULAR ELECTRONICS

Fig. 64-4

A 4049 Hex inverter drives an IRF731 hex FET. The 4049 is configured as an oscillator. Q1 should be heatsinked. T1 is an auto ignition coil.

STROBE POWER SUPPLY

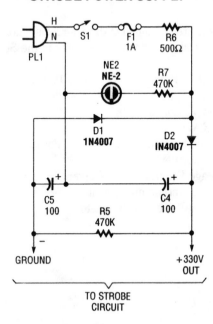

TO STROBE
CIRCUIT

RADIO-ELECTRONICS

Fig. 64-5

This 330-V power supply is a simple voltage doubler which provides 330 Vdc for a strobe circuit. This supply is not isolated from the power lines and extreme caution is advised.

65

Power Supplies (Variable)

The sources of the following circuits are contained in the Sources section, which begins on page 674. The figure number in the box of each circuit correlates to the entry in the Sources section.

2.5-A/1.25-to 25-V REGULATED POWER SUPPLY

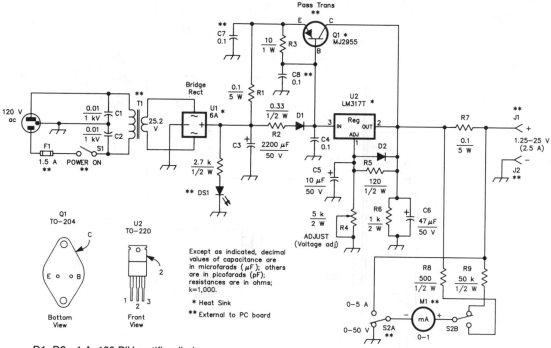

D1, D2—1-A, 100-PIV rectifier diode.
DS1—Red LED.
F1—1.5-A, 3AG fuse in chassis-mount holder.
J1, J2—Standard five-way binding post, one red, one black.
M1—Milliammeter, 0-1 mA dc.
Q1—NPN power transistor MJ2955 (Radio Shack) or equiv device with a +70-V, 10-A, 150-W rating in a TO-204 case.
R1, R2, R7—5-W wire-wound resistor. See Notes 3 and 4 for source. or, use 17 inches of no. 28 enam wire, single-layer wound, on a 10-kΩ, 1-W carbon-composition resistor for R1 and R7. For R2, use 36 inches of no. 30 enam wire on a 10-kΩ, 1-W carbon composition resistor (scramble wound).

R-4—Panel-mount, 5-kΩ, 2-W or 5-W potentiometer, carbon or wire wound (See Note 8).
R8, R9—See text.
S1—SPST toggle switch.
S2—DPDT toggle or rotary wafer switch.
T1—25.2-V, 2.75-A power transformer (see text).
U1—6-A, 200 PIV bridge rectifier with heat sink. See text.
U2—LM317T +1.25- to 30-V, 1.5-A TO-220 regulator. Use an LM317HVK (TO-204 case) for dc output voltage greater than 40. See text.

QST

Fig. 65-1

This power supply uses an LM317J adjustable regulator and an MJ2955 pass transistor. Q1 and U2 as well as U1 should be heatsinked. A suitable heatsink would typically be 4″ × 4″ × 1″ fins, extruded type, because up to 65 W dissipation can occur. R8 and R9 should be 1% types or selected from 5% film types with an accurate ohmmeter. Capacitors are disc ceramic except for those with polarity marked, which are electrolytic.

DUAL 0-to 50-V/5-A UNIVERSAL POWER SUPPLY

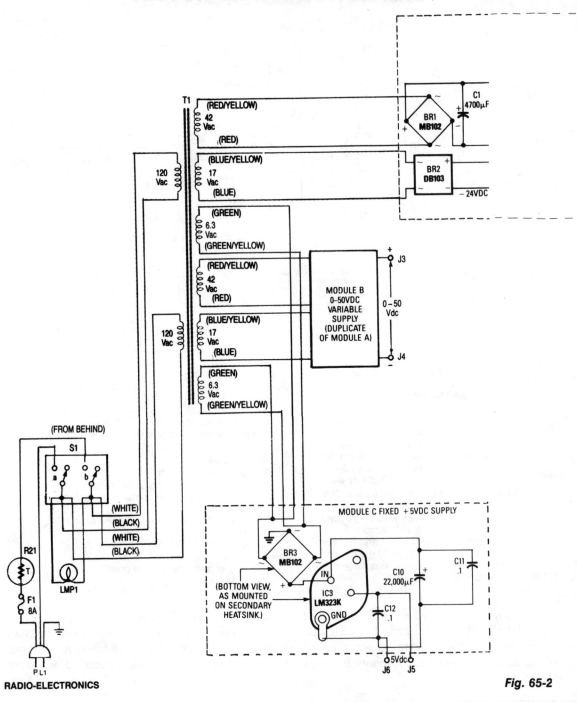

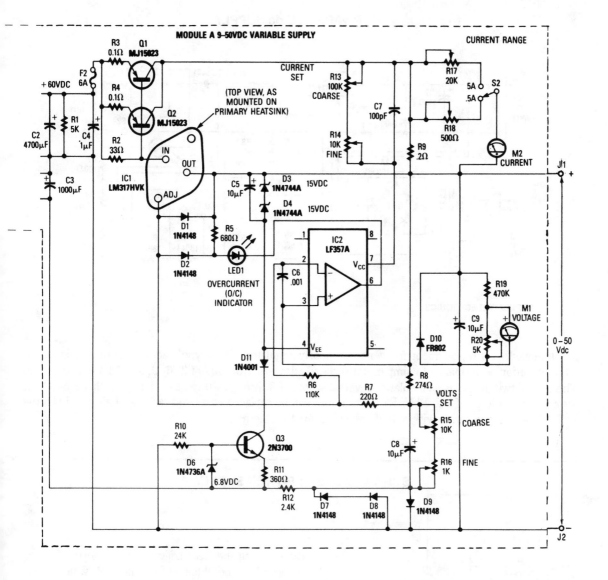

T1 has two primaries and six secondaries; the two 120-VAC primaries and 6.3-VAC secondaries are in parallel. Modules A and B are identical; hence, only Module A's parts are called out. Module C is wired point-to-point on the IC3 heatsink.

STEP VARIABLE dc SUPPLY

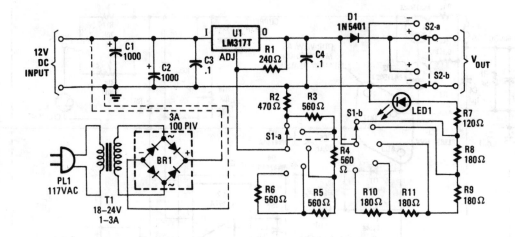

POPULAR ELECTRONICS

Fig. 65-3

Intended as a replacement for generally poorly regulated "wall-type" ac/dc adapters, this circuit offers superior performance to simple, unregulated adapters. Voltages of 3, 6, 9, and 12 V are available. The DPDT switch serves as a polarity-reversal switch. R2 through R6 can be replaced with a 2.5-kΩ pot for a variable voltage of 1 to 12 V. R7 through R10 can be replaced by a fixed resistor of about 1 kΩ if the LED1 brightness variation with output voltage is not a problem.

SCR VARIABLE dc POWER SUPPLY

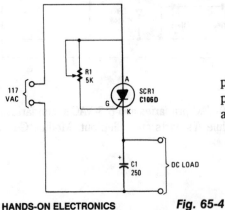

By adjusting the SCR trigger point, a form of phase control can be obtained from an SCR, which produces a dc output, depending on the conduction angle.

HANDS-ON ELECTRONICS

Fig. 65-4

SWITCH-SELECTED FIXED-VOLTAGE POWER SUPPLY

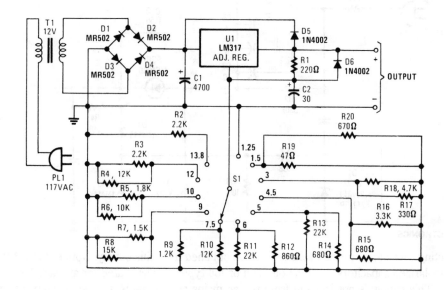

POPULAR ELECTRONICS

Fig. 65-5

This supply can serve as a battery eliminator for various devices (such as tape recorders, small radios, clocks, etc.).

S1 selects a resistance that is predetermined to provide a preselected output voltage. In this circuit, various commonly used supply voltages produced by batteries were chosen, but any voltages up to the rating of T1 (approximately) can be produced by choosing an appropriate resistor.

The resistor value is given by:

$$R_X = R_1 \left(\frac{V_{OUT}}{1.25} - 1 \right)$$

$$R_1 = 220 \ \Omega$$

Remember to provide adequate heatsinking for U1.

TRANSFORMERLESS POWER SUPPLY

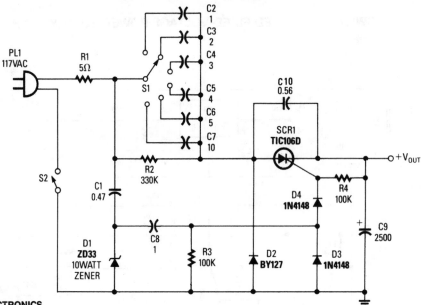

POPULAR ELECTRONICS

Fig. 65-6

By selecting capacitors, various voltages can be obtained from this supply. Notice that C2 through C7 must be nonpolarized capacitors, such as oil-filled or foil types (Mylar) rated for at least 250 Vac.

Warning: This supply is not isolated from the ac mains and presents a serious safety hazard if body contact is made anywhere to this circuit or anything that is powered by it. Use only for applications where contact is avoided or impossible.

VOLTAGE-PROGRAMMABLE CURRENT SOURCE

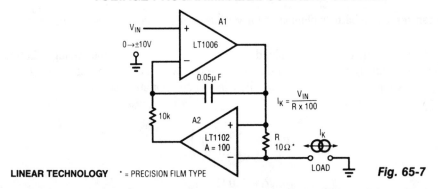

LINEAR TECHNOLOGY   * = PRECISION FILM TYPE

Fig. 65-7

This circuit is a programmable current source in which op amp LT1102 (Linear Technology Corp.) is used in conjunction with LT1006 op amp. A1, biased by V_{in}, drives current through R (10 Ω) and the load. A2 senses this current and controls A1. The 10-kΩ resistor and 0.05-μF capacitor sets the frequency response of the circuit.

SIMPLE DARLINGTON REGULATOR

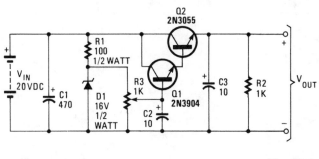

POPULAR ELECTRONICS

Fig. 65-8

A Darlington pair is used as an emitter-follower that produces about 1.2 V less than the wiper voltage of R3. Output voltage for this circuit will range from close to zero to about 14.5 V.

0-TO-50-V VARIABLE REGULATOR

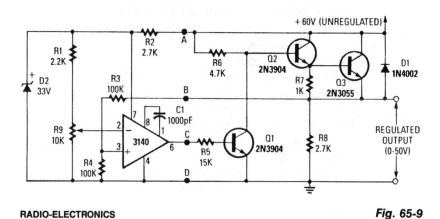

RADIO-ELECTRONICS

Fig. 65-9

A CA3140 op amp compares the regulator output to a reference voltage, depending on the setting of R9. The output voltage will be nominally twice the voltage between the plus input of the CA3140 and ground. R1 and R9 allow 0 to 50 V.

66

Power-Supply Monitors

The sources of the following circuits are contained in the Sources section, which begins on page 674. The figure number in the box of each circuit correlates to the entry in the Sources section.

ISOLATED VOLTAGE SENSOR

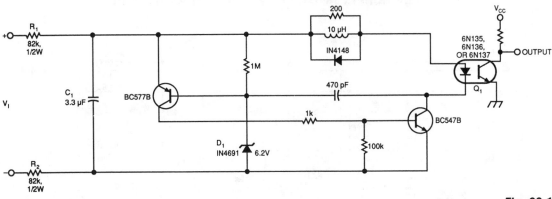

EDN

Fig. 66-1

A simple voltage-controlled oscillator (VCO), coupled to your instrumentation by an optoisolator, allows you to measure high voltages. The component values suit a 0- to 600-V input range (power dissipation in R1 and R2 set a limit on the input-voltage range). The circuit's linearity is not an issue, because you can linearize its output in software.

The input voltage (V_1), charges capacitor C1 until zener diode D1 conducts. Then, the zener diode triggers an "avalanche" circuit that discharges C1 into optocoupler Q1. After C1 discharges, the charging cycle repeats. C1 also averages the sensed-voltage level, which thereby provides noise immunity.

The optocoupler's output is a pulse train whose frequency increases with increasing input voltage. To develop a linearizing equation for the circuit, measure its output at two convenient, widely spaced input voltages. Then plug the resulting periods into this second-order polynomial approximation and solve the two simultaneous equations for the two constants, k_1 and k_2:

$$V_1 = [k_1/T20 + (k_2/T) + V_z]$$

V_z is the zener voltage of D1.

CIRCUIT BREAKER TRIPPER

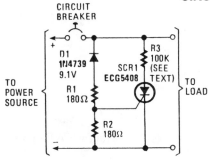

This tripper is designed to protect against overvoltages. D1 conducts over 9.1 V and triggers SCR1. R3 is chosen to draw enough current to trip the breaker.

POPULAR ELECTRONICS

Fig. 66-2

BACKUP SUPPLY ACTIVATES BY DROP-IN MAIN SUPPLY

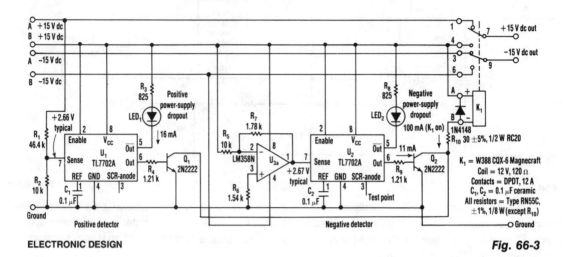

ELECTRONIC DESIGN

Fig. 66-3

A supply monitor using two TL7702A chips monitors the ±15-V supplies and activates the backup supply in case of a voltage drop. Although the chips are intended for use as reset controllers in microprocessor systems, they work well in this application.

CONSTANT-CURRENT TEST LOAD

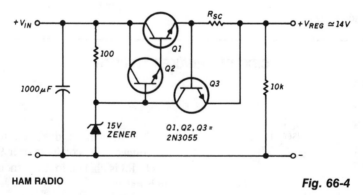

HAM RADIO

Fig. 66-4

This circuit will supply a constant load of 500 mA to 1.5 A. R4 controls the current while R3 provides fine adjustment.

POWER BUFFER BOOSTS REFERENCE CURRENT

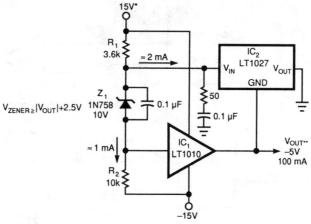

EDN

NOTES:
* = BYPASS SUPPLIES WITH 10 μF AND 0.1 μF CAPACITORS
** = OUTPUT WILL OSCILLATE WITH LOW ESR CAPACITORS

Fig. 66-5

A method of boosting the output current of a reference and also protecting against overloads is shown in Fig. 66-5. IC1 acts as a power buffer. The LT1027 forces the output of V_{OUT} and ground to be 5 V. The RC damper (50 Ω and 0.1 μF) provides loop stability. The output might oscillate if low ESR capacitors are connected to it, so use aluminum electrolytic or tantalum capacitors instead of ceramic or mylar.

POWER SUPPLY MONITOR/MEMORY PROTECTOR

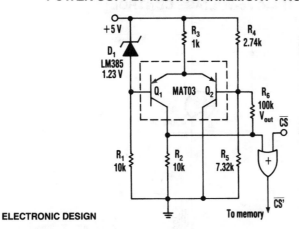

ELECTRONIC DESIGN

Fig. 66-6

This circuit detects low-voltage supply conditions, down to 0.6 V. D1 sets the trip point of the circuit. The circuit is useful to protect memory circuits from accidental writes in the event of power-supply low-voltage conditions, which cause other circuits to turn off, etc. Response time is about 700 ns. R6 provides some hysteresis to ensure clean transitions.

TUBE AMPLIFIER ISOLATES HIGH VOLTAGES

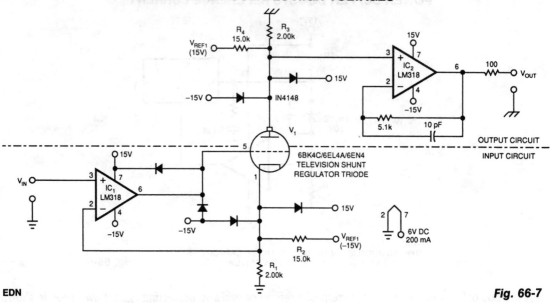

EDN

Fig. 66-7

This amplifier can transfer dc-to 5-MHz signals across a potential difference of 25 000 V. This circuit can be used in CRT displays, high-voltage applications, etc. Notice that the tube must be shielded because the tube will generate X-rays. Typically, about 0.1″-thick sheet metal would be used.

TRIAC ac-VOLTAGE CONTROL

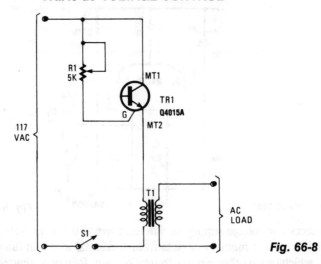

HANDS-ON ELECTRONICS

Fig. 66-8

By using a variable resistor in the gate of TR1, variable conduction angles can be achieved via R1.

POLARITY-PROTECTION RELAY

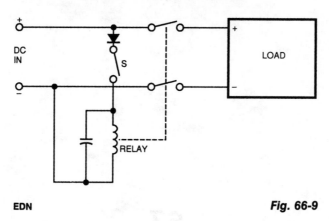

EDN

Fig. 66-9

A diode prevents the relay from applying power if polarity is reversed.

67

Probes

The sources of the following circuits are contained in the Sources section, which begins on page 675. The figure number in the box of each circuit correlates to the entry in the Sources section.

3-In-1 Test Set (Logic Probe, Signal Tracer, and Injector)
Logic Tester
Universal Test Probe
4-Way Logic Probe
Active RF Detector Probe
Single-IC Logic Probe
Logic Probe
Logic Probe

3-IN-1 TEST SET (LOGIC PROBE, SIGNAL TRACER, AND INJECTOR)

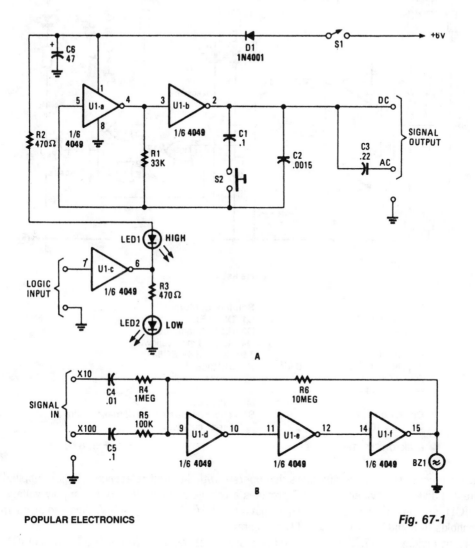

POPULAR ELECTRONICS

Fig. 67-1

This circuit for a test set contains a signal injector (U1A/U1B) and associated components, a logic probe (U1C) and an audio amplifier. S1 selects either 10-kHz or 100-Hz output. U1D, U1E, and U1F form an audio amplifier that drives a piezo sounder element without an internal driver so that it functions as a piezoelectric speaker.

LOGIC TESTER

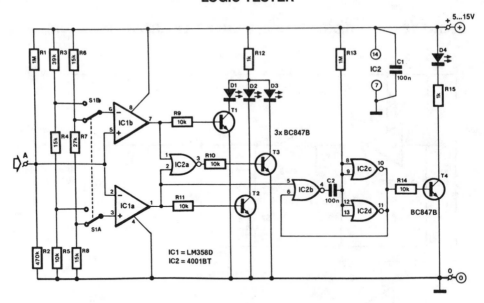

Parts list

Resistors:
R1, R13 = 1 MΩ
R2 = 470 kΩ
R3 = 39 kΩ
R4, R6, R8 = 15 kΩ
R5, R9, R10, R11, R14 = 10 kΩ
R7 = 27 kΩ
R12, R15 = 1 kΩ

Capacitors:
C1, C2 = 100 nF

Semiconductors:
D1, D2 = LED, 3 mm, green
D3 = LED, 3 mm, red
D4 = LED, 3 mm, yellow
T1, T2, T3, T4 = BC847
IC1 = LM358
IC2 = 4001

Miscellaneous:
S1 = sub-miniature switch, 2 make-before-
    break contacts

ELEKTOR ELECTRONICS   *Fig. 67-2*

The input consists of two comparators that operate with different reference voltages supplied by separate potential dividers. Divider R3/R4/R5 provides a voltage of about 40% of the supply voltage, U_{CC}, to pin 6 of IC1B and one of about 16% of U_{CC} to pin 3 of UC1A. When $U_{CC} = 5$ V, these voltages are exactly the thresholds (0.8 and 2.0 V) of the TTL comparators.

Similarly, divider R6/R7/R8 provides voltages of 23% of U_{CC} and 73% of U_{CC} to pin 3 of IC1A and pin 6 of IC1B respectively; these levels correspond to the standard threshold for CMOS comparators.

The voltage to be measured U_a, is applied to pin 5 of IC1B and pin 2 of IC1A and compared with the respective reference. The output of comparator IC1B goes high when U_a exceeds the reference, whereas the output of IC1A goes high when U_a lies below the voltage at pin 3.

The comparators are followed by driver stages, T1 and T2, for the LED display (D1 for "high" and D2 for "low") and also NOR gate IC2A that switches on T3 when the output of both comparators is low, that is, when it is undefined. This state is indicated by D3.

The remaining three gates in IC2 form a monostable. During quiescent operation, U_{CC} is present at the input of inverter IC2C. The output of the inverter is then low, T4 is off and D4 is out. Pin 4 of IC2B is also high, but this state changes when a pulse arrives at pin 5. The output of IC2B then goes low, C2 discharges, the inverter toggles, T4 is switched on, and D4 lights. This state is unstable, however, because C2 recharges via R13. Although the pulse at pin 5 might be very short, the time constant R_{13}/C_2 lengthens it to about 100 ms.

The supply voltage can lie between 5 and 15 V. At 5 V, the circuit draws a current of about 15 mA. The input impedance of the tester is of the order of 330 kΩ.

UNIVERSAL TEST PROBE

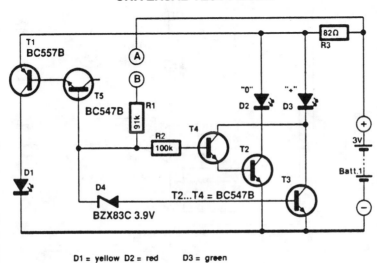

D1 = yellow   D2 = red   D3 = green

ELEKTOR ELECTRONICS

Fig. 67-3

The compact test probe provides rapid "measurement" of voltage levels at digital gates, fuses, diodes, batteries, and others. It does not provide absolute values, but rather it provides a good indication of correct operation or otherwise.

Measurements are carried out with pins A and B. If the potential difference between A (the reference pin) and B is 1.9 to 2.0 V, D2 will light. If the voltage at B is ≤ 1.4 V higher than that at A, D3 will light. Finally, if the potential at B is ≥ 11 V, with respect to that at A, D1 will light.

Transistor T5 is used as a zener diode. The probe allows the measurement of alternating voltage. The maximum input voltage is highly dependent on the dissipation allowed in R1. For example, when this resistor is a 0.5-W type, the input voltage can be as high as 200 Vrms.

The current drawn by the circuit depends on the number of lighting LEDs: it is not more than 10 mA at a supply voltage of 3 V. In quiescent operation, the current is so low (about 5 μA) that an on/off switch is not necessary.

4-WAY LOGIC PROBE

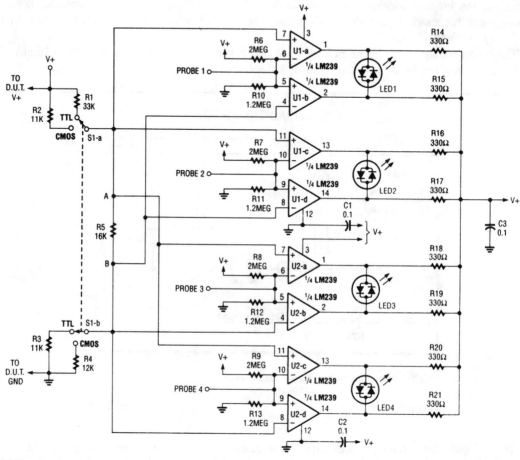

POPULAR ELECTRONICS

Fig. 67-4

This logic probe has four channels and uses two IF quad comparators to drive four bicolor LEDs. S1 and S1B program the comparator trip levels for TTL and CMOS. R6 through R13 bias the probe inputs to prevent the probe from indicating a HIGH for an open circuit. An open circuit will produce an OFF indication on the LED. The LEDs will indicate one color for high, the other color for low, and intermediate colors for pulsing (assuming a duty cycle between 30 and 70%). The color that corresponds to HIGH or LOW depends on how you connect the LEDs.

ACTIVE RF DETECTOR PROBE

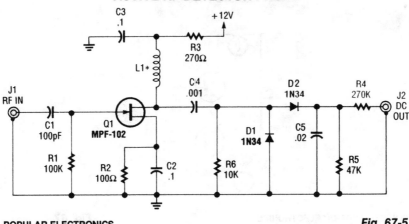

POPULAR ELECTRONICS

Fig. 67-5

An MFP102 FET is used as a wideband amplifier. L1 can be 100 μH for 30 to 100 MHz or 1 000 μH for 2 to 30 MHz. For LF, (less than 3 MHz), use a 2.5-μH RF choke. This probe will work as a relative indicator device, but it can be calibrated over a frequency range, if needed.

SINGLE-IC LOGIC PROBE

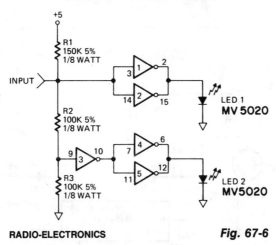

RADIO-ELECTRONICS

Fig. 67-6

This logic probe uses a CD4009 CMOS hex inverter. The characteristic high-input impedance of CMOS gives the advantage of not loading the circuit being tested. Because the output of the inverters is not specified at either a high or low level with a floating input, an input-bias network produces lows at both input inverter pairs if the input is open or at less than 2 V. Resistor R3 holds the input of inverter 3 low which makes the output of inverters 4 and 5 low and will not permit LED 2 to light. At the same time, R1 holds the inputs of inverters 1 and 2 high so that their output is low and LED 1 will not light. If the probe input is touched to a logic low, the output of inverter 3 is held high by R3 and inverter 1 and 2 are brought low, which causes their outputs to go high and turns on LED 1. With no input, both LEDs should be off.

LOGIC PROBE

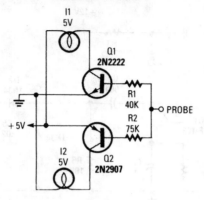

Fig. 67-7

By connecting this circuit to a logic device that's under power, you can get an indication as to its status. If the circuit is open, neither of the test lamps will light. If the circuit is grounded, the low (or zero) lamp will light. If 3 to 6 V are present, the high-voltage lamp will light.

Other than its application in logic testing, the probe is also convenient for checking supply voltages and grounds. You can select resistors to turn the lamps on at any desired threshold voltage within the component limits.

LOGIC PROBE

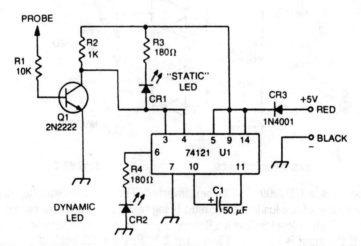

Fig. 67-8

The "static" LED indicates a logic level of 1 when lit. The "dynamic" LED indicates a logic 1 *pulse*. A 4-V, 100-ns pulse will be stretched to about 50 ms so that it can be easily seen.

68

Pulse Circuits

The sources of the following circuits are contained in the Sources section, which begins on page 675. The figure number in the box of each circuit correlates to the entry in the Sources section.

Inexpensive Pulse Generator for Logic Troubleshooting
Negative Pulse Stretcher
Low-Power Ring Counter (less than 6 mW)
Transistor Pulse Generator
Free-Running Pulse Generator
Stable Start-Stop Oscillator
Positive Pulse Pulse Stretcher
555 Pulse Generator
Fast Low Duty-Cycle Pulse Oscillator
Simple Pulse Stretcher
Delayed Pulse Generator

INEXPENSIVE PULSE GENERATOR FOR LOGIC TROUBLESHOOTING

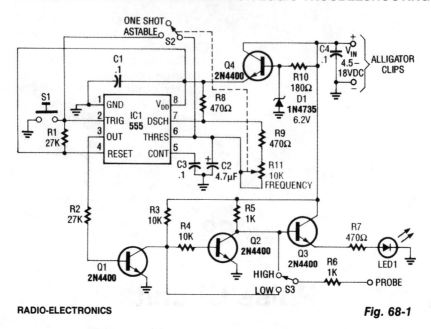

RADIO-ELECTRONICS

Fig. 68-1

Built around a 555 timer IC, this pulse generator can be built into a probe for logic troubleshooting. R11 is frequency controlled and gives a range of about 5 to 200 Hz. C2 can be reduced for higher frequencies. S2 selects one shot or pulse operators. Q1 and Q2 provide a fast rise-time pulse, which acts as a clipper amplifier. Q4 acts as a regulator. The supply can be anything from about 4.5 to 18 Vdc. LED1 is an indicator that shows circuit operation.

NEGATIVE PULSE STRETCHER

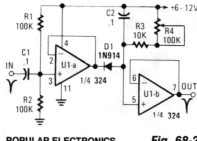

POPULAR ELECTRONICS

Fig. 68-2

U1A acts as an amplifier, which drives D2 and charges C2. U1B acts as a voltage follower. R3 and R4 determine the amount of stretch that the input pulse receives. C2 can be charged to accommodate different pulse rates.

LOW-POWER RING COUNTER (LESS THAN 6 mW)

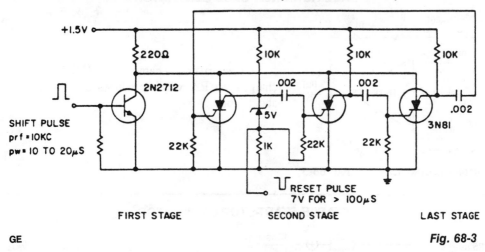

GE

Fig. 68-3

The ring counter operates from 1.0 to 6.0 V and requires only 6 mW at 1.5 V. The reset pulse turns on the first stage with its trailing edge. The maximum shift pulse width increases with voltage and approaches 70 μs for a 6.0-V supply. Minimum pulse width is 10 μs.

TRANSISTOR PULSE GENERATOR

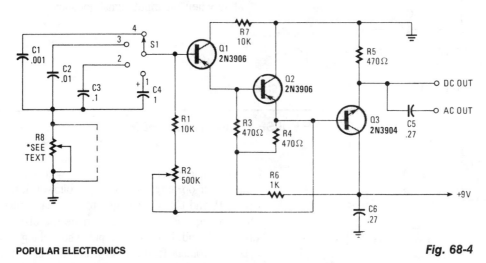

POPULAR ELECTRONICS

Fig. 68-4

Seven-V narrow pulses from 2 Hz to 50 kHz are produced by this circuit. C1 through C4 provide frequency ranges in decode steps. R1 and R2 control the charging time of C1 through C4. R2 is a potentiometer used to set the frequency. R8 controls pulse width. Pulse width varies from 7 μs to 10 ms. Depending on the frequency, R8 can be deleted if it is not needed.

FREE-RUNNING PULSE GENERATOR

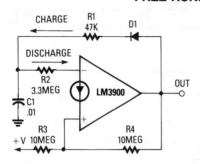

C1 alternately charges via R1/D1 and discharges via R2, which produces a duty cycle of about 1:60.

RADIO-ELECTRONICS   **Fig. 68-5**

STABLE START-STOP OSCILLATOR

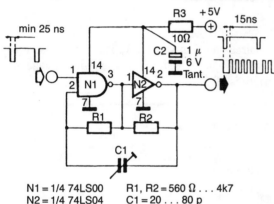

N1 = 1/4 74LS00   R1, R2 = 560 Ω . . . 4k7
N2 = 1/4 74LS04   C1 = 20 . . . 80 p

Oscillators that generate a predetermined number of pulses are often required in applications such as video work. This oscillator starts 13 ms after the control signal goes high and stops immediately when the input signal goes low.

ELEKTOR ELECTRONICS   **Fig. 68-6**

POSITIVE PULSE PULSE STRETCHER

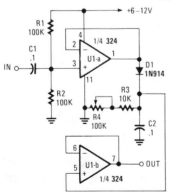

A simple pulse stretcher built with two sections of an op amp uses voltage follower U1A to drive D1 and C2. C2 charges to the peak value of pulse voltage. R3 and R4 determine the discharge time of C2 and therefore the pulse stretching. U1B acts as a voltage follower. Typically this circuit can stretch a pulse by a factor of 50. C2 can be charged to accommodate different pulse rates.

POPULAR ELECTRONICS   **Fig. 68-7**

555 PULSE GENERATOR

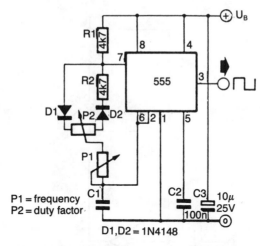

P1 = frequency
P2 = duty factor

D1,D2 = 1N4148

ELEKTOR ELECTRONICS   **Fig. 68-8**

This approach to using a NEC555 timer uses two diodes to set the charge and discharge timer of capacitor, which gives the circuit a variable duty factor.

FAST LOW DUTY-CYCLE PULSE OSCILLATOR

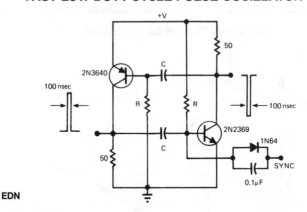

EDN   **Fig. 68-9**

This simple and symmetrical free-running generator has a 50-Ω output impedance, a pulse width of 100 ns and complementary outputs that swing essentially from ground to the power-supply voltage. Moreover, it functions with a power supply range from <1 to >15 V and maintains a low voltage and temperature drift while consuming little power.

For oscillation to occur, each transistor must have a gain greater than unity. This restricts the value of R to a range of 1 kΩ to 1 MΩ, because the gain will be less than unity when the transistor is saturated or when beta is low as a result of small collector currents. The two RC timing networks do not have to match because the RC with the longest time constant will determine the frequency of oscillation.

SIMPLE PULSE STRETCHER

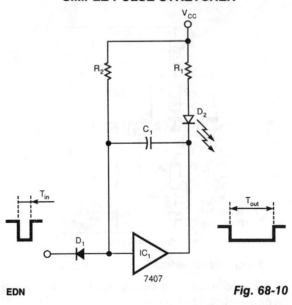

EDN

Fig. 68-10

A single gate (open collector, noninverting) produces a simple one-shot to produce a pulse that stretches equal to the pulse duration, plus the R_1C_1 time constant. R2 is a pull-up resistor to keep the gate's input high while waiting for a pulse.

DELAYED PULSE GENERATOR

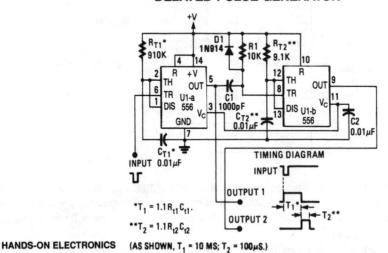

HANDS-ON ELECTRONICS   (AS SHOWN, T_1 = 10 MS; T_2 = 100μS.)

Fig. 68-11

This circuit produces a delayed pulse width of $1.1(R_{t2}C_{t2})$. The delay is given by $T = 1.1(R_{t1}C_{t1})$. A 556 dual timer can be used.

69

Radar Detectors

The sources of the following circuits are contained in the Sources section, which begins on page 675. The figure number in the box of each circuit correlates to the entry in the Sources section.

Radar Detector
Deluxe Radar Detector

RADAR DETECTOR

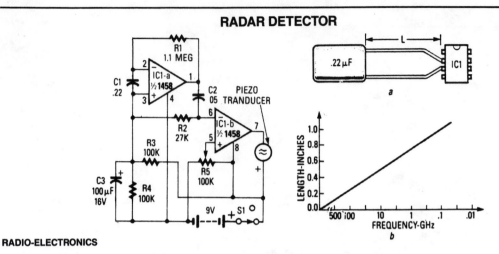

RADIO-ELECTRONICS

Fig. 69-1

A simple detector of radar signals uses the junctions of the input transistors of an LM1458 op amp as RF detectors. The leads of C1 act as an antenna and should be about 0.4″ long measured from the IC package. Detected audio components are further amplified by IC1A and IC1B, and drive a piezo transducer. Mount the circuit so that incident RF energy will not be blocked.

DELUXE RADAR DETECTOR

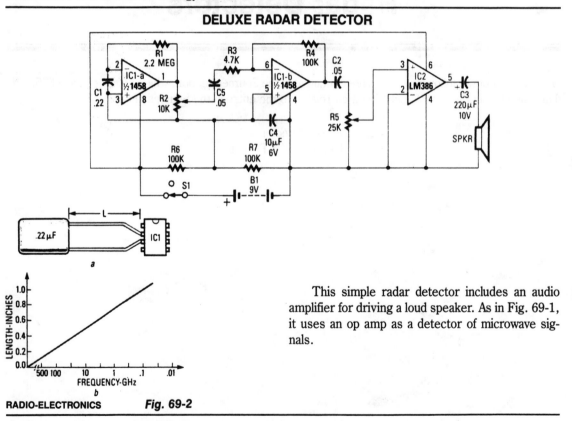

This simple radar detector includes an audio amplifier for driving a loud speaker. As in Fig. 69-1, it uses an op amp as a detector of microwave signals.

RADIO-ELECTRONICS

Fig. 69-2

70

Ramp and Staircase Generators

The sources of the following circuits are contained in the Sources section, which begins on page 675. The figure number in the box of each circuit correlates to the entry in the Sources section.

Digital Sawtooth Generator
Simple Staircase Generator
Ramp Generator
Sawtooth Generator for Sweep Generators
Stepped Waveform Generator

DIGITAL SAWTOOTH GENERATOR

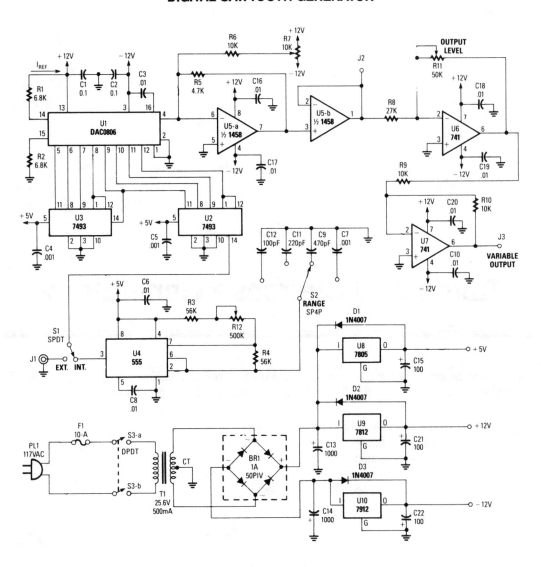

POPULAR ELECTRONICS

Fig. 70-1

This straightforward circuit uses an NE555 timer as an oscillator, a 7493 counter chain and a DAC0806 D/A converter. The counter output feeds the DAC, producing a linearly increasing voltage, which drives an op amp and a dc-level insertion circuit. The clock frequency must be 256 times the desired sawtooth output frequency.

SIMPLE STAIRCASE GENERATOR

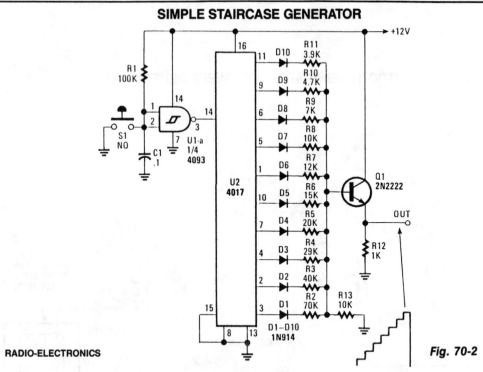

RADIO-ELECTRONICS

Fig. 70-2

U2 is a decade counter /divider. U1 is used as a switch debouncer. For a self-generating system, connect a resistor between pins 2 and 3 of a U1 value that should be between 10 kΩ and several MΩ, depending on desired frequency. C1 can also be varied to change frequency. Also, S1 can be omitted in the self-generating version.

RAMP GENERATOR

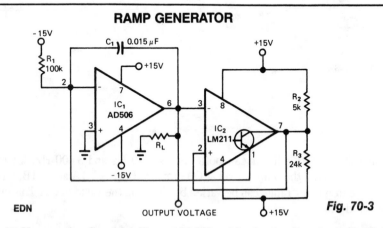

EDN

Fig. 70-3

Providing a 0-to 10-V excursion from 0.4 Hz to 100 kHz, this circuit offers both simplicity and small size. The negative current through R1 produces the ramp's positive slope and causes the output of IC1 to increase linearly toward the +15-V rail.

445

SAWTOOTH GENERATOR FOR SWEEP GENERATORS

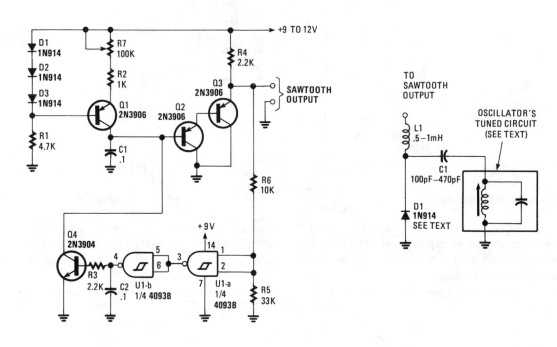

RADIO-ELECTRONICS

Fig. 70-4

This circuit will generate a linear sawtooth between 30 Hz and 3 000 Hz. Q1 is a constant-current source that charges C1 until the output level at Q3 emitter triggers U1A and U1B, which turns on Q4 and discharges C1. The frequency range can be varied by changing the value of C1. This circuit should be good to several tens of kHz.

STEPPED WAVEFORM GENERATOR

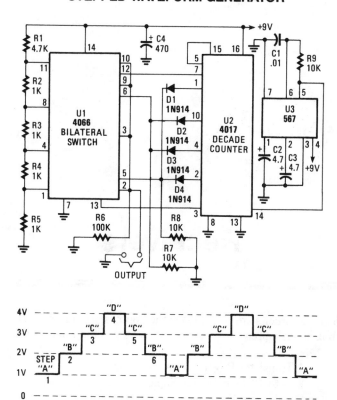

The output of the stepped waveform generator is made up of 3-up and 3-down steps in 1-volt increments (much like the output of a digital-to-analog converter). Switch trigging for the 4066 is controlled by a 4017 decade counter/divider (U2), which is clocked by a square-wave generator built around a 567 PLL (phase-locked loop).

POPULAR ELECTRONICS

Fig. 70-5

A decode counter (U2) is used to perform sequential switching via a CD 4066. Analog switch is to generate a waveform. The clock is a 567 PLL or other VCO chip.

71

Receivers

The sources of the following circuits are contained in the Sources section, which begins on page 675. The figure number in the box of each circuit correlates to the entry in the Sources section.

1-Transistor Regenerative Receiver
3.5- to 10-MHz Simple Superheterodyne Receiver
Simple Low-Frequency Receiver
Reflex Radio Receiver
TRF Radio
Old-Time Radio
Pulse-Frequency Modulated Receiver
Shortwave Receiver
Simple AM Radio

1-TRANSISTOR REGENERATIVE RECEIVER

RECEIVER

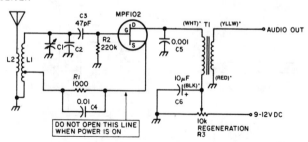

T1 = ANY AUDIO TRANSFORMER AS LONG AS
PRIMARY IS GREATER THAN 1kΩ

$T50-2 \quad N = 100\sqrt{L/50}$
N = NUMBER OF TURNS
L = INDUCTANCE IN μH

*COLOR CODED CONNECTIONS FOR 1:1
AUDIO TRANSFORMER

L1 & C1 = RESONANT FREQUENCY
USE FORMULA

$$f = \frac{10^6}{2\pi\sqrt{LC}}$$

TAP 25% FROM COLD END

L2 = 3 TURNS AROUND L1

COIL FORM CAN BE ANYTHING
EVEN T50-2 TOROID

$$f = \frac{10^6}{2\pi\sqrt{LC}} \qquad L = \frac{10^6}{39.5 \cdot f^2 \cdot C} \qquad C = \frac{10^6}{39.5 \cdot f^2 \cdot L}$$

f = FREQUENCY IN MHz
L = INDUCTANCE IN μH
C = CAPACITANCE IN pF

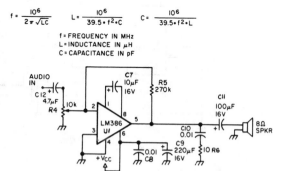

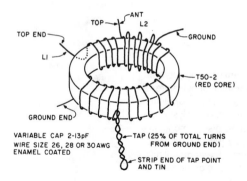

VARIABLE CAP 2-13pF
WIRE SIZE 26, 28 OR 30 AWG
ENAMEL COATED

Using an MPF102 FET, this one-transistor radio receiver is suitable for experimental use. It can tune 4 to 26 MHz. For speaker operation, a simple IC amplifier can be used.

Frequency (MHz)	C2 (pF)	C18, C19 (pF)	C23 (pF)	L1, L2 (Ant)	L3 (Osc)
				(# of turns on T-37-2 core)	
5	100	120	68	5, 41	45
6	100	120	68	4, 30	34
7	82	100	47	4, 26	29
8	82	100	47	3, 22	24
10	82	100	47	3, 17	19
12	82	100	47	2, 15	17
14	68	82	33	2, 14	15
15	68	82	33	2, 13	14

Fig. 71-1

3.5- TO 10-MHz SIMPLE SUPERHETERODYNE RECEIVER

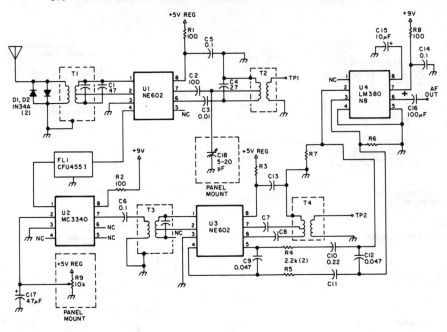

SupeRX Parts List

Part	Value	Type
C1	47 pF	silver-mica or polystyrene
C2	100 pF	silver-mica or polystyrene
C3	0.01µF	polystyrene or monolithic
C4	27 pF	silver-mica or polystyrene
C5	0.1µF	ceramic disc or monolithic
C6	0.1µF	polystyrene or monolithic
C7	0.022µF	polystyrene or monolithic
C8	0.1µF	polystyrene or monolithic
C9	0.047µF	monolithic
C10	0.22µF	monolithic
	0.15µF*	(alternate)
C11	0.22µF	monolithic
	0.15µF*	(alternate)
C12	0.047µF	monolithic
C13, C14	0.1µF	ceramic disc or monolithic
C15	10µF	electrolytic, 16V
C16	100µF	electrolytic, 16V
C17	47µF	electrolytic, 16V
C18	5–20 pF	panel mounted tuning capacitor
C19, C20	220µF	electrolytic, 16V
D1, D2	1N34A	germanium diode or equivalent
R1, R2, R3, R8	100 ohm, ¼W	carbon composition
R4, R5	2.2k, ¼W	carbon composition.
	1.5k	(alternate)

3.5- TO 10-MHz SIMPLE SUPERHETERODYNE RECEIVER (*Cont.*)

R6, R7	10k, ¼W	carbon composition	
R9	10k	potentiometer	
T1, T2	10.7 MHz	microminiature (7mm) IF transformer, green core	Mouser PN 42IF223
T3, T4	455 kHz	microminiature (7mm) IF transformer, black core	Mouser PN 42IF203
U1, U3	NE602	double-balanced mixer	
U2	MC3340	variable attenuator	
U4	LM380N-8	audio amplifier	
U5	78L05	100 mA miniature +5V regulator	

* C10 and C11 can range from 0.1 to 0.22µf. Valves greater than 0.33 cause distortion.

Other: Printed stripboard, DSE PN H5614 or equivalent, cabinet, plastic stick-on feet, 4-40 hardware, etc.

In this circuit, U1 is a frequency converter that feeds the 455-kHz IF stage U2 and detector U3. U4 is the audio output stage. R9 is a gain control that varies the gain of U2. Coil data is given in the part list.

SIMPLE LOW-FREQUENCY RECEIVER

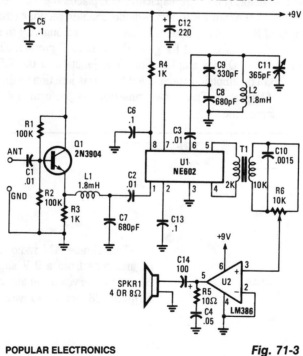

POPULAR ELECTRONICS

Fig. 71-3

Using an NE602 heterodyne detector and U1 as an RF amplifier, this receiver tunes the middle portion of the low-frequency spectrum from 150 to 250 kHz. U2 is a loudspeaker amplifier.

REFLEX RADIO RECEIVER

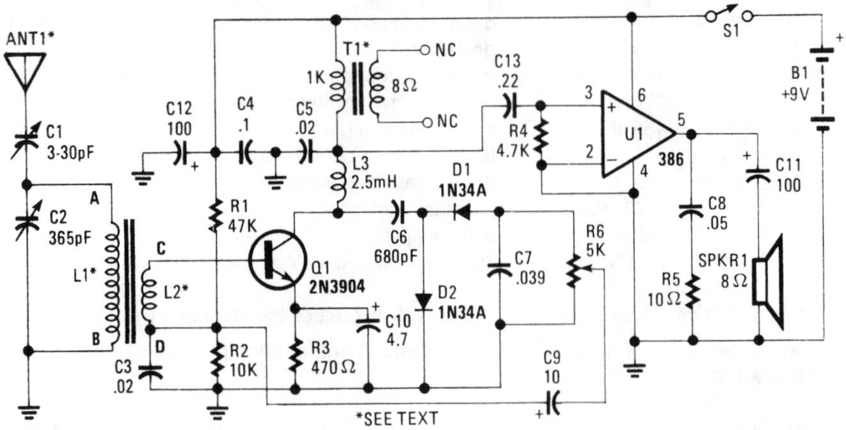

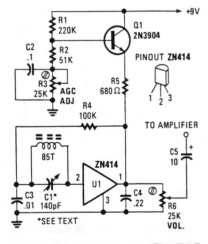

POPULAR ELECTRONICS *Fig. 71-4*

The RF signal is passed from the antenna through C1 to the tuned circuit made up of L1 and C2. One end of L2 feeds the RF signal to the base of Q1 for amplification and the other end ties to the junction of R1 and R2 to supply bias to the transistor. A 0.02-μF capacitor, C3, places the "D" end of L1 at RF ground.

The amplified RF signal is fed through C6 to a two-diode doubler/rectifier circuit and then on to the volume control, R6. The wiper of R6 feeds the detected audio signal through C9 to the junction of R1, R2, and the "D" end of L2. The "D" end of L2 is at RF ground, but not AF ground, allowing the AF signal to be passed through L2 to the base of Q1 for amplification. The junction of the 2.5-mH choke and T1 is placed at RF ground through C5. The amplified audio is fed from this junction to the input of the 386 audio amplifier, U1, to drive the 4″ 8-Ω speaker. The single transistor has performed a dual duty by amplifying the RF and AM signals at the same time.

TRF RADIO

POPULAR ELECTRONICS *Fig. 71-5*

This simple AM radio uses a Ferranti ZN414 IC and runs from a 9-V supply. A 5/16″-diameter ferrite rod serves as an antenna, and uses about 85 turns of #28 enamelled wire.

OLD-TIME RADIO

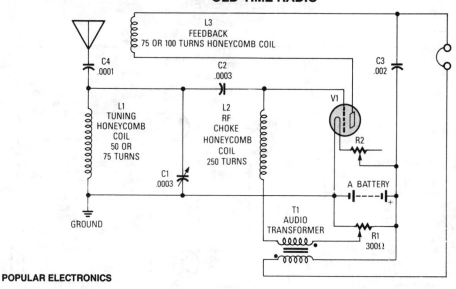

POPULAR ELECTRONICS

Fig. 71-6

This circuit was used in the early days of radio to receive signals. Almost any battery-operated triode, such as a type 30, can be used. "A" battery is 3 V, R2 is a 100-Ω rheostat. Coils are typically 2" to 3" diameter honeycomb wound.

PULSE-FREQUENCY MODULATED RECEIVER

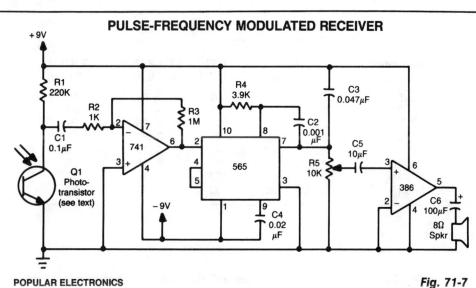

POPULAR ELECTRONICS

Fig. 71-7

This receiver uses an IR-sensitive phototransistor (Clairex, HP, etc.) mounted in a light-tight enclosure with an aperture for the incoming IR beam. An optical system can be used with this receiver for increased range. A 741 amplifies the pulsed IR signal and a 565 PLL FM demodulator recovers the audio, which drives an LM386 audio amplifier and speaker.

SHORT WAVE RECEIVER

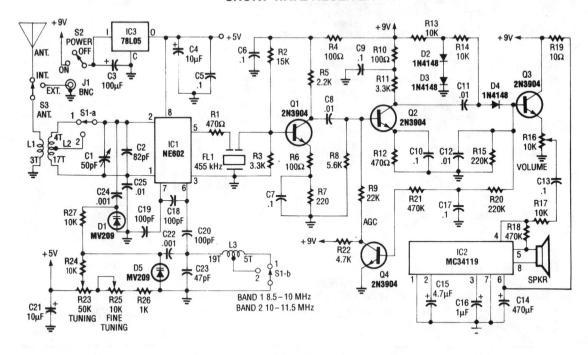

TABLE

Frequency (MHz)	C2 (pF)	C18, C19 (pF)	C23 (pF)	L1, L2 (Ant) (# of turns on T-37-2 core)	L3 (Osc)
5	100	120	68	5, 41	45
6	100	120	68	4, 30	34
7	82	100	47	4, 26	29
8	82	100	47	3, 22	24
10	82	100	47	3, 17	19
12	82	100	47	2, 15	17
14	68	82	33	2, 14	15
15	68	82	33	2, 13	14

RADIO-ELECTRONICS

Fig. 71-8

Using a Signetics NE602 in a varactor-tuned front end, the circuit of a shortwave receiver can be very simple and yet give high performance. This circuit also uses a ceramic filter as a sensitivity-determining device, two IF stages, AGC, and an audio amplifier. It has a sensitivity of under $1\mu V$. The table shows coil data for the frequencies from 5 to 16 MHz. The values C_{18}, C_{19}, and C_{23} depend on the frequency range chosen.

SIMPLE AM RADIO

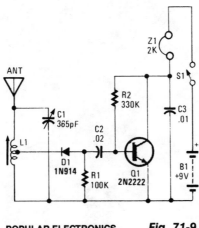

POPULAR ELECTRONICS   *Fig. 71-9*

An AM radio can be built of a simple diode detector and an audio amplifier. A random length of wire always serves as an antenna. L1 is an adjustable ferrite loopstick of the type used in transistor radios.

455

72

Receiving Circuits

The sources of the following circuits are contained in the Sources section, which begins on page 676. The figure number in the box of each circuit correlates to the entry in the Sources section.

HF Transceiver Mixer
AGC System For CA3028 IF Amplifier
Receiver IF Amplifier
30-MHz IF Preamp
Carrier-Operated Relay

HF TRANSCEIVER MIXER

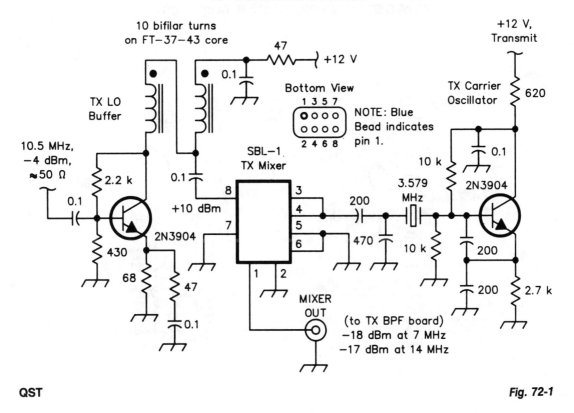

QST

Fig. 72-1

The transceiver mixer and carrier oscillator in the band-imaging (7- and 14-MHz) CW transceiver. Careful selection of drive levels, and use of a spectrally clean carrier oscillator, assure low spurious-signal content in the transmitter output. This transceiver mixer should prove useful in HF and VHF CW or SSB applications. A Mini-Circuits SBL1 low-cost mixer is used with a 3.579-MHz crystal oscillator that uses a low-cost TV color-burst crystal. By paying careful attention to drive leads, good performance and low spurious content can be obtained.

AGC SYSTEM FOR CA3028 IF AMPLIFIER

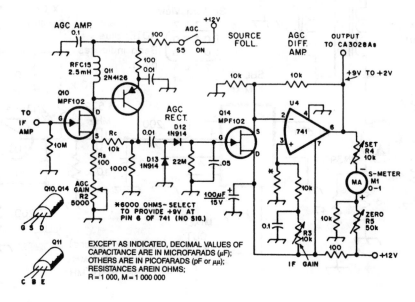

Fig. 72-2

An MPF102 amplifier feeds IF signals to a 2N4126. A potentiometer in the MPF102 source acts as a gain control. This voltage is rectified by an 1N914 doubling detector, and drives a 741 op amp via a source follower (Q14). S-meter and IF-gain controls are provided.

RECEIVER IF AMPLIFIER

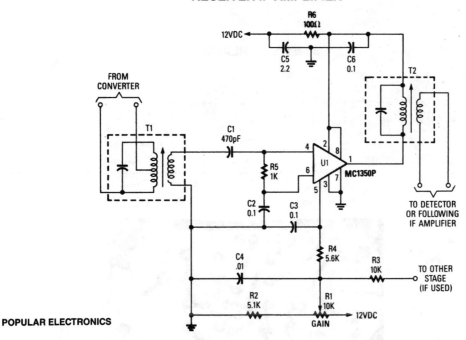

POPULAR ELECTRONICS

Fig. 72-3

T1 is tuned to converter-output frequency U1 to provide 45-to-50-dB gain, depending on the design of T1 and T2. C2, C3, C4, C5, and C6 are bypass capacitors. R5 is a bias resistor. Gain is set by R1, which controls the voltage on pin 5 of U1. T1 and T2 should provide source and load impedance of 1-kΩ and 3- to 10-kΩ, respectively. R3 supplies dc bias to other stages, if required.

IF AMPLIFIER

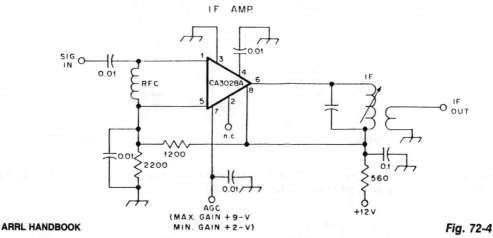

ARRL HANDBOOK

Fig. 72-4

Using a CA3028A, this circuit is useful up to about 15 or 20 MHz.

30-MHz IF PREAMP

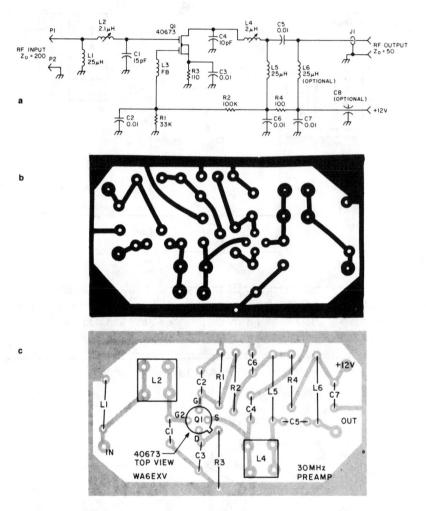

Fig. 72-5

This preamp for 30 MHz is useful for IF applications used in microwave work, etc. A 40673 MOSFET is used and typically the gain at 30 MHz will be 20 to 25 dB.

CARRIER-OPERATED RELAY

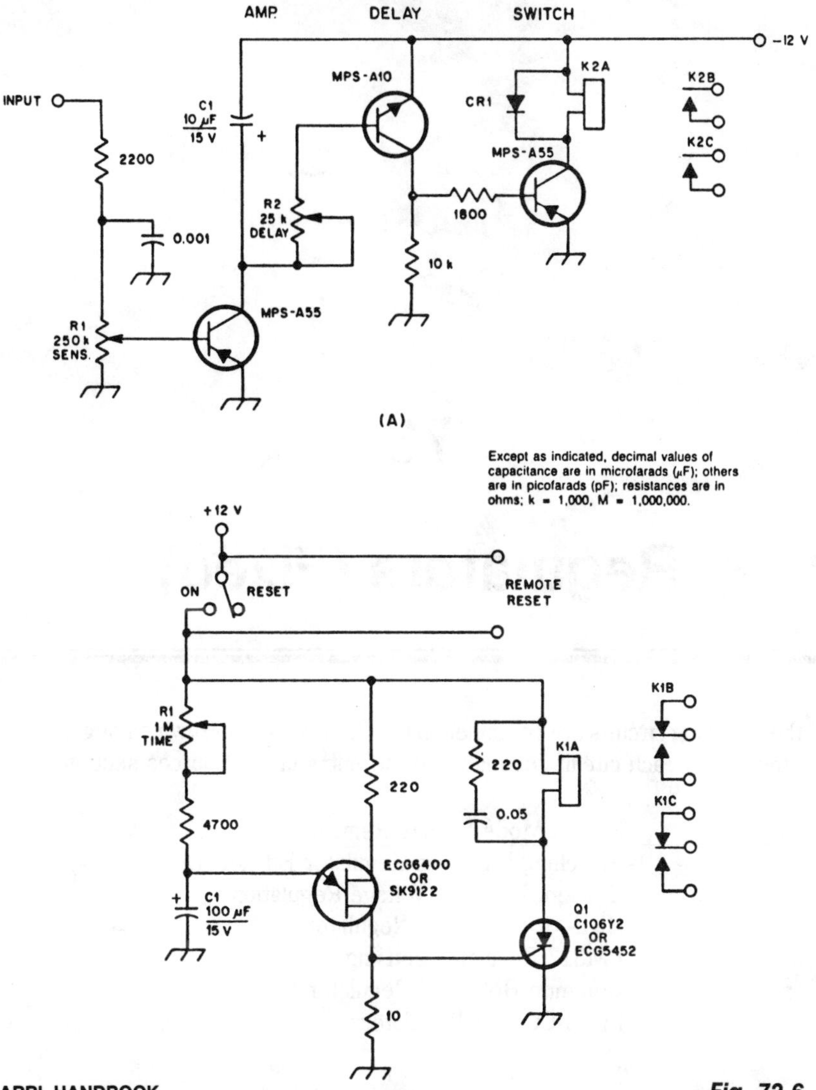

AMP.   DELAY   SWITCH

INPUT

C1
10 µF
15 V

MPS-A10

CR1

K2A

MPS-A55

K2B

K2C

−12 V

2200

0.001

R2
25 k
DELAY

1800

10 k

R1
250 k
SENS.

MPS-A55

(A)

Except as indicated, decimal values of
capacitance are in microfarads (µF); others
are in picofarads (pF); resistances are in
ohms; k = 1,000, M = 1,000,000.

+12 V

ON   RESET

REMOTE
RESET

R1
1 M
TIME

220

K1A

220

K1B

0.05

K1C

4700

ECG6400
OR
SK9122

C1
100 µF
15 V

Q1
C106Y2
OR
ECG5452

10

ARRL HANDBOOK

(B)

Fig. 72-6

A shows a COR/CAS circuit for repeater use. CR1 is a silicon diode. K2 may be any relay with a 12-V coil (a long-life reed relay is best). R2 sets the length of time that K2 remains closed after the input voltage disappears (hang time). B shows a timer circuit. Values shown for R1 and C1 should provide timing up to four minutes or so. C1 should be a low-leakage capacitor; Q1 is a silicon-controlled rectifier, ECG-5452 or equivalent. K1 may be any miniature relay with a 12-volt coil. The timer is reset when the supply voltage is momentarily interrupted. The switch must be in the RESET position for the remote reset to work. This circuit operates from the detector output of a receiver. A delay circuit is included so that the relay stays closed for a time period after the carrier output from the receiver disappears.

73

Regulators (Fixed)

The sources of the following circuits are contained in the Sources section, which begins on page 676. The figure number in the box of each circuit correlates to the entry in the Sources section.

Switch-Mode Voltage Regulator
Switching Improves Regulator Efficiency
Efficient Negative Voltage Regulation
High- or Low-Input Regulator
LM317 Regulator Sensing
Common Hot-Lead Regulator
Fixed-Current Regulator

SWITCH-MODE VOLTAGE REGULATOR

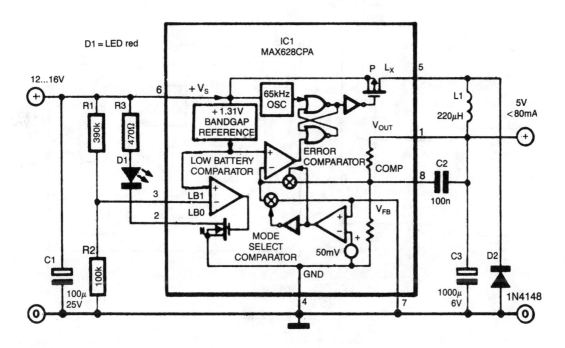

ELEKTOR ELECTRONICS

Fig. 73-1

Switch-mode power supplies offer the benefit of a much greater efficiency than obtainable with a tradi-tional power supply. The switch-mode regulator presented here has an efficiency of around 85%.

An input voltage of 12 to 16 Vdc is converted into a direct voltage of exactly 5 V. The use of a MAX638CPA enables the design and construction of the regulator to be kept fairly simple: only nine addi-tional components are needed to complete the circuit.

Resistors R1 and R2 are used to indicate when the battery voltage becomes low: as soon as the volt-age on pin 3 becomes lower than 1.3 V, D1 lights. With values as shown for the potential divider, this corresponds to the supply voltage getting lower than about 6.5 V. The output of the IC is shunted by a simple LC filter formed by L1, C3 and D2.

The oscillator on board the IC generates a clock frequency of around 65 kHz and drives the output transistor via two NOR gates. The built-in error detector, the "battery low" indicator or the voltage com-parator can block the clock frequency, which causes the transistor to switch off.

The IC compares the output voltage of 5 V with a built-in reference (FET). Depending on the load, the FET will be switched on for longer or shorter periods. The maximum current through the FET is 375 mA, which corresponds with a maximum output current of 80 mA.

SWITCHING IMPROVES REGULATOR EFFICIENCY

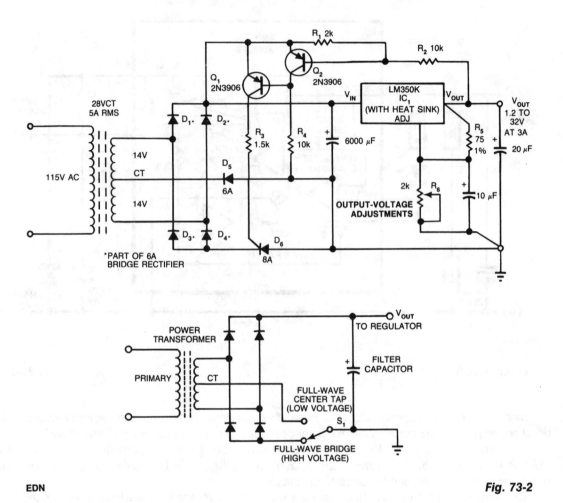

EDN

Fig. 73-2

In this circuit, a full-wave bridge is switched to a full-wave center tap to reduce regulator dissipation. SCR D6 switches between configurations. When D6 is off, the circuit is an FWCT rectifier using D1, D2, and D5. It applies 17 V plus ripple to the regulator input. The drop across the regulator supplies base drive to Q2. If Q2 is on, Q1 is off, and D6 is off. If the regulator voltage drops below about 3 V, Q2 turns off, and turns Q1 on, which turns on D6. This changes the circuit to an FW bridge using D1 through D4.

EFFICIENT NEGATIVE VOLTAGE REGULATION

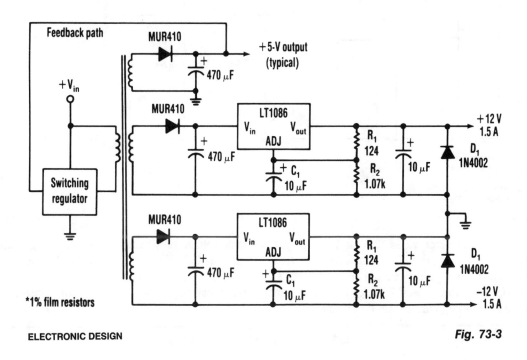

ELECTRONIC DESIGN

Fig. 73-3

Many applications require highly efficient negative-voltage post regulators with low dropout voltage in switch-mode supplies.

A way to provide good negative-voltage regulation is with a low-dropout positive-voltage regulator operating from a well-isolated secondary winding of the switch-mode transformer. The technique works with any positive-voltage regulator, although highest efficiency occurs with low-dropout types.

In the circuit, two programming resistors, R1 and R2, set the output voltage to 12 V, and the LT1086s servo the voltage between the output and its adjusting (ADJ) terminals to 1.25 V. Capacitor C1 improves ripple rejection, and protection diode D1 eliminates common-load problems.

Because a secondary winding is galvanically isolated, a regulator's 12-V output can be referenced to ground. Therefore, in the case of a negative-voltage output, the positive-voltage terminal of the regulator connects to ground, and the -12-V output comes off the anode of D1. The V_{in} terminal floats at 1.5 V or more above ground. This arrangement is the equivalent of connecting the positive terminal of a battery to ground and taking the output from the negative terminal.

HIGH- OR LOW-INPUT REGULATOR

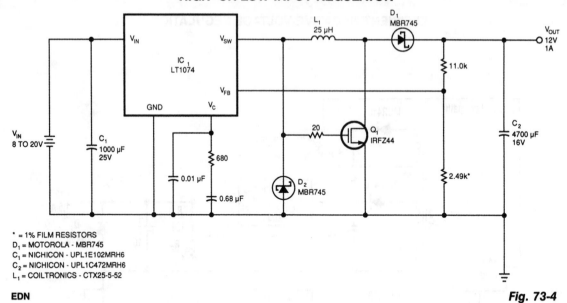

* = 1% FILM RESISTORS
D_1 = MOTOROLA - MBR745
C_1 = NICHICON - UPL1E102MRH6
C_2 = NICHICON - UPL1C472MRH6
L_1 = COILTRONICS - CTX25-5-52

EDN

Fig. 73-4

This regulator provides 12 V at 1 A out with an input voltage of 8 to 20 V. Output voltage can be changed by charging the 11-kΩ and 2.49-kΩ resistors to provide 2.21 V at the V_{FB} pin of IC1, if desired. If you need to handle a higher input voltage, make sure to clamp the gate of Q1 below its 20-V max. rating.

Efficiency can exceed 70% for output currents greater than 0.5 A; above 15-V input voltage, more than 2 A of output current can be obtained.

LM317 REGULATOR SENSING

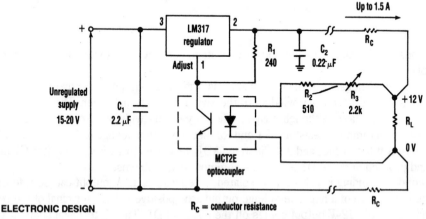

ELECTRONIC DESIGN

R_c = conductor resistance

Fig. 73-5

The optocoupler (as shown) provides load sensing for a 3-terminal regulator, such as the LM317 series. R1 sets a current of 5 mA through the optocoupler transistor and R3 is adjusted for 12 V across the load.

COMMON HOT-LEAD REGULATOR

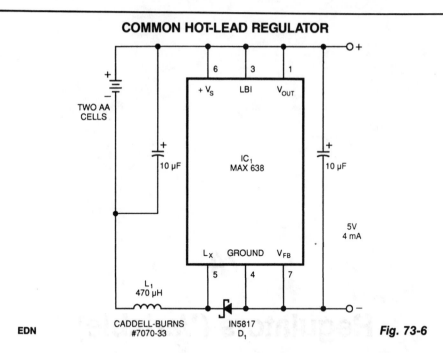

EDN

Fig. 73-6

This circuit derives 5 Vdc from 2-AA cells—even at their end-life voltages of 1.05 V, and is approximately 80% efficient, providing 5 V at 4 mA from 2.1 V at 11 mA. IC1 is manufactured by Maxim Integrated Products, Inc.

FIXED-CURRENT REGULATOR

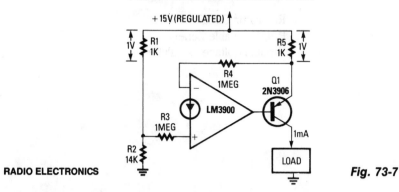

RADIO ELECTRONICS

Fig. 73-7

This fixed 1-mA current source delivers a fixed current to a load connected between Q1's collector and ground; the load can be anywhere in the range from 0 Ω to 14 Ω. The circuit is powered from a regulated 15-V supply, and the R1/R2 voltage divider applies a 14-V reference to R3. The op amp's output automatically adjusts to provide an identical voltage at the junction of R4 and R5. That produces 1 V across R5, resulting in an R5 current of 1 mA. Because that current is derived from Q1's emitter, and the emitter and collector currents of a transistor are almost identical, the circuit provides a fixed-current source. The output current can be doubled by halving the value of R5.

74

Regulators (Variable)

The sources of the following circuits are contained in the Sources section, which begins on page 676. The figure number in the box of each circuit correlates to the entry in the Sources section.

Regulator Circuit
Programmable Zener
Variable Voltage Regulator

REGULATOR CIRCUIT

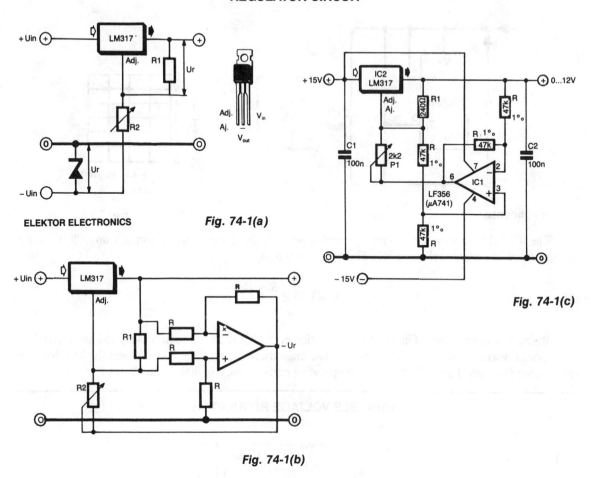

ELEKTOR ELECTRONICS

Fig. 74-1(a)

Fig. 74-1(c)

Fig. 74-1(b)

The special characteristic of this regulator is that the output voltage can be adjusted down to 0 V. The regulation is provided by an integrated regulator Type LM317. As is normal in supplies that can be adjusted to 0 V, this IC is used in conjunction with a zener diode. This diode provides a reference voltage that is equal, but of opposite sign, to the reference voltage (U_r) of the regulator, as shown in Fig. 74-1(a). Potential divider R1/R2 enables the output voltage to be adjusted.

In this circuit, the negative reference voltage is derived in a different manner: from the regulator with the aid of an op amp (Fig. 74-1(b)). The op amp is connected as a differential amplifier that measures the voltage across R1 and inverts this voltage to U_r. An additional advantage of this method is that at low-output voltages, a change in the reference voltage has less effect on the output voltage than the circuit in Fig. 74-1(a). The prototype, constructed as shown in Fig. 74-1(c), gave very satisfactory results.

The op amp need not meet any special requirements: a μA741 works fine, although an LF356 gives a slightly better performance. The negative supply for the op amp can be obtained with the aid of a center-tapped mains transformer.

469

PROGRAMMABLE ZENER

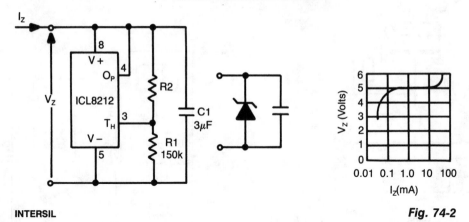

INTERSIL

Fig. 74-2

The ICL8212 is connected as a programmable zener diode. Zener voltages from 2 V up to 30 V can be programmed by suitably selecting R2. The zener voltage is:

$$V_Z = 1.15 \times \frac{R_1 + R_2}{R_1}$$

Because of the absence of internal compensation in the ICL8212, C1 is necessary to ensure stability. Two points worthy of note are the extremely low-knee current (less than 300 μA) and the low dynamic impedance (typically 4 to 7 ohms) over the operating current range of 300 μA to 12 mA.

VARIABLE VOLTAGE REGULATOR

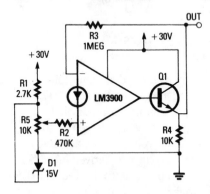

RADIO-ELECTRONICS

Fig. 74-3

The op amp is wired as a ×2 noninverting dc amplifier with a gain that is determined by the R3/R2 ratio. The input voltage to the op amp is variable between 0 and 15 V via R5. The output voltage is therefore variable over the approximate range from 0.5 to 30 V. The available output current has been boosted by adding transistor Q1 to the output.

75

Relay Circuits

The sources of the following circuits are contained in the Sources section, which begins on page 676. The figure number in the box of each circuit correlates to the entry in the Sources section.

LATCHING ac SOLID-STATE RELAY

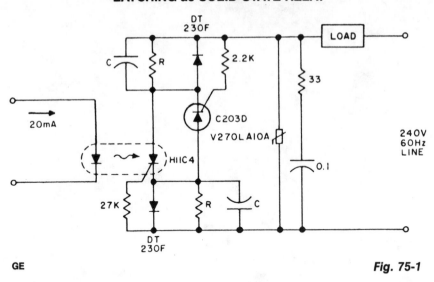

GE

Fig. 75-1

Latching is obtained by storing the gate trigger energy from the preceding half cycle in the capacitors. Power must be interrupted for more than one full cycle of the line to ensure turn-off. Resistor R and capacitor C are chosen to minimize dissipation, while assuring triggering of the respective SCRs for each cycle. A pulse of current, over 10 ms duration into the H11C4 IRED, ensures triggering the latching relay into conduction.

BIDIRECTIONAL SWITCH

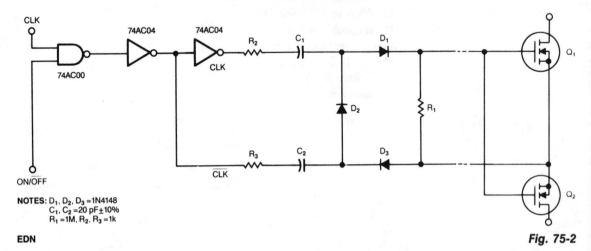

NOTES: D_1, D_2, D_3 = 1N4148
$\quad\quad$ C_1, C_2 = 20 pF±10%
$\quad\quad$ R_1 =1M, R_2, R_3 =1k

EDN

Fig. 75-2

Using voltage doublers, this simple switch circuit uses a clock signal and a control signal to switch MOSFETs.

LOW-CONSUMPTION MONOSTABLE RELAY

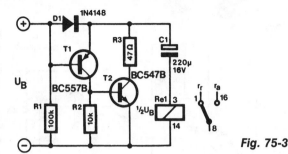

ELEKTOR ELECTRONICS

Fig. 75-3

A monostable relay has two states: *operative* when a large enough current flows through its coil and *quiescent* when no current flows. A relay contact that assumes a certain position after the supply voltage has been switched on is required in many applications. Of course, many relays operate in that manner.

However, most of these relays require an energizing current of 50 mA or more and that normally precludes a battery supply. The circuit presented here, which uses a bistable relay, can solve that problem.

The contact of a bistable relay normally remains in the position it is in after the supply is switched off. This circuit, however, makes the bistable relay behave like a monostable type, at a modest current.

When the supply voltage is switched on, C1 charges via D1 and the relay coil. The current then flowing through the coil causes the relay contact to assume one of two positions. The forward drop across D1 ensures that the base of T1 (in this condition) is more positive than its emitter so that T1, and thus T2, is switched off.

When the supply voltage is switched off, the emitter of T1 is connected to the positive terminal of C1, while the base is connected to the negative terminal of the capacitor via R1 and the relay coil. This results in T1, and thus T2, switching on so that C1 discharges via T4 and the relay. The current flows through the relay coil, then flows in an opposite direction and this causes the contact to change over.

The bistable relay thus behaves exactly as a monostable with the advantage, however, that the operational current is determined by R1, which amounts to only 130 μA.

To ensure reliable operation, the rating of the relay coil should be 65 to 75% of the supply voltage. In the prototype, a 9-V relay was used with a battery supply voltage of 12 V.

DELAY-OFF RELAY CIRCUIT

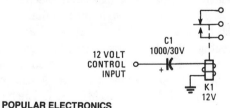

POPULAR ELECTRONICS

Fig. 75-4

When voltage is applied to the capacitor, it charges. While it's charging, the relay remains latched. When the charging current falls below the level needed to hold the relay down, the relay unlatches. The higher the value of the capacitor, the longer the relay will remain latched.

SOLID-STATE RELAY

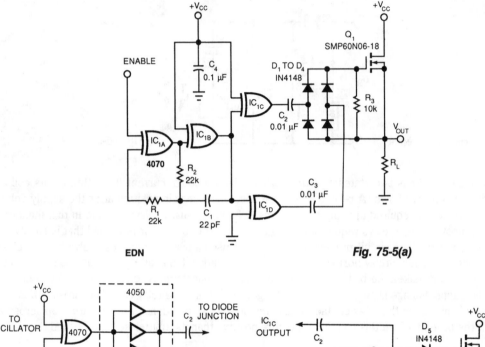

EDN

Fig. 75-5(a)

Fig. 75-5(b)

Fig. 75-5(c)

A power MOSFET and a quad exclusive-OR oscillator makes an effective solid-state relay. Figure 75-5(a)'s capacitively isolated drive circuit provides gate drive to turn on the n-channel device. This consists of a gated oscillator (IC1A and IC1B running at 500 kHz, set by R1, R2, and C1).

The diode bridge (D1 through D4) rectifies the charge transferred through C2 and C3. When you disable the oscillator, R3 discharges the stored gate charge, thereby turning off the MOSFET. R3 needs to allow fast turn-off times without loading the gate's enhancement voltage. A value of 10-kΩ is sufficient to produce a turn-off time of 800 μs with an 18-mΩ SMP60N06-18 MOSFET. The measured turn-on time is 150 μs.

You can reduce the turn-off time to 100 μs by using a pnp transistor as a diode-steering emitter-follower in the MOSFET gate circuit (Fig. 75-5(b)). Adding a hex buffer to Fig. 75-5A's circuit increases the drive capability of the complementary outputs (Fig. 75-5(c)).

OPTOISOLATOR

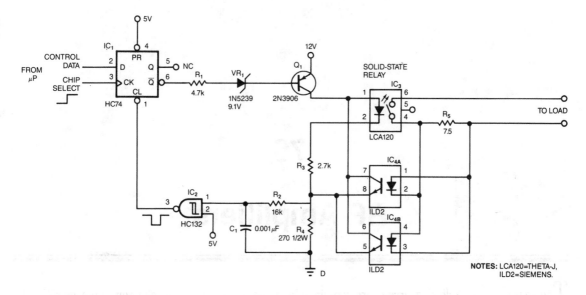

Fig. 75-6

The circuit protects a solid-state relay from overloads. The circuit limits current, automatically disconnects the load after detecting a short circuit, and develops a fault-condition output signal.

In normal operation, the controlling μP sets the flip-flop, IC1, which turns on transistor Q1. When Q1 turns on, current flows through the solid-state relay's input , thus activating the relay.

If an overcurrent or fault condition occurs, the excessive load current flowing through the relay develops enough potential across sense resistor R5 to turn on one of the optoisolators, IC4A or IC4B. The optoisolator's output transistor diverts current around the solid-state relay's input, which limits the current that the relay's output can pass.

If the overload is severe enough, the optoisolator pulls the input of the Schmitt trigger above its threshold, thus clearing the flip-flop and turning off the solid-state relay. R2 has two functions: It keeps the input of the Schmitt trigger below 5 V max. to prevent latchup, and it forms an RC filter in conjunction with C1. The RC filter prevents spurious triggering of the Schmitt trigger. You can use the output of the flip-flop to signal overload conditions to the controlling μP.

76

RF Amplifiers

The sources of the following circuits are contained in the Sources section, which begins on page 676. The figure number in the box of each circuit correlates to the entry in the Sources section.

4-W AMPLIFIER FOR 900 MHz

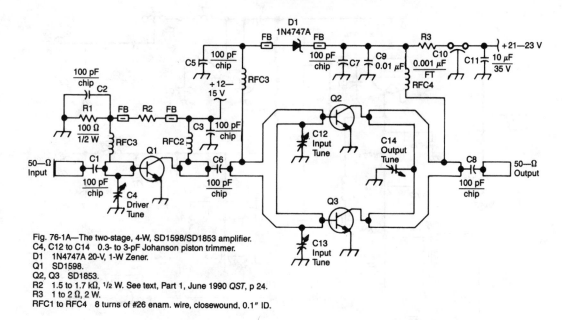

Fig. 76-1A—The two-stage, 4-W, SD1598/SD1853 amplifier.
C4, C12 to C14   0.3- to 3-pF Johanson piston trimmer.
D1   1N4747A 20-V, 1-W Zener.
Q1   SD1598.
Q2, Q3   SD1853.
R2   1.5 to 1.7 kΩ, ½ W. See text, Part 1, June 1990 *QST*, p 24.
R3   1 to 2 Ω, 2 W.
RFC1 to RFC4   8 turns of #26 enam. wire, closewound, 0.1" ID.

Fig. 76-1B—Parts-placement diagram for the
2 × SD1853 amplifier. The PC-board
edges are not shown. All components
mount to the trace side of the PC board
(except those mounted to the enclosure).

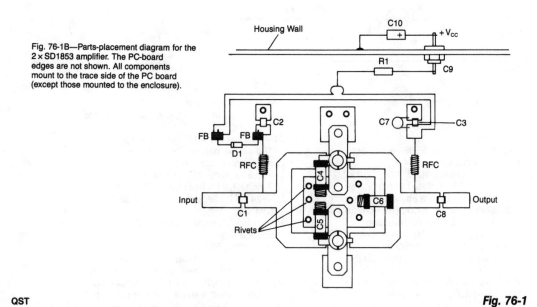

Fig. 76-1

Using Wilkinson power dividers in the base and collector circuits of Q2 and Q3, two SD1853 devices are paralleled for twice the power output of the 2-W amplifier.

1500-W RF AMPLIFIER

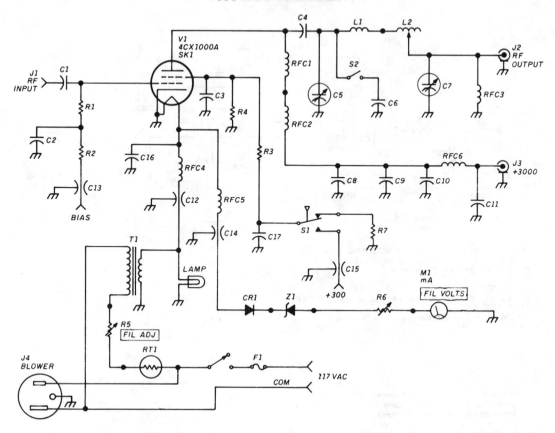

C1,2—0.01 μF silver mica transmitting
  capacitor
C3—1500 pF screen grid bypass (internal to
  tube socket)
C4—2000 pF, 5 kV, two 858S capacitors in
  parallel, Centralab
C5—Vacuum variable, 10 to 300 pF, 10 kV,
  Jennings UCS-300 (plate tuning)
C6—300 pF, 5 kV, Centralab-type 858S
C7—Vacuum variable, 3000 pF maximum,
  3 kV, Jennings UCSL-3000 (loading)
C8—0.0014 μF, 10 kV (EM)
C9—0.005 μF, 15 kV oil capacitor
C10,11—500 pF, 20 kV "TV doorknob,"
  Centralab
C12—0.05 μF, 20 A feedthrough, Sprague
C13,14,15—1000 pF feedthrough
C16—0.1 μF, 100 volts DC disc ceramic
D1—Silicon diode, 1N4148 or 1N914
F1—3 A fuse
J1—BNC chassis mount

J2—N chassis mount
J3—HN chassis mount, used with RG213 for
  3 kV feed
J4—Chassis-mount AC outlet, Amphenol
  160-2N, AL part no. 713-5202
Lamp—No. 47 6.3 volt lamp
L1—3.5 turns no. 10 silver-plated wire,
  1 inch ID, 2 inches long, self-supporting
L2—24 H roller inductor, Johnson 226-1
  (EM, CC)
M1—1 mA DC meter movement, filament
  voltage
R1—50 ohm, 60 watt noninductive resistor,
  thirty 1500 ohm 2 watt carbon composition
  resistors connected in parallel (see text)
R2—1000 ohm, 2 watt carbon composition
R3—100 ohm, 2 watt carbon composition
R4—220 K, 2 watt carbon composition
R5—25 ohm, 25 watt wirewound adjustable
R6—1000 ohm, 2 watt carbon or wirewound
  pot

1500-W RF AMPLIFIER (Cont.)

R7—10,000 ohm, 2 watt carbon composition
RFC1—28 turns no. 18 wire solenoid wound on 0.5 inch OD by 2.5 inch ceramic form (plate choke)
RFC2—Plate choke, surplus part (suggest B&W 801)
RFC3—74 turns no. 20 wire solenoid wound on 0.75 OD by 3 inch ceramic form
RFC4—9 μH 15 A RF choke, surplus part (suggest Dale no. IH15, HF no. 18-105)

RFC5—1.0 mH RF choke
RFC6—10 H, 1 A RF choke
RT1—Surgistor, GC 25-933-S
S1—Air flow switch, Rotron 2A-1350
SK1—Eimac socket, Sk-810B; chimney SK-806 (BY)
T1—Transformer, 6.3 volts AC at 10 A second—ary, Thordarson 21F12, AL 704-2019
Z1—Zener diode, 6.2 volts, 1N473

CQ

Fig. 76-2

The frequency range of this amplifier is 1.8 to 54 MHz. The amount of RF drive required for full output is about 30 W. The grid compartment (R1, R2, RFC4, RFC5) should be shielded from the other circuitry—especially the output circuitry.

WIDEBAND RF AMPLIFIER

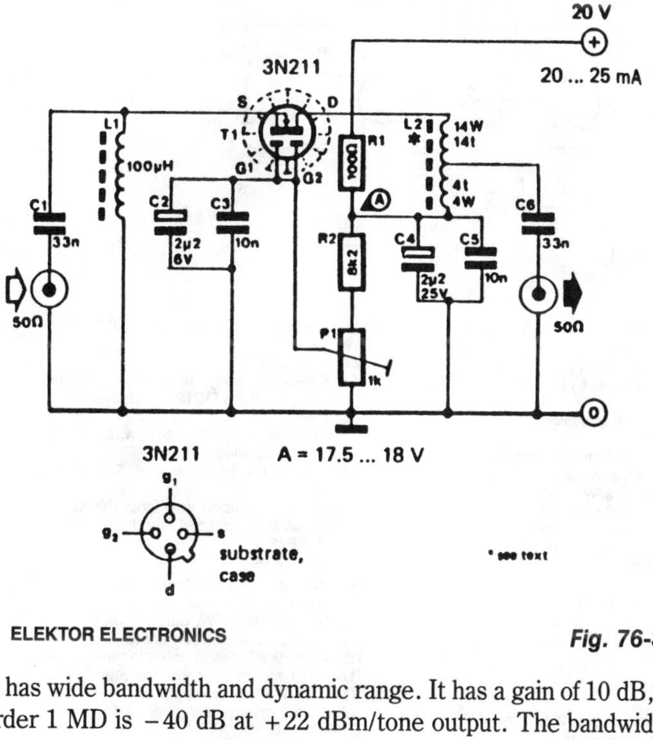

ELEKTOR ELECTRONICS

Fig. 76-3

This RF amplifier has wide bandwidth and dynamic range. It has a gain of 10 dB, noise figure less than 5 dB, and the third-order 1 MD is −40 dB at +22 dBm/tone output. The bandwidth is 4 to 55 MHz.

6-m 100-W LINEAR AMPLIFIER

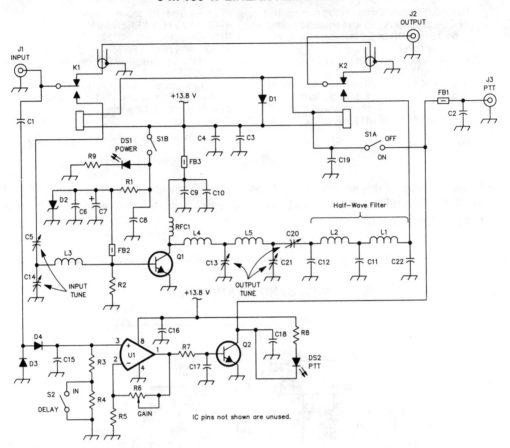

C1—10 pF, ceramic disc.
C2—0.01 μF, polyester film.
C3, C10—0.001 μF, ceramic disc.
C4, C6, C8, C9, C15-C19—0.1 μF,
  polyester film.
C5—9- to 180-pF mica compression
  trimmer, Arco no. 463.
C7—1000 μF, 6.3 V, aluminum electrolytic.
C11—120 pF, 100 V, silver mica.
C12, C22—62 pF, 100 V, silver mica.
C13, C14, C20—50- to 380-pF mica
  compression trimmer, Arco no. 465.
C21—25- to 280-pF mica compression
  trimmer, Arco no. 464.
D1—1N4002.
D2—1N1200 stud-mount diode.
D3, D4—1N4148.
DS1—Green LED.
DS2—Red LED.

FB1, FB3—Ferrite bead, Amidon no.
  FB43-901 or similar.
FB2—VK-200 wide-band choke, 2½ turns
  no. 24 solid wire on Amidon no.
  FB43-5111 ferrite core.

J1, J2—Female RF connector (UHF, BNC,
  N, etc).
J3—Phono jack.
K1, K2—SPDT relay, 12-V dc coil, Omron
  no. G5L112P-PS-DC12. Available from
  Digi-Key.
L1, L2—4 turns no. 14 enam wire, 7/16 in.
  diam, 3/8 in. long.
L3—2 turns no. 14 enam wire, 3/8 in. diam,
  ½ in. long.
L4—no. 14 U-shaped wire loop, 3/8 in.
  diam, 9/16 in. finished length.
Q1—MRF492.
Q2—2N2222A.

R1—5-W bias resistor (see text).
R2—10 Ω, ½ W, carbon comp.
R3, R5—10 kΩ, ¼ W, carbon comp.
R4—1 MΩ, ¼ W, carbon comp.
R6—100-kΩ PC-board potentiometer.
R7-R9—1 kΩ, ¼ W, carbon comp.
S1—DPDT miniature toggle.
S2—SPST miniature toggle.
U1—LM358.

QST

Fig. 76-4

6-m 100-W LINEAR AMPLIFIER (Cont.)

Miscellaneous
Suitable die-cast-aluminum enclosure and
 heat sink.
Pair of binding posts or other suitable dc-
 supply connector.

Two 1- × 3-in. strips of double-sided foam
 tape.
Two LED holders.
Seven no. 4-40 × ¼ in. machine screws.

100 W output at 50 MHz is available from this circuit. U1 and Q2 form a T-R relay driver, switching the amplifier on when RF input at J1 is sensed. During receive periods, J1 and J2 are directly connected. A 13.8-V supply is required for this amplifier.

10- to 15-W ATV LINEAR AMPLIFIER

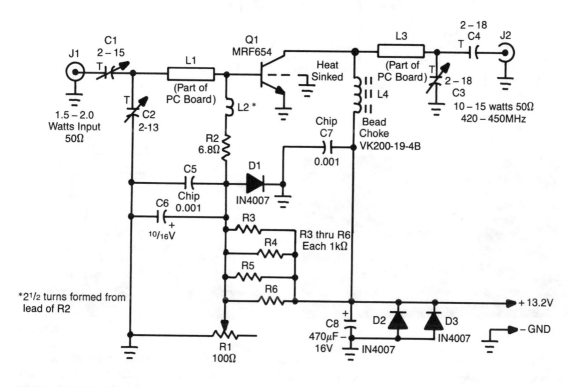

NORTH COUNTRY RADIO

Fig. 76-5

This amplifier is useful for applications where a 10- to 15-W peak-envelope-power (PEP) signal is needed in the 420- to 520-MHz range. C1, C2, and L1 form a matching network for amplifier Q1. L3, C3, and C4 form an output matching network for a 50-Ω load. L2, R2, D1, and R3 through R6 (with bias adjust R1) form a biasing network for Q1. D2 and D3 provide polarity protection for Q1, which must be heat-sinked. A kit of all parts including case and PC board is available from North Country Radio, P.O. Box 53, Wykagyl Station, New Rochelle, NY 10804.

481

UHF TV-LINE AMPLIFIER

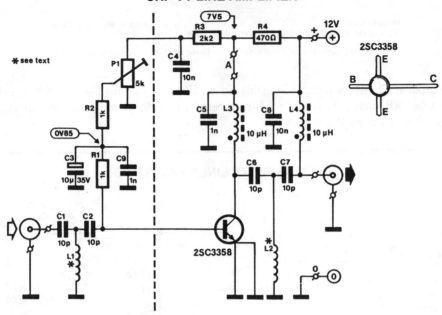

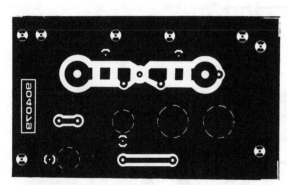

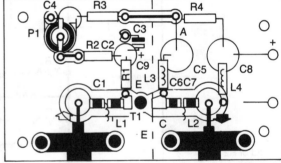

ELEKTOR ELECTRONICS

Fig. 76-6

This circuit offers 10- to 15-dB gain from 400 to 850 MHz and is therefore eminently suitable for situations where the television signal is weak. Moreover, the filters can be adapted to the individual needs of users. Construction is simplicity itself if the ready-made PC board shown on the next page is used. The tracks should be tinned or silvered for optimum performance and long life.

The opening at the center of the board is intended to accommodate the transistor. This device has two emitter pins, both of which should be connected to ground. The drawings show that the board is divided into two by a small piece of tin plate, which should have a small cut-out for the transistor.

The input and output terminals are made from small cable clamps and M3 nuts and bolts. One side of disc capacitors C4, C5, C8, and C9 is soldered direct to the board.

Input and output capacitors, C1/C2 and C6/C7 respectively are surface-mount types. C1/C2/L1 form an input filter and C6/C7/L2 form an output filter. The value of the capacitors might have to be lowered to

3.9 pF to obtain the correct frequency range. The amplifier can be housed in a watertight case and then mounted near the antenna at the top of the mast (if used).

The power is obtained from a simple stabilized 12-V supply: a mains adapter with a 78L12 will do nicely. This can be kept indoors, of course. The amplifier can be powered via the coaxial feeder cable, for which purpose a 10- to 100-μH choke is inserted in the supply line. Calibrating the amplifier is straightforward: set P1 to the center of its travel and then adjust it for optimum picture quality.

In practice, the collector current of the transistor is then 5 to 15 mA. This may be checked by temporarily replacing jump lead A by one suitable meter.

PARTS LIST

R1, R2 = 1 kΩ

R3 = 2 kΩ

R4 = 470 Ω

P1 = 5 kΩ preset potentiometer

C1, C2, C6, C7 = 10 pF surface mount

C3 = 10 μF; 35 V

C4, C8 = 1 nF disc

C5, C9 = 1 nF disc

L1, L2 = air core, 2 turns of 3mm diameter enamelled copper wire

L3, L4 = 10-μH choke or 10 turns of 0.2 mm diameter enamelled copper wire on a ferrite bead.

T1 = 2SC3358

DOUBLE-TUNED JFET PRESELECTOR

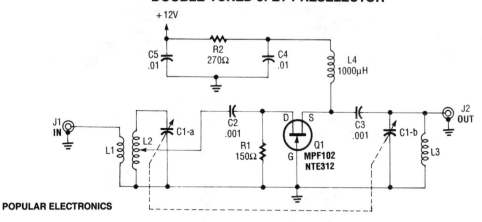

POPULAR ELECTRONICS

Fig. 76-7

This circuit uses an MPF102 JFET and a double-tuned common-gate amplifier. Gain is typically 10 to 15 dB:

$$L_2 \text{ and } L_3 = \frac{1}{(2\pi f)^2 C_1}$$

L1 $\approx$ 10% turns on L2 or 1 turn (larger of these two).

903-MHz LINEAR AMPLIFIERS

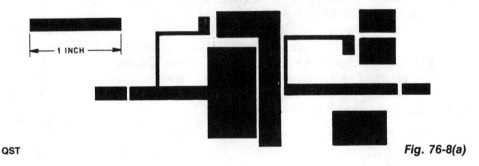

QST

Fig. 76-8(a)

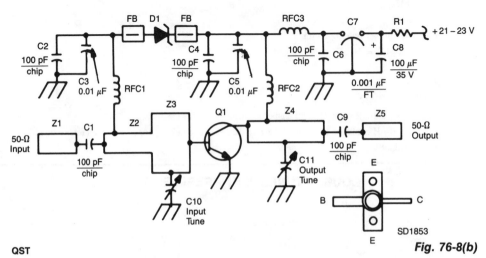

QST

Fig. 76-8(b)

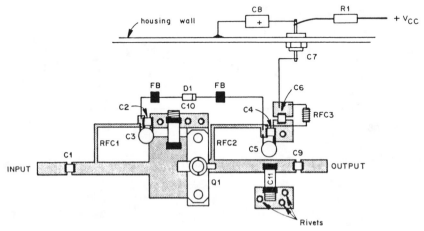

QST

Fig. 76-8(c)

903-MHz LINEAR AMPLIFIERS (*Cont.*)

The 2-W SD1853 amplifier. RFC1 and RFC2 are implemented as PC-board traces.
C10,C11   0.3- to 3-pF Johanson piston trimmer.
D1   1N4747A 20-V, 1-W Zener.
Q1   SD1853.
R1   2 – 3 Ω, 1 W.
RFC3   8 turns of #26 enam wire, closewound, 0.1″ ID.
Z1 – Z5   Microstriplines. See text and Fig. 18.

Two W output can be produced by this circuit designed for the 902- to 928-MHz amateur band. As shown in the figures, much of the circuitry is in the form of PC board traces.

SIMPLE JFET PRESELECTOR

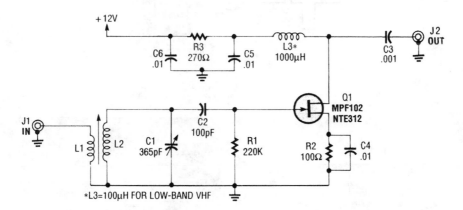

POPULAR ELECTRONICS

Fig. 76-9

This preselector will improve the performance of simple shortwave receivers. L2 is tuned with Q1. The inductance of L2 is:

$$L_2 = \frac{1}{(2\pi f)^2 C_1}$$

and L1 is found around the cold end of L2. Typically, L1 has 10% of the turns in L2, or one turn, whichever is larger.

1296-MHz RF AMPLIFIER

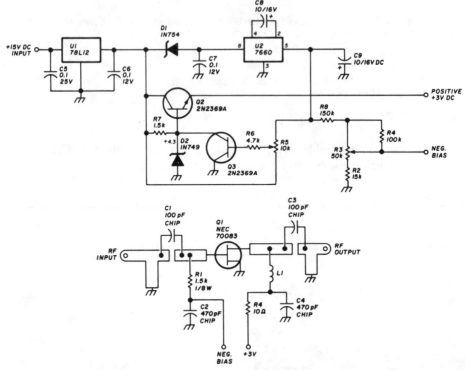

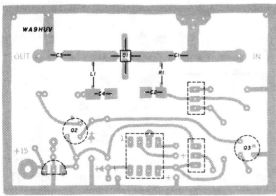
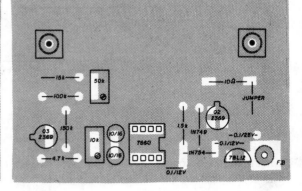

HAM RADIO

Fig. 76-10

Using an NEC70083, this 1296-MHz amplifier delivers about 17-dB gain and around 1- to 1.5-dB noise. It is constructed on a G-10 epoxy fiberglass PC board. Use the layout shown because this is important for correct performance.

The power supply/regulator delivers the regulated 3-Vdc for the drain circuit and U2 produces a negative bias for the gate circuit. R5 sets the drain voltage to +3 V and R3 sets the gate bias. Typically, the drain current is about 10 mA.

486

BROADCAST BAND BOOSTER

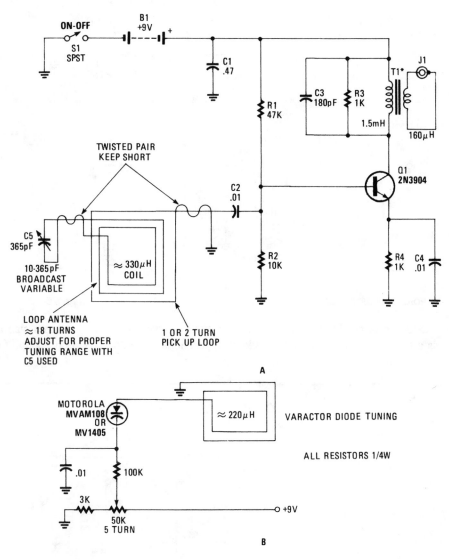

POPULAR ELECTRONICS

Fig. 76-11

The use of a loop antenna of large size (12 × 18″) provides a large signal pickup at AM broadcast frequencies. It has about 18 turns. T1 is a toroidal transformer of about 3:1 turns ratio (not critical). The primary winding should have about 1.5 mH inductance. By using the circuit at (B), a varactor diode can be used in place of the 10- to 365-pF variable capacitor.

VARACTOR-TUNED PRESELECTOR

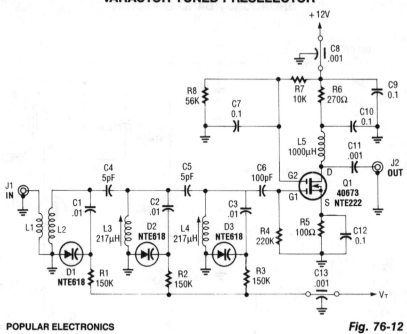

POPULAR ELECTRONICS

Fig. 76-12

Varactor diodes replace the conventional 365-pF tuning capacitors, which reduces size and weight when used as a turned RF stage for AM broadcast applications. Selectivity is good enough for use as a TRF receiver if a detector is connected to J2.

CASCODE RF AMPLIFIER

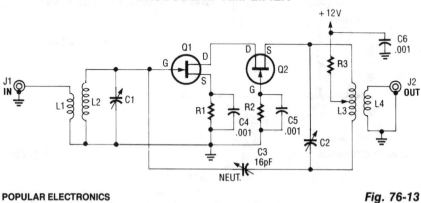

POPULAR ELECTRONICS

Fig. 76-13

A cascode amplifier using two MOSFETs is shown in the diagram. L2C1 and L3C2 resonate to the frequency in use. The circuit has the advantage of good gain, low NF, and excellent linearity. Q1 and Q2 can be MPF102 or 2N4416.

DUAL-GATE MOSFET RF-AMP STAGE

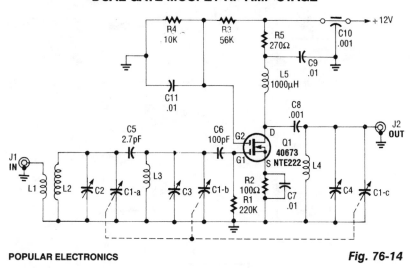

POPULAR ELECTRONICS

Fig. 76-14

The use of a double-tuned input and a single-tuned output yield superior RF selectivity to that of equivalent single-tuned designs. AGC, if required, can be added to gate 2 of Q1, and should drive gate 2 negative for decreased gain.

WIDEBAND AMPLIFIER

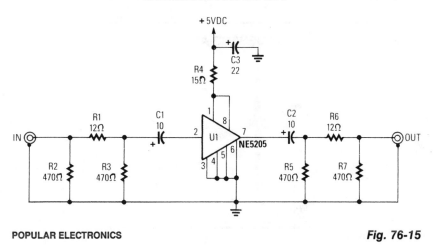

POPULAR ELECTRONICS

Fig. 76-15

Using a Signetics NE5205, this circuit gives about +16-dB gain from LF to 600 MHz. The π minimizes load impedance and source-impedance variations and aids stability.

BUFFER AMPLIFIER WITH MODULATOR

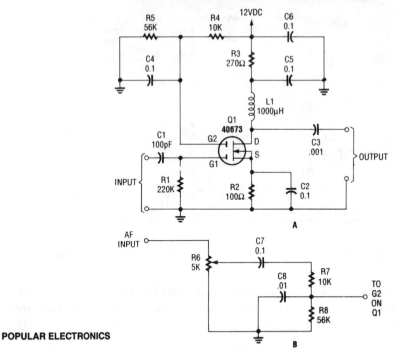

POPULAR ELECTRONICS

Fig. 76-16

A 40673 MOSFET is used as a wideband buffer amplifier (to 20 MHz). If desired, the amplifier can be modulated, considering that the gate 2 voltage of a MOSFET can be used to vary the gain of the stage.

WIDEBAND AMPLIFIER

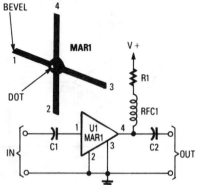

POPULAR ELECTRONICS

Fig. 76-17

This wideband amplifier uses a MAR1 IC, which is a gain block. The device is manufactured by Mini-Circuits Lab and offers +13-dB gain from dc to 1 000 MHz. R1 is selected to provide 17-mA current to U1 at +5 V. For 12-V supply $R_1 = 470 \ \Omega$. RFC1 is typically 1 to 5 μH.

TWO CA3100 WIDEBAND OPERATIONAL AMPLIFIERS

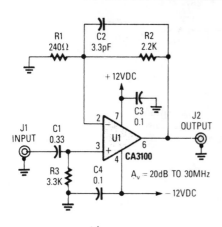

POPULAR ELECTRONICS   *Fig. 76-18(a)*

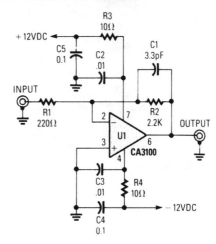

Fig. 76-18(b)

These circuits use the RCA CA3100 wideband amplifier IC. The gain is:

$$A_v = 1 + \frac{R_2}{R_1}$$

These circuits are useful for video applications to 10 MHz. The 3.3-pF capacitors are compensation capacitors. Capacitors on pins 7 and 4 are bypass capacitors to prevent self-oscillations. Figure 76-18(a) is noninverting and Fig. 76-18(b) is an inverting configuration.

500-MHz AMPLIFIER

C1, C2, C3, C4, ARCO #400, 1→7pF
L1, L2, 1/2″ length #12 wire
L3, 4 turns #22 enameled wire close wound on 1/4″ diameter

MOTOROLA   *Fig. 76-19*

This amplifier provides 1-W output at 500 MHz with a DV1202W FET. About 6- to 8-dB power gain can be expected.

HF WIDEBAND AMPLIFIER

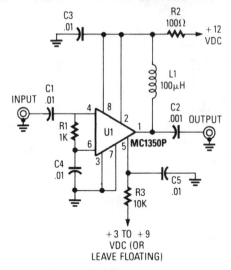

POPULAR ELECTRONICS   *Fig. 76-20*

About 20-dB gain can be obtained using this IC. The gain can be controlled by varying the voltage applied to R3.

MOSFET WIDEBAND AMPLIFIER

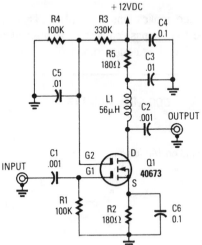

POPULAR ELECTRONICS   *Fig. 76-21*

For high-impedance (7 500 Ω) applications, this amplifier will provide a voltage gain of approximately $-gm/Z_L$ where Z_L is the load impedance in ohms and gm is $\approx 12 \times 10^{-3}$ for the 40673 FET. The G2 voltage can be used to control the gain.

JFET WIDEBAND AMPLIFIER

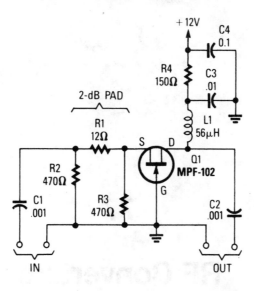

POPULAR ELECTRONICS

Fig. 76-22

Using an MPF102 JFET, this circuit has a 50-Ω input impedance. Load impedance should be about 470 Ω. A 3:1 matching transformer can be used to get the impedance back to 50 Ω. This circuit will show about 6- to 8-dB gain at VHF frequencies.

77

RF Converters

The sources of the following circuits are contained in the Sources section, which begins on page 677. The figure number in the box of each circuit correlates to the entry in the Sources section.

Radio Beacon
Low-Noise 420-MHz ATV Receiver/Converter
VLF Converter
2-Meter Converter
Receiver Frequency-Converter Stage
10-MHz WWV to 80-Meter SW Converter
Shortwave Converter for AM Car Radios
220-MHz Receiving Converter
RF Upconverter For TVRO Subcarrier Reception

RADIO BEACON CONVERTER

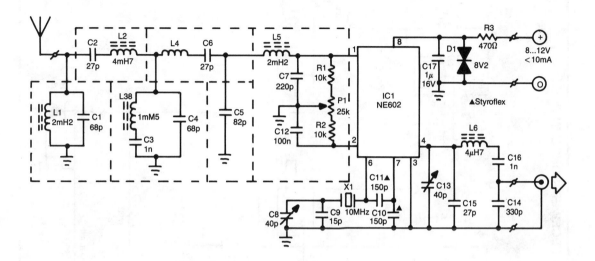

ELEKTOR ELECTRONICS

Fig. 77-1

The radio beacon band extends from 280 to 516 kHz. Each beacon has its own characteristic AM-modulated morse-coded callsign that is transmitted on a specific frequency. To be able to receive distant beacons, the aerial signal is passed through a band-pass filter that effectively suppresses longwave and mediumwave signals. The filter also converts the aerial impedance, Z_{in}, from about 10 kΩ to the input impedance of mixer IC1, which is about 1 kΩ.

The mixer adds or subtracts the received signal to/from the local oscillator signal so that the beacon signal can be received on a normal shortwave receiver. The resulting frequencies are from 9.72 to 9.48 MHz or from 10.280 to 10.516 MHz. In the construction of the converter, some components must be surrounded by a metal shield, as indicated by dashed lines on the PC board layout.

The circuit is aligned with the aid of an SSB receiver, to which the output of the converter is connected. Tune the receiver to 10 MHz and adjust the oscillator frequency of the converter with C8 for zero beat. Next, detune the receiver slightly until you hear a pleasant whistle, which is adjusted for minimum level with the aid of P1. Finally, tune to a beacon transmitting at or about 300 kHz and adjust C13 for maximum sound output.

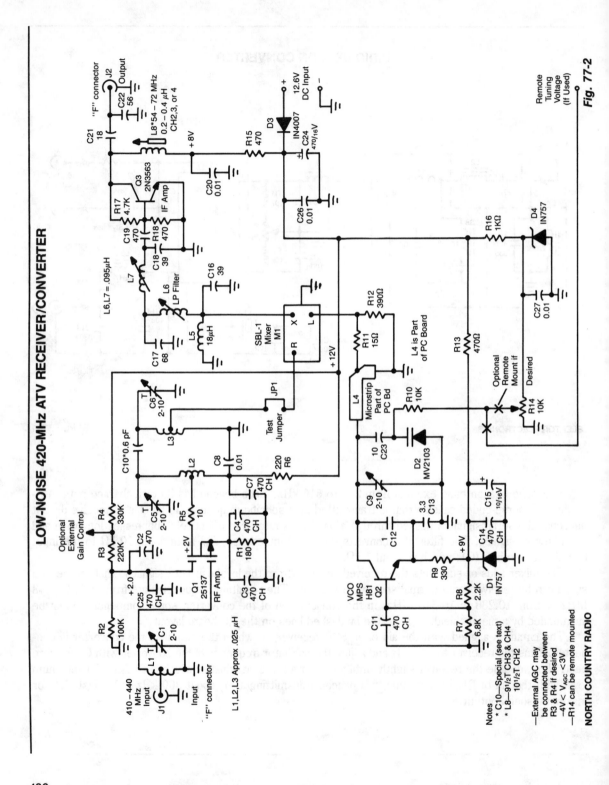

LOW-NOISE 420-MHz ATV RECEIVER/CONVERTER

Fig. 77-2

NORTH COUNTRY RADIO

Notes

—External AGC may
be connected between
R3 & R4 if desired
—4V < V$_{AGC}$ <3V
—R14 can be remote mounted

* C10—Special (see text)
* L8—9½T CH3 & CH4
10½T CH2

LOW-NOISE 420-MHz ATV RECEIVER/CONVERTER *(Cont.)*

   L1, Q1, L2, and L3 compose an RF amplifier stage that feeds M1, a doubly balanced mixer. Q4 is a local oscillator stage in the 375-MHz range. Signals in the 420- to 450-MHz range from Q1 are mixed in M1 and fed through filter L6/L7/C17, where only the 60- to 70-MHz (CH3/CH4) signals pass. The IF signal is passed to Q3, an IF amplifier. The overall gain is 25 dB and the noise figure less than 2 dB. A kit of all parts, including the PC board, is available from North Country Radio, P.O. Box 53, Wykagyl Station, New Rochelle, NY 10804.

VLF CONVERTER

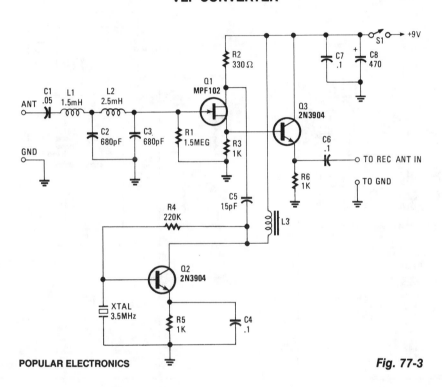

POPULAR ELECTRONICS   *Fig. 77-3*

   The VLF Converter can be used to pick up signals for the general coverage of shortwave receivers. A number of unusual signals can be heard on frequencies below 15 kHz.

   This converter will convert frequencies from 0 to 250 kHz to 3 500 to 3 750 kHz so that the LF- and VLF-band segments can be received on an amateur or shortwave receiver that covers 3 500 to 4 000 kHz. Signals from a short whip antenna (8 to 10 feet) are coupled through low-pass filter L1/L2/C2/C3 to RF amp Q1. Q3 mixes these signals with a 3.5-MHz signal from Q2 and associated components C4, R5, R4, and 3.5-MHz XTAL. L3 is an RF choke that presents an inductive load to Q3. It should be resonant somewhat above 3.5 MHz when placed in the circuit. An adjustable coil of about 30 to 100 μH should be sufficient. The converter output is taken from the emitter of Q3 through C6.

2-METER CONVERTER

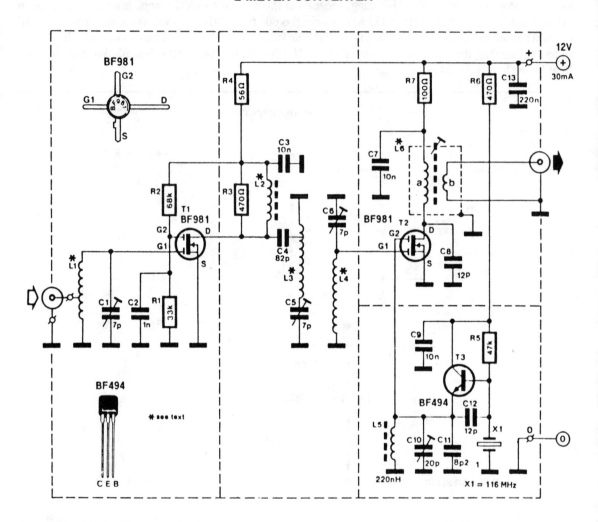

ELEKTOR ELECTRONICS

Fig. 77-4

This converter enables a receiver that tunes 28 to 32 MHz to receive the 144- to 148-MHz amateur band. A BF981 dual-gate MOSFET provides RF gain and feeds mixer T2, another BF981. T3 is a 116-MHz crystal oscillator used to provide L.O. injection to T2. Coils are wound on a 6-mm form. L1, L3, and L4 are 8 turns of 1-mm diameter silver-plated copper wire. L2 is 4 turns of 0.2-mm wire through a ferrite lead. L6 has 19 turns on the primary and 3 turns on the secondary.

RECEIVER FREQUENCY-CONVERTER STAGE

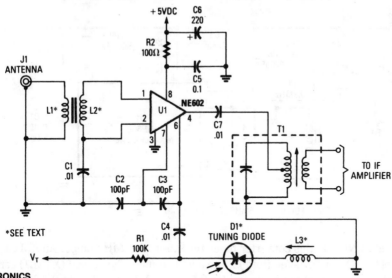

POPULAR ELECTRONICS

Fig. 77-5

L1, L2   1:12 Turns Ratio Toroid (Broadband).
L3   Resonates to L.O. Frequency with D1 capacitance.
LO FREQ   Desired received frequency ± IF frequency.

In this case, the NE602 is used in this superhet front-end configuration. U1 serves as a frequency converter. L1/L2 is a broadband toroidal transformer. A tuned transformer may be used instead. The supply voltage is +5 to +9 Vdc. T1 is tuned to the IF frequency. The typical IF frequency is 455 kHz. This circuit, depending on L1, L2, and L3, should be usable in the frequency range from audio to 30 MHz. The varactor tuning diode can be replaced with an air-variable capacitor, if desired.

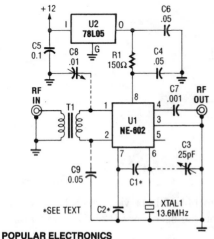

POPULAR ELECTRONICS   Fig. 77-6

10-MHz WWV TO 80-METER SW CONVERTER

This converter is useful where reception of WWV is desirable and only a ham-band receiver is available. U1 acts as a mixer/oscillator. The values of C_1 and C_2 are given by:

$$C1 = \frac{100 \text{ pF}}{\sqrt{f_{MHz}}}$$

$$C2 = 10 \times C_1$$
$$f_{MHz} = \text{crystal frequency}$$

SHORTWAVE CONVERTER FOR AM CAR RADIOS

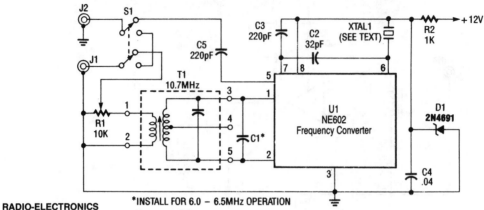

RADIO-ELECTRONICS *INSTALL FOR 6.0 – 6.5MHz OPERATION **Fig. 77-7**

Using a Signetics NE602, this converter tunes the 9.5- to 9.8-MHz range. An AM car radio is used as a tunable IF amplifier. Output is taken from J2, the auto antenna. The crystal (XTAL1) can be a frequency about 1 MHz below the desired tuning range; for 9.5 to 9.8 MHz, an 8.5- to 8.8-MHz crystal should be used.

220-MHz RECEIVING CONVERTER

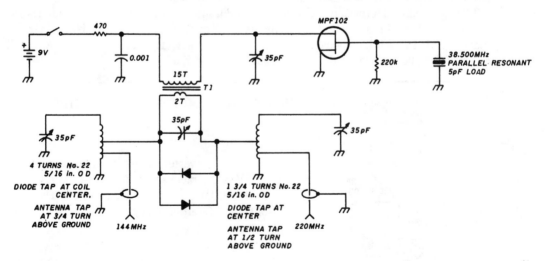

HAM RADIO **Fig. 77-8**

A simple circuit using a single transistor converts 220 MHz to 144 MHz or vice versa, because the mixer is bilateral. T1 has 15 turns on the primary, and 2 turns on the secondary (#24 AWG wire) on a 0.375" ID SF-material toroidal coil.

RF UPCONVERTER FOR TVRO SUBCARRIER RECEPTION

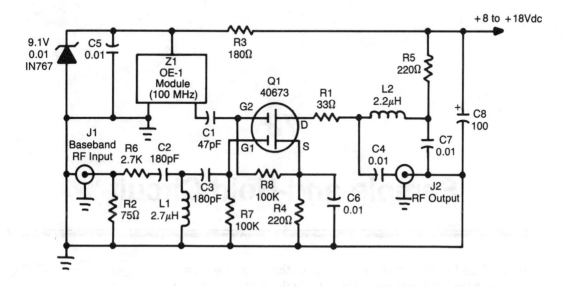

RADIO-ELECTRONICS

Fig. 77-9

This converter uses a 40673 MOSFET to heterodyne the 5.5- to 8-MHz TVRO subcarriers to the FM broadcast band, where a stereo receiver can be used for high-fidelity stereo reception of TV sound subcarriers. Z1 is a prepackaged 100-MHz oscillator module, which is available from International Crystal Corporation, Oklahoma City, OK.

78

Sample-and-Hold Circuit

The source of the following circuit is contained in the Sources section, which begins on page 677. The figure number in the box correlates to the entry in the Sources section.

Sample and Hold

SAMPLE AND HOLD

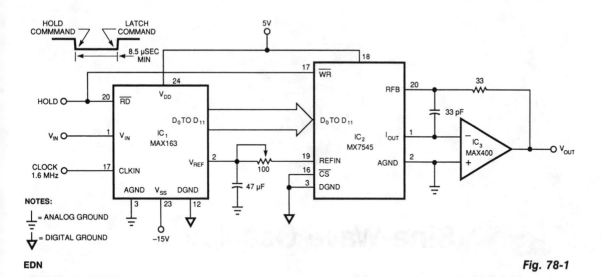

EDN

Fig. 78-1

By using the ADC and DAC, this circuit uses a negative pulse to latch data into the ADC. The hold pulse width should be at least 8.5 μs. This circuit has zero drop and infinite hold time.

79

Sine-Wave Oscillators

The sources of the following circuits are contained in the Sources section, which begins on page 677. The figure number in the box of each circuit correlates to the entry in the Sources section.

Sine-Wave Generator
Pure Sine-Wave Generator
LC Sine-Wave Generator
60-Hz Sine-Wave Generator
Two-Transistor Sine-Wave Oscillator
VLF Audio Tone Generator
Very Low Distortion Oscillator
Low-Frequency LC Oscillator

Three-Decade 15-Hz to 15-kHz Wien-Bridge
  Oscillator
Phase-Shift Oscillator for Audio Range
Wien-Bridge Oscillator
Low-Frequency Sine-Wave Generator
Sine- and Square-Wave TTL Oscillator
Wien-Bridge-Based Oscillator with Very Low
  Distortion

SINE-WAVE GENERATOR

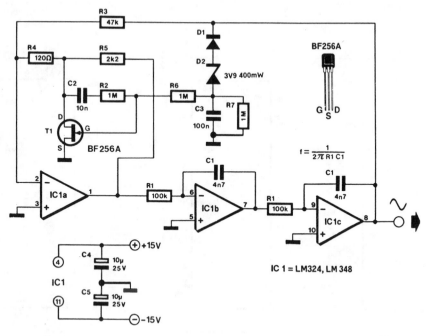

ELEKTOR ELECTRONICS

Fig. 79-1

The frequency of the generator is determined by integrators IC1B and IC1C. An integrator has two properties that are used in this design. Firstly, a phase shift of 90° is between the input and output (ignoring, for the moment, the nonideal behavior of the op amp), and secondly, its amplification is −1 (i.e., the signal inverts), provided the frequency:

$$f = \frac{1}{2\pi R_1 C_1}$$

Cascading two identical integrators will thus result in an overall phase shift of 180° and an amplification of unity (provided that the frequency is $1/2\pi R_1 C_1$): an ideal basis for an oscillator. The two integrators are connected in the feedback circuit of an amplifier whose gain is determined by the amplitude of the output signal. Consequently, the generator has reasonably stable output voltage (at a level of about 4.5 Vpp).

With the values of C1 (C1′) and R1 (R1′), as shown in the diagram, the output has a frequency of about 300 Hz. The frequency can be varied by replacing R1 and R1′ with a stereo potentiometer. To keep the frequency setting within bounds, the overall range of this potentiometer should not exceed a decade.

The maximum attainable frequency is about 5 kHz. Distortion is not greater than 0.1%. The current drawn by the generator is only a few milliamperes. Finally, the LM348 is a quadruple 741; it is thus possible to construct the generator from four 741s.

PURE SINE-WAVE GENERATOR

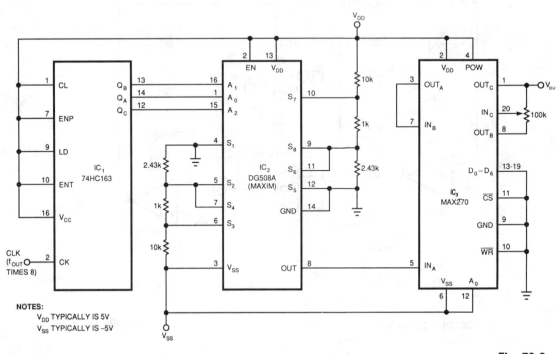

EDN

Fig. 79-2

This circuit produces a pure, -80-dB THD sine wave with a frequency that is equal to the f_c of IC3's filter. A TTL counter, an 8-channel analog multiplexer, and a fourth-order low-pass filter can generate 1- to 25-kHz sine waves with a THD of better than -80 dB. The circuit cascades the two second-order, continuous-time Sallen-Key filters within IC3 to implement the fourth-order low-pass filter. Two resistive dividers connected from ground to V_{DD} and ground to V_{SS} provide bipolar dc inputs to the multiplexer.

To operate the circuit, you first must choose the filter's cutoff frequency, f_c, by tying IC3's D_0 through D_6 inputs to 5 V or ground. The cutoff frequency can be at 128 possible levels between 1 and 25 kHz, depending on those 7 digital input levels. Because this figure ties D_0 through D_6 to ground, f_c equals 1 kHz. The 100-kHz potentiometer adjusts the output level anywhere from 1.5 V below V_{DD} to 1.5 V above V_{SS}.

The clock input frequency must be 8 times higher than the filter's f_c. The multiplexer then produces an 8× oversampled staircase approximation of a sine wave. 8× oversampling greatly simplifies the smoothing requirements of the low-pass filter by pushing the first significant harmonic out to 7× the fundamental. All higher-order harmonics are removed by IC3, which includes an uncommitted amplifier for setting the output level.

LC SINE-WAVE GENERATOR

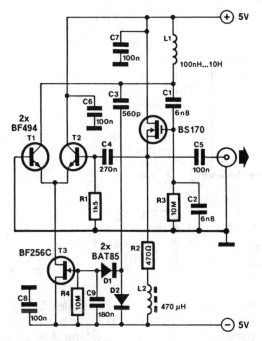

ELEKTOR ELECTRONICS

Fig. 79-3

This compact LC oscillator offers a frequency range of about 1 kHz to almost 9 MHz and a low-distortion sine-wave output. The heart of the circuit is series-resonant circuit L1/C2/C3 in the feedback loop of amplifiers T1/T2. Transistor T2, which is connected as an emitter follower, serves as impedance converter, whereas T1, connected to a common base circuit, is a voltage amplifier whose amplification is determined by the impedance of L1 in its collector circuit and the emitter current.

The feedback loop runs from the collector of T1 via the junction of capacitive divider C1/C2, source-follower BS170, and the input impedance is formed by R1/C4. The whole is strongly reminiscent of a Colpitts circuit. The signal is also taken to the output terminal via C5.

Of particular interest is the amplitude control by the current source. The signal is rectified by two Schottky diodes, smoothed by C9, then used to control the current through T3. The gain of amplifier T1 is therefore higher at low input levels than at higher ones. This arrangement ensures very low distortion, because the amplifier cannot be overdriven.

The resonant frequency can be calculated from:

$$f = \frac{1}{2\pi}\left(\frac{L_1\,C_1\,C_2}{C_1+C_2}\right)$$

With values as shown, it extends from 863 Hz ($L_1 = 10$ H) to 8.630 MHz ($L_1 = 100$ H).

The unit can be used to measure the Q of inductors. To that end, a potentiometer is connected in parallel with L1 and adjusted so that the current through the amplifier is doubled. The Q is then calculated from:

$$Q = \frac{R_p}{2\pi f\,L}$$

60-Hz SINE-WAVE GENERATOR

A chip by Micro Linear and two CMOS (40165 and 4060) chips generate sine waves at 60 Hz.

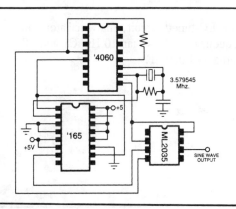

MICRO LINEAR

Fig. 79-4

TWO-TRANSISTOR SINE-WAVE OSCILLATOR

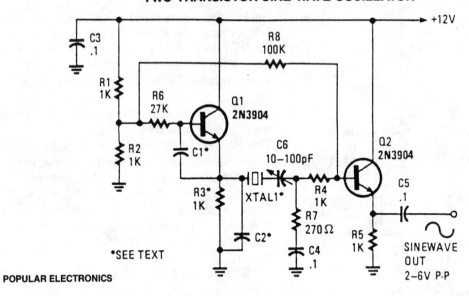

POPULAR ELECTRONICS

Fig. 79-5

This oscillator uses two transistors and operates the crystal in the fundamental mode. C_1 and C_2 should be about 2 700 pF for 1 MHz, 680 pF for 5 MHz, and 330 pF for 10 MHz. 150 pF can be used for up to 20 MHz. The output is a near perfect sine wave. Try varying C_1 and C_2 for best waveform. About 2 to 6 Vpp is available.

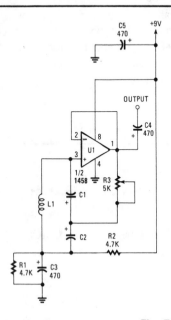

VLF AUDIO TONE GENERATOR

Using an LC-tuned circuit, this oscillator can produce frequencies of less than 10 Hz. C1 and C2 can be as large as 1 000 μF.

POPULAR ELECTRONICS

Fig. 79-6

VERY LOW DISTORTION OSCILLATOR

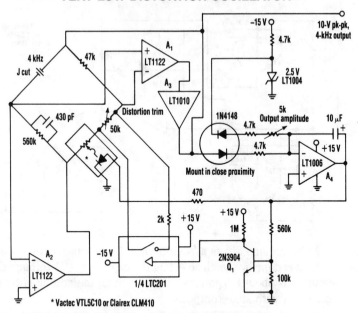

ELECTRONIC DESIGN

* Vactec VTL5C10 or Clairex CLM410

Fig. 79-7

This oscillator uses a bridge circuit with an optoisolator as a gain-control device. The resultant distortion can be held to 9 ppm (.000 9%) with proper adjustment.

LOW-FREQUENCY LC OSCILLATOR

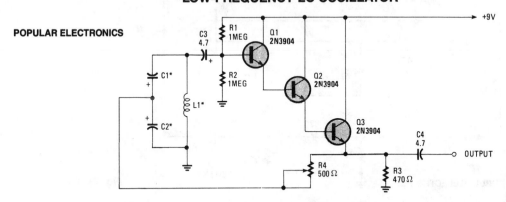

POPULAR ELECTRONICS

Fig. 79-8

Basically a Hartley oscillator using a triple-emitter follower, this oscillator can be used at audio and low radio frequencies. The frequency is given by:

$$f = \frac{1}{2\pi\sqrt{L_1 C_T}}, \text{ where } C_T = \frac{C_1 C_2}{C_1 + C_2}$$

At 1 kHz, typically C would be 4.7 μF tantalums, but this is only a guide as to convenient values to use.

THREE-DECADE 15-Hz TO 15-kHz WIEN-BRIDGE OSCILLATOR

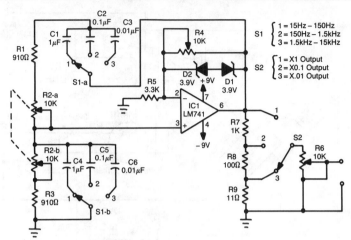

RADIO-ELECTRONICS

Fig. 79-9

In this circuit, an LM741 op amp drives a Wien-bridge network using two zener diodes as an amplitude limiter. Range selection is done by switch selecting the capacitors (C1 through C6) and tuning is done via a ganged pot. The output is about 8 Vpp max., depending on the setting of S2 and R6. R4 is set for maximum distortion consistent with stable output.

PHASE-SHIFT OSCILLATOR FOR AUDIO RANGE

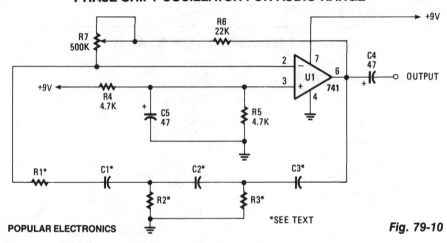

POPULAR ELECTRONICS

Fig. 79-10

This phase-shift oscillator is useful for audio oscillator applications. Adjust R7 for a good sine wave. An amplifier gain of 29 is required for oscillation. If $C = C_1 = C_2 = C_3$ and $R = R_1 = R_2 = R_3$:

$$f = \frac{1}{2\pi\sqrt{6RC}}$$

Typically, R will be 1 to 100 kΩ and C will be 0.1 μF down to 100 pF, in most practical circuits. As a start, choose $R = 10$ kΩ and $C = 0.006\,8$ μF (for $\approx$ 1-kHz range).

WIEN-BRIDGE OSCILLATOR

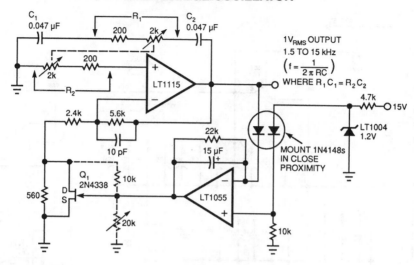

1V$_{RMS}$ OUTPUT
1.5 TO 15 kHz
$\left(f = \dfrac{1}{2\pi RC}\right)$
WHERE R$_1$C$_1$ = R$_2$C$_2$

MOUNT 1N4148s
IN CLOSE
PROXIMITY

EDN

Fig. 79-11

This complex oscillator circuit uses a photocell and common-mode-suppression circuitry to achieve distortion of 0.0003%.

WIEN-BRIDGE OSCILLATOR

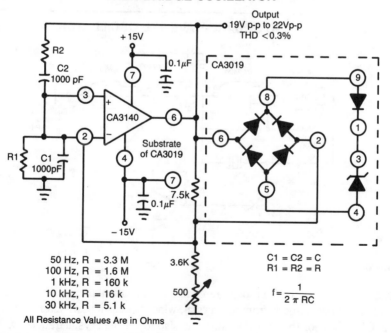

Output
19V p-p to 22Vp-p
THD <0.3%

50 Hz, R = 3.3 M
100 Hz, R = 1.6 M
1 kHz, R = 160 k
10 kHz, R = 16 k
30 kHz, R = 5.1 k

C1 = C2 = C
R1 = R2 = R

$f = \dfrac{1}{2\pi RC}$

GE/RCA   All Resistance Values Are in Ohms

Fig. 79-12

This circuit makes excellent use of high input impedance, high slew rate, and high-voltage qualities of CA3140 BiMOS op amp, in combination with CA3019 diode array.

LOW-FREQUENCY SINE-WAVE GENERATOR

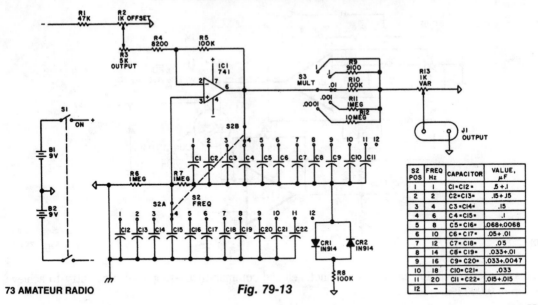

S2 POS	FREQ Hz	CAPACITOR	VALUE, μF
I	I	CI•CI2•	.5 +.I
2	2	C2•CI3•	.I5+.I5
3	4	C3•CI4•	.I5
4	6	C4•CI5•	.I
5	8	C5•CI6•	.068+.0068
6	IO	C6•CI7•	.05+.0I
7	I2	C7•CI8•	.05
8	I4	C8•CI9•	.033+.0I
9	I6	C9•C20•	.033+.0047
IO	I8	CIO•C2I•	.033
II	20	CII•C22•	.0I5+.0I5
I2	–	–	–

Fig. 79-13

Using a Wien-bridge oscillator, this circuit generates switch-selected frequencies from 1 to 20 Hz. A 741 op amp is used as the active element.

SINE- AND SQUARE-WAVE TTL OSCILLATOR

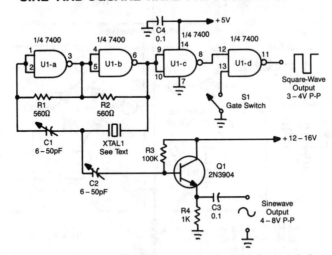

Fig. 79-14

Using a Quad NAND Gate, the TTL oscillator can use fundamental crystals between 1 and 10 MHz. The sine-wave output is taken directly from the crystal, which acts as its own filter, and yields a fairly clean sine wave. Adjust C1 and C2 for the best sine wave and also to set the crystal frequency. TTL square wave can be taken from U1D. The gate switch can be replaced with a logic gate to electronically control the output.

WIEN-BRIDGE-BASED OSCILLATOR WITH VERY LOW DISTORTION

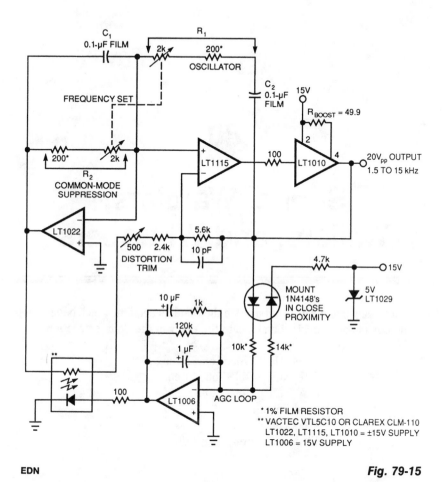

Fig. 79-15

This complex oscillator circuit uses a photocell and common-mode-suppression circuitry to achieve distortion of 0.0003%. This oscillator circuit replaces the lamp in the traditional Wien bridge with an electronic equivalent.

80

Sirens, Warblers, Wailers, and Whoopers

The sources of the following circuits are contained in the Sources section, which begins on page 677. The figure number in the box of each circuit correlates to the entry in the Sources section.

"Hee-Haw" Electronic Siren
Alternate Warble-Tone Siren
6-W Warble-Tone Siren
Low-Cost Siren
Whooper
Electronic Siren

"HEE-HAW" ELECTRONIC SIREN

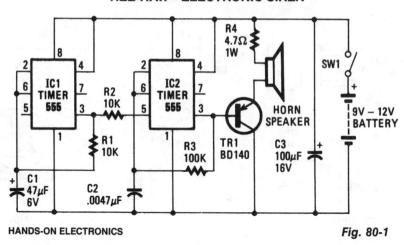

HANDS-ON ELECTRONICS

Fig. 80-1

The oscillator based on IC2 is responsible for producing the sound. Its output is connected to the base of TR1, which amplifies it to drive the speaker. Resistor R4 is included in the circuit to limit the current through TR1 to a safe and reasonable level. The oscillation frequency of IC2 is partially dependent on the values of R_3 and C_2. Another factor that governs the frequency of oscillation is the magnitude of voltage fed to pin 5 of IC2. If a voltage of varying magnitude is fed to pin 5, the internal circuitry of the IC is forced to reset at a different rate, which changes the frequency.

IC1 is also connected as an oscillator, but it runs much slower than IC2: around 1 Hz. Each time the IC triggers, the voltage at pin 3 goes high. As pin 3 is connected to pin 5 of IC2, this forces IC2 to change its note. That produces the "hee-haw" sound of the siren.

ALTERNATE WARBLE-TONE SIREN

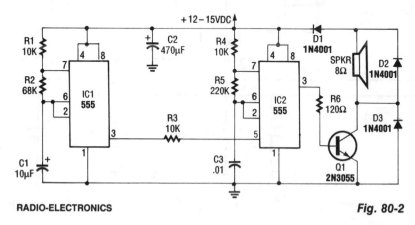

RADIO-ELECTRONICS

Fig. 80-2

This circuit uses two NE555 timers to generate a warble tone. IC1 frequency shifts IC2 by feeding a square wave to pin 5, the modulation input of IC2. IC1 runs at about 1 Hz.

6-W WARBLE-TONE SIREN

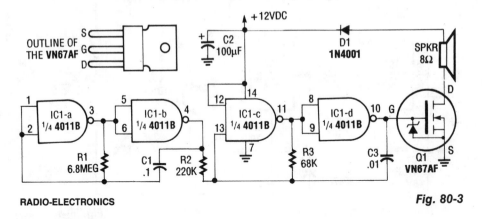

RADIO-ELECTRONICS

Fig. 80-3

This circuit uses a CMOS chip and a VMOS FET amplifier for 6 W of audio output. 18 W of audio can be generated using a +24-Vdc supply. IC1A and IC1B are used as a 1-Hz oscillator. IC1C and IC1D form a 1-kHz multivibrator that is gated by the 1-Hz signal from IC1A and IC1B.

LOW-COST SIREN

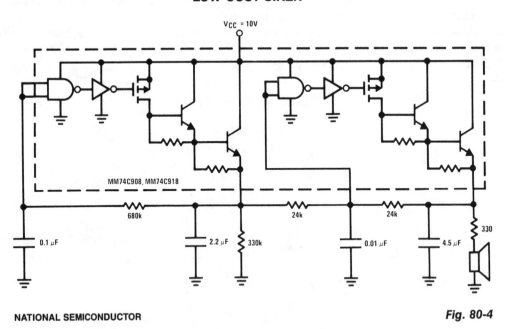

NATIONAL SEMICONDUCTOR

Fig. 80-4

This low-cost 1-package siren has one VCO, and the other oscillator generates the voltage ramp to vary the frequency at the VCO output. All components within the dotted line are part of the IC.

WHOOPER

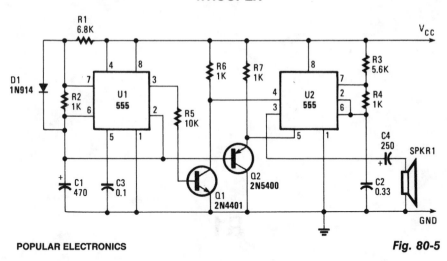

POPULAR ELECTRONICS

Fig. 80-5

Integrated circuit U1 is connected as a low-frequency asymmetrical oscillator. Its output is inverted by Q1 and fed to the reset terminal of U2 at pin 5. Integrated circuit U2 is configured as an audio oscillator and is enabled when the output of U1 is low. With the voltage at pin 5 of U2 constant, the circuit just "bleeps."

The voltage across capacitor C1 is fed to the base of Q2, which turns it on and grounds pin 5 of U2. When the frequency of the reset signal on pin 4 falls, the output frequency of U2 rises. The output then becomes a whoop, starting low in frequency and ending high. Resistor R1 sets the repetition rate and R2 determines the time duration of the whoop. Resistors R3 and R4 set the center-operating frequency.

ELECTRONIC SIREN

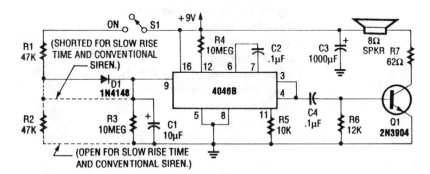

RADIO-ELECTRONICS

Fig. 80-6

For normal wailing tone, short D1 and open R2. For fast rise and slow fall in frequency, include D1 and R2. Use of a CD4046B with a diode-RC network as shown produces a siren tone, using a VCO.

517

81

Sound Effects

The sources of the following circuits are contained in the Sources section, which begins on page 677. The figure number in the box of each circuit correlates to the entry in the Sources section.

Guitar Compressor
Single-Chip Melody Generator
Electronic Bagpipe
Electronic Music Maker
Musical Doorbell
Octave Shifter for Musical Effects
Phasor Sound Generator
Single-Chip Chime

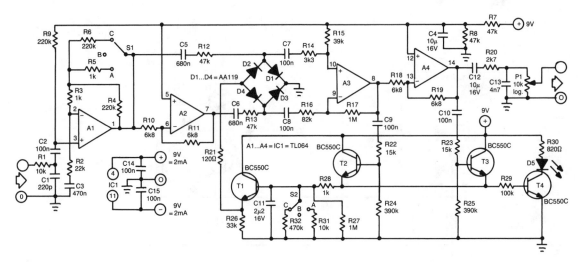

Fig. 81-1

The control of this compressor is based on the dependence of the dynamic resistance of a diode on the current flowing through it. The heart of the present circuit is the diode bridge (D1 through D4), which behaves as a variable resistance controlled by the current flowing in T1.

The input signal is applied to preamplifier state A1 via low-pass filter R1/C1 that removes any HF noise from the input.

Switch S1 in the feedback loop of A1 sets the amplification to 1 (position A), 6 (C) or 11 (B). The amplified signal is applied directly to the diode bridge via R12 and C5, and inverted via inverter A2, capacitor C6 and resistor R13.

The two signals are summed by the bridge, amplified (in A3), then split again into two, one of which is inverted by A4. The positive half cycles of the two signals are used to switch on T2 and T3, respectively. Capacitor C11 is then charged via R12. When the potential across this capacitor reaches a certain level, T1 is also switched on, after which a control current flows through the bridge via R21. This current lowers the resistance of the bridge so that the signal is attenuated (compressed). At the same time, the LED lights to indicate that the signal is being compressed. Capacitor C12 prevents any dc voltage from reaching the output.

The output signal is taken from the wiper of P1. Low-pass section R20/C13 limits its bandwidth to 12 kHz. Switch S2 enables the selection of various decay times from C11. The values shown in the diagram are the most useful. Nevertheless, these values are subjective and can be altered to personal taste and requirements.

SINGLE-CHIP MELODY GENERATOR

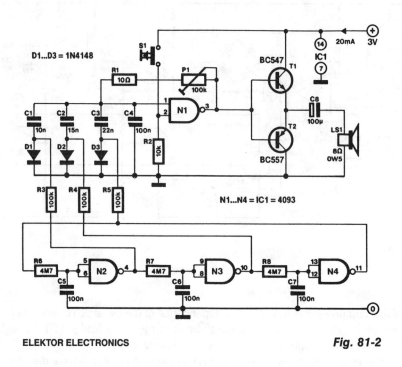

ELEKTOR ELECTRONICS   *Fig. 81-2*

   This melody generator, based on a 4093 CMOS Schmitt trigger, can be used in alarms, doorbells, and cars (audible reverse gear or lights-on indicator).
   Three of the four NAND gates in the 4093 are connected in series by RC networks. Oscillation is affected by feedback of the output signal of N4 to the input of N2. The logic-high levels produced by the cascaded gates in the oscillator circuit are used to bias one of associated diodes D1, D2, or D3). The relevant diode connects one of the frequency-determining capacitors (C1 through C3) to tone oscillator N1. The audio signal available when S1 is pressed is applied to complementary transistor pair (T1/T2) that drives the loudspeaker. The frequency of the emitted tone can be adjusted to individual taste with preset P1.

ELECTRONIC BAGPIPE

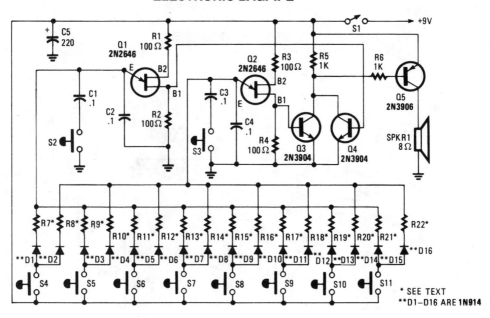

POPULAR ELECTRONICS

Fig. 81-3

The electronic bagpipe mimics the sound of real instruments. This circuit uses two UJT oscillators and an amplifier (Q3, Q4, and Q5). R7 through R22 are selected for tonal range desired (typically 3 300 Ω). Each key selects resistors for the two oscillator circuits Q1 and Q2. S2 and S3 vary the tonal range of S4 through S11.

ELECTRONIC MUSIC MAKER

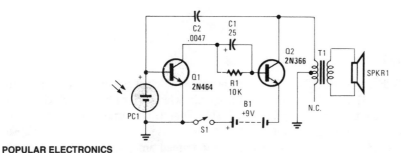

POPULAR ELECTRONICS

Fig. 81-4

This electronic music maker uses an astable oscillator circuit that is controlled by a photocell. The light falling on the photo cell controls the tone. By mounting the circuit in a box, you can control light-reading PC1 with your hand.

MUSICAL DOORBELL

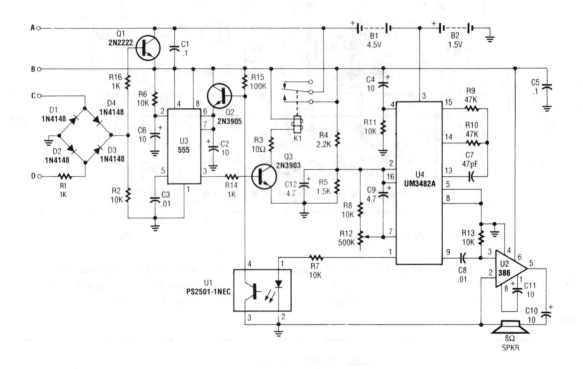

Fig. 81-5

8 to 15 Vac is applied to terminals C and D, which produces a dc voltage across R2, and turns on Q1. This connects the batteries B1 and B2 to the rest of the circuit, which activates it. Latch U3 is triggered, it remains on until Q2 turns on, charges C2, and turns off U2. When U3 is turned on, Q3 is forward-biased, which energizes K1, powers up U4. At the time the K1 contacts close, C4 couples a positive spike to pin 4 of U4, a melody synthesizer chip. U4 generates a pre-programmed tune, at the end of which pin 1 of U4 goes positive. This activates optocoupler U1 and turns off Q2, which drops out the relay. U2 acts as an audio amplifier, which drives an 8-Ω speaker.

OCTAVE SHIFTER FOR MUSICAL EFFECTS

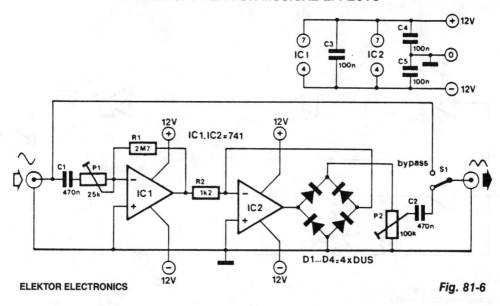

ELEKTOR ELECTRONICS

Fig. 81-6

This musical special-effect device is basically a frequency doubler. The input audio is amplified and doubled by using a full-wave rectifier, which has a dc output plus a twice the ac frequency component. The $2\times$ frequency component is fed to the output.

PHASOR SOUND GENERATOR

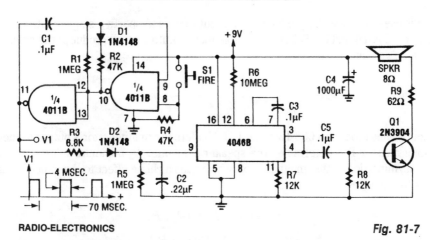

RADIO-ELECTRONICS

Fig. 81-7

The 4011B astable is gated by S1 to produce 4-ms pulses at 70-ms intervals. Each pulse charges C2 via D2, producing a high tone that decays slowly as C2 discharges through R5. The process repeats for each pulse.

SINGLE-CHIP CHIME

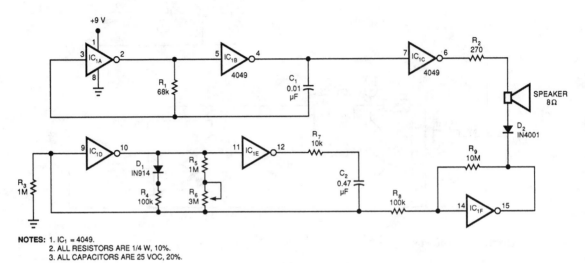

NOTES: 1. IC_1 = 4049.
2. ALL RESISTORS ARE 1/4 W, 10%.
3. ALL CAPACITORS ARE 25 VOC, 20%.

EDN

Fig. 81-8

This circuit uses only one IC, produces a pleasant tone, and sports a single control for adjusting the tone's chiming rate. IC1A and IC1B form an astable multivibrator, which produces the circuit's basic tone. The multivibrator's frequency is:

$$f = \frac{1}{2.2 \times R_1 \times C_1}$$

The component values produce a 668-Hz tone. IC1C buffers the multivibrator's output to the 8-Ω speaker. Current-limiting resistor R2, determines the speaker's volume. R_2 minimum value is 220 Ω. IC1D and IC1E form an asymmetric, astable multivibrator, which adds a chime effect to the circuit's basic tone. The chime effect's frequency is:

$$t_{LO} = 1.1 C_2 (R_4 \parallel (R_5 + R_6))$$

$$t_{HI} = 1.1 C_2 (R_5 + R_6)$$

R_7 gives this rate multivibrator a slowly varying output signal to produce a pleasant decay for the chime effect. IC1F is an inverting amplifier for the chime multivibrator.

82

Sound Operated Circuits

The sources of the following circuits are contained in the Sources section, which begins on page 678. The figure number in the box of each circuit correlates to the entry in the Sources section.

Sound-Activated Switch
Sound-Operated Switch
Microphone-Controlled Voice-Activated Switch
Gain-Controlled Amplifier

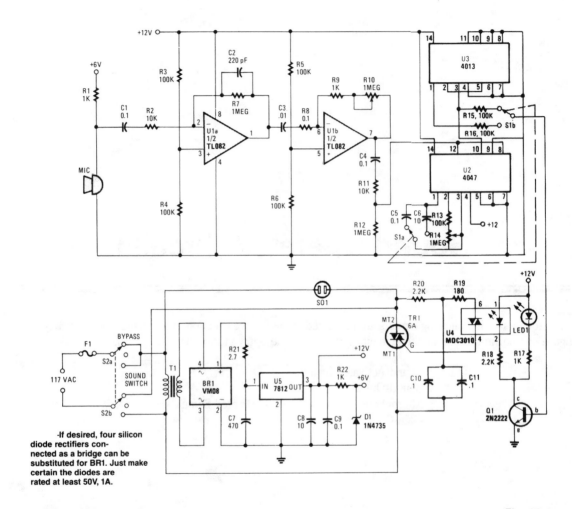

RADIO-ELECTRONICS

Fig. 82-1

This circuit provides either latched switching or timed switching. U1A and U1B provide audio amplification from the microphone. U2 is a retriggerable monostable multivibrator. S1A and S1B select either U3, a flip-flop, or U2. R13 and R14 allow a 6- to 60-second timer delay after the sound ceases, in the timed mode. BR1, U5, and associated components form a power supply. Q1 drives optocoupler U4 and triggers triac TR1.

SOUND-OPERATED SWITCH

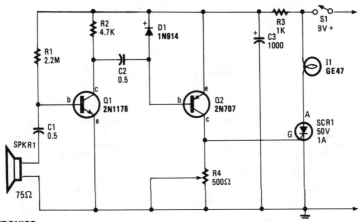

HANDS-ON ELECTRONICS

Fig. 82-2

This sound-operated switch will sense the ring of the phone and translate this to a lamp that will go on and off. The amplified signal across R2 reaches D1 through capacitor C2. The rectified audio signals provide a negative bias for Q2, a pnp transistor. This causes Q2 to conduct so the current that triggers SCR1 is provided at the gate. Potentiometer R4 sets the sensitivity. R3 and C3 delay the operating voltage for Q1 so that the circuit will not be triggered on by the sound of the on/off switch, S1 or by the current surge.

Set the lamp atop a TV receiver, turn it on and set the potentiometer so that a finger snap at two feet will trigger the lamp on. Place the speaker close to the telephone and give it a try.

MICROPHONE-CONTROLLED VOICE-ACTIVATED SWITCH

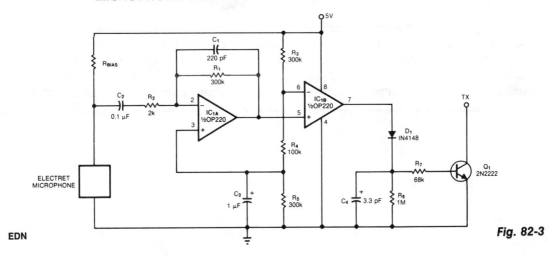

EDN

Fig. 82-3

An electret microphone feeds a bandpass filter circuit (IC1A), then feeds a comparator, which in turn drives Q1. Q1 is a switch that conducts when audio from IC1B causes D1, C4, R6, and R7 to bias it ON.

GAIN-CONTROLLED AMPLIFIER

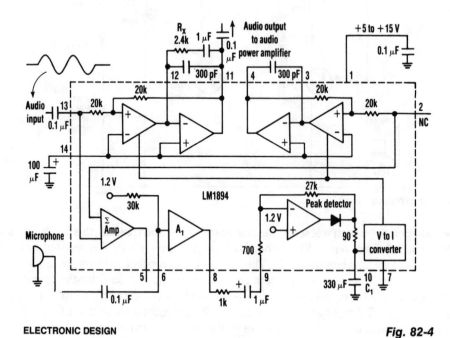

ELECTRONIC DESIGN

Fig. 82-4

This single-chip circuit adjusts its audio gain according to the ambient noise picked up by the microphone. When operating in a quiet environment, the audio output is quiet, while a noisy environment results in a louder audio output. Audio to pin 13 is amplified by the variable-gain amplifier within the LM1894 IC. Audio from the microphone connected through 0.1-μF capacitor to pin 6 controls the audio gain of the variable-gain amplifier. The output appears on pin 11 and is taken off through an 0.1-μF capacitor.

83

Square-Wave Generators

The sources of the following circuits are contained in the Sources section, which begins on page 678. The figure number in the box of each circuit correlates to the entry in the Sources section.

SIMPLE TTL, LSTTL, CMOS, SQUARE-WAVE GENERATORS

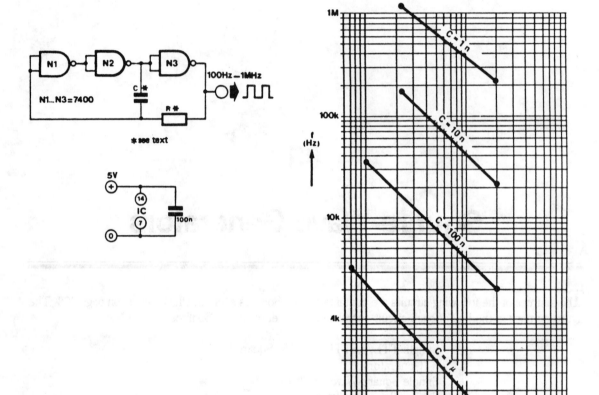

Fig. 83-1(a)

These three circuits for TTL, LSTTL, and CMOS logic use three gates each, and one or two resistors and capacitor as a square-wave oscillator. The circuits are useful for clock oscillators, etc. *R* and *C* are determined from the nomographs.

SIMPLE TTL, LSTTL, CMOS, SQUARE-WAVE GENERATORS (*Cont.*)

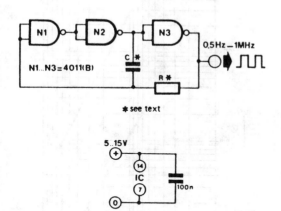

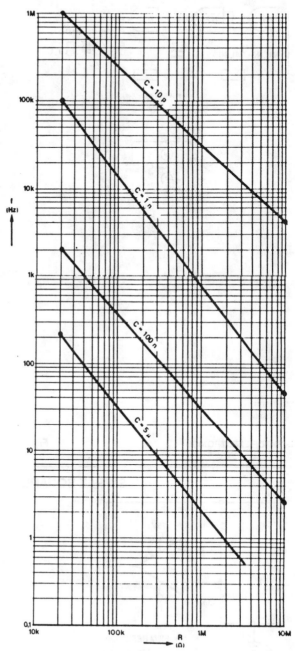

Fig. 83-1(b)

SIMPLE TTL, LSTTL, CMOS, SQUARE-WAVE GENERATORS (*Cont.*)

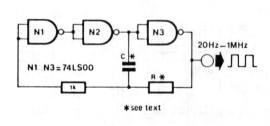

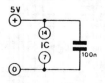

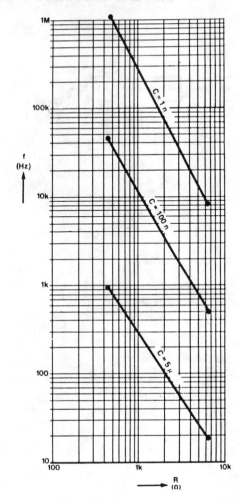

Fig. 83-1(c)

SIMPLE SQUARE-WAVE OSCILLATOR

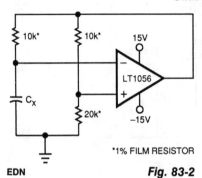

*1% FILM RESISTOR

Fig. 83-2

Using only four components, this circuit generates a square wave. Oscillation frequency is $\approx 1/RC_X$ Hz, $R = M\Omega$, $C_X = \mu F$ (in this case, $R = 10$ kΩ).

SQUARE-WAVE OSCILLATOR

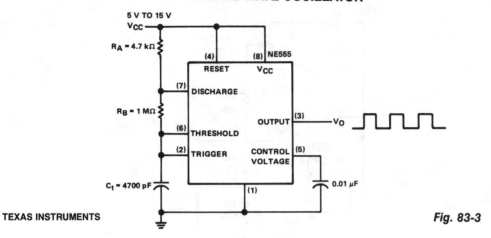

TEXAS INSTRUMENTS

Fig. 83-3

The NE555 is connected in the astable mode and uses only three timing components (RA, RB, and Ct. A 0.01-μF bypass capacitor is used on pin 5 for noise immunity. The operating restrictions of the astable mode are few. The upper frequency limit is about 100 kHz for reliable operation, as a result of internal storage times. Theoretically, it has no lower frequency limit, only that which is imposed by R_t and C_t limitations. The frequency for the circuit can be calculated as:

$$f = \frac{1.44}{(R_A + 2R_B)C_t}$$

$$= \frac{1.44}{(4.7 \text{ k}\Omega + 2 \text{ M}\Omega)\,(0.0047 \ \mu\text{F})}$$

$$= \frac{1.44}{9.42209 \times 10^{-3}}$$

$$f = 152.8 \text{ Hz}$$

VARIABLE DUTY-CYCLE SQUARE-WAVE GENERATOR

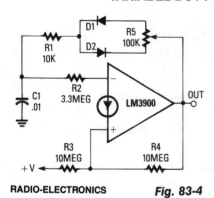

C1 alternately charges via R1/D1 and the upper half of R5, and discharges via R1/D2 and the lower half of R5. The duty cycle can be varied over the range from 1:10 to 10:1 via R5.

RADIO-ELECTRONICS

Fig. 83-4

SQUARE-WAVE GENERATOR

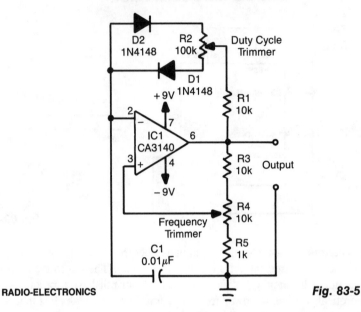

RADIO-ELECTRONICS

Fig. 83-5

This relaxation oscillator circuit uses diodes to produce charge and discharge paths for C1. The duty cycle is set via R2 and the frequency via R4. C1 can be varied to vary the frequency range, which, for this circuit is approximately 300 to 3 000 Hz.

SQUARE-WAVE ASTABLE CIRCUIT

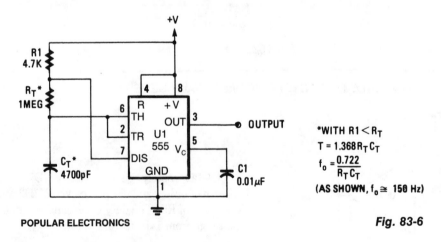

POPULAR ELECTRONICS

*WITH R1 < R_T
$T = 1.368 R_T C_T$
$f_o = \dfrac{0.722}{R_T C_T}$
(AS SHOWN, $f_o \cong 150$ Hz)

Fig. 83-6

This 555 circuit produces a square wave. The frequency depends on the values of R_T and C_T, as per the design equations.

4-DECADE SQUARE-WAVE GENERATOR

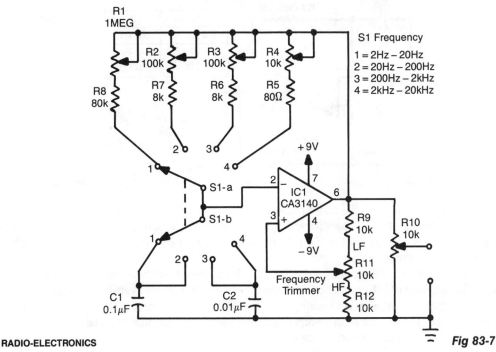

S1 Frequency
1 = 2Hz – 20Hz
2 = 20Hz – 200Hz
3 = 200Hz – 2kHz
4 = 2kHz – 20kHz

RADIO-ELECTRONICS

Fig 83-7

This circuit will generate a square wave of 2 Hz to 20 kHz. The circuit uses an op amp in a relaxation-oscillator configuration. The output is about 15 Vpp. R1 through R4 are calibration controls for each of the four frequency ranges, as selected by S1-a and S1-b. R10 adjusts the output level.

SIMPLE VARIABLE-FREQUENCY SQUARE-WAVE GENERATOR

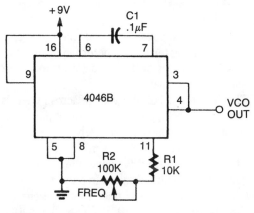

A CD4046B PLL is used as a simple generator. The range is 200 Hz to 2 kHz, but it can be changed by changing C1.

RADIO-ELECTRONICS

Fig. 83-8

BASIC MULTIVIBRATOR

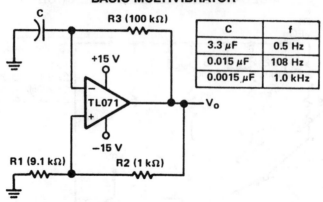

C	f
3.3 µF	0.5 Hz
0.015 µF	108 Hz
0.0015 µF	1.0 kHz

TEXAS INSTRUMENTS

Fig. 83-9

When this circuit is turned on, the natural offset of the devices serves as an automatic starting voltage. Assume that output voltage V_O goes positive and the positive feedback through R2 and R1 forces the output to saturate. The high-voltage level at V_O, then charges C through R3, until the voltage at the inverting input exceeds that at the noninverting input.

As the inverting input exceeds the noninverting input level, the output switches to the negative saturation voltage. This action starts the capacitor discharging toward the new noninverting input level. When the capacitor reaches that level, the op amp switches back to the positive saturation voltage, and the process starts again. With the TL071, the positive and negative output levels are nearly equal, which results in a 50% duty cycle. The total time period of one cycle will be:

$$t_T = 2\ (R_3)C\ \ln\!\left(\frac{1+2R_1}{R_2}\right)$$

1-kHz SQUARE-WAVE GENERATOR

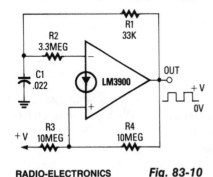

RADIO-ELECTRONICS   Fig. 83-10

When the output is high, R3 and R4 are in parallel, and C1 charges via R1 until the current in R2 equals that at the noninverting terminal. This action occurs when C1's voltage rises to 2/3 of the supply voltage. At that point, the circuit switches regeneratively. The output switches low and C1 starts to discharge via R1.

Now, R4 is effectively disabled and the current to the noninverting terminal is determined solely by R3, so C1 discharges until the current through R2 falls slightly below that of R3.

This happens when the voltage across C1 falls to about 1/3 of the supply voltage. At that point, the circuit again switches regeneratively, and the output again goes high.

This circuit is useful for generating symmetrical square waves with maximum frequencies of only a few kHz. Because of the poor slew-rate characteristics of the LM3900 (0.5 V/µs), the output waveforms have rather slow rise and fall times.

84

Switching Circuits

The sources of the following circuits are contained in the Sources section, which begins on page 678. The figure number in the box of each circuit correlates to the entry in the Sources section.

Electronic Antenna Selector
$(N+1)$ Wires Connect N Hall-Effect Switches
Audio/Video Switcher
Precision Narrow-Band Tone Switch
Diode RF Switch
Satellite TV Audio Switcher

On/Off Switch
Switching Circuit
Inexpensive VHF/UHF Diode RF Switch
Mechanically Controlled Bistable
Bounce-Free Auto-Repeat Switch
Transistor Turns Op Amp On or Off

ELECTRONIC ANTENNA SELECTOR

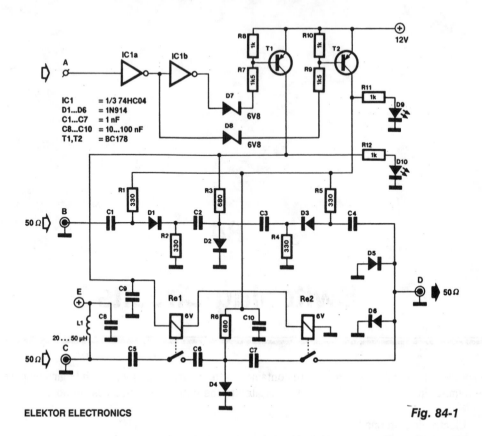

ELEKTOR ELECTRONICS

Fig. 84-1

The electronic antenna selector is intended to switch between two FM antennas via a logic signal. Gates IC1A and IC1B ensure a clean switching action and at the same time form the interface between the 5-V logic level (probably available from the receiver) and the 12-V supply voltage for the selector. Depending on the type of gate used, a digital TTL or CMOS control signal is available in direct and inverted form at the outputs of IC1.

When input A is logic high, the output of IC1A is low and that of IC1B is high. The current then flows from the positive supply line to IC1A via T2, R9, and D8. T2 is switched on and D9 lights.

Because direct currents flow through R1/D1/R2 and R5/D3/R4, diodes D1 and D3 conduct and pass the VHF signal from input A to output D. At the same time, a direct current flows through R6/D4 so that D4 conducts. This arrangement ensures that any VHF signal at input C cannot reach the output via the parasitic capacitances of the relay contacts and the wiring.

When A is logic low, and IC1B is therefore low, the current flows from the positive supply line to IC1B via T1, R7, and D17. T1 is then switched on and D10 lights. At the same time, the two series-connected relays, Re1 and Re2, are energized, their contacts close, and the VHF signal at input C is fed to output D. Moreover, a direct current flows through R3/D2 so that D2 conducts. Any signal at input B is then shorted to ground via D2.

ELECTRONIC ANTENNA SELECTOR (Cont.)

All resistors should be carbon-film types because these have a higher parasitic series inductance than metal-film resistors. Thus, the attenuation of the VHF signal caused by them is reduced to a minimum.

The attenuation losses caused by the diode junctions (5-10 dB) are somewhat larger than those caused by the relays. It is thus advisable to connect the antenna that provides the weaker signal (normally the domestic one) to input C. If the domestic antenna is equipped with an antenna amplifier, it can be supplied via terminal E. Diodes D5 and D6 protect the circuit against high-voltage spikes that occur during the on and off switching. The selector draws a current of approximately 65 mA.

(N + 1) WIRES CONNECT N HALL-EFFECT SWITCHES

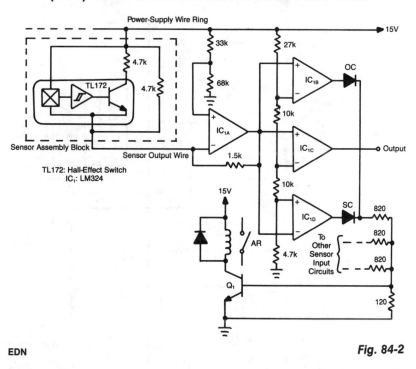

Fig. 84-2

Hall-effect switches have several advantages over mechanically and optically coupled switches. Their major drawback is that they require three wires per device. This circuit, however, reduces this wire count to $N+1$ wires for N devices.

Amplifier IC1A is configured as a current-to-voltage converter. It senses the sensor assembly's output current. When the Hall-effect switch is actuated, the sensor's output current increases to twice its quiescent value. Amplifier IC1B, configured as a comparator, detects this increase. The comparator's output goes low when the Hall-effect switch turns on.

The circuit also contains a fault-detection function. If any sensor output wire is open, its corresponding LED will turn on. If the power-supply line opens, several LEDs will turn on. A short circuit will also turn an LED on. Every time an LED turns on, Q1 turns on and the alarm relay is actuated.

AUDIO/VIDEO SWITCHER

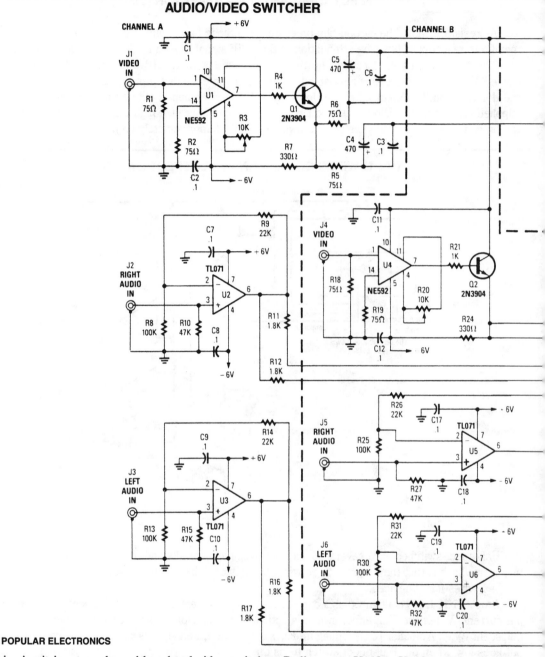

POPULAR ELECTRONICS

This circuit is a two-channel baseband video switcher. Buffer amps U1/Q1, U4/Q4, and associated components produce a buffered 75-Ω video signal, which is routed to switching network K1/K2/K3. Relay K1 selects either of these two video amplifiers and feeds J7. K2 also routes the output of either video amplifier to K3, which passes the selected video channel to J8 or connects J9 to J8.

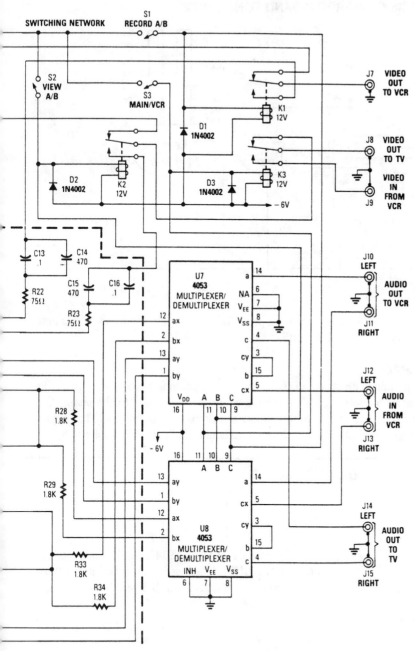

Fig. 84-3

U2 and J3 are audio amplifiers, which drive U7 and J8 (CD4053 analog switches) to route audio from J1 and J2 to either J14/J15, or J10/J11. Also, audio from J12/J13 can be routed to these jacks.

PRECISION NARROW-BAND TONE SWITCH

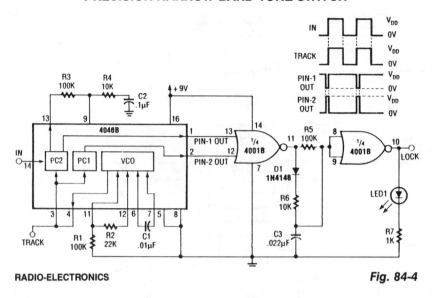

RADIO-ELECTRONICS

Fig. 84-4

This signal tracker and lock detector combine to make a precision tone switch. Filter R3/R4/C2 determines signal capture and tracking range, as well as settling time.

Max. VCO frequency: R_1C_1

Min. VCO frequency: $(R_1 + R_2)C_2$

Pin 9 voltage affects both. The minimum at pin 9 is 0 V and the maximum at pin 9 is V_{DD}. In the lock detector, the PC (phase comparator) outputs are pulses whose width is proportional to the phase difference between the two PC inputs. At lock up, the two PC outputs are almost mirror images. The output of IC1A remains low and IC1B is high. This lights LED1. If the loop is unlocked, the LED will not light.

DIODE RF SWITCH

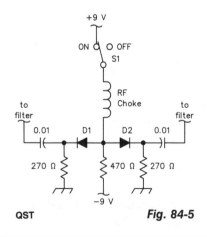

QST

Fig. 84-5

D1 and D2 can be IN914B- or HP2800-series diodes (for UHF). The loss is over 60 dB in the OFF state, and less than 3 dB at 3.5 to 30 MHz (using common IN914B diodes).

SATELLITE TV AUDIO SWITCHER

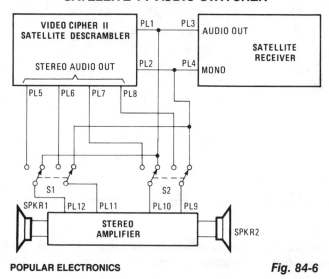

POPULAR ELECTRONICS

Fig. 84-6

Some channels offer a separate audio mode (SAP) in a second language. It is usually transmitted on 6.8 MHz, which is the frequency used for unscrambled channels. The audio in the scrambled channels is transmitted along with the picture, so when the descrambler descrambles the signal, it also descrambles the audio in stereo. When the channel offers SAP, you'll find it on 6.8 MHz.

The switches are a pair of DPDT switches that have the toggle handles tied together. In one position, you hear the audio in stereo and in the other position, you hear the SAP. Just turn down the volume level on the TV and you can connect it to a stereo amplifier and a pair of speakers.

ON/OFF SWITCH

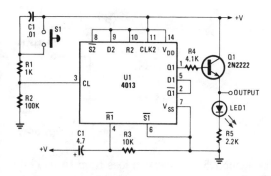

A CD 4013 dual-D flip-flop is used to drive an emitter-follower. This circuit can be used where a simple pushbutton on/off is desired.

POPULAR ELECTRONICS

Fig. 84-7

SWITCHING CIRCUIT

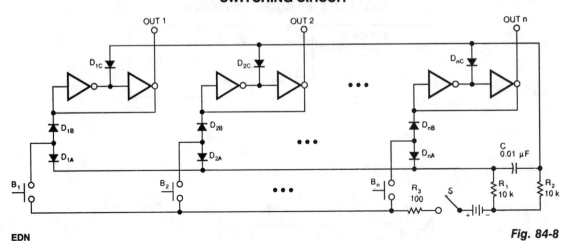

EDN

Fig. 84-8

This switching circuit acts like a bank of interlocked mechanical switches; pushing one of the buttons latches its corresponding output and unlatches a previously selected output. A pair of inverters forms a latch for each output.

Pressing button B1, for example, applies a positive pulse, via resistor diode D1B, to the input of the first output's, OUT1, latch. This positive pulse will set OUT1 high. Feedback locks OUT1's pair of converters in this HIGH state. Meanwhile, the pulse will also pass through diode D1A to the differentiator that is formed by C and R2. The differentiator will shorten the pulse.

The shortened pulse goes to all the latches and resets all of them, except the latch that sees the longer setting pulse. Obviously, if you press more than one button at once, more than one output will latch at once.

INEXPENSIVE VHF/UHF DIODE RF SWITCH

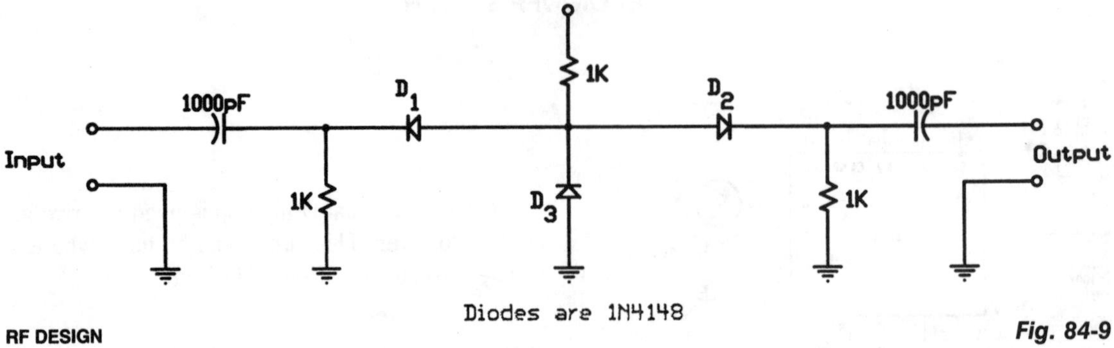

Diodes are 1N4148

RF DESIGN

Fig. 84-9

This circuit uses low-cost IN4148 diodes and exhibits about 1.5 dB insertion loss from 10 to 1 000 MHz with a few volts of negative bias. D3 conducts and D1/D2 are cut off, which results in 30 to 50 dB isolation. When a few volts of positive bias are applied, D1 and D2 are biased on and D3 is cut off. This circuit should be useful in applications where a low-cost RF switch is necessary.

MECHANICALLY CONTROLLED BISTABLE

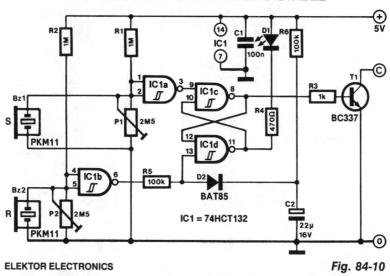

ELEKTOR ELECTRONICS

Fig. 84-10

Applications for this mechanically set and reset bistable are found, among others, in antitheft devices and model railway crossings.

The transducers are formed by buzzer BZ1, which sets the bistable, and BZ2, which resets it. Their sensitivity is set with P1 and P2, respectively. The presets are adjusted correctly if the output of buffers IC1A and IC1B just toggles from high to low or vice versa.

If all have been set correctly, a slight tap of BZ1 will set the bistable. This tap causes T1 to switch on, which enables, for instance, a relay to be energized. At the same time, D1 lights. A tap on BZ2 or on its mounting resets the bistable, whereupon D1 goes out and T1 is switched off.

BOUNCE-FREE AUTO-REPEAT SWITCH

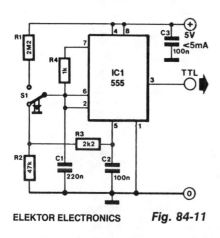

ELEKTOR ELECTRONICS   *Fig. 84-11*

This switch that keeps pulsing as long as it is pressed is often required. The circuit here used the well known Type 555 for this purpose. Its output is a TTL-compatible signal.

Pin 5 of the timer has a potential of 67% of the supply voltage, U_{CC}. In the quiescent condition (switch not pressed), C1 charges via R2 and R3 to a voltage that is lower than that at pin 5, and thus is also lower than the toggle voltage.

When the switch is pressed, C1 is rapidly charged via R1 to the toggle voltage, upon which the timer emits a pulse. At the same time, the capacitor is discharged again via R4. As long as the switch is pressed, the circuit functions as an astable toggle and produces pulses. When it is released, the capacitor cannot charge to the toggle voltage.

TRANSISTOR TURNS OP AMP ON OR OFF

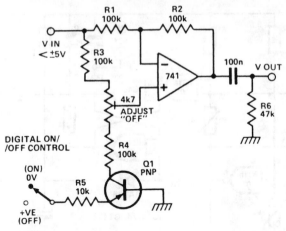

ELECTRONICS TODAY INTERNATIONAL   *Fig. 84-12*

When transistor Q1 is switched off, the circuit behaves as a voltage follower. By applying a positive voltage to the emitter of Q1 via a 10 kΩ resistor, the transistor is made to turn on and go into saturation. Thus, the lower end of R4 is connected to ground. The circuit has not changed into that of a differential amplifier, except that the voltage difference is always 0 V. As long as the resistor ratios in the two branches around the op amp are in the same ratio, the output should be zero. A 47-kΩ resistor is used to null out any ratio errors so that the OFF attenuation is more than 60 dB. The high common-mode rejection ratio of a 741 enables this large attenuation to be obtained.

85

Tape Recorder Circuits

The sources of the following circuits are contained in the Sources section, which begins on page 678. The figure number in the box of each circuit correlates to the entry in the Sources section.

Audio-Powered Tape Recorder Controller
12-V Auto-Powered Circuit For Cassette Recorders

AUDIO-POWERED TAPE RECORDER CONTROLLER

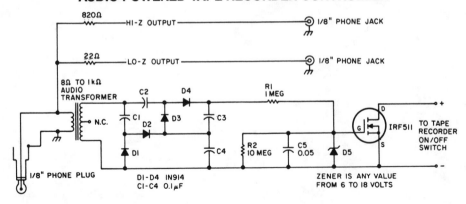

73 AMATEUR RADIO

Fig. 85-1

A tape recorder can be controlled by rectifying the audio input and driving an IRF511 power MOSFET to switch a tape recorder on when audio is present. This circuit was used with a communications receiver to record intermittent transmissions (such as aircraft, repeater output, etc.).

12-V AUTO-POWERED CIRCUIT FOR CASSETTE RECORDERS

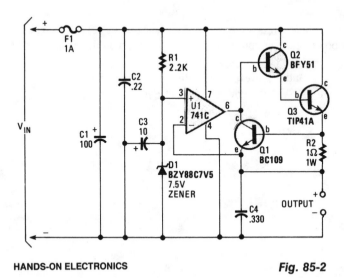

HANDS-ON ELECTRONICS

Fig. 85-2

A regulator allows you to power a 7.5-V cassette recorder or other device from a 12-Vdc auto system. About 600 mA is available from the circuit. Q3 should be heatsinked because it dissipates up to 4 W. F1 should be a slow-blow fuse so that the surge caused by C1 does not cause unnecessary fuse failures.

86

Telephone-Related Circuits

The sources of the following circuits are contained in the Sources section, which begins on page 678. The figure number in the box of each circuit correlates to the entry in the Sources section.

Telephone-Operated ac Power Switch
Telephone Toll Totalizer
Remote-Controlled Telephone/Fax Machine Switch
Telemonitor for Recording Phone Calls
Duplex Audio Link
Low-Power Touch-Tone Decoder
Telephone Speaker Amplifier
Telephone Ringer
Phone Message Flasher
Telephone Silencer
Telephone Intercom
Telephone Auto Record

Tell-A-Bell
Answering Machine Beeper
Telephone Visual Ring Indicator
Phone-In-Use Indicator
Telephone Amplifier
Extension Phone Ringer
Telephone Visual Ring Indicator
Call Tone Generator
Remote Ringer
Telephone Message Taker
Telephone Line-In-Use Indicator
Simple Ring Detector

TELEPHONE-OPERATED ac POWER SWITCH

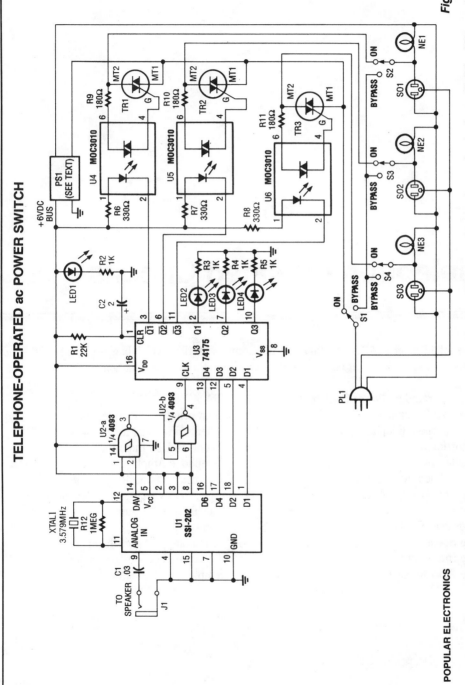

Fig. 86-1

Tones (from the DTMF) on the telephone line are detected by U1. When a valid tone is received, pin 14 (DAV) of U1 produces a positive pulse that is used to drive NAND gates U2A and U2B and then to latch binary data from U1 into Quad-D flip-flop U3 (U1 could be decoded into 16 bits, if required). The Q outputs of U3 drive optoisolators that control triacs TR1, TR2, and TR3. PS1 is a 6-V 150-mA dc adapter that operates from 120 Vac.

550

TELEPHONE TOLL TOTALIZER

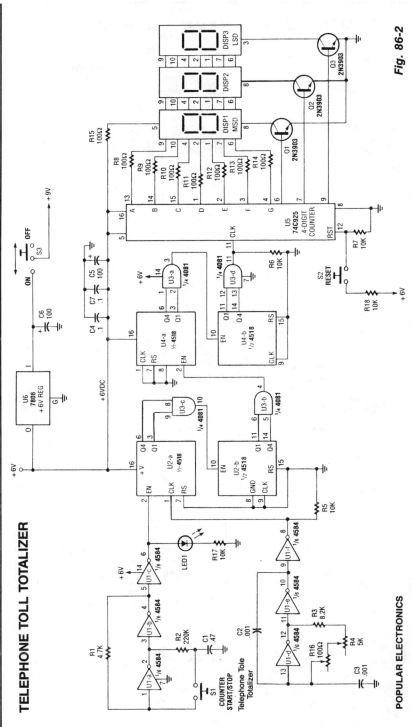

Fig. 86-2

POPULAR ELECTRONICS

The Telephone Toll Totalizer—built around two 4 518 dual synchronous up counters, a 74C925 4-digit counter, a 4 584 hex inverting buffer, and a 4 081 quad 2-input AND gate—is fairly simple.

Approximate toll charges can be calculated with this counter. It is started when dialing and stopped (manually) on hang-up. It is actually a counter that measures the time you are on the telephone. By calibrating it to the average cost/second of calls (get this from calculations you have done on your monthly phone bill), you can closely estimate your phone bill.

The circuit consists of an oscillator running at the 100 000 × frequency into the main counter (74C925). Typically, cost of telephone calls is 15 to 25 cents/minute so that the clock frequency (U1) is in the 25- to 40-kHz range. U2 and U4 with gates U3 form a ÷ 100 000 counter. The approximate cost in dollars and cents is read out on the multiplexed display, DISP 1, 2, 3. S2 resets the counter to zero after each use.

551

REMOTE-CONTROLLED TELEPHONE/FAX MACHINE SWITCH

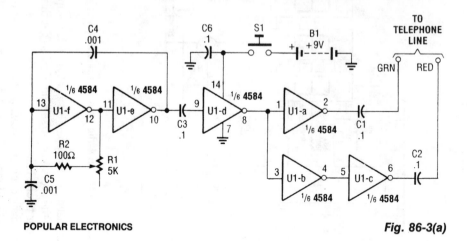

POPULAR ELECTRONICS

Fig. 86-3(a)

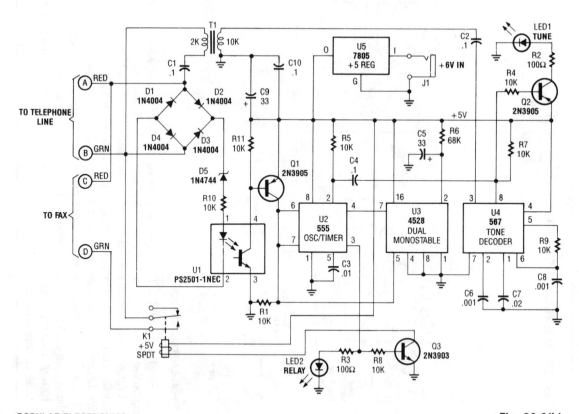

POPULAR ELECTRONICS

Fig. 86-3(b)

REMOTE-CONTROLLED TELEPHONE/FAX MACHINE SWITCH (*Cont.*)

This system uses a transmitter at around 100 kHz (see Fig. 86-3(a)) to control a remote receiver. A line splitter can be used to connect the transmitter to the telephone line in use. The transmitter is a CMOS oscillator and has output buffer stages to drive the telephone line.

When the receiver (Fig. 86-3(b)) detects the off-hook condition (the line voltage drops from about 48 V to less than 10 V). Optocoupler U1 has the LED extinguished. This enables timer U2. When the transmitter is activated, tone decoder U4 detects the 100-kHz signal and outputs a low signal, which lights indicator LED1. LED1 is also used to set the transmitter frequency. Also, U2 is triggered. U2 is configured for the latching condition. U2 feeds the base of Q3, turns it on and energizes relay K1, which switches in the fax machine to the telephone line.

U3 prevents U2 from being accidentally triggered by transients on the telephone line. When the phone is lifted off-hook and it resets U2 after about one or two seconds of delay, transients are allowed to subside before U2 is reset, and it waits for a negative pulse on pin 2 to turn on. When U2 turns on, Q3 is biased on, which activates change-over relay K1.

TELEMONITOR FOR RECORDING PHONE CALLS

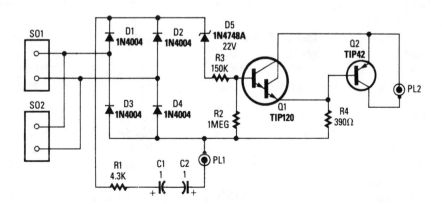

POPULAR ELECTRONICS

Fig. 86-4

This circuit switches a tape recorder via PL2. When on-hook, D5 conducts, turns on Q1, and cuts off Q2. Ringing voltage will also cause D5 to conduct; C2 and C1 should be rated 150 V or higher. When phone is off-hook, the 10 V or so present on the line will not break down D5, and therefore Q1 is off and Q2 is biased on. PL2 connects to the remote control jack on the tape recorder. Audio is taken from PL1.

Caution: Use either a battery tape recorder or an FCC/CSA/UL-approved ac adapter-powered tape recorder. This precaution is to avoid inadvertent 120 Vac on the telephone line.

DUPLEX AUDIO LINK

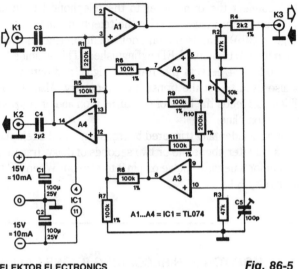

ELEKTOR ELECTRONICS

Fig. 86-5

Duplex communication is, of course, not a new technique: it has been used, for instance, in telephone systems for many years. Those systems, however, use transformers to achieve duplex—this circuit does it with the aid of electronics.

The principle is fairly simple. Two senders impose signals (U_1 and U_2, respectively) on to the audio cable. The voltage across the cable is then $(U_1 + U_2)/2$. The receivers at both sides of the cable deduct their side's sender signal from the cable signal: the result is that the signal is sent from the other end of the cable. This principle is the basis of the circuit shown. Notice that a similar circuit is required at either end of the link.

Op amp A1 is connected as a buffer amplifier and serves as sender. The send signal is imposed on the cable via R4. Terminating the cable by R4 results in the voltage across the cable being only half the voltage output of A1. This does not detract from the operation of the circuit, however. At the same time, R4 ensures that signals emanating from the other end of the link cannot get to the output of A1; if they could, they would be short-circuited by the output.

The receiver is a differential amplifier consisting of op amps A2 through A4. The quality of the differential amplifier depends largely on the resistors used with the op amps; 1% types are, therefore, essential.

The cable signal, $(U_1 + U_2)/2$, is applied to one input of the differential amplifier and the (halved) output signal of A1 to the other. Because the differential amplifier has a gain of 6 dB, the received signal applied to K2 has the same level as the original signal.

The circuit is calibrated by connecting the cable to it and to its twin circuit, then injecting a 1-kHz sinusoidal signal and a 5-Vrms level to its input. The input bus of the other circuit must be short-circuited during the calibration. Adjust P2 for minimum signal at K2. Next, increase the frequency of the input signal to 10 kHz and adjust C5 for a minimum signal at K2. Repeat the procedure with the other circuit. The signal suppression at 1 kHz is of the order of 80 dB; at 20 kHz, it is approximately 60 dB.

LOW-POWER TOUCH-TONE DECODER

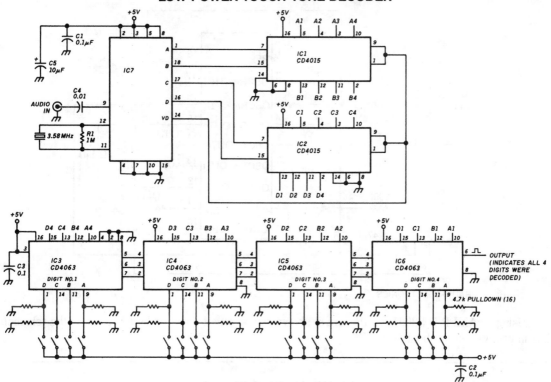

HAM RADIO

Fig. 86-6

This decoder will respond to a preselected 4-digit DTMF number. IC7 is a Radio Shack IC device (part #276-1303). The logic is all CMOS. The digits are selected by SW1 and SW2, a pair of 8-position DIP switches.

TELEPHONE SPEAKER AMPLIFIER

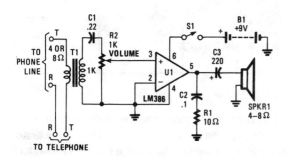

This simple telephone amplifier (which can be switched off for privacy) allows everyone in the room to listen to your telephone conversations.

POPULAR ELECTRONICS

Fig. 86-7

TELEPHONE RINGER

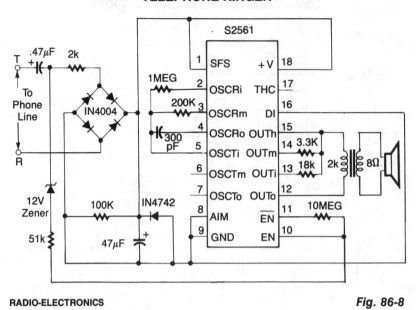

RADIO-ELECTRONICS

Fig. 86-8

Using an AMI P/N S2561 IC, the circuit shown can be either powered by a battery or the telephone line in use. Output is about 50 mW.

PHONE MESSAGE FLASHER

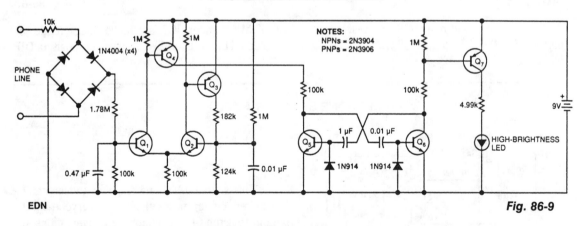

EDN

Fig. 86-9

This circuit flashes an LED to indicate that your phone rang during your absence. A differential amplifier with hysteresis (Q1, Q2, and Q3) detects high line voltage (ringing), which turns on Q4, multivibrator Q5/Q6, and flashes the LED via Q7. Q1 and Q2 remain on until the phone-line voltage drops to less than 9 V, which indicates an off-hook condition.

TELEPHONE SILENCER

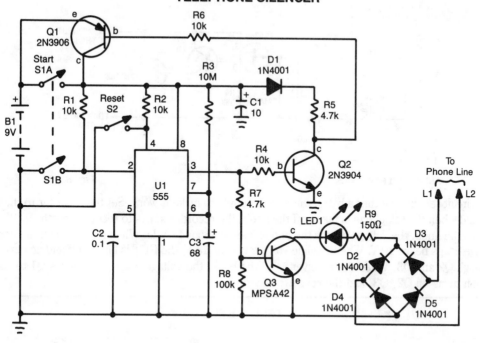

WELS' THINK TANK

Fig. 86-10

If you are busy and cannot answer or do not wish to answer your phone, this circuit will give a busy signal without you having to leave the phone off the hook. After a predetermined time, the circuit is deactivated. U1 forms an astable multivibrator that can be set for a time up to 10 minutes by values of R3 and C3. When S1A is depressed, U1 starts, and Q1 latches, which powers the circuit. At the end of a time interval determined by R3 and C3, Q2 and Q3 cut off and remove power from the circuit. During the operation, S3 throws a 150-Ω resistor across the phone line, which simulates an off-hook condition.

TELEPHONE INTERCOM

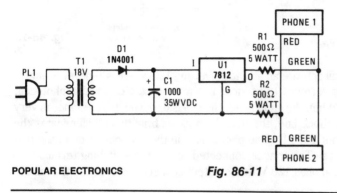

POPULAR ELECTRONICS

Fig. 86-11

Two telephones can be used as an intercom setup with this simple power-supply arrangement. The 500-Ω resistors maintain line balance.

557

TELEPHONE AUTO RECORD

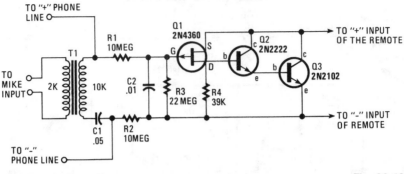

HANDS-ON ELECTRONICS

Fig. 86-12

The circuit requires neither a battery nor an ac supply to make it work. Set the recorder to the record position and when the telephone is taken off the hook, the recorder starts to record everything. When the phone is on the hook, the voltage across the phone lines is about 48 Vdc. When it is taken off the hook, the line voltage drops below 10 V. When the line voltage is near 48 V, the FET is biased off and no current can flow through Q2 and Q3. When the receiver is off the hook, the voltage drops, and allows Q1 to conduct. This action turns on Q2, Q3, and the cassette recorder.

TELL-A-BELL

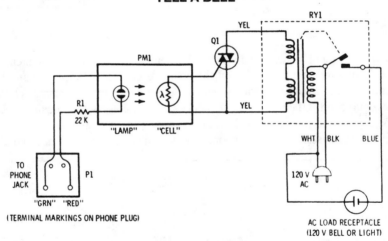

TAB BOOKS

Fig. 86-13

This accessory connects to phone and will activate an attention-getting 120-V bell whenever a ringing voltage appears on the phone line. The four-terminal transducer has a neon bulb close to a photocell, which are both enclosed in a light-tight tube. When the lamp is off, the cell is dark and its resistance is very high. When a ringing voltage appears on the phone line, the neon bulb glows brightly and illuminates the photocell whose resistance then drops to about $1\,000\ \Omega$. The photocell is in the gate circuit of a triac that will turn on whenever the cell resistance drops. The triac is connected across the switching terminals of the isolation relay. Thus, the relay closes and applies 120 V to the output socket whenever the triac is on.

ANSWERING MACHINE BEEPER

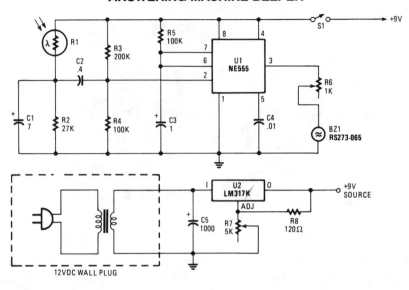

POPULAR ELECTRONICS

Fig. 86-14

When the light on the answering machine blinks, the resistance of photoresistor R1 charges, triggers the timer (U1), and generates 0.2-s pulse that activates BZ1. R1 is optically coupled to the LED on the answering machine and is properly light shielded.

TELEPHONE VISUAL RING INDICATOR

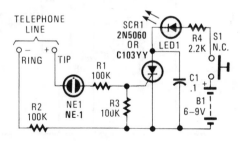

POPULAR ELECTRONICS

Fig. 86-15

This circuit will indicate the receipt of a call. When the telephone rings, a 100- to 120-V ring signal breaks over NE1, and causes SCR1 to trigger. This causes LED1 to light until the SCR1 is turned off by depressing S1.

PHONE-IN-USE INDICATOR

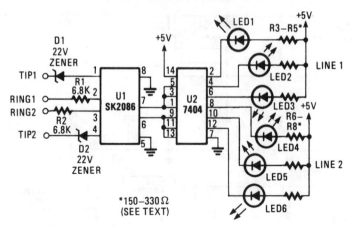

POPULAR ELECTRONICS

Fig. 86-16

The circuit receives its power from a 15-V wall adapter (not shown). The circuit takes advantage of the fact that the phone line voltage drops from 48 to 10 V when an extension is taken off the hook. When the voltage on a line drops, the optoisolator/coupler is turned off so that the inputs to the line-1 hexinverters (U2 pins 1, 3, and 5) float high. The corresponding outputs (U2 pins 2, 4, and 6) go low and light the line-1 LEDs.

TELEPHONE AMPLIFIER

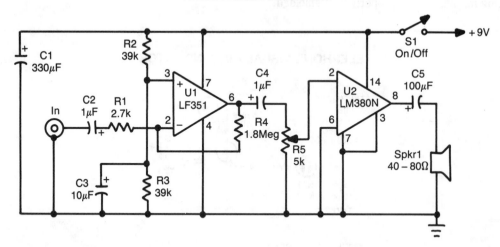

POPULAR ELECTRONICS

Fig. 86-17

This amplifier can be used in telephone work or where a simple speech amplifier is required. The frequency response can be varied by the value of C_2, C_4, and addition of a capacitor across R4 (≈ 33 pF for voice band) to limit the HF response.

EXTENSION PHONE RINGER

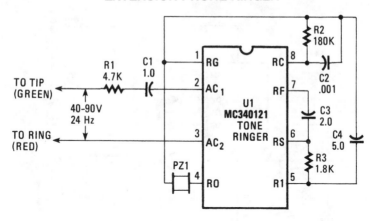

HANDS-ON ELECTRONICS

Fig. 86-18

The ac ringing voltage (typically 40 to 90 V at 26 Hz) is rectified by U1, the tone-ringer IC, and is used to drive that IC's internal tone-generator circuitry. The tone-generator IC includes a relaxation oscillator (with a base frequency of 500, 1 000, or 2 000 Hz) and frequency dividers that produce the high- and low-frequency tones, as well as the tone-warble frequency. An on-board amplifier feeds a 20-Vpp signal to the transducer.

TELEPHONE VISUAL RING INDICATOR

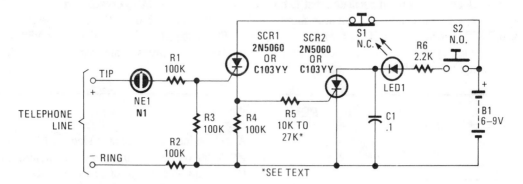

POPULAR ELECTRONICS

Fig. 86-19

In this circuit, the ringing voltage on a telephone line causes NE-1 to break over, triggering SCR1, which in turn triggers SCR2. If a call has been received, depressing S2 will cause LED1 to light. Depressing S1 resets the circuit. This circuit has the advantage of lower battery drain because LED1 is not left on continuously after a ring signal, but only when S2 is depressed.

CALL TONE GENERATOR

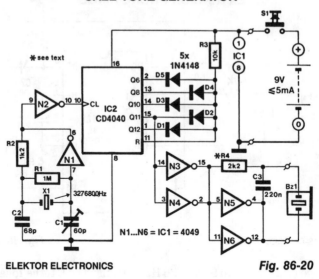

ELEKTOR ELECTRONICS

Fig. 86-20

Amateur VHF relay stations are normally actuated by a 1750-Hz call tone. This might give problems if the relevant sending equipment has no internal call-tone generator, if it does not have sufficiently accurate frequencies, or if the tone duration is not long enough to securely energize the relevant relay.

These problems can be overcome by the stand-alone generator described here. Simply placed in front of the microphone, it makes absolutely certain that the relay station is actuated. The generator consists of a quartz oscillator, a frequency counter and a buffer-amplifier—all contained in just two CMOS ICs. It is powered by a 9-V (p-p) battery, from which it draws a current of around 5 mA.

Gates N1 and N2 form an oscillator that is controlled by a 3.276 80-MHz crystal and provides clock pulses to IC2, which is connected as a programmable scaler. Diodes D1 through D5 determine the divide factor of 1 872. Counter output Q1 provides the wanted 1 750-Hz signal, which is buffered by N3 through N6 before being applied to a piezoelectric buzzer. Capacitor C3 suppresses any harmonics, while R4 determines the volume of the output signal.

REMOTE RINGER

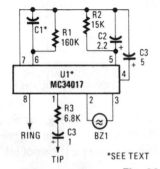

POPULAR ELECTRONICS   Fig. 86-21

A telephone bell circuit using a Motorola MC34017 can be built from a few components. C1 and U1 depend on the type of ring required.

U1	C1	Ring
MC34017-1	1 000 pF	1 kHz
MC34017-2	500 pF	2 kHz
MC34017-3	2 000 pF	500 Hz

Select the version of MC34017 and C1 from this table.

TELEPHONE MESSAGE TAKER

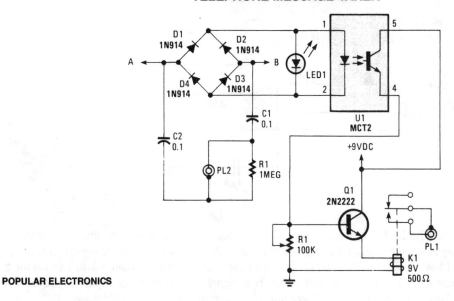

Fig. 86-22

This circuit operates on the ringing voltage of the telephone to trigger a tape recorder to record messages. K1 can be made to latch using extra contacts if the tape recorder requires a constant-contact closure.

TELEPHONE LINE-IN-USE INDICATOR

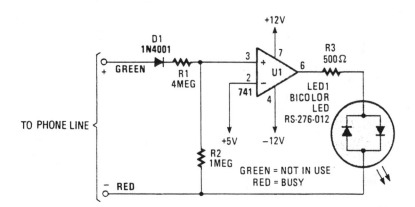

Fig. 86-23

When a telephone line is not in use, about 48 V appears across the line, which drops to about 10 V or less when the line is in use. This circuit switches a bicolor LED as an indicator.

SIMPLE RING DETECTOR

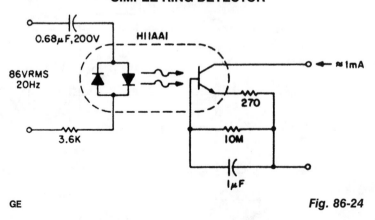

GE   *Fig. 86-24*

This circuit detects the 20-Hz approximately 86-Vrms ring signal on telephone lines and initiates action in an electrically isolated circuit. Typical applications include automatic answering equipment, interconnect/interface and key systems. The detector is the simplest and provides about a 1-mA signal for a 7-mA line, which loads for 0.1 s after the start of the ring signal. The time-delay capacitor provides a degree of dial-tap and click suppression, and filters out the zero crossing of the 20-Hz wave.

87

Temperature Controls

The sources of the following circuits are contained in the Sources section, which begins on page 679. The figure number in the box of each circuit correlates to the entry in the Sources section.

Temperature Controller with Defrost Cycle
Thermocouple Temperature Control
Temperature Controller

TEMPERATURE CONTROLLER WITH DEFROST CYCLE

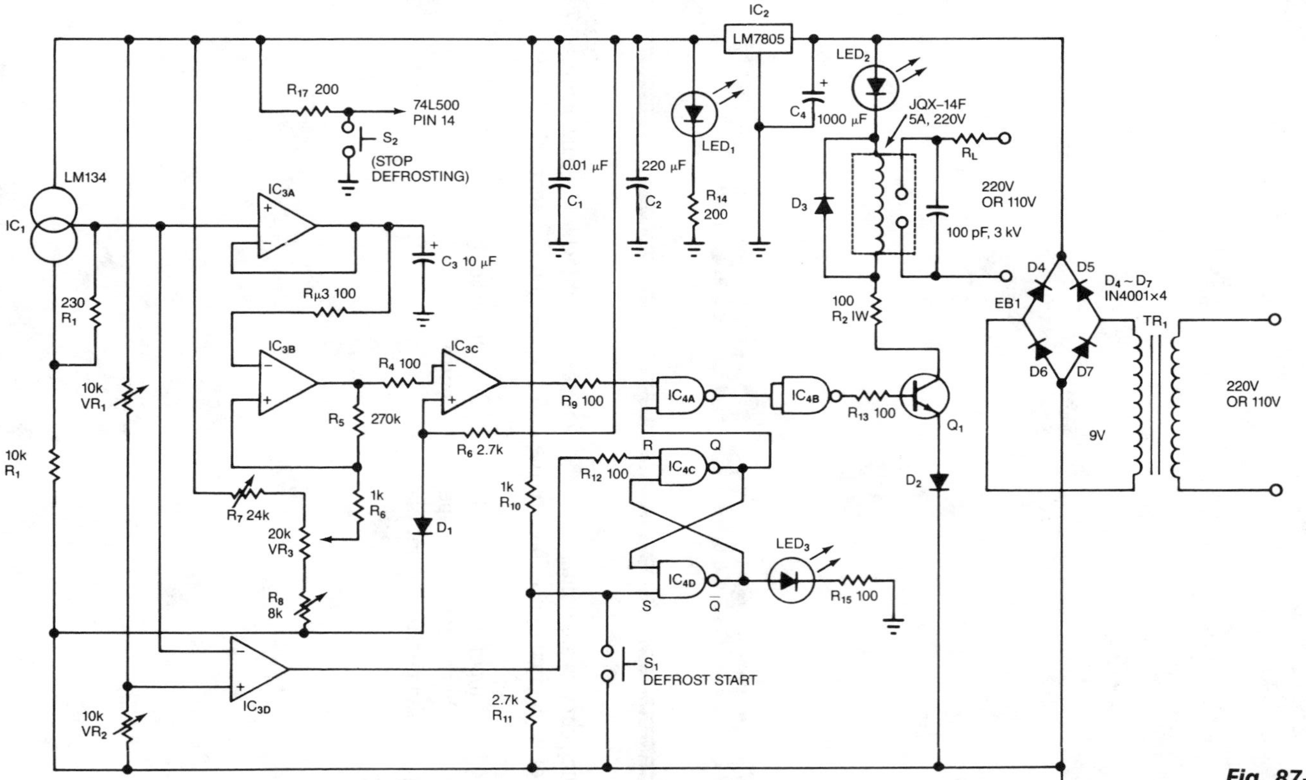

Fig. 87-1

NOTE: D_1, D_2, AND D_3 =1N4148; IC_3 =LM324; IC_4 =74LS00; EB_1 =1A/ 50V; Q_1 =C59013; TR_1 ≥3W.

EDN

This temperature controller has a range of – 50 to + 150°C and permits defrosting. R7, VR3, and R8 set the controller's trip point. S1 initiates defrosting, S2 cancels defrosting. VR1 and VR2 set the defrost-temperature trip point.

The LM134, IC1, is a thermal sensor. One section of IC3, a quad op amp, buffers the sensor's output. The other section functions as Schmitt trigger and buffer for the normal-cycle circuitry and as a comparator for the defrost-cycle circuitry. If you wish to control a heater rather than a refrigerator, omit the final inverter in Q1's base circuit. You must select LM7805s that have outputs between 4.95 and 5.05 V, or the circuit might not work.

THERMOCOUPLE TEMPERATURE CONTROL

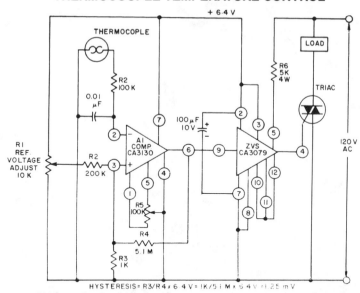

GE/RCA

HYSTERESIS= R3/R4 x 6.4 V = 1K/5.1 M x 6.4 V = 1.25 mV

Fig. 87-2

This control, with zero-voltage load switching, uses a CA3130 BiMOS op amp and a CA3079 zero-voltage switch. The CA3130, used as a comparator, is ideal because it can "compare" the low voltages generated by the thermocouple to the adjustable reference voltage over the range of 0 to 20 mV.

TEMPERATURE CONTROLLER

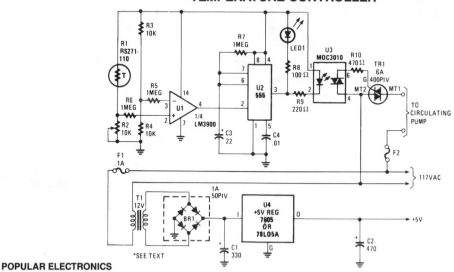

POPULAR ELECTRONICS

Fig. 87-3

A thermistor (R1) is compared with a reference (R2) in a Wheatstone-bridge circuit. Comparator U1's output goes high, which triggers U2. U2 is a delay of about 25 s. After 15 s, LED1 lights, U3 actuates, triac TR1 triggers, and turns on a hot water pump. This system was used with a hot-water heater.

567

88

Temperature Sensors

The sources of the following circuits are contained in the Sources section, which begins on page 679. The figure number in the box of each circuit correlates to the entry in the Sources section.

THERMAL MONITOR

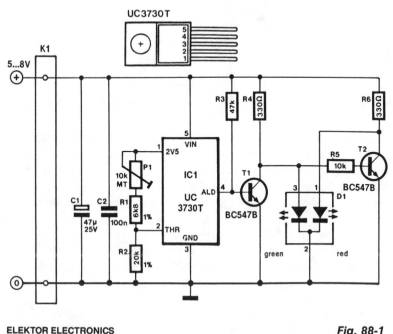

ELEKTOR ELECTRONICS

Fig. 88-1

Unitrode's UC1730 family of integrated circuits is designed for use in a number of thermal monitoring applications. Each IC combines a temperature transducer, a precision reference, and a temperature comparator to allow the device to respond with a logic output if temperatures exceed a predetermined level.

The monitor presented here is based on a UC3730T and it is intended to be fitted to a heatsink. Although the supply to the device can be as high as 40 V, 5- to 8-V is chosen here, because it is normally readily available in the equipment where the monitor will be used (power amplifiers, power supplies, etc.).

The threshold temperature, T_t, in °C, is determined by:

$$T_t = \frac{2.5 R_2}{0.005(R_1 + R_2 + P_1) - 273.15}$$

The temperature can be preset with $P1$ to values between $-1°C$ and $+100°C$. The indicator is formed by a bicolor LED and controlled by transistors T1 and T2. Resistors R4 and R5 limit the current through the LED. When the temperature of the heatsink is below the threshold temperature, the ALD (alarm delay) output, pin 4, is logic low so that T1 is switched off and the green LED lights.

SIMPLE TEMPERATURE INDICATOR

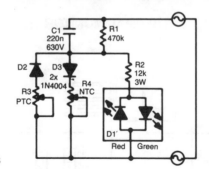

ELEKTOR ELECTRONICS

Fig. 88-2

For the absolute measurement of temperatures, a thermometer is indispensable. However, in many situations, an absolute value is not needed and a relative indication is sufficient. It would be a further advantage if a green light would indicate that all is well as far as temperature is concerned. As the temperature rises, the light should change color slowly to indicate that the equipment is getting too hot.

This circuit does this and works directly from the mains. The indicator proper is a two-color LED (D1), while the sensor is a combination of a negative-temperature coefficient (NTC) and a positive-temperature coefficient (PTC) resistor (R4 and R3, respectively).

At a relatively low temperature, the value of R_3 is low and that of R_4 is high. During the positive half cycle of the mains voltage, a voltage will exist across R3/D2 that is sufficiently high to cause the green section of D1 to light. The value of R_3 has been chosen to ensure that during the negative half cycle of the mains voltage, the potential across it is too low to cause the red section of D1 to light.

If the temperature rises, the value of R_4 diminishes and that of R_3 rises. Slowly, but surely, the green section will light with lesser and lesser brightness. At the same time, the red section lights with greater and greater brightness until ultimately only the red section will light.

Resistor R2 and capacitor C1 ensure that the current drawn by the LEDs does not become too large. This arrangement keeps the dissipation relatively low. Both R3 and R4 should be of reasonable dimensions—approximately 6 mm in diameter, not less. At 25°C, the NTC must be 22 to 25 kΩ and the PTC must be 25 to 33 Ω. The circuit should be treated with great care because it carries the full mains voltage.

UNDER-TEMPERATURE SWITCH

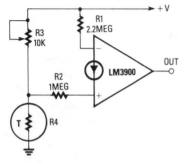

RADIO-ELECTRONICS

Fig. 88-3

The reference current is fed from the supply voltage via R1, to the inverting terminal, and the variable (noninverting) current is supplied from the junction of R3 and R4. Because the value of R_1 is approximately double that of R_2, and generates a current that is proportional to the supply voltage, the trip temperature (preset via R3) is independent of the supply voltage.

TEMPERATURE SENSOR

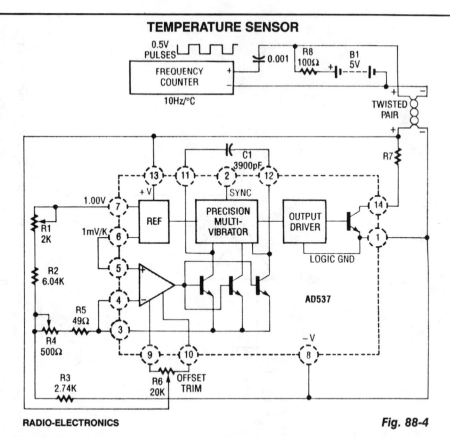

RADIO-ELECTRONICS

Fig. 88-4

The AD537 uses its two reference outputs—one fixed at 1 V, the other of which varies with temperature (1 mV per °K). At 0°C, the 1-V reference multiplied by 0.273 will balance this voltage and produce a zero output. The scale in this circuit is 10 Hz/°C. Output from, as well as power to, the circuit, is fed via a two-wire twisted pair. A frequency counter is used as a readout.

OVER-TEMPERATURE SWITCH

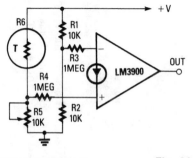

RADIO-ELECTRONICS

Fig. 88-5

The output goes high when a preset temperature is exceeded. A fixed half-supply reference voltage feeds a reference current to the inverting input, and a variable current is fed to the noninverting input. Resistor R6 is a negative-temperature-coefficient (NTC) thermistor, so the potential at the junction of R5 and R6 rises with temperature. The op amp will switch high when that voltage exceeds the half-supply value. The trip temperature can be preset via R5.

TRANSISTOR SENSOR TEMPERATURE MEASURER

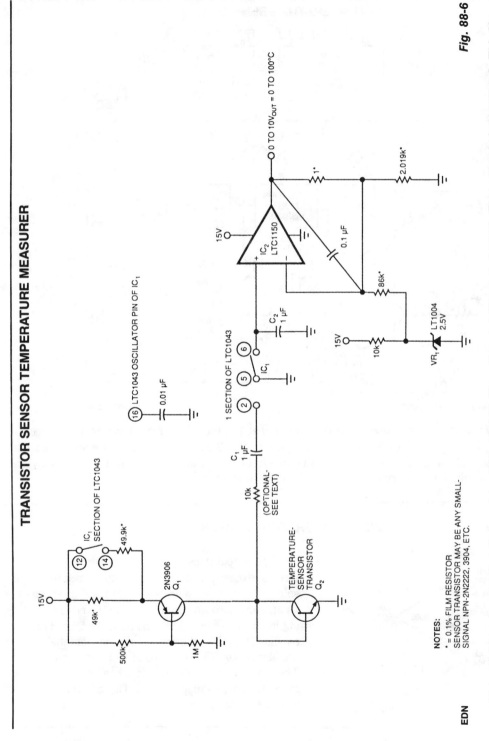

Fig. 88-6

NOTES:
* = 0.1% FILM RESISTOR
SENSOR TRANSISTOR MAY BE ANY SMALL-
SIGNAL NPN-2N2222, 3904, ETC.

EDN

Using the fact that the V_{BE} of a transistor shifts 59.16 mV per decade of current at 25°C. This V_{BE}-vs.-current relationship holds true regardless of the V_{BE} absolute value. IC, an LT1043, acts as an oscillator and switches a current source (Q1) at a 10:1 ratio. The stepped 10:1 current drive is translated to temperature by IC2, Q2, and the associated components. Accuracy is ±1%. No compensation is needed if Q2 is changed.

89

Thermometer Circuits

The sources of the following circuits are contained in the Sources section, which begins on page 679. The figure number in the box of each circuit correlates to the entry in the Sources section.

INEXPENSIVE LINEAR THERMOMETER

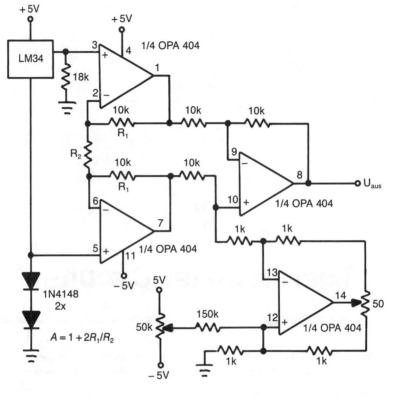

ELECTRONIC ENGINEERING

Fig. 89-1

As a sensor in the design of this linear thermometer, the LM34 is used. The output is the difference between two base-emitter voltages ΔV_{EB} of two transistors operated at different collector-current densities. The voltage difference (ΔV_{EB}) is:

$$\Delta V_{EB} = V_{EB1} - V_{EB2}$$
$$= \frac{kT}{q} \ell n \ (I_{C1}/I_{C2})$$

where the current densities are I_{C1} and I_{C2}, k is Boltzmann's constant, and q is the electron charge. Because all factors, including the ratio I_{C1}/I_{C2} are constant, the output of the LM34 (National Semiconductor) is a linear function of T in the range over $-50°$ to $300°F$, which provides the output voltage of 10 mV/°F with a max. nonlinearity of $\pm 0.35°F$.

The output of the LM34 is amplified by a three-op-amp instrumentation amplifier. The fourth op amp gives the possibility to control the offset voltage of the amplifier. The gain A of the instrumentation amplifier can be set to any desired value by the choice of the resistance R_2 only.

ELECTRONIC THERMOMETER

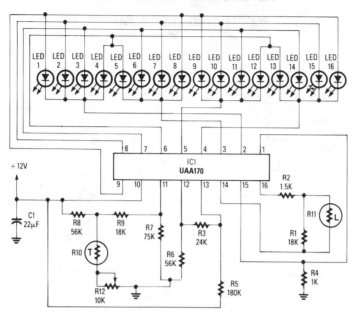

RADIO-ELECTRONICS

Fig. 89-2

This thermometer uses an NTC thermistor (R10) to produce a dc voltage that decreases with temperature, to drive an IC that lights one of the 16 LEDs as a function of this voltage. R11 is a light-dependent resistor that adjusts the LED brightness as a function of ambient light.

SINGLE-dc SUPPLY THERMOMETER

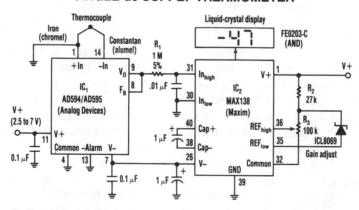

ELECTRONIC DESIGN

Fig. 89-3

Using a J-type thermocouple, this circuit can indicate temperatures from −350° to 400° with a 6-V supply or −50 to +100° with a 3-V lithium battery. The AD954 produces 10 mV/°C output to the MAX 138 digital voltmeter chip, which drives the LCD display.

ELECTRONIC THERMOMETER

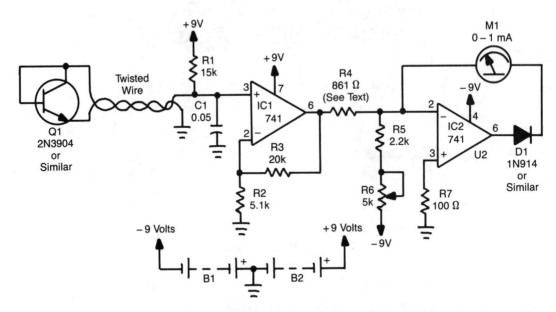

GERNSBACK PUBLICATIONS

Fig. 89-4

This thermometer is capable of measuring temperatures from −30 to +120°F. A diode-connected 2N3904 transistor forms a voltage divider with R1. The transistor is used as the temperature sensor and, for best results, it should be connected to the rest of the circuit with twisted wire, as shown. As temperature increases, the voltage drop across the transistor changes by approximately −1.166 mV/°F. As a result, the current at pin 3 of IC1, a 741 op amp with a gain of 5, decreases as the temperature measured by the sensor increases. A second 741 op amp, IC2, is configured as an inverting amplifier.

Resistors R5 and R6 are used to calibrate the current. At a temperature of about −30°F, the current through R4 (formed by connecting a 910- and a 1 600-Ω resistor in parallel) should equal the current through R5 and R6. A temperature of −30°F will result in a meter reading of 0 mA, while a temperature of 120°F will result in a meter reading of 1 mA. Divide the scale between those points into equal segments and mark the divisions with the appropriate corresponding temperatures. If you divide it into 150 equal segments, for instance, each division will equal one degree. Calibration is completed by placing the sensor in an environment with a known temperature, such as in an ice-point bath. The freezing point of water is approximately 32°F. Verify that the temperature is indeed 32°F using another thermometer that is known to be accurate. Then, simply place the sensor in the bath and adjust R6 until you get the correct meter reading.

HIGH-ACCURACY THERMOMETER

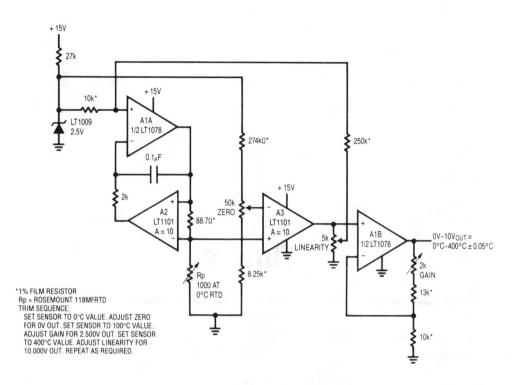

LINEAR TECHNOLOGY

Fig. 89-5

This circuit combines a current source and a platinum RTD bridge to form a complete high-accuracy thermometer. The ground-referred RTD sits in a bridge that is composed of the current drive and the LT1009 biased resistor string. The current drive allows the voltage across the RTD to vary directly with its temperature-induced resistance shift. The difference between this potential and that of the opposing bridge leg forms the bridge output. The RTD's constant drive forces the voltage across it to vary with its resistance, which has a nearly linear positive temperature coefficient. The nonlinearity could cause several degrees of error over the circuit's 0°C – 400°C operating range.

The bridge's output is fed to instrumentation amplifier A3, which provides differential gain, while simultaneously supplying nonlinearity correction. The correction is implemented by feeding a portion of A3's output back to A1's input via the 10- to 250-kΩ divider. This causes the current supplied to R_P to slightly shift with its operating point, compensating sensor nonlinearity to within ± 0.05°C. A1B, providing additional scaled gain, furnishes the circuit output. To calibrate this circuit, follow the procedure given in the diagram.

90

Timers

The sources of the following circuits are contained in the Sources section, which begins on page 679. The figure number in the box of each circuit correlates to the entry in the Sources section.

Mains-Powered Timer
Transmit Time Limiter
Long-Interval Programmable Timer
Programmable Timer for Long Intervals
Appliance Cutoff Timer
SCR Timer
Watchdog Timer/Alarm
10-Minute ID Timer
Adjustable Timer
Long-Duration Time Delay
Time-Out Circuit

MAINS-POWERED TIMER

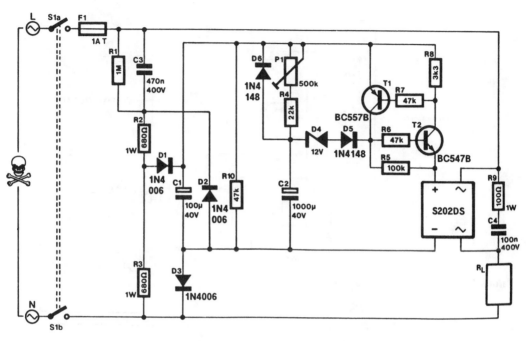

ELEKTOR ELECTRONICS

Fig. 90-1

This timer can be inserted in a power line to provide a controllable delay before a load is energized. The mains voltage is reduced by C3 and rectified to give about 30 V across C1. This potential charges C2 slowly via R4/P1. When U_{C2} reaches about 14 V, electronic switch T1/T2 actuates a solid-state relay (a Sharp S202DS). When the mains voltage is removed, C2 discharges rapidly via D6 and R10. The delay extends from 15 s (P1 set to minimum resistance) to 5 min (P1 set to maximum resistance).

The solid-state relay needs cooling in accordance with the current drawn by the load: at up to 1 A, no heatsink is required; at 1 to 3 A (max), a 5×5 cm heatsink is advisable.

During the building of the circuit, consideration must be given to safety because many parts will be at mains potential. For instance, fitting the unit in an ABS or other man-made fiber enclosure is a must. If a potentiometer is used for P1, its spindle should be insulated. If a preset is used, it must not be accessible through a hole in the enclosure.

Switch S1 is a DPST that disconnects the circuit from the mains. Nevertheless, the only way to safely work on the circuit is to unplug the mains socket and allow C3 sufficient discharge time.

TRANSMIT TIME LIMITER

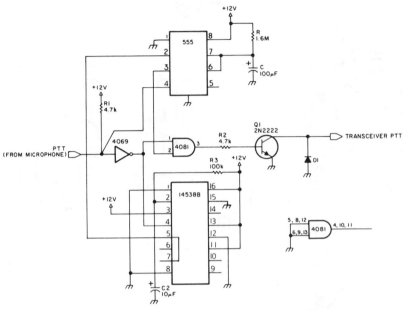

Values for Diverse Duty Cycles

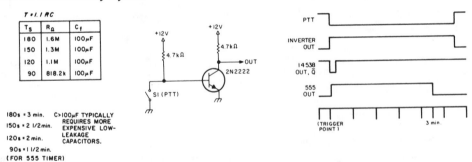

$$T = 1.1 RC$$

T_S	R_Ω	C_f
180	1.6M	100μF
150	1.3M	100μF
120	1.1M	100μF
90	818.2k	100μF

180s = 3 min. C>100μF TYPICALLY
150s = 2 1/2 min. REQUIRES MORE
                   EXPENSIVE LOW-
120s = 2 min. LEAKAGE
               CAPACITORS.
90s = 1 1/2 min.
(FOR 555 TIMER)

Fig. 90-2

This circuit prevents making transmissions that are too long, which "time out" repeaters and/or tie up a communications channel too long. On transmit, the PTT (push to talk line) from the microphone is at ground. This causes one input of the AND gate (4081) to go high. The 555 is held in the reset mode. An MC1453B monostable multivibrator generates a pulse to the 555 and causes it to produce a gate of length approximately $1.1RC$, where R is selected for desired time delay and C is 100 μF. The output of pin 3 of the 555 causes the 4081 gate to go high, turns on Q1, and keys the transmitter. At the end of the cycle ($1.1RC$), the AND gate will lose one input, which turns off Q1 and unkeys the transmitter.

LONG-INTERVAL PROGRAMMABLE TIMER

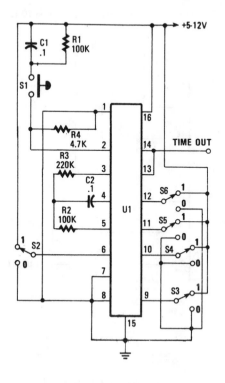

BINARY CODE INPUT					SELECTED STAGE FOR OUTPUT
8 BYPASS PIN 6	D PIN 12	C PIN 11	B PIN 10	A PIN 9	
0	0	0	0	0	9
0	0	0	0	1	10
0	0	0	1	0	11
0	0	0	1	1	12
0	0	1	0	0	13
0	0	1	0	1	14
0	0	1	1	0	15
0	0	1	1	1	16
0	1	0	0	0	17
0	1	0	0	1	18
0	1	0	1	0	19
0	1	0	1	1	20
0	1	1	0	0	21
0	1	1	0	1	22
0	1	1	1	0	23
0	1	1	1	1	24

A

BINARY CODE INPUT					SELECTED STAGE FOR OUTPUT
8 BYPASS PIN 6	D PIN 12	C PIN 11	B PIN 10	A PIN 9	
1	0	0	0	0	1
1	0	0	0	1	2
1	0	0	1	0	3
1	0	0	1	1	4
1	0	1	0	0	5
1	0	1	0	1	6
1	0	1	1	0	7
1	0	1	1	1	8
1	1	0	0	0	9
1	1	0	0	1	10
1	1	0	1	0	11
1	1	0	1	1	12
1	1	1	0	0	13
1	1	1	0	1	14
1	1	1	1	0	15
1	1	1	1	1	16

B

These truth tables can be used to set the desired output frequency. Refer to the text for details.

Fig. 90-3

Using an RC oscillator, an up to 24-stage ripple counter ($\div$) 16777216 or 2^{24}, and a 0.1-Hz count-rate with $R_2 = 39$ kΩ, $C_2 = 0.001$ μF, and $R_4 = 220$ kΩ for example, the count cycle would take about 654 s. This example shows the capabilities of this time circuit, using the Motorola MC14536 timer. A low-frequency oscillator can be used for longer time periods.

PROGRAMMABLE TIMER FOR LONG INTERVALS

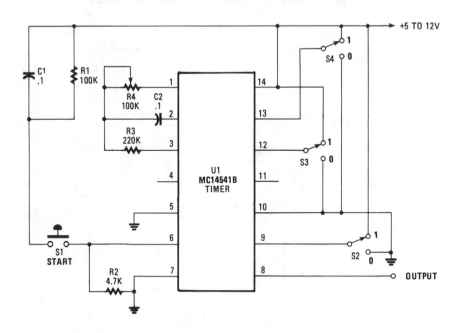

COUNTER SELECTOR CHART

PIN 12	PIN 13	NUMBER OF COUNTER STAGES (N)	COUNT 2^N
0	0	13	8192
0	1	10	1024
1	0	8	256
1	1	16	65536

The inputs at pins 12 and 13 of the MC14415 programmable counter determine the number of counter stages selected, and therefore the count.

POPULAR ELECTRONICS

Fig. 90-4

By using an RC oscillator and a programmable divider, this counter can run for hours. An interval oscillator runs at a frequency given by (see figure schematic):

$$f = \frac{1}{2.3\ R_4 C_2} \text{ and } R_3 \approx 2R_2$$

By using, for example, $R_4 = 390$ kΩ and $C_2 = 10$ μF and R_2, the oscillator can run at 0.1 Hz. Divided by 65536, this is a cycle of approximately 655 000 s (182 hours, slightly more than a week).

APPLIANCE CUTOFF TIMER

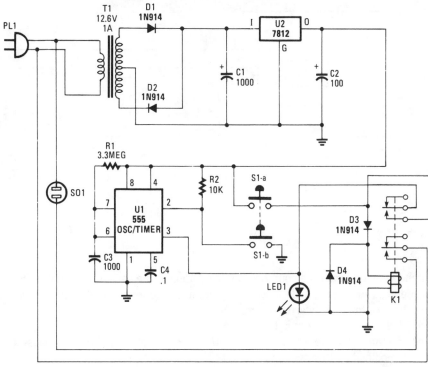

POPULAR ELECTRONICS

Fig. 90-5

Suitable for cutting off an appliance or other ac load, this timer will cut the ac power after a period determined by R1/C3, as shown, for about 40 minutes. K1 is a relay that should handle about 10 A. S1A and S1B is a momentary switch that starts the timer cycle.

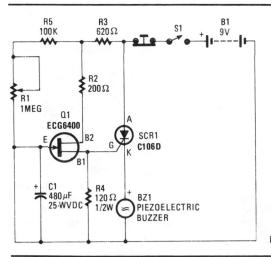

SCR TIMER

Depending on the R1 (adjustable) C1 time constant, when C1 charges up to a certain level, Q1 conducts, triggers SCR1, and sounds BZ1. A pushbutton resets the circuit.

HANDS-ON ELECTRONICS

Fig. 90-6

WATCHDOG TIMER/ALARM

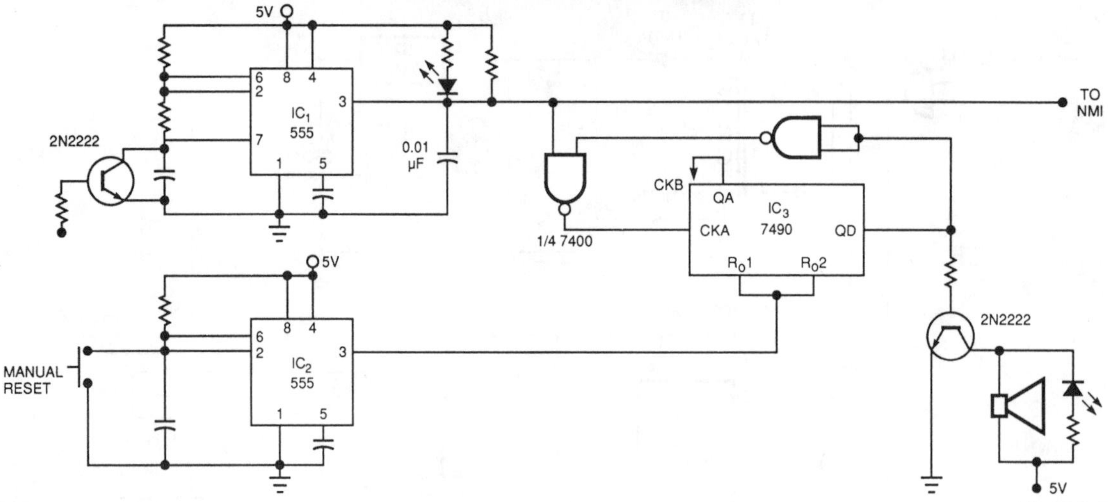

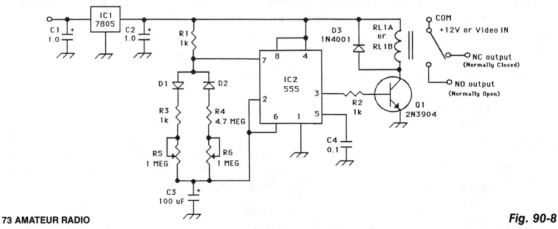

EDN

Fig. 90-7

The watchdog timer contains a counter, IC3, in addition to the usual retriggerable 555 timer, IC1. The counter will sound an audible alarm if the watchdog timer trys to reset the μP a certain number of times (8, in the case of the counter). The alarm indicates that despite numerous resets, the system μP has failed to restart successfully, and the system is truly dead.

A second 555 timer, IC2, resets the counter, 1C3, for the duration of the manual system restart. The design could be modified so that system μP resets the counter.

10-MINUTE ID TIMER

73 AMATEUR RADIO

Fig. 90-8

Designed to automatically identify a transmitter every 10 minutes, this 555 circuit has adjustable charge and discharge paths. The IC should be a standard 555 type, not a CMOS type. C3 should be tantalum. The relay is a small 5-V reed type.

ADJUSTABLE TIMER

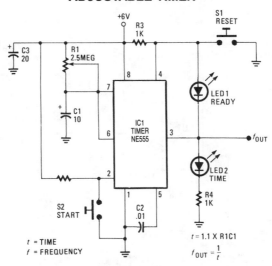

t = TIME
f = FREQUENCY

$t = 1.1 \times R1C1$

$f_{OUT} = \dfrac{1}{t}$

Fig. 90-9

LEDs indicate at a glance what the status of the circuit is at any given moment. Once the reset switch, S1, makes contact, the timer remains in that state until the start switch, S2, is pressed. When either switch is activated, LED1 (ready) and the time indicator, LED2, keep track of the situation. Although not necessary, the two LEDs should be of different colors (for example, red for ''ready'' and green for ''time'').

LONG-DURATION TIME DELAY

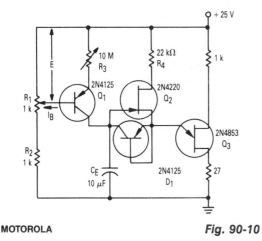

Fig. 90-10

Transistor Q1 and resistors R1, R2, and R3 form a constant current source and the charge current might be adjusted to be as low as a few nanoamperes. This current would, of course, not be sufficient to fire the UJT where $I_P = 0.2$ μA, unless the peak current was supplied from another source. Field-effect transistor Q2, acting as a source follower, supplies the current flowing into the emitter lead prior to firing and diode D1 provides a low-impedance discharge path for C_E. D1 must be selected to have a leakage that is much lower than the charge current.

Because I_B is small, the delay time will vary linearly with R3. The voltage (E), applied across R3 and the base-emitter junction of Q1, is set by the variable resistor R1. Time delays up to 10 hours are possible with this circuit. Resistor R4, in series with the FET drain terminal, must be large enough not to allow currents in excess of I_V to flow when the UJT is on. Otherwise, the UJT will not turn off and the circuit will latch up.

585

TIME-OUT CIRCUIT

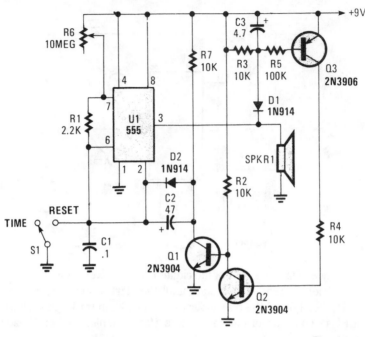

POPULAR ELECTRONICS

Fig. 90-11

This circuit operates in the astable mode and at the end of the first period (up to several minutes), it produces a tone. When S1 is placed in the time position, Q3 is cut off because pin 3 of U1 is high and D1 holds Q3 in cutoff. Q2 is off, and Q1 is on, which grounds the negative end of D2 and C2. Therefore, C1 and C2 are returned to ground.

After a time of about $1.1\,R_6\,(C_1 + C_2)$, the timer cycle completes and pin 3 U1 goes low. This turns on Q3 and Q2, cuts off Q1, and effectively disconnects C2. Now, the circuit oscillates with a period determined by R7 and C1, because D2 is forward-biased. A tone is then generated and can be heard from SPKR1. Closing S2 resets the circuit.

91

Tone-Control Circuits

The sources of the following circuits are contained in the Sources section, which begins on page 680. The figure number in the box of each circuit correlates to the entry in the Sources section.

Baxandall Tone-Control Audio Amplifier
Active Tone Control
Tremolo Circuit

BAXANDALL TONE-CONTROL AUDIO AMPLIFIER

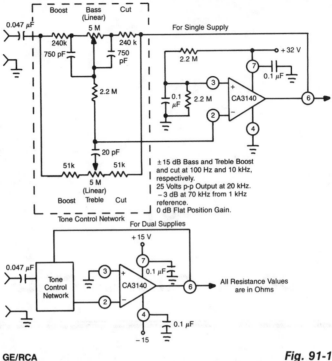

GE/RCA

Fig. 91-1

This circuit exploits the high slew rate, high input impedance, and high output-voltage capability of CA3140 BiMOS op amp. It also provides mid-band unity gain with standard linear potentiometers.

ACTIVE TONE CONTROL

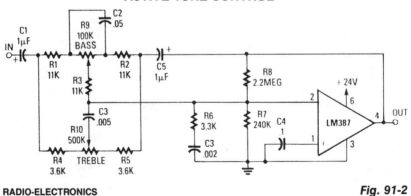

RADIO-ELECTRONICS

Fig. 91-2

The use of a low noise LM387 in a feedback circuit provides 20-dB boost or rejection of treble and bass. The supply voltage is +24 V.

TREMOLO CIRCUIT

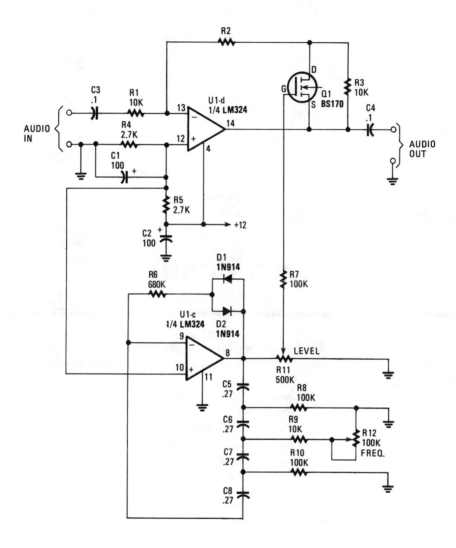

Fig. 91-3

This circuit adds a VLF AM component to an audio signal. This effect is widely used in musical instruments. U1C, a phase-shift oscillator operating at a few Hz applies a signal to Q1, which modulates the gain of U1D. R11 varies the level of the effect, while R12 varies the frequency.

92

Touch Controls

The sources of the following circuits are contained in the Sources section, which begins on page 680. The figure number in the box of each circuit correlates to the entry in the Sources section.

SENSOR SWITCH AND CLOCK

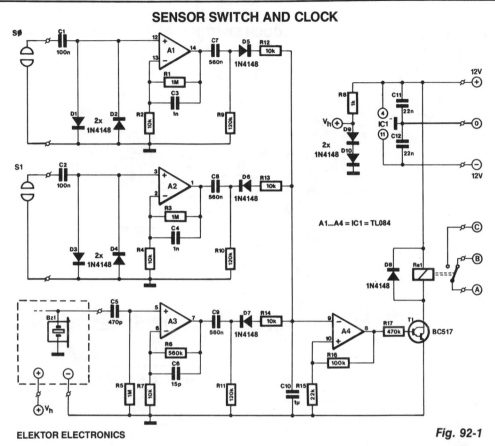

ELEKTOR ELECTRONICS

Fig. 92-1

One TL084 IC and an old quartz watch enable the construction of a deluxe on/off switch. Two of the four op amps contained in the TL084 (A1 and A2) are used to amplify the input signals from the sensors by one hundredfold (with the component values as shown in the diagram). Just touching the sensors with a finger causes a good 50-Hz input signal (hum). Notice that the amplification drops rapidly with rising frequency.

Diodes D5 and D6 rectify (single-phase) the 50-Hz signal. Because the diodes are connected in anti-phase, touching the "off" sensor causes a positive potential across C10, whereas touching the "on" sensor produces a negative potential across C10.

Op amp A4 is connected as an inverting bistable so that a negative potential across C10 causes relay Re1 to be energized. Because of feedback resistor R16, this state is maintained until the other sensor is touched.

The relay can also be energized at a predetermined time with the aid of a quartz watch. The 1.2-V supply for the watch is derived from the voltage drop across diodes D9 and D10; it can be increased to 1.8 V by adding a third diode.

The piezo buzzer in the watch is connected to the input of A3 via C5. As soon as the alarm goes off (the hour signal must be off), the voltage across C10 becomes negative, the relay is energized, and the load is switched on. The circuit, excluding the relay, draws a current of about 20 mA.

TOUCH SWITCH I

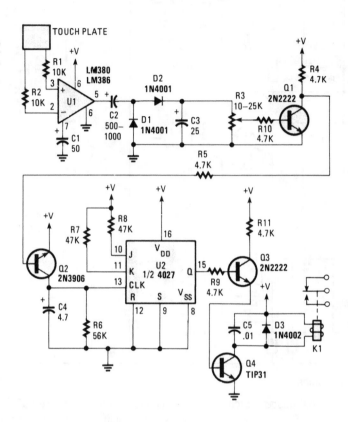

POPULAR ELECTRONICS   *Fig. 92-2*

This switch reacts to the touch of a finger to turn lights and/or appliances on or off. The device uses the human body as an antenna to pick up 60-Hz hum, which is applied to a metal plate by your finger. The signal is fed to the input of U1, an LM380 audio-power amplifier. An LM386 should work as well.

The 60-Hz output from the amplifier is rectified by D1 and D2, then filtered by C3. Potentiometer R3 sets the trigger voltage used to saturate Q1. When Q1 turns on, the collector end of R4 goes almost to ground and provides the needed voltage to turn on Q2. Transistor Q2 turns on and clocks. The flip-flop is configured for toggle-mode operation, so its output switches states with each clock pulse.

The 4027 (U2) is wired to toggle by tying the J and K inputs high and the set and resets low. Transistor Q3 is connected to the Q output through the 4.7-kΩ resistor. Transistor Q3 drives Q4, the relay driver. Be sure that the load does not exceed the relay ratings.

ON/OFF TOUCH SWITCH

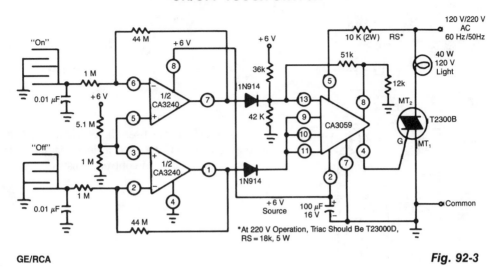

GE/RCA

Fig. 92-3

This circuit uses a CA3240 dual BiMOS op amp to sense small currents flowing between the contact points on a touch plate. The high input impedance of the CA3240 allows the use of 1-MΩ resistors in series with the touch plates to ensure user safety. A positive output on either pin 7 (ON) or pin 1 (OFF) of the CA3240 actuates the CA3059 zero-voltage switch, which then latches the triac on or turns it off. The internal power supply of the CA3059 powers the CA3240.

TOUCH SWITCH II

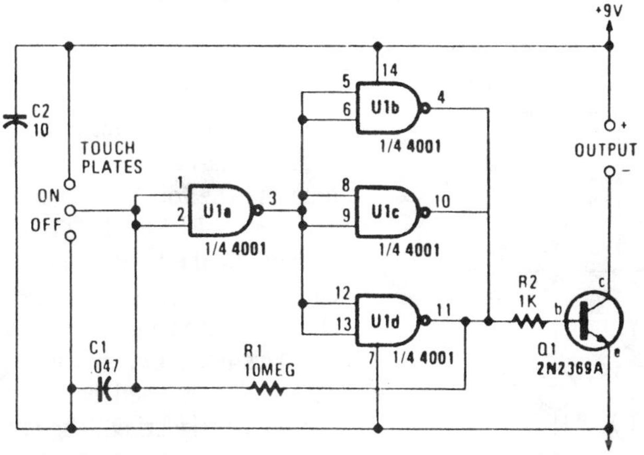

WELS' THINK TANK

Fig. 92-4

U1A and U1B/U1C/U1D form a bistable multivibrator that drives Q1, which switches the load. Touching the two upper contacts makes Q1 conduct; the two lower contacts cause Q1 to cut off.

HUM-DETECTING TOUCH SENSOR

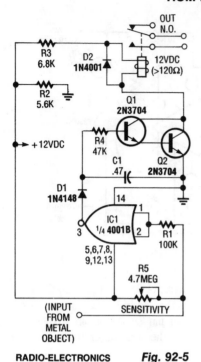

This touch sensor uses the 60-Hz hum pickup by the human body to drive a detector and relay drivers, Q1 and Q2. R5 controls sensitivity of the circuit.

RADIO-ELECTRONICS *Fig. 92-5*

TIME-ON TOUCH SWITCH

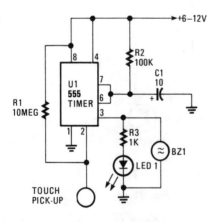

The circuit is built around a 555 oscillator (U1), which is turned on when a trigger is applied by touching the touch terminal to pin 2 of U1. When activated, LED1 and BZ1 (a piezoelectric buzzer) turn on for the time period set by the values of R_2 and C_1. The ON-time of the touch circuit can be altered by changing the values of C_1 and R_2.

This touch switch can be powered from batteries so that it need not be near a 60-Hz power source for triggering. The extremely small amount of current supplied to the trigger input through the 10-MΩ resistor. R1, makes the input circuitry very sensitive to any external loading, and it is easily triggered by touching the pickup.

HANDS-ON ELECTRONICS *Fig. 92-6*

93

Transmitters and Transceivers

The sources of the following circuits are contained in the Sources section, which begins on page 680. The figure number in the box of each circuit correlates to the entry in the Sources section.

20-m CW TRANSCEIVER

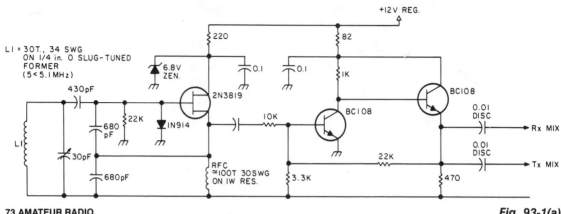

73 AMATEUR RADIO

Colpitts VFO circuit.

Fig. 93-1(a)

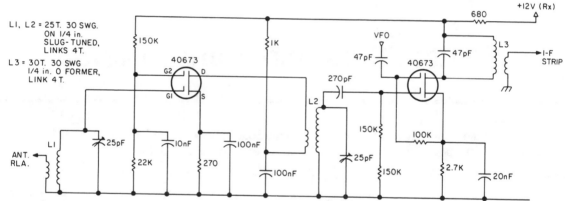

73 AMATEUR RADIO

Fig. 93-1(b)

The receive front-end. The associated tuned circuits are peaked to the center of the CW band trimmers.

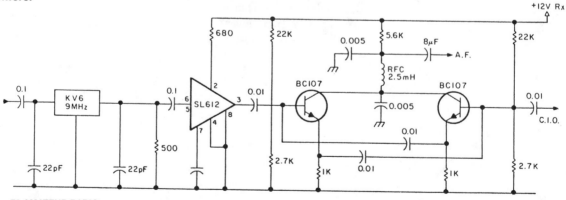

73 AMATEUR RADIO

Fig. 93-1(c)

Receive filter, IF, and product detector.

20-m CW TRANSCEIVER (*Cont.*)

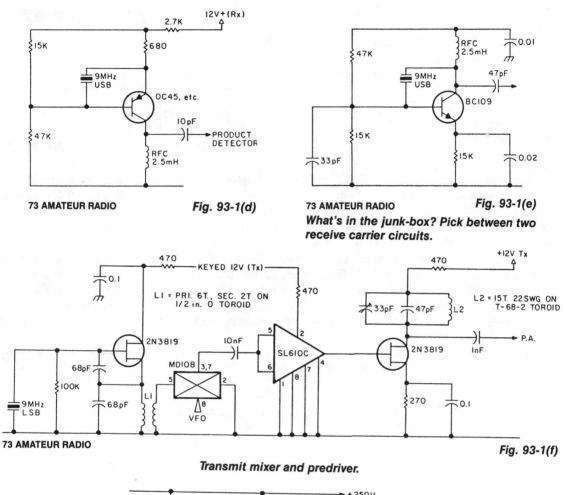

Fig. 93-1(d)

Fig. 93-1(e)

What's in the junk-box? Pick between two receive carrier circuits.

Fig. 93-1(f)

Transmit mixer and predriver.

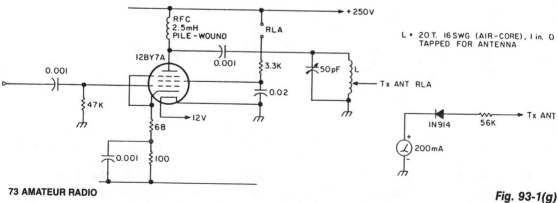

Fig. 93-1(g)

A simple diode/meter circuit to measure relative power output.

20-m CW TRANSCEIVER (Cont.)

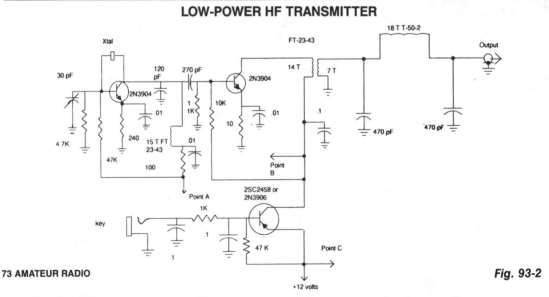

R*: SELECT FOR DESIRED TONE (10 < 50K)
R≠: VARY INPUT RES. FOR REQUIRED OUTPUT.

73 AMATEUR RADIO

Fig. 93-1(h)

73 AMATEUR RADIO

Fig. 93-1(i)

The Colpitts VFO circuit (Fig. 93-1(a)) has gate clamping to improve stability. It is followed by two buffers: the second provides individual outputs to the transmitter and receiver. An RIT circuit operates on receive. The front end uses 40673 dual-gate MOSFETs (Fig. 93-1(b)). The tuned circuits are peaked to the center of the CW band with the trimmer. The mixer output is link-coupled to a KVG 9-MHz SSB filter whose output is amplified by an SL612 IF amplifier IC (Fig. 93-1(c)).

The product detector uses two BC 107 transistors. Carrier reinjection is from a crystal oscillator using the USB crystal supplied with the KVG filter. Figure 93-1(d) and 93-1(e) contain: an FET oscillator and a pnp bipolar oscillator. Use either or alter the circuit polarities to suit an npn transistor. The transmitter mixer is an MD108 (Fig. 93-1(f)), fed from the VFO and the LSB carrier-injection oscillator.

A simple diode/meter circuit measures the relative power output (Fig. 93-1(g)). Sidetone is provided by an NE555 circuit (Fig. 93-1(h)).

LOW-POWER HF TRANSMITTER

73 AMATEUR RADIO

Fig. 93-2

This transmitter runs up to ½ W on the 40-m amateur band. Coils are wound on T25 and FT23 toroids, respectively. Point A is connected to point B to key the oscillator, along with the final (point C) to leave the oscillator running constantly. Use ¹/₁₀-W resistors and VERY small components.

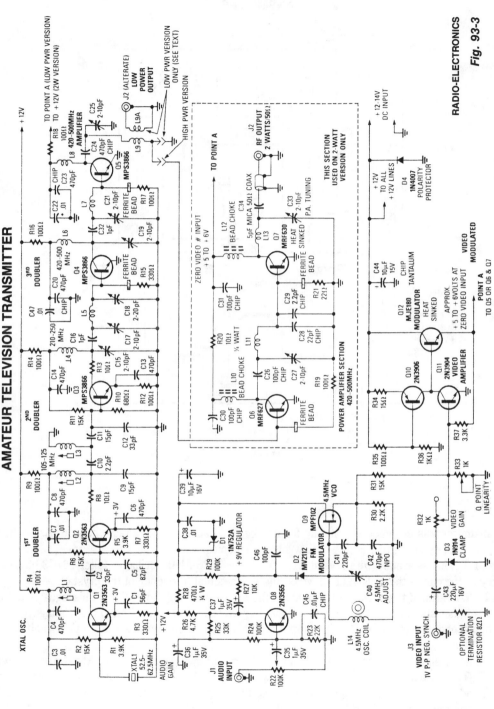

AMATEUR TELEVISION TRANSMITTER

Although the unit was designed for 2 W operation, Q6 and Q7 stages and associated components can be omitted and about 1 to 30 mW of RF can be obtained by link coupling to L9. A complete set of parts, including PC board is available from North Country Radio, P.O.Box 53, Wykagyl Station, New Rochelle, NY 10804.

RADIO-ELECTRONICS

Fig. 93-3

599

2-m TRANSMITTER

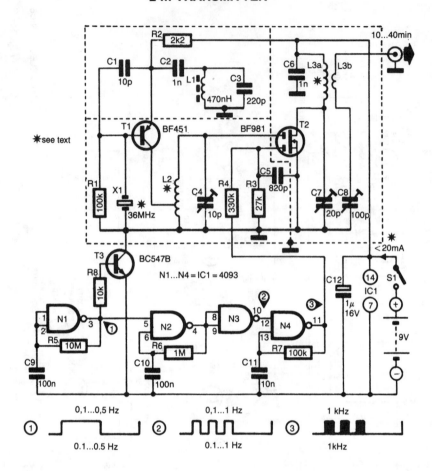

ELEKTOR ELECTRONICS

Fig. 93-4

The transmitter was designed specifically for use by radio amateurs as a radio beacon. As such, it provides a good-quality signal that is free of unwanted harmonics.

Transistor T1, in association with crystal X1, operates as a 36-MHz oscillator. Filter L1/C3 obviates any tendency of the circuit to oscillate at 12 MHz (the fundamental frequency of the crystal).

Circuit L2/C4 is tuned to the fourth harmonic of the oscillator signal (144 MHz). This signal is fed to the aerial via a buffer stage that consists of T2, a double-gated FET. The (amplitude) modulating signal is applied to the second gate of the buffer. The output power of the transmitter has been kept low, about 10 to 40 mW.

The modulating signal is generated by N1, an oscillator that switches the transmitter on and off via transistor T3. The switching rate lies between 0.1 and 0.5 Hz. When the output of N1 is low, T3 is switched off, and the transmitter is inoperative because the supply is disabled. When the output of N1 is high, T3 is on and the transmitter operates normally.

2-m TRANSMITTER (Cont.)

The digital pattern at the gate of T2 shapes the modulating signal. Gate N2 generates a square wave at a frequency of 0.1 to 1 Hz. As long as the output of T3 is high, N4 oscillates at a frequency of about 1 kHz. At the relevant gate of N2, there is, therefore, a periodic burst-signal at 1 kHz, and this signal is used to modulate the transmitter.

The digital pattern at the relevant gate of T2 can be varied to individual requirements by altering the values of the feedback resistors in the digital chain. The transmitter is calibrated by setting trimmers C4, C7, and C8 for maximum output power.

Inductors L2 and L3 are wound from 0.8-mm diameter enamelled copper wire: L2 = 5 turns with a tap of 1 turn from ground; L3A = 3 turns and L3B = 2 turns. The coupling between L3A and L3B should be arranged for maximum output power. The circuit draws a current of only 20 mA, which enables the transmitter to be operated from a 9-V battery for several hours.

HF LOW-POWER CW TRANSMITTER

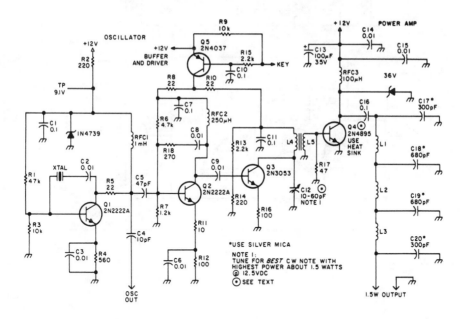

Fig. 93-5

Suitable for amateur use, this 1.5-W transmitter runs on a 12-V supply. Q1 is an oscillator using a surplus FT243 crystal. Q2 is a buffer driver and is keyed via keying transistor Q5. Q3 acts as a driver for Q4 (which should be heatsinked). Q4 develops about 1.5-W output. Coil data is given in the parts list. C12 is adjusted for best power output.

5-W 80-m CW TRANSCEIVER

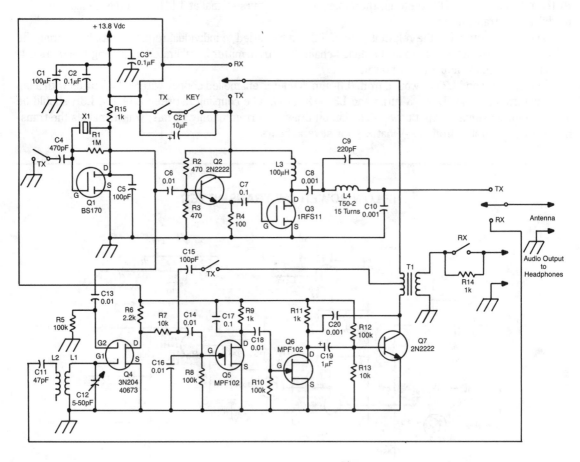

Fig. 93-6

This transceiver has a 3-stage transmitter and a direct-conversion receiver. Q1 is the transmitter's oscillator, and the frequency is controlled by X1, which also serves as the receiver local oscillator. Buffer Q2 drives final amplifier Q3 to about 5 W output. The B+ lead to these stages is keyed. The receiver consists of mixer Q4 followed by high gain amplifiers Q5/Q6/Q7. The audio signal appears at the secondary of Q7. In the transmit mode, Q5/Q6/Q7 serve as a sidetone oscillator. A 6PDT switch is required for the T/R switching.

LOW-COST BEACON TRANSMITTER

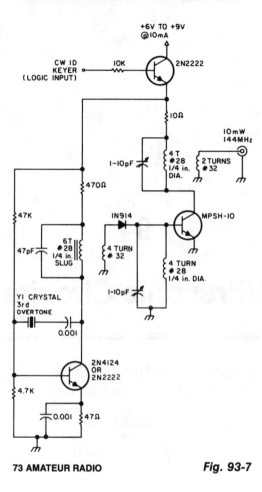

73 AMATEUR RADIO

Fig. 93-7

    This transmitter uses a 48-MHz crystal oscillator to drive a diode tripler to the 144-MHz amplifier. The output is 5 to 10 mW.

94

Ultrasonic Circuits

The sources of the following circuits are contained in the Sources section, which begins on page 680. The figure number in the box of each circuit correlates to the entry in the Sources section.

Ultrasonic Sound Source
Ultrasonic Pest Repeller I
Ultrasonic Pest Repeller II

ULTRASONIC SOUND SOURCE

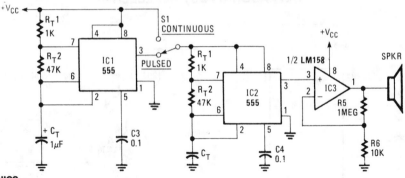

RADIO-ELECTRONICS

Fig. 94-1

Using two NE555 timer IC devices, this circuit generates either pulsed or continuous ultrasonic signals. The sound frequency is:

$$f = \frac{1.44}{C_T(R_{T1} + R_{T2})}$$

The values of C_T for both pulse rate and ultrasonic frequencies can be calculated this way. SPKR is a small hi-fi tweeter.

ULTRASONIC PEST REPELLER I

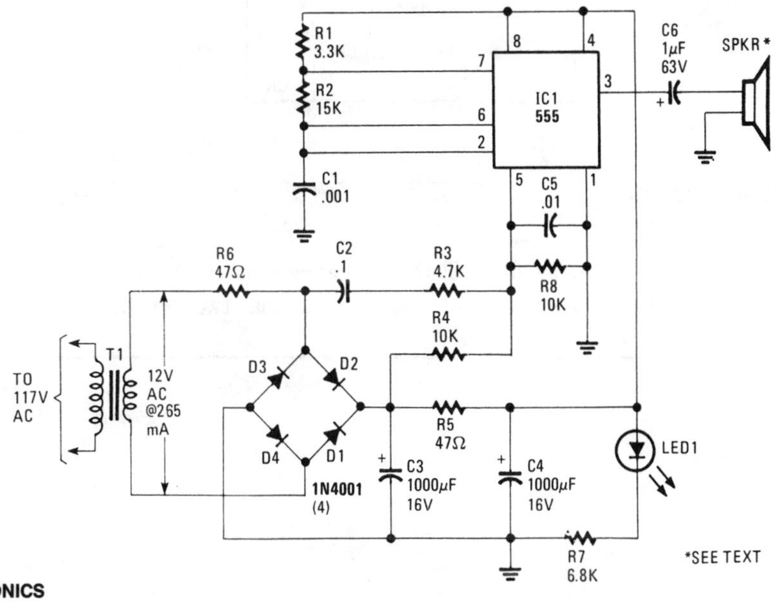

RADIO-ELECTRONICS

Fig. 94-2

An NE555 timer is used to generate an ultrasonic signal in the 20- to 65-kHz range. The speaker is a small piezoelectric tweeter with response above 20 kHz. These frequencies are said to be annoying to rats, mice, and insects.

ULTRASONIC PEST REPELLER II

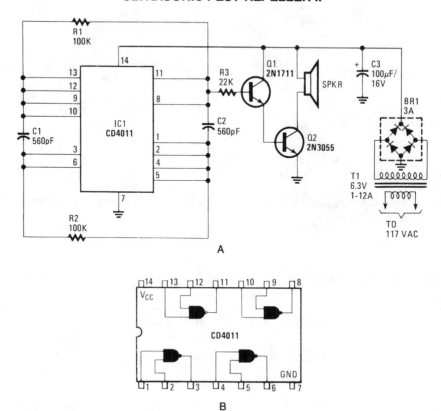

A

B

RADIO-ELECTRONICS

Fig. 94-3

A CD4011 Quad NAND gate acts as an oscillator, operating around 40 kHz. The small amount of filtering used modulates this with 120-Hz hum. The speaker is a small tweeter for hi-fi applications.

95

Video Circuits

The sources of the following circuits are contained in the Sources section, which begins on page 680. The figure number in the box of each circuit correlates to the entry in the Sources section.

ANALOG VOLTAGE CAMERA-IMAGE TRACKER

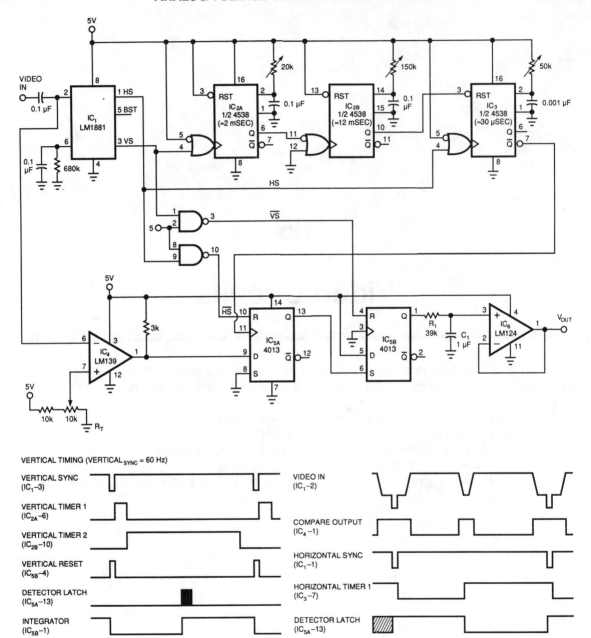

NOTE: ASSUME TARGET IS IN MIDDLE OF SCREEN

EDN

Fig. 95-1

ANALOG VOLTAGE CAMERA-IMAGE TRACKER (*Cont.*)

By using a low-cost RS-170 camera and this circuit, a voltage that trades the position of an object in the field of view of a camera is generated. IC2A and IC2B form a valid video gate that holds IC3 in reset during the internal vertical blanking to prevent false interpretation of the UBI as black video. IC4 is a black level detector. The circuit tests for a black object in the middle of each video line. IC5 latches the comparator's output and produces a square wave whose duty cycle depends on where the black level is detected in the video field. R1, C1, and IC6 integrate and buffer the analog output voltage.

HIGH-PERFORMANCE VIDEO MIXER

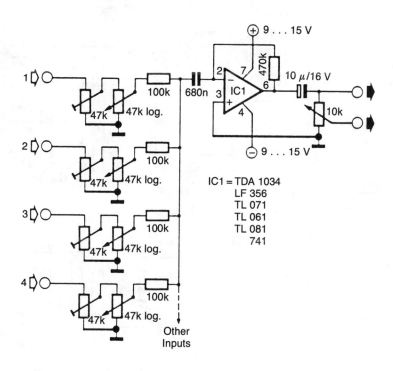

ELEKTOR ELECTRONICS

Fig. 95-2

This circuit mixes H synch, V synch, and actual video. T2 mixes the synch, while T1 serves as an emitter-follower. Bandwidths of up to 25 MHz are typical for this circuit.

VIDEO A/D—D/A CONVERTER

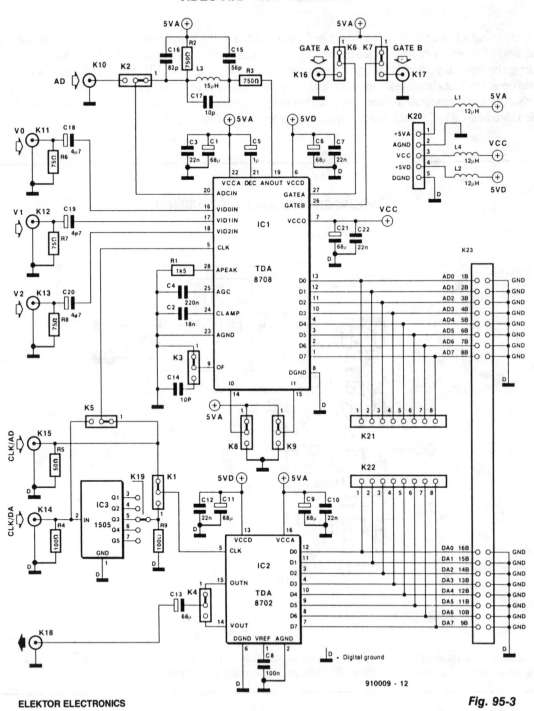

Fig. 95-3

910009 - 12

VIDEO A/D—D/A CONVERTER (*Cont.*)

This circuit is useful for digital video experiments and for interfacing video with a computer that has a TDA8708 (Philips). The A/D converter provides 8-bit digitized video to k23 socket and k21 socket. A TDA8702 (Philips) D/A converter recovers analog video. IC3 is a 1- to 5-ms delay line (1505) to delay clock pulses in 1-ms steps to the D/A converter. Clock speeds can be up to 30 MHz. Three video inputs are provided for three analog channels (e.g., R, G, B video).

RGB/NTSC CONVERTER

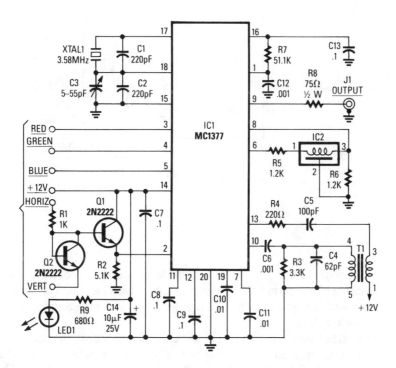

RADIO-ELECTRONICS

Fig. 95-4

Using a Motorola MC1377, this circuit produces NTSC video from an RGB source. Components are not critical, except for R7 and CB, which should be 1% and 2% tolerance, respectively.

TV LINE PULSE EXTRACTOR

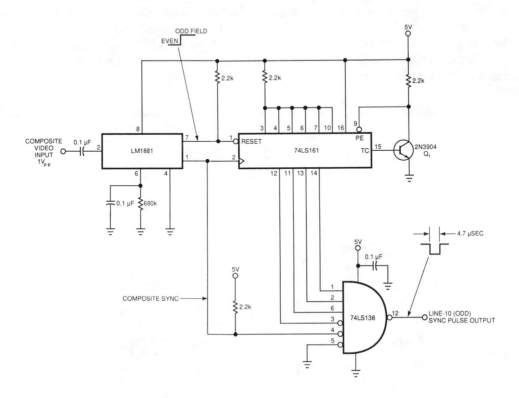

EDN

Fig. 95-5

This circuit uses a sync-generator chip, a counter, and a decoder to detect the horizontal sync pulse that occurs at the beginning of line 10 in field 1 of an NTSC television picture. You can use this circuit to compare the time delay between sync signals at various locations, and to determine and correct for any drift between the two master clocks.

The output of the LM1881 sync separator is the key to detecting line 10; the odd/even line goes high on the leading edge of the first equalizing pulse in the middle of line 4. Thus, you can use this knowledge to find virtually any other line in the field. This particular circuit locates line 10 of field one. The circuit resets the 74LS161 counter until the odd/even line goes high. Then, 74LS161 counts the positive transitions of the sync signal. After 11 positive transitions, the sync pulse drives pin 4 of the 74LS138 decoder low, and the line-10 sync pulse appears at pin 12 of the decoder. (The circuit counts to 11, as opposed to the 6 you might expect—because the composite sync signal contains more than 1 pulse per line). The counter remains in its maximum-count state until the sync separator causes a reset because Q1 feeds the inverted terminal-count output back to the parallel-enable input.

NTSC/RGB VIDEO DECODER

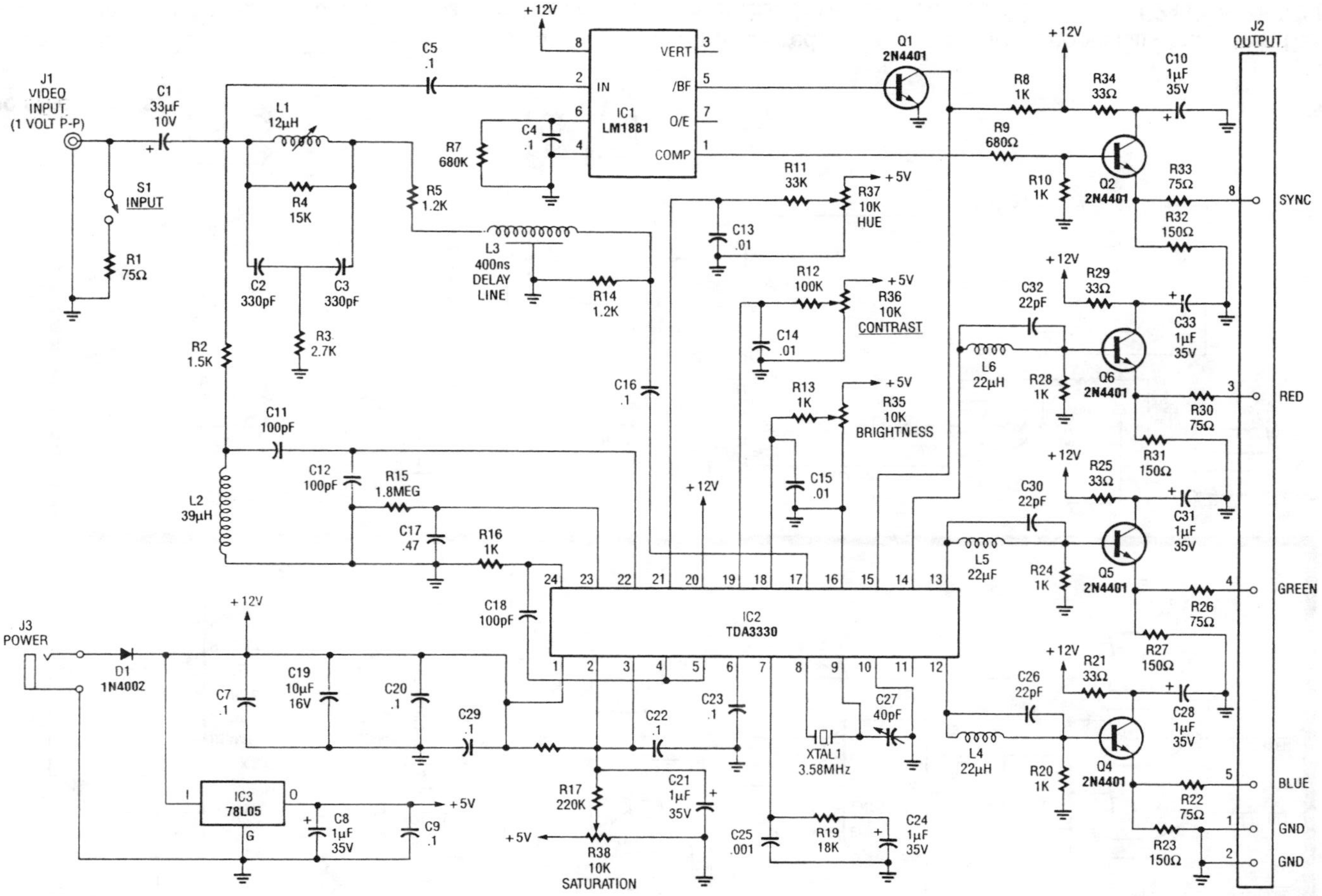

Fig. 95-6

RADIO-ELECTRONICS

An NTSC/RGB decoder is shown here. Using a TDA3330, 1-V input video is broken down into its R, G, B components, and composite synch. U1 is an integrated synch separator (LM1881). This circuit should be useful for interfacing RGB monitors to NTSC video systems.

COLOR-BAR GENERATOR

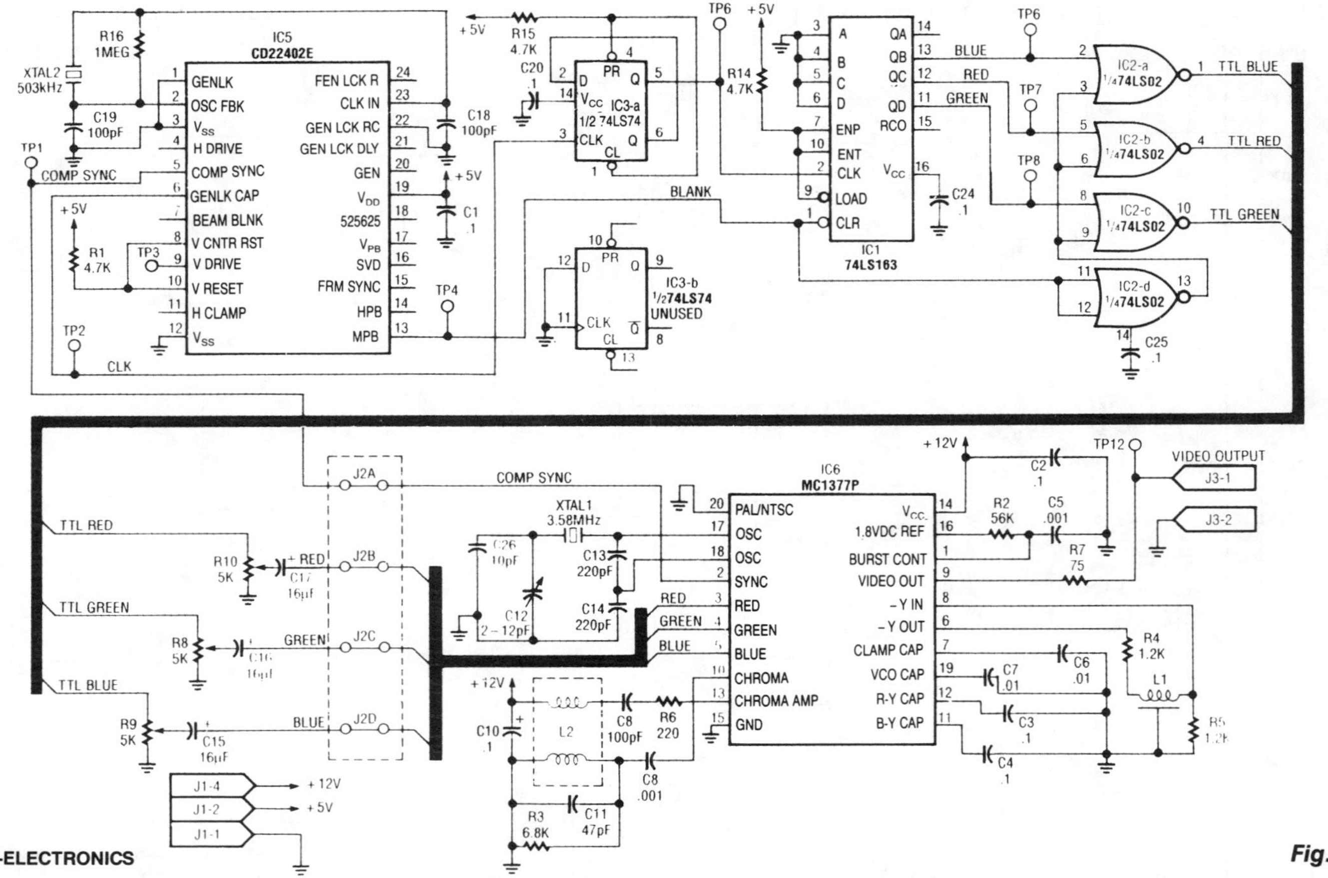

Fig. 95-7

RADIO-ELECTRONICS

IC5 generates RS-170 NTSC synch. IC1, IC2, and IC3 make up the red, green, and blue video signals that drive the video encoder section of IC6 to make up the color bars. IC3 is an $a \div 2$ counter that drives 4-bit counter IC1. Gates IC2a through IC2b form the R, B, and G, video signals. IC6 encodes these, plus synch, to form an NTSC video output signal, which appears at TP12.

VIDEO OP AMP CIRCUITS

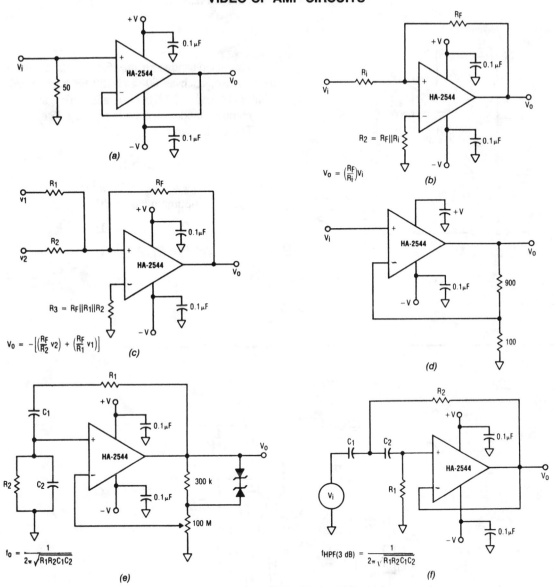

$$V_o = \left(\frac{R_F}{R_i}\right)V_i$$

(b)

$$V_o = -\left[\left(\frac{R_F}{R_2}v_2\right) + \left(\frac{R_F}{R_1}v_1\right)\right]$$

(c)

$$f_0 = \frac{1}{2\pi\sqrt{R_1R_2C_1C_2}}$$

(e)

$$f_{HPF(3\ dB)} = \frac{1}{2\pi\sqrt{R_1R_2C_1C_2}}$$

(f)

ELECTRONIC PRODUCTS

Fig. 95-8

These 6 circuits use the Harris HA2544 op amp. Component values are obtained by using the equations in the figure.

The HA-2544 can be used in any number of standard op amp configurations, including a voltage follower (a), an inverting amplifier (b), an inverting summer (c), a buffer amplifier with a gain of 10 (d), a Wien-bridge oscillator with zener diode adaptive feedback (e), and a second-order, high-pass active filter (f).

VIDEO LOOP-THRU AMPLIFIER

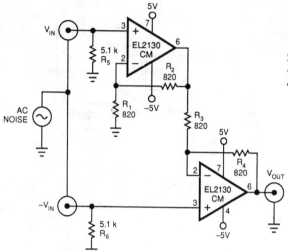

This video bandwidth amplifier rejects common-mode noise, such as 60- and 120-Hz hum. Bandwidth is typically 60 MHz for a differential gain of 2. The common-mode rejection is typically 45 dB. The design equations are:

$$A_{DIFF} \geq 1,$$

$R_2 = R_4$ (select both for optimum bandwidth)

$$R_1 = R_2 (A_{DIFF} - 1)$$

$$R_3 = R_4 (A_{DIFF} - 1)$$

EDN

Fig. 95-9

SYNC SEPARATOR

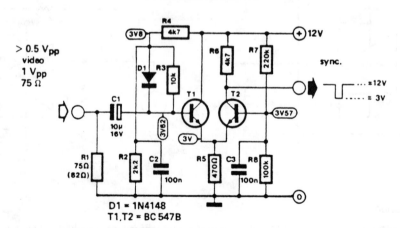

ELEKTOR ELECTRONICS

Fig. 95-10

This circuit separates the synch from the composite video signal. It uses a two-transistor comparator. Output is 9 Vpp with a 0.5-Vpp (minimum) video input signal.

SIMPLE MONOCHROME TV-PATTERN GENERATOR

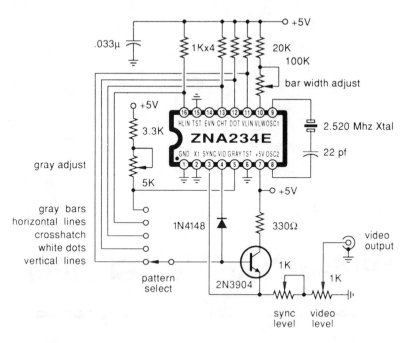

RADIO-ELECTRONICS

Fig. 95-11

Using a Plessey ZNA234E IC, this generator produces sync, blanking, gray bars, lines, dots, and crosshatch patterns.

96

Video-Switching Circuits

The sources of the following circuits are contained in the Sources section, which begins on page 680. The figure number in the box of each circuit correlates to the entry in the Sources section.

Remote-Selection Video Switch
Remote-Controlled Switcher

REMOTE SELECTION VIDEO SWITCH

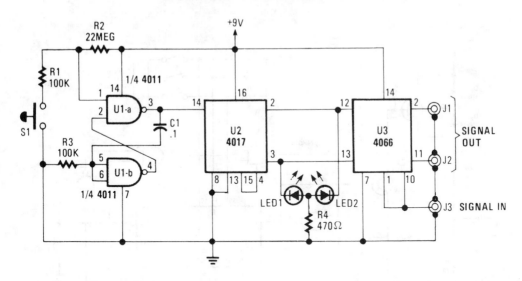

POPULAR ELECTRONICS

Fig. 96-1

The A/B Switch circuit consists of three ICs and a handful of resistors. Two gates from a 4011 quad 2-input NAND gate (U1A and U1B) are configured as a monostable multivibrator that, when switch S1 is pressed, triggers a 4017 decade counter/divider, which has been set to recycle after a count of two. The outputs of U2 at pins 2 and 3 are fed to the control inputs of U3 (a 4066 quad bilateral switch) at pins 12 and 13. Depending on which control input is high, either the J1 or J2 output is selected.

With a little modification, the switch could be set to trigger at a set rate (automatically). With the addition of another 4066, it could have as many as 8 channels. One possible application would be in a security surveillance system.

REMOTE-CONTROLLED SWITCHER

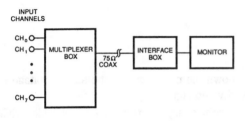

This 1-cable system carries composite video (NTSC, PAL, or SECAM), power, and channel-select signals.

EDN

Fig. 96-2(a)

619

REMOTE-CONTROLLED SWITCHER (Cont.)

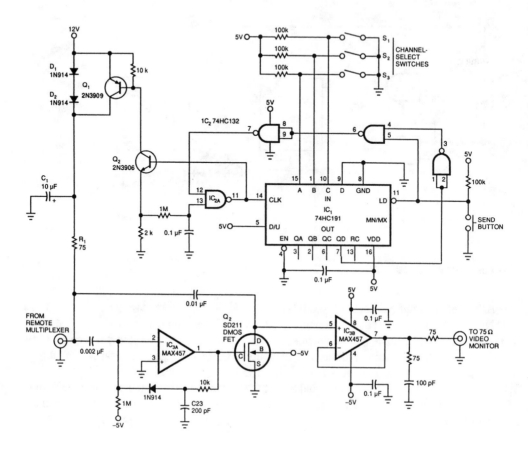

EDN

Fig. 96-2(b)

The interface end of Fig. 96-2(a)'s circuit deliver 10 V down the cable, pulses the supply voltage to transmit channel-change commands, and buffers the received video signal.

The multiplexer circuit in Fig. 96-2(a) receives power and control signals over the coaxial cable, while driving the cable with the currently selected video signal.

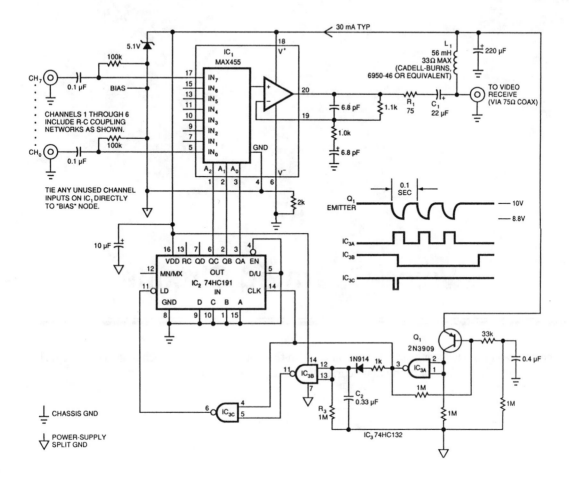

EDN

Fig. 96-2(c)

The interface circuit (Fig. 96-2(b)) delivers 10 V to the cable and pulses the supply voltage to select one of 8 channels. When the send button is depressed, a digital burst of 1.2 V amplitude (negative) is superimposed on the 12-V line (as a voltage drop). This does not affect the video signal. The multiplexer circuit (Fig. 96-2(c)) consists of a multiplexer and an amplifier. The multiplexer is a Maxim MAX455. The digital code on the supply line is picked off by A1, IC3A, and interfaced to counter IC2, which drives the multiplexer to select the desired video channel.

97

Voice-Operated Circuits

The sources of the following circuits are contained in the Sources section, which begins on page 681. The figure number in the box of each circuit correlates to the entry in the Sources section.

Simple VOX
Scanner Voice Squelch

SIMPLE VOX

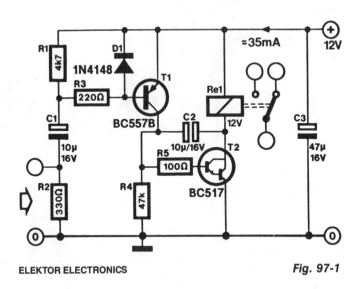

ELEKTOR ELECTRONICS

Fig. 97-1

A *VOX* is a voice-operated switch that is often used as a substitute for the press-to-talk switch on a microphone. This VOX can be connected to almost any audio equipment that has a socket for an external loudspeaker. The actuation threshold is set by the volume control on the AF amplifier that drives the VOX.

The (loudspeaker) signal across R2 is capacitively fed to the base of T1. Resistor R3 limits the base current of this transistor when the input voltage exceeds 600 mV. Diode D1 blocks the positive excursions of the input signal, so that V_{eb} cannot become more negative than about 0.6 V.

The output relay is driven by Darlington T2. Resistor R4 keeps the relay disabled when T1 is off. The value of bipolar capacitor C2 allows it to serve as a ripple filter in conjunction with T2. Resistor R5 limits the base current of T2 to a safe level.

The switching threshold of the VOX is about 600 mV across R2. The maximum input voltage is determined by the maximum permissible dissipation of R2 and R3. As a general rule, the input voltage should not exceed 40 Vpp. The current drawn by the VOX is mainly the sum of the currents through the relay coil and through R5. The resistor can carry up to 100 mA when the VOX is overdriven.

SCANNER VOICE SQUELCH

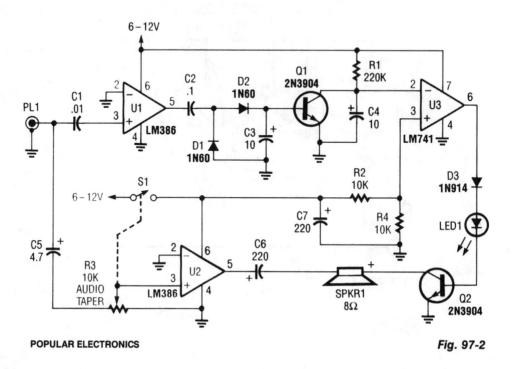

POPULAR ELECTRONICS

Fig. 97-2

This circuit detects the presence of audio (voice) on the output of a scanner. If the scanner stops on a "dead carrier" or noise, the circuit mutes the speaker to avoid annoying noise.

U1 amplifies speech and drives rectifier D1/D2 and switch Q1. Comparator U3 drives speaker switch Q1 and indicator LED1. Q2 completes the speaker path to ground. U2 is an audio amplifier to drive the speaker. R3 is a volume control. PL1 connects to the scanner speaker or to the headphone jack.

98

Voltage-Controlled Oscillators

The sources of the following circuits are contained in the Sources section, which begins on page 681. The figure number in the box of each circuit correlates to the entry in the Sources section.

SIMPLE AUDIO-FREQUENCY VCO

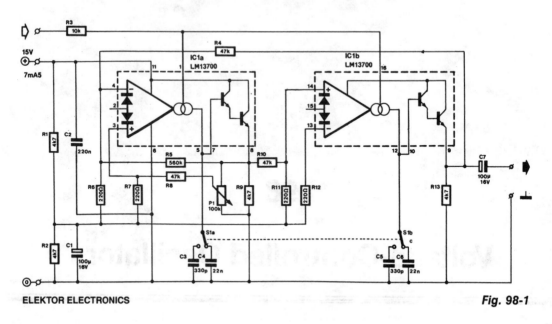

ELEKTOR ELECTRONICS

Fig. 98-1

The frequency of this sine-wave oscillator is determined by a direct voltage, U_c, of 0 to 15 V. The distortion on output signals of up to 10 Vpp is not greater than 1%. When the output is reduced with the aid of P1 to 1 Vpp, the distortion drops to below 0.1%. It is not recommended to use output signals below 1 Vpp, because the oscillator then becomes unstable and temperature-dependent.

The oscillator consists of two operational transconductance amplifiers (OTAs) contained in one package. Their Amp-bias inputs, pins 1 and 16, are connected in parallel. These inputs can drive the output currents at pins 5 and 12 to a peak value of up to 0.75 mA.

Switch S1 enables the oscillator output to be set to two ranges: 6.7 to 400 Hz and 400 Hz to 23.8 kHz. The overall range needs a control voltage that varies from 1.34 to 15 V. When the frequency is changed by a variation of U_C and the setting of P1 is not altered, the output signal might be distorted. In other words, the amplitude of the signal must be adapted to the frequency.

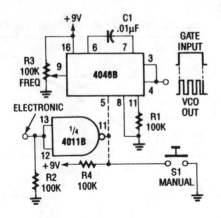

GATED WIDE-RANGE VCO

A CD4046 can be gated either with a switch or electronically, as shown in the figure. Frequency range of this circuit is to 1.5 kHz; use another C_1 for higher frequencies.

RADIO-ELECTRONICS

Fig. 98-2

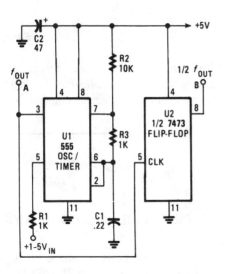

SIMPLE 555 VCO

The VCO has an output frequency that ranges from 1 500 Hz at $V_{in}=1$ V to 300 Hz at $V_{in}=5$ V. R_1 or C_1 can be varied to change this range. U2 provides a symmetrical square-wave output of half the timer frequency.

POPULAR ELECTRONICS

Fig. 98-3

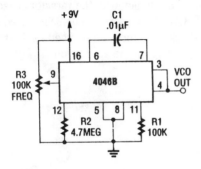

RESTRICTED-RANGE VCO

This VCO is adjustable from 60 Hz to 1.4 kHz. C_1 can be changed for other ranges.

RADIO-ELECTRONICS

Fig. 98-4

LINEAR VCO

220 pF

6   7

HC4046

4   F_OUT

5

9   10   12

12V

V_IN

LF356

+

−

−12V

10k

VCO CHARACTERISTIC OF 4046 COMBINED WITH LF356

Frequency (Hz) vs. V_{IN} (V)

EDN   *Fig. 98-5(a)*

EDN   *Fig. 98-5(b)*

This VCO uses an LF356 op amp to produce a linear frequency vs. voltage characteristic using the CMOS HC4046. The frequency range can be changed by changing the capacitor connected between pins 6 and 7 of the HC4046. Using the HC4046's internal transistor instead of an external component achieves the linearization in Fig. 98-5(b).

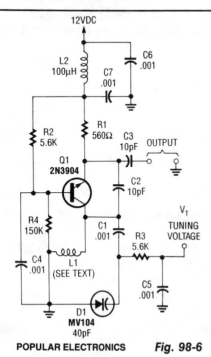

12VDC

L2 100µH

C6 .001

C7 .001

R1 560Ω

R2 5.6K

C3 10pF   OUTPUT

Q1 2N3904

C2 10pF

V_T

R4 150K

C1 .001

R3 5.6K

TUNING VOLTAGE

C4 .001

L1 (SEE TEXT)

C5 .001

D1 MV104 40pF

POPULAR ELECTRONICS   *Fig. 98-6*

VOLTAGE-TUNED VHF OSCILLATOR

This VHF VCO circuit is suitable for 30 to 200 MHz. Q1 can be replaced by a 2N3563 for operation above 100 MHz. L1 is chosen to resonate to the desired frequency with the varactor capacitance of 40 pF. Other varactors can be substituted or two back-to-back varactors can be used for better linearity, depending on the application.

VOLTAGE-CONTROLLED CURRENT SINK

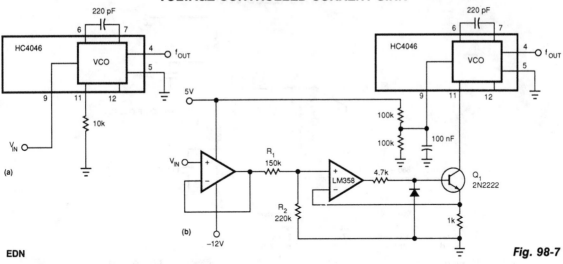

EDN

Fig. 98-7

This circuit widens the linear frequency span of an HC4046 from one decade to nearly three decades. An LM358 is used as a constant-current sink to replace the frequency-determining resistor (10 kΩ) from pin 9 to ground. Pin 9 is held at a fixed 2.5 V for this application.

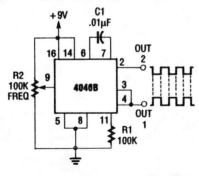

RADIO-ELECTRONICS   **Fig. 98-8**

BIPHASE WIDE-RANGE VCO

Using a CD4046B, this circuit generates a biphase signal. The frequency range is below 100 Hz to about 1.5 kHz.

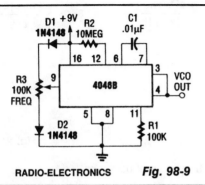

RADIO-ELECTRONICS   **Fig. 98-9**

WIDE-RANGE VCO

This circuit covers 0 to 1.4 kHz. C_1 can be changed to cover other ranges, as desired.

VARACTORLESS VCO

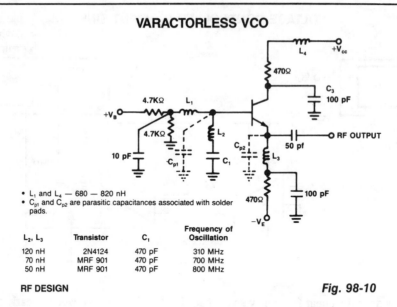

- L_1 and L_4 — 680 — 820 nH
- C_{p1} and C_{p2} are parasitic capacitances associated with solder pads.

L_2, L_3	Transistor	C_1	Frequency of Oscillation
120 nH	2N4124	470 pF	310 MHz
70 nH	MRF 901	470 pF	700 MHz
50 nH	MRF 901	470 pF	800 MHz

RF DESIGN

Fig. 98-10

The varactorless VCO utilizes a modified Clapp-oscillator configuration, together with some data on the type of transistor, circuit values, and operating frequencies.

VCO

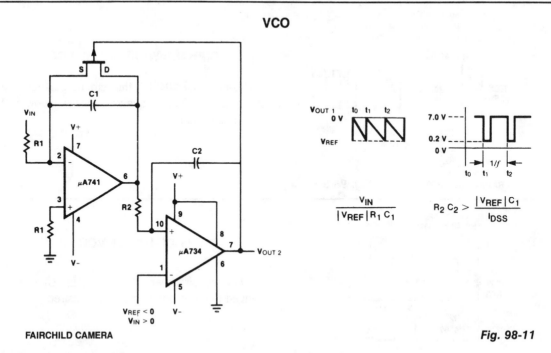

$$\frac{V_{IN}}{|V_{REF}| R_1 C_1}$$

$$R_2 C_2 > \frac{|V_{REF}| C_1}{I_{DSS}}$$

FAIRCHILD CAMERA

Fig. 98-11

Q1, an FET, is used as a variable resistance to control frequency of oscillator.

99

Voltage Multiplier Circuits

The sources of the following circuits are contained in the Sources section, which begins on page 681. The figure number in the box of each circuit correlates to the entry in the Sources section.

VOLTAGE MULTIPLIER

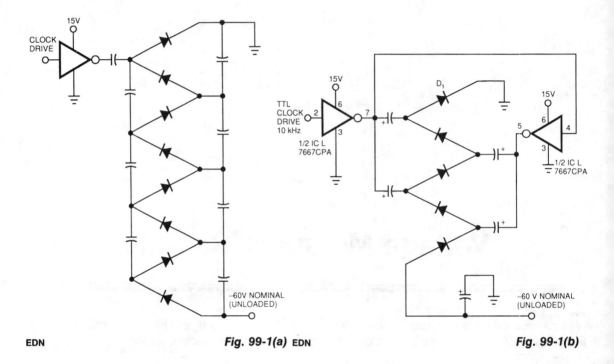

EDN **Fig. 99-1(a)** EDN   **Fig. 99-1(b)**

Figure 99-1(a)'s circuit exhibits a high-output impedance as a result of the small effective capacitance of the series-connected capacitors, and it exhibits considerable voltage loss due to all of the diode drops. Further, this circuit requires $2n$ diodes and $2n$ capacitors to produce a dc output voltage approximately n times the rail voltage.

Figure 99-1(b)'s circuit multiplies more effectively using fewer diodes and capacitors. The parallel arrangement of the capacitors lets you use smaller capacitors than those required in Fig. 99-1(a). Alternatively, when using the same capacitor values of Fig. 99-1(a), the output impedance will be lower.

Whereas the clock source directly drives only one of the two strings of capacitors in Fig. 99-1(a), Fig. 99-1(b)'s clock drives both strings with opposite phases. This drive scheme doubles the voltage per stage of two diodes. A final diode is necessary to pick off the dc output voltage because both strings of capacitors now carry the p–p ac input-voltage waveform. The ICL7667 dual-FET driver accepts a TTL drive swing and provides a low-impedance push-pull drive to the diode string. This low impedance is particularly helpful when using a long string to raise output voltage to more than 100 V, starting from a low rail voltage.

CORONA WIND GENERATOR

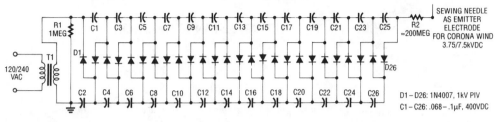

RADIO-ELECTRONICS

Fig. 99-2

This 25-stage voltage doubler will generate "corona wind." It delivers 3.75 kVdc when powered from 120 Vac, or 7.5 kVdc when powered from 240 Vac.

10 000-Vdc SUPPLY

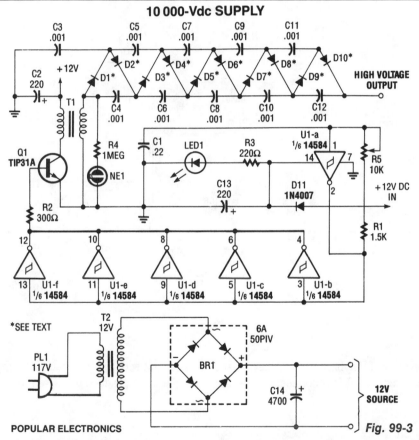

POPULAR ELECTRONICS

Fig. 99-3

A CMOS oscillator (U1A) drives. U1B through U1F, which drives Q1, which generates a 12-Vpp square wave across the primary of T1. This square wave is applied to a rectifier-multiplier circuit consisting of D1 through D10 (each is two 1N4007 diodes in series) and C3 through C12. About 10 kV is available.

HIGH-VOLTAGE NEGATIVE-ION GENERATOR

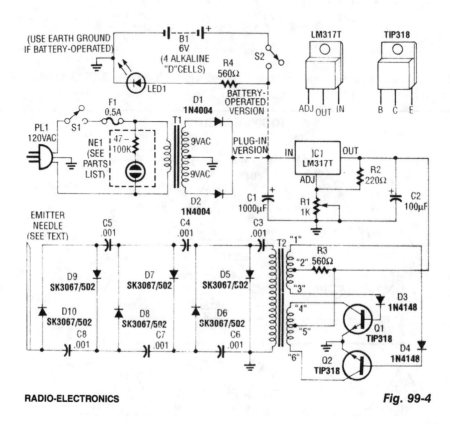

Fig. 99-4

A modified B/W TV flyback transformer is used in this circuit with a voltage multiplier to produce 9- to 14-kV negative voltage. This is connected to a discharge needle to produce negative ions.

VOLTAGE DOUBLERS

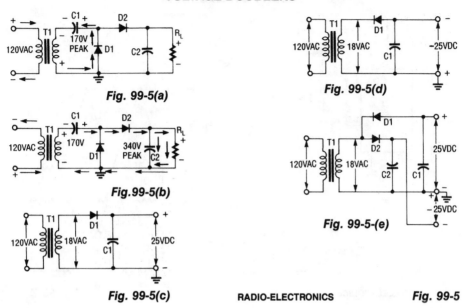

Fig. 99-5(a)

Fig. 99-5(d)

Fig.99-5(b)

Fig. 99-5-(e)

Fig. 99-5(c)

RADIO-ELECTRONICS

Fig. 99-5

During the first half-cycle (Fig. 99-5(a)), D1 conducts, D2 cuts off, C1 charges to 170 V peak, and C2 discharges through R_L. For the second half-cycle (Fig. 99-5(b)), the input polarity is reversed, and both the input and C1 are in series, which produces 340 V (peak). Now, D1 cuts off, while D2 conducts, and the current divides between C2 and R_L; the cycle then repeats. Two Half-Wave Rectifiers, one with a positive output (Fig. 99-5(c)) and one negative (Fig. 99-5(d)), combine to make a full-wave voltage doubler (Fig. 99-5(e)).

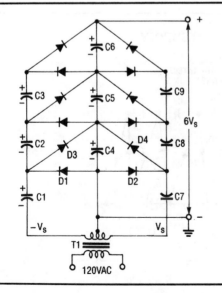

COCKCROFT-WALTON
CASCADED VOLTAGE DOUBLER

A center-tapped transformer of secondary voltage. Two V_s can be used to power a voltage multiplier. For higher voltages, simply add more sections.

RADIO-ELECTRONICS

Fig. 99-6

LASER POWER SUPPLY

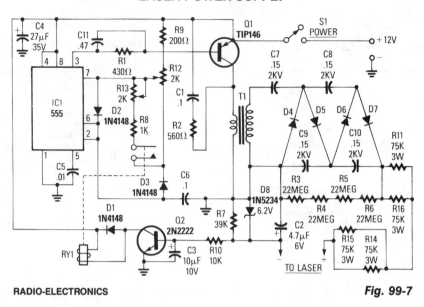

RADIO-ELECTRONICS

Fig. 99-7

IC1 is a 555 timer running at about 16 kHz. This IC drives Q1, a TIP146, which produces a 12-V square wave across T1 primary. This produces between 800 and 2,000 V across the secondary, which is doubled to 3 to 5 kV. When the load (laser) on the power supply increases, current Q2 is turned on, which energizes RY1. This changes the duty cycle of the 555 timer. To adjust this supply, set R12 and R13 at the center. Adjust R12 until the laser tube triggers, and make sure that the relay pulls in. If the relay chatters, adjust R12. If the full-clockwise adjustment of R12 fails to ignite the tube, adjust R13.

2,000-V LOW-CURRENT POWER SUPPLY

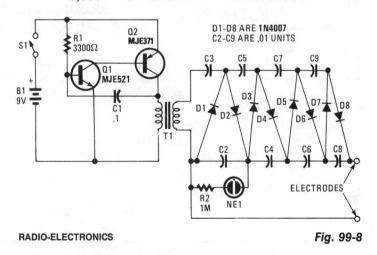

RADIO-ELECTRONICS

Fig. 99-8

2,000-V LOW-CURRENT POWER SUPPLY (*Cont.*)

In this circuit Q1, Q2, R1, and C1 form a multivibrator. The square wave that results from the oscillation of this circuit (20 to 30 Vpp) is stepped up by T1 (an audio transformer of the type used in radios or small TVs). An 8- to 1,200-Ω impedance ratio equates to a turn ratio of 12:1. The ac from the secondary of T1 is applied to the multiplier circuit (D1 to D8 and C2 to C9). NE1/R2 are used as an operating indicator. The circuit will supply about 2,000 V. C2 to C9 should have a 400-V or higher voltage rating.

LOW-CURRENT VOLTAGE TRIPLER

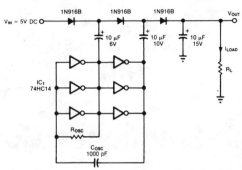

This circuit generates a 15V output by tripling V_{IN}.

(a)

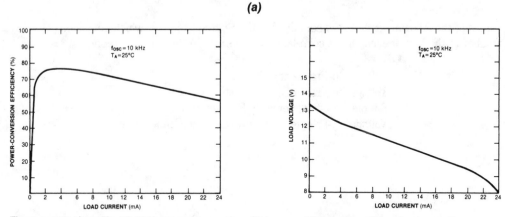

These curves show Fig. 99-9(a)'s power-conversion efficiency and laod voltage (b) vs load current.

(b)   **(c)**

EDN   **Fig. 99-9**

Using a 74HC14 operating at 350 kHz, this voltage tripler delivers approximately 12 V from a 5-V supply.

100

Voltage-To-Frequency Converters

The sources of the following circuits are contained in the Sources section, which begins on page 681. The figure number in the box of each circuit correlates to the entry in the Sources section.

VOLTAGE-TO-FREQUENCY CONVERTER I

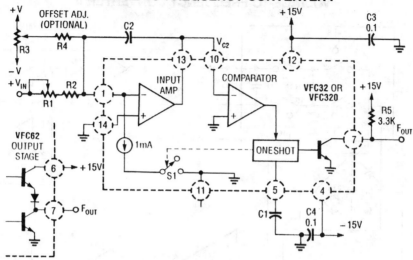

Full-Scale Input	Full-Scale Frequency	C1	C2	R1 (ohms)	R2 (ohms)
1 V	10 kHz	3300 pF	0.01 μF	1K	3K
10 V	10 kHz	3300 pF	0.01 μF	10K	30K
1 V	100 kHz	300 pF	0.001 μF	1K	3K
10 V	100 kHz	300 pF	0.001 μF	10K	30K

CIRCUIT VALUES

RADIO-ELECTRONICS

Fig. 100-1

(a)

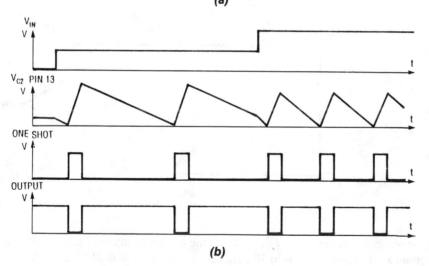

(b)

Using a Burr-Brown VFC 32 IC, this voltage-to-frequency converter uses few components. The circuit values are shown in the figure.

This charge-balanced V/F converter uses a VFC32 or a VFC320 IC. The positive charge from the 1-mA balances the negative charge from the input. V/F converter waveforms are shown in Fig. 100-1(b).

VOLTAGE-TO-FREQUENCY CONVERTER II

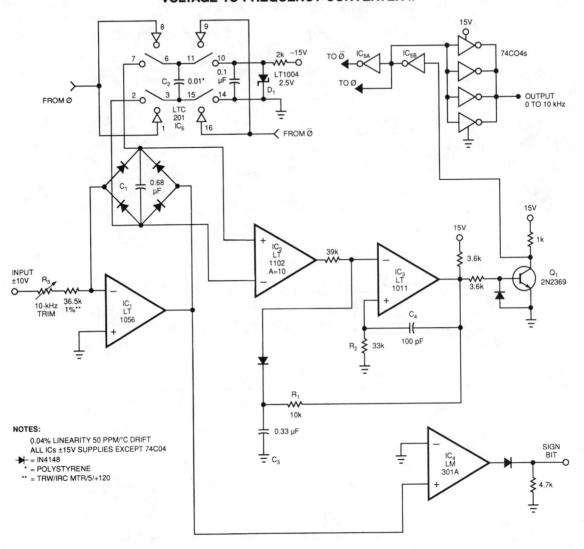

Fig. 100-2

This voltage-to-frequency converter (VFC) accepts the bipolar-ac inputs. For −10- to +10-V inputs, the converter produces a proportional 0- to 10-kHz output. Linearity is 0.04%, and temperature coefficient (TC) measures about 50 ppm/°C.

To understand the circuit, assume that its input sees a bipolar square wave. During the input's positive phase, IC1's output swings negative and drives current through C1 via the full-wave diode bridge. IC1's current causes C1's voltage to ramp up linearity. Instrumentation amplifier IC2 operates at a gain of 10 and measures the differential voltage across C1.

VOLTAGE-TO-FREQUENCY CONVERTER II (Cont.)

IC2's output biases comparator IC3's negative input. When IC2's output crosses zero, IC3 fires ac positive feedback to IC3's positive input and hangs up IC3's output for about 20 μs. The Q1 level shifter drives ground-referred inverters IC5A and IC5B to deliver biphase drive to LT1004 switch IC6.

IC6, configured as a charge pump, places C2 across C1 each time the inverters switch, which resets C1 to a lower voltage. The LT1004 reference (D1), along with C2's value, determines how much charge the charge pump removes from C1 each time the charge pump cycles. Thus, each time IC2's output tries to cross zero, the charge pump switches C2 across C1, which resets C1 to a small negative voltage and forces IC1 to begin recharging C1.

The frequency of this oscillatory behavior is directly proportional to the input-derived current into IC1. During the time that C1 is ramping toward zero, IC6, places C2 across the reference diode (D1), and prepares C2 for the next discharge cycle.

The action is the same for negative-input excursions, except that IC1's output phasing is reversed. IC2, looking differentially across IC1's diode bridge, sees the same signal as it does for positive inputs; therefore, the circuit's action is identical. IC4, detecting IC1's output polarity, provides a signal bid output.

SIMPLE LOW-FREQUENCY V/F CONVERTER

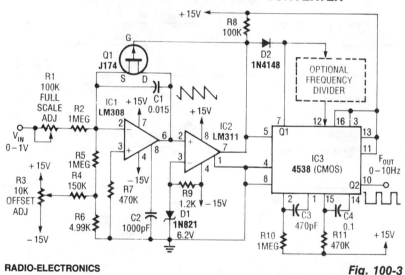

RADIO-ELECTRONICS

Fig. 100-3

In this circuit, C1 is charged to a fixed reference level, then discharged. Integrator IC1 circuit charges C1 until IC1 has -6.2-V output, when comparator IC2 outputs a low. FET Q1, triggers one-section monostable multivibrator IC3, pulls pin 3 low for 470 μs, ensuring that Q1 completely discharges C1. The other section of IC3 produces a longer pulse of about 47 ms.

Full scale of this circuit is 10 Hz. For lower output pulse rates, a counter circuit can be inserted between the sections of IC3. Notice that because C1 does not integrate while Q1 is biased on, this circuit has an error in the output period, which must be as short as possible. Therefore, the circuit's use is limited to low frequencies.

VOLTAGE-TO-FREQUENCY CONVERTER WITH OPTOCOUPLER

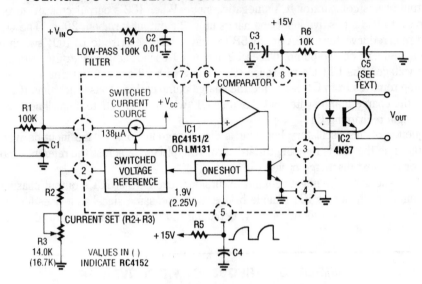

CIRCUIT VALUES

Full-Scale Input	Full-Scale Frequency	C1	C4	R5 (ohms)
10 V	1 kHz	10 μF	0.1 μF	6.8K
10 V	10 kHz	1 μF	0.01 μF	6.8K
10 V	100 kHz	0.1 μF	0.001 μF	6.8K

RADIO-ELECTRONICS

Fig. 100-4

In this circuit, a Raytheon RC4151 or National LM131 is used in conjunction with an optocoupler for applications where input-to-output isolation is desirable. Circuit values are shown in the figure for various applications.

101

Volume/Level-Control Circuit

The source of the following circuit is contained in the Sources section, which begins on page 681. The figure number in the box correlates to the entry in the Sources section.

Digital Volume Control

DIGITAL VOLUME CONTROL

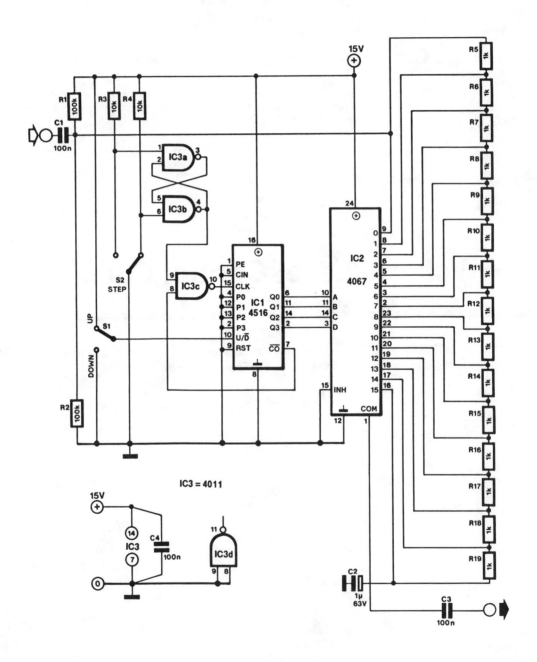

IC3 = 4011

Fig. 101-1

DIGITAL VOLUME CONTROL (*Cont.*)

The heart of the digitally operated volume control is IC2, a 4067 16-channel analog multiplexer. Depending on the logic state on pins A, B, C, and D of the multiplexer, one of its 16 inputs or outputs is connected to pin 1, which is the "wiper" of the control.

Because a 1-kΩ resistor has been connected between each input and output, the multiplexer can be considered a linear potentiometer with 16 fixed steps. Its overall resistance is 15 kΩ. It is, of course, possible to use a different value for each of the resistors to obtain a different characteristic.

The setting of the potentiometer is controlled by counter IC1. Dependent on the position of switch S1, the counter moves one step up or down when switch S1 is changed over. Circuits IC3A and IC3B debounce S2.

A jump from 0000 to 1111 or the other way around is not possible, because further count pulses are suppressed with the aid of the $\overline{CO}$ line. This line is logic low when both the counter state and signal U/$\overline{D}$ are 0.

When U/$\overline{D}$ is high and the counter state is 15, $\overline{CO}$ again becomes logic low. It is then necessary to reverse the logic state at U/$\overline{D}$, and thus the direction of counting. The volume control draws a current of around 1 mA.

102

Wave-Shaping Circuits

The sources of the following circuits are contained in the Sources section, which begins on page 681. The figure number in the box of each circuit correlates to the entry in the Sources section.

PHASE SHIFTER

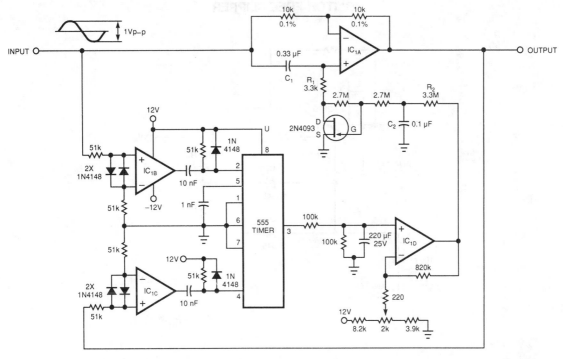

EDN

Fig. 102-1

This circuit adds 120 degrees of phase shift to a 50- or 60-Hz input, regardless of the frequency and amplitude fluctuations of that input. The circuit configures a 2N4093 JFET as a voltage-controlled resistor whose value is proportional to the phase difference between the input and the output. The values of C_1, R_1, and r_{DS} determine the amount of phase shift (120° this case.)

A 555 timer implements a phase detector whose two inputs are related to the input and output. The input and output, respectively, drive IC1B and IC1C, which operate as zero-crossing detectors. D1 and D2 limit the positive-going pulses at the 555 inputs. Thus, the falling edges of IC1B and IC1C's outputs control the 555 timer. The timer's output signal stays low for a time that is proportional to the phase shift between the circuit's input and output.

The average value of the timer's output and an offsetting voltage drive IC1D. R2 and C2 filter IC1D's output. The resultant signal controls the JFET. The potentiometer sets the control at a value for which the phase shift between input and output is equal to 120 degrees when the input signal frequency is 50 or 60 Hz. Any differences between the input and output changes the 555 output's average value, thus ultimately modifying the control voltage and the JFET's resistance.

To calibrate the circuit, apply a 50-Hz sine wave with an amplitude of less than 1 Vpp to the input and adjust the potentiometer until the phase shift reads 120° on a digital phase meter. For input frequency variations between 40 and 60 Hz, the phase shift changed by a maximum of $\pm 0.17\%$ (equivalent to an offset of only 0.02°/Hz). The average value at IC1D's noninverting input is 3.864 V.

GLITCH-FREE CLIPPER

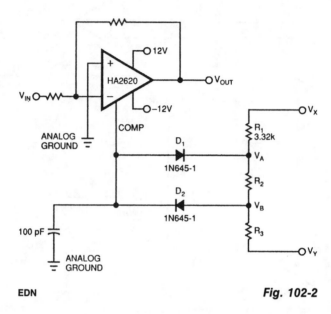

EDN

Fig. 102-2

Adding a simple clamping circuit to a Harris 2620 high-speed op amp produces a glitch-free amplifier/clipper. The op amp pin that controls the device's bandwidth is a high-impedance, isolated input. This pin also tracks the device's output voltage.

Therefore, D1, D2, R1, R2, and R3 will clamp the amplifier's output voltage only when the amplifier's input voltage exceeds your clamping-voltage limits. V_D is the diode drop of D1 or D2. The two clamp voltages, $V_A + V_D$ and $V_B - V_D$, are:

$$V_A = V_X \left(\frac{R_2 + R_3}{R_1 + R_2 + R_3} \right) + V_Y \left(\frac{R_1}{R_1 + R_2 + R_3} \right)$$

$$V_B = V_X \left(\frac{R_3}{R_1 + R_2 + R_3} \right) + V_Y \left(\frac{R_1 + R_2}{R_1 + R_2 + R_3} \right)$$

where V_X and V_Y are the clamping circuit's bias voltages. Choosing R_1 lets you determine the values of R_2 and R_3. Try a value for R_1 around 3 kΩ.

One example of this circuit had clamping voltages of ±3.7 V and exhibited THD below −75 dB for a sinusoidal, 30-kHz input signal. When the input signal increased beyond the ±3.7-V clamping voltage, the clipper symmetrically clamped the output voltage with no glitches in the waveform.

SIGNAL CONDITIONER

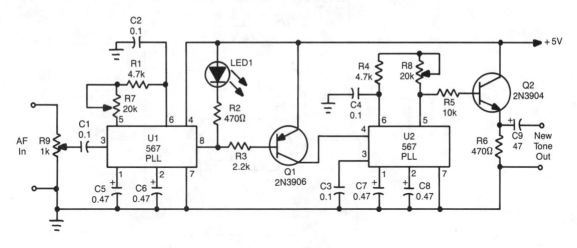

POPULAR ELECTRONICS

Fig. 102-3

This circuit takes audio from a receiver that might have a weak CW or tone signal and uses a PLL (U1) to recover the weak signal. U1 produces a low on receipt of a tone or note of frequency, determined by R1, R7, and C2. The output of U1 (pin 8) goes low, keys tone generator U2, and produces a new tone. The circuit is useful in cleaning up CW reception in static, noise, etc.

HARMONIC GENERATOR

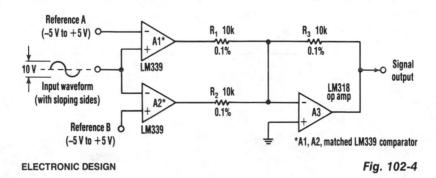

ELECTRONIC DESIGN

Fig. 102-4

This circuit can extract harmonics from various waveforms. With a sloped input waveform, the comparator produces a pulse width that is proportional to a reference plus input amplitude. As the pulse width changes, the harmonics spectrum changes. Combining the two comparator outputs eliminates some harmonics, depending on the duty cycle. Adjusting the references can create virtually any harmonic.

A1 and A2 should be matched, R1 and R2 should be equal to 0.1%, and A3 should have good common-mode rejection and high slew rate.

FULL-WAVE RECTIFIER (TO 10 MHz)

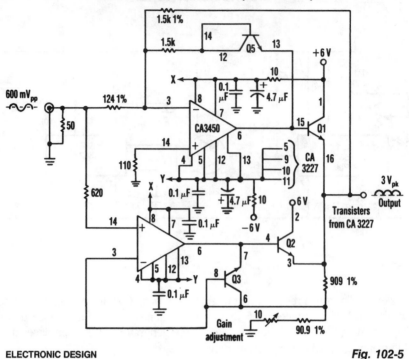

ELECTRONIC DESIGN

Fig. 102-5

Using two CA3450 op amps and a CA3227 transistor array, this circuit will accurately full-wave rectify signals to 10 MHz. Two of the CA3227 transistors drive the output, two are in the feedback circuits. Two transistors serve as clamping diodes, limit the negative-going signals of each amplifier, and keep both amplifiers active during the end cycle. The maximum output is determined by the slew rate of 300 V/μs at highest frequencies. This output equals 300 V/ms ÷ $6\pi V$ peak.

CAPACITOR ALLOWS HIGHER SLEW RATES

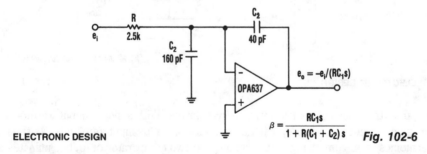

ELECTRONIC DESIGN

Fig. 102-6

In this circuit, a Burr-Brown op amp supplies a slew rate of 135 V/μs. The addition of C2 charges the high-frequency feedback factor to less than unity, and allows higher slew-rate amplifiers to be compensated for greater-than-unity gain.

S/R FLIP-FLOP

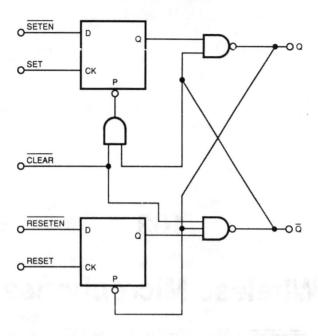

EDN   *Fig. 102-7*

This circuit combines the characteristics of an asynchronous S/R flip-flop and an edge-triggered JK flip-flop. It changes state on the leading edges of its inputs, and ignores the levels at all other times.

In operation, outputs of both D flip-flops are normally high, going low for brief periods after seeing an edge at their respective clock inputs.

103

Wireless Microphones

The sources of the following circuits are contained in the Sources section, which begins on page 682. The figure number in the box of each circuit correlates to the entry in the Sources section.

Wireless FM Microphone
Wireless Microphone

WIRELESS FM MICROPHONE

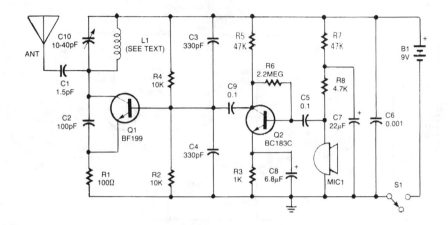

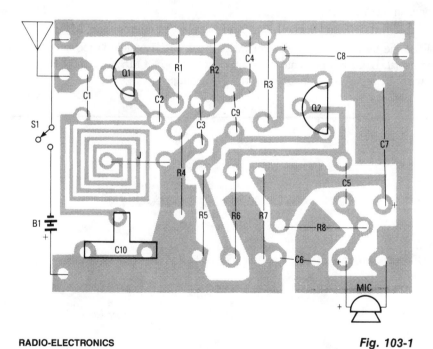

RADIO-ELECTRONICS

Fig. 103-1

A simple FM wireless microphone uses a single BC183C transistor as an audio amplifier. A 2N3565 can be substituted. Q1 is an oscillator that is FM modulated by the signal from Q1. Other transistors can be substituted, but the modulation characteristics should be checked.

WIRELESS MICROPHONE

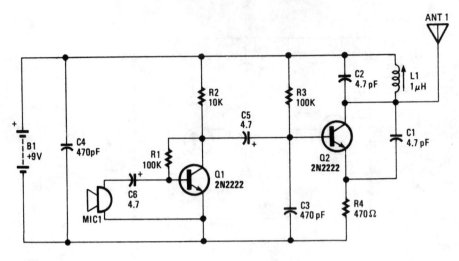

POPULAR ELECTRONICS

Fig. 103-2

Q1 amplifies the output from an electret microphone MIC1. Audio is fed into oscillator Q2, which modulates the signal. L1C1 is a tank circuit for operation in the 88-MHz region. The antenna is a 6- to 8-inch piece of wire. L1 is a variable inductor in the 1-μH range.

104

Window Circuits

The sources of the following circuits are contained in the Sources section, which begins on page 682. The figure number in the box of each circuit correlates to the entry in the Sources section.

Low-Cost Window Comparator
Window Generator
Window Comparators
Window Detector
Voltage Comparators

LOW-COST WINDOW COMPARATOR

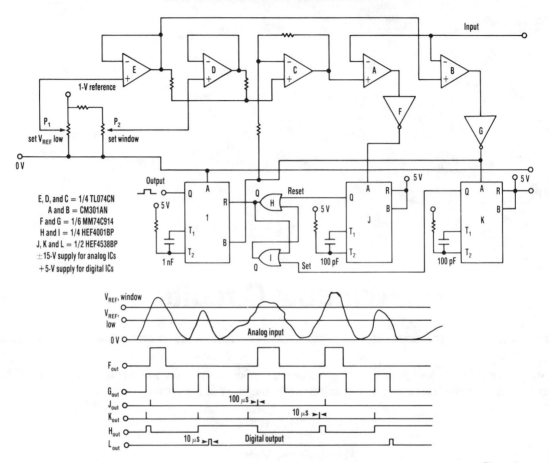

ELECTRONIC DESIGN

Fig. 104-1

This circuit outputs a TTL-compatible 100-µs pulse whenever the signal falls within the limits set by potentiometers and can be varied to suit the application.

Op amps, E, D, and C are used with the two potentiometers to supply reference voltages, derived from a 1-V precision source, for two other op amps (A and B) configured as voltage comparators. The input signal is taken to the negative inputs of both these comparators. C is wired as a noninverting summing amp, used to derive the higher reference voltage. Consequently, the acceptance window is set. Because the voltage across potentiometer P2 is 0.5 V, the window can be set between 0 and 0.5 V above the value chosen for the lower reference value. The lower value is set by P1.

The outputs from the comparators are sent to the inputs of inverting Schmitt triggers F and G. Although these triggers operate from a 5-V supply, they have an extended input-voltage range and are capable of handling the comparator's output voltage swings.

The two monostables, J and K, are triggered on the rising edge of the Schmitt outputs. J and K set and reset the bistable latch formed by two NOR gates. Latch output Q controls the reset of output

656

LOW-COST WINDOW COMPARATOR *(Cont.)*

monostable L, which can only be triggered if its B input goes low, taken from the output of G, while its reset is high.

From the timing diagram, when the signal input exceeds the lower reference level, the latch is set and Q goes high. When the higher reference level is exceeded, the latch is reset and Q goes low. If only the lower reference is exceeded, the latch will be set and L is triggered when G's output goes low. When both levels are exceeded, the latch is set, then reset. When G's output goes low, it won't trigger the monostable because its reset is low.

The circuit, as shown, was used for a bandwidth-limited, 500- to 5 000 Hz input signal with an amplitude range of ± 1 V. Modifications can easily be made to cover other frequency and amplitude ranges.

WINDOW GENERATOR

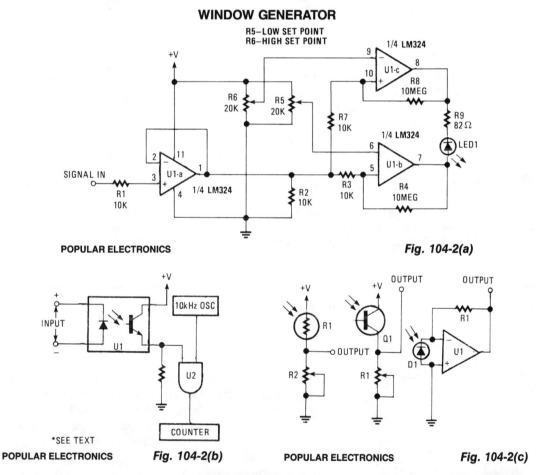

POPULAR ELECTRONICS

Fig. 104-2(a)

*SEE TEXT

POPULAR ELECTRONICS   *Fig. 104-2(b)*

POPULAR ELECTRONICS   *Fig. 104-2(c)*

This window generator uses a single LM324 op amp and features two adjustable set points. When the two comparators formed by U1B go high, LED1 lights. When U1C goes high, the LED extinguishes. Some hysteresis is provided by the 10-mΩ resistors. As shown in Fig. 104-2(b), LED1 can be replaced by an optoisolator (etc.) for use in several applications.

WINDOW COMPARATORS

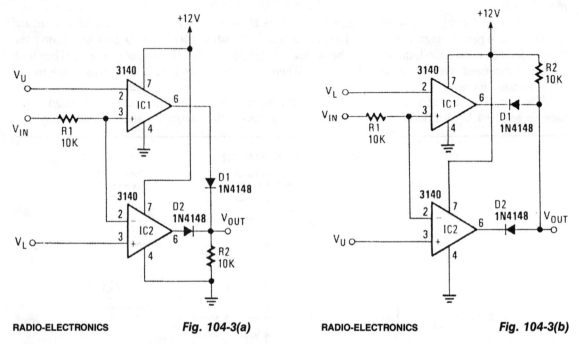

RADIO-ELECTRONICS **Fig. 104-3(a)**

RADIO-ELECTRONICS **Fig. 104-3(b)**

In Fig. 104-3(a), when V_{IN} is between reference voltages V_L and V_U, output V_{OUT} goes low. If $V_{IN} > V_L$, IC2 produces a low. Because IC1 outputs low, if $V_{IN} > V_U$, both outputs are low and V_{OUT} is low. If $V_{IN} < V_L$, both IC1 and IC2 are low. Figure 104-3(b) operates the reverse of this; it produces a high when $V_L < V_{IN} < V_U$.

WINDOW DETECTOR

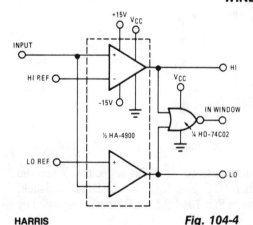

HARRIS **Fig. 104-4**

The high switching speed, low offset current and low offset voltage of the HA-4900 series make this window detector extremely well-suited for applications that require fast, accurate decision-making. This circuit is ideal for industrial-process system-feedback controllers, or "out-of-limit" alarm indicators.

VOLTAGE COMPARATORS

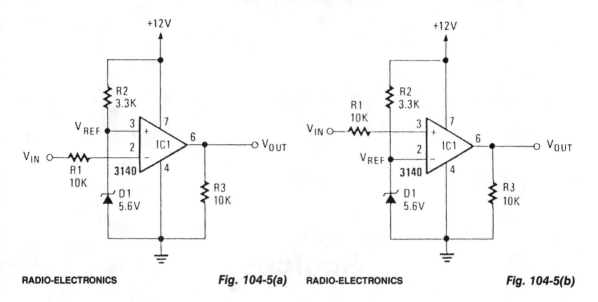

RADIO-ELECTRONICS   *Fig. 104-5(a)*   RADIO-ELECTRONICS   *Fig. 104-5(b)*

These two comparators are over- and under-voltage comparators. In Fig. 104-5(a), if V_{IN} exceeds the reference voltage, the output of IC1 goes low. In Fig. 104-5(b), if the V_{IN} exceeds the reference, V_{OUT} goes high.

Sources

Chapter 1

Fig. 1-1. Reprinted with permission from Popular Electronics, 11/90, p. 85. © Copyright Gernsback Publications, Inc., 1990.

Fig. 1-2. Reprinted with permission from Popular Electronics, 11/90, © Copyright Gernsback Publications, Inc., 1990.

Fig. 1-3. Elektor Electronics USA, 5/91, p. 51 – 53.

Fig. 1-4. Reprinted with permission from R-E Experimenter's Handbook, 1990, p. 88. © Copyright Gernsback Publications, Inc., 1990.

Chapter 2

Fig. 2-1. Reprinted from EDN, 2/91, p. 184. © 1991 Cahners Publishing Co., a division of Reed Publishing USA.

Chapter 3

Fig. 3-1. Hands-On Electronics, 8/87, p. 94.

Fig. 3-2. Reprinted with permission from Popular Electronics Fact Card No. 185. © Copyright Gernsback Publications, Inc.

Fig. 3-3. Reprinted with permission from Radio-Electronics, 9/87, p. 65. © Copyright Gernsback Publications, Inc., 1987.

Fig. 3-4. Reprinted with permission from Popular Electronics, 12/77, p. 82 – 83. © Copyright Gernsback Publications, Inc., 1977.

Fig. 3-5. Reprinted with permission from Popular Electronics, 1/91, p. 81. © Copyright Gernsback Publications, Inc., 1991.

Chapter 4

Fig. 4-1. Elektor Electronics USA, 6/91, p. 42.

Fig. 4-2. Reprinted with permission from Popular Electronics, 7/90, p. 80 – 81. © Copyright Gernsback Publications, Inc., 1990.

Chapter 5

Fig. 5-1. Elektor Electronics, 12/90, p. 523 – 524.

Fig. 5-2. Hands-On Electronics, 8/87, p. 93.

Fig. 5-3. QST, 5/90, p. 22 – 24.

Fig. 5-4. Elektor Electronics, 7/89 Supplement, p. 5 – 6.

Fig. 5-5. Electronics Today International, 4/79, p. 70.

Fig. 5-6. Elektor Electronics, 301 Circuits, p. 30 – 31.

Fig. 5-7. Electronics Today International, 4/78, p. 31.

Fig. 5-8. Reprinted with permission from Popular Electronics, 4/90, p.37 – 38. © Copyright Gernsback Publications, Inc., 1990.

Fig. 5-9. Electronics World, 4/66.

Fig. 5-10. Reprinted with permission from Electronic Design, 12/89, p. 71–72. Copyright 1989 Penton Publishing.

Chapter 6

Fig. 6-1. Reprinted with permission from Popular Electronics, 7/91, p. 70–71. © Copyright Gernsback Publications, Inc., 1991.

Fig. 6-2. Hands-On Electronics, 8/87, p. 90.

Chapter 7

Fig. 7-1. Reprinted with permission from Popular Electronics, 7/91, p. 40–42. © Copyright Gernsback Publications, Inc., 1991.

Fig. 7-2. Reprinted with permission from R-E Experimenter's Handbook, 1990, p. 90. © Copyright Gernsback Publications, Inc., 1990.

Fig. 7-3. Reprinted with permission from Wels' Think Tank, Gernsback Publications Inc., p. 23.

Fig. 7-4. Reprinted with permission from Popular Electronics, 3/90. © Copyright Gernsback Publications, Inc., 1990.

Fig. 7-5. Reprinted with permission from Radio-Electronics, 1/90, p. 40. © Copyright Gernsback Publications, Inc., 1990.

Fig. 7-6. Reprinted with permission from Popular Electronics, 10/90, p. 90. © Copyright Gernsback Publications, Inc., 1990.

Fig. 7-7. Reprinted with permission from Radio-Electronics, 9/89, p. 41. © Copyright Gernsback Publications, Inc., 1989.

Fig. 7-8. Reprinted with permission from Radio-Electronics, 1/89, p. 71. © Copyright Gernsback Publications, Inc., 1989.

Chapter 8

Fig. 8-1. Elektor Electronics, 12/90 Supplement, p. 36–37.

Fig. 8-2. Reprinted with permission from Radio-Electronics, 2/90, p. 56. © Copyright Gernsback Publications, Inc., 1990.

Fig. 8-3. Reprinted with permission from Radio-Electronics, 2/90, p. 56. © Copyright Gernsback Publications, Inc., 1990.

Fig. 8-4. Reprinted with permission from Radio-Electronics, 2/90, p. 57. © Copyright Gernsback Publications, Inc., 1990.

Fig. 8-5. Reprinted with permission from Radio-Electronics, 2/90, p. 57. © Copyright Gernsback Publications, Inc., 1990.

Fig. 8-6. Reprinted with permission from Popular Electronics, 8/90, p. 82. © Copyright Gernsback Publications, Inc., 1990.

Fig. 8-7. Reprinted with permission from Radio-Electronics, 2/90, p. 57. © Copyright Gernsback Publications, Inc., 1990.

Fig. 8-8. Reprinted with permission from Radio-Electronics, 2/90, p. 57. © Copyright Gernsback Publications, Inc., 1990.

Fig. 8-9. Reprinted with permission from Popular Electronics, 3/90, p. 72. © Copyright Gernsback Publications, Inc., 1990.

Fig. 8-10. Elektor Electronics, 12/90, p. 514.

Fig. 8-11. Elektor Electronics, 12/90, p. 511.

Fig. 8-12. Reprinted with permission from Electronic Design, 7/91, p. 139. Copyright 1991, Penton Publishing.

Fig. 8-13. Reprinted with permission from Electronic Design, 9/90, p. 90. Copyright 1990, Penton Publishing.

Fig. 8-14. Reprinted with permission from Popular Electronics, Fact Card No. 173. © Copyright Gernsback Publications, Inc.

Fig. 8-15. QST, 10/90, p. 41.

Fig. 8-16. Reprinted with permission from Popular Electronics, Fact Card No. 185. © Copyright Gernsback Publications, Inc.

Chapter 9

Fig. 9-1. Reprinted with permission from Radio-Electronics, 7/90, p. 33. © Copyright Gernsback Publications, Inc., 1990.

Fig. 9-2. Reprinted with permission from Radio-Electronics, 7/90, p. 36. © Copyright Gernsback Publications, Inc., 1990.

Fig. 9-3. Elektor Electronics, 12/90 Supplement, p. 43.

Fig. 9-4. Reprinted with permission from Radio-Electronics, 7/90, p. 36. © Copyright Gernsback Publications, Inc., 1990.

Fig. 9-5. Reprinted with permission from Radio-Electronics, 7/90, p. 35–36. © Copyright Gernsback Publications, Inc., 1990.

Fig. 9-6. Reprinted with permission from Popular Electronics, 6/89, p. 23. © Copyright Gernsback Publications, Inc., 1989.

Fig. 9-7. Reprinted with permission from Radio-Electronics, 8/90, p. 42. © Copyright Gernsback Publications, Inc., 1990.

Fig. 9-8. Reprinted with permission from Radio-Electronics, 7/90, p. 32. © Copyright Gernsback Publications, Inc., 1990.

Fig. 9-9. Reprinted with permission from Radio-Electronics, 7/90, p. 36. © Copyright Gernsback Publications, Inc., 1990.

Chapter 10

Fig. 10-1. Reprinted with permission from Popular Electronics, 2/90, p. 24–26. © Copyright Gernsback Publications, Inc., 1990.

Fig. 10-2. Reprinted with permission from Popular Electronics, 2/90, p. 27. © Copyright Gernsback Publications, Inc., 1990.

Fig. 10-3. Reprinted with permission from Radio-Electronics, 7/90, p. 57. © Copyright Gernsback Publications, Inc., 1990.

Fig. 10-4. Reprinted with permission from Popular Electronics, 8/89, p. 24. © Copyright Gernsback Publications, Inc., 1989.

Fig. 10-5. Reprinted with permission from Popular Electronics, 12/90, p. 24. © Copyright Gernsback Publications, Inc., 1990.

Fig. 10-6. Reprinted with permission from Popular Electronics, 1/90, p. 25. © Copyright Gernsback Publications, Inc., 1990.

Fig. 10-7. Reprinted with permission from Radio-Electronics, 5/90, p. 65. © Copyright Gernsback Publications, Inc., 1990.

Fig. 10-8. Reprinted from EDN, 9/84, p. 290. © 1991 Cahners Publishing Company, a division of Reed Publishing USA.

Fig. 10-9. Reprinted with permission from Radio-Electronics, 5/90, p. 64. © Copyright Gernsback Publications, Inc., 1990.

Fig. 10-10. Reprinted with permission from Popular Electronics, 11/90, p. 24. © Copyright Gernsback Publications, Inc., 1990.

Chapter 11

Fig. 11-1. Elektor Electronics, 7/89 Supplement, p. 15.

Fig. 11-2. Elektor Electronics, 12/90, p. 527–528.

Fig. 11-3. Reprinted with permission from Popular Electronics, 12/90, p. 25. © Copyright Gernsback Publications, Inc., 1990.

Fig. 11-4. Reprinted with permission from Popular Electronics, 9/89, p. 22. © Copyright Gernsback Publications, Inc., 1989.

Fig. 11-5. Reprinted with permission from Hands-On Electronics/Popular Electronics, 12/88, p. 24. © Copyright Gernsback Publications, Inc., 1988.

Fig. 11-6. Elektor Electronics, 7/89, p. S41.

Fig. 11-7. Reprinted with permission from Popular Electronics, 7/89, p. 76–77. © Copyright Gernsback Publications, Inc., 1989.

Fig. 11-8. Reprinted with permission from Popular Electronics, 8/89, p. 23. © Copyright Gernsback Publications, Inc., 1989.

Fig. 11-9. Hands-On Electronics, 9–10/86, p. 26.

Fig. 11-10. Reprinted with permission from Popular Electronics, 12/90, p. 25. © Copyright Gernsback Publications, Inc., 1990.

Chapter 12

Fig. 12-1. Reprinted with permission from Radio-Electronics, 4/90, p. 47. © Copyright Gernsback Publications, Inc., 1990.

Fig. 12-2. Elektor Electronics, 12/90, p. 57.

Fig. 12-3. Reprinted with permission from Radio-Electronics, 3/89, p. 51–52. © Copyright Gernsback Publications, Inc., 1989.

Fig. 12-4. Elektor Electronics, 12/90, p. 526.

Fig. 12-5. Reprinted with permission from Radio-Electronics, 8/91, p. 8. © Copyright Gernsback Publications, Inc., 1991.

Chapter 13

Fig. 13-1. Elektor Electronics, 12/90, p. 522.

Fig. 13-2. Hands-On Electronics, 11/87, p. 92.

Fig. 13-3. Reprinted with permission from Popular Electronics, Fact Card No. 134. © Copyright Gernsback Publications, Inc.

Fig. 13-4. Reprinted with permission from Popular Electronics, Fact Card No. 95. © Copyright Gernsback Publications, Inc.

Fig. 13-5. Reprinted with permission from Popular Electronics, 3/90, p. 38. © Copyright Gernsback Publications, Inc., 1990.

Fig. 13-6. Reprinted from EDN, 2/7/85, p. 243. © 1991 Cahners Publishing Company, a division of Reed Publishing USA.

Chapter 14

Fig. 14-1. Elektor Electronics, 12/90, p. 519.

Fig. 14-2. Reprinted from EDN, 5/91, p. 165. © 1991 Cahners Publishing Company, a division of Reed Publishing USA.

Fig. 14-3. Reprinted from EDN, 6/91, p. 168. © 1991 Cahners Publishing Company, a division of Reed Publishing USA.

Fig. 14-4. Reprinted from EDN, 5/90, p. 148. © 1991 Cahners Publishing Company, a division of Reed Publishing USA.

Fig. 14-5. Reprinted from EDN, 2/91, p. 108. © 1991 Cahners Publishing Company, a division of Reed Publishing USA.

Fig. 14-6. Elektor Electronics, 5/91, p. 33.

Fig. 14-7. Reprinted with permission from Popular Electronics, 8/90, p. 40–41. © Copyright Gernsback Publications, Inc., 1990.

Fig. 14-8. 42 New Ideas, 1984, p. 14, Gernsback Publications, Inc.

Fig. 14-9. Electronic Engineering, 11/78, p. 24.

Fig. 14-10. Maxim, Seminar Applications Book, 1988/89, p. 35.

Fig. 14-11. Reprinted with permission from Wels' Think Tank, Gernsback Publications Inc., p. 19.

Chapter 15

Fig. 15-1. Reprinted from EDN, Design Ideas Special Issue, Vol. III, 2/89, p. 47. © 1991 Cahners Publishing Company, a division of Reed Publishing USA.

Fig. 15-2. Reprinted from EDN, Design Ideas Special Issue, Vol. IV, 7/20/89, p. 26. © 1991 Cahners Publishing Company, a division of Reed Publishing USA.

Fig. 15-3. 73 Amateur Radio, 4/91, p. 11–12.

Chapter 16

Fig. 16-1. Reprinted with permission from Radio-Electronics, 5/90, p. 64. © Copyright Gernsback Publications, Inc., 1990.

Fig. 16-2. Reprinted with permission from Popular Electronics, 8/90, p. 92. © Copyright Gernsback Publications, Inc., 1990.

Fig. 16-3. Reprinted with permission from Popular Electronics, 12/90, p. 41–42. © Copyright Gernsback Publications, Inc., 1990.

Fig. 16-4. Reprinted with permission from Popular Electronics, Fact Card No. 173. © Copyright Gernsback Publications, Inc.

Fig. 16-5. Reprinted with permission from Radio-Electronics, 5/90, p. 65. © Copyright Gernsback Publications, Inc., 1990.

Fig. 16-6. Reprinted with permission from Radio-Electronics, 5/90, p. 62. © Copyright Gernsback Publications, Inc., 1990.

Fig. 16-7. Reprinted with permission from Radio-Electronics, 5/90, p. 62. © Copyright Gernsback Publications, Inc., 1990.

Chapter 17

Fig. 17-1. Reprinted from EDN, 2/91, p. 105. © 1991 Cahners Publishing Company, a division of Reed Publishing USA.

Fig. 17-2. Reprinted from EDN, 11/90, p. 282. © 1991 Cahners Publishing Company, a division of Reed Publishing USA.

Fig. 17-3. Hands-On Electronics, Fact Card No. 89.

Chapter 18

Fig. 18-1. Reprinted with permission from R-E Experimenter's Handbook, 1990, p. 158. © Copyright Gernsback Publications, Inc., 1990.

Fig. 18-2. Reprinted with permission from R-E Experimenter's Handbook, 1990, p. 162. © Copyright Gernsback Publications, Inc., 1990.

Chapter 19

Fig. 19-1. Reprinted from EDN, 1/91, p. 160–161. © 1991 Cahners Publishing Company, a division of Reed Publishing USA.

Fig. 19-2. Elektor Electronics, 12/90, p. 534–535.

Chapter 20

Fig. 20-1. Reprinted from EDN, 3/91, p. 161. © 1991 Cahners Publishing Company, a division of Reed Publishing USA.

Fig. 20-2. Elektor Electronics, 6/89, p. S42.

Fig. 20-3. Reprinted from EDN, 9/89, p. 168. © 1991 Cahners Publishing Company, a division of Reed Publishing USA.

Fig. 20-4. Reprinted from EDN, 5/91, p. 166. © 1991 Cahners Publishing Company, a division of Reed Publishing USA.

Fig. 20-5. Electronic Engineering, 2/90, p. 29.

Fig. 20-6. Reprinted from EDN, 2/90, p. 135. © 1991 Cahners Publishing Company, a division of Reed Publishing USA.

Fig. 20-7. Reprinted with permission from Popular Electronics, 8/91, p. 20. © Copyright Gernsback Publications, Inc., 1991.

Fig. 20-8. Reprinted with permission from Popular Electronics, 1/91, p. 23–24. © Copyright Gernsback Publications, Inc., 1991.

Fig. 20-9. Reprinted with permission from Electronic Design, 9/89, p. 93. Copyright 1989, Penton Publishing.

Fig. 20-10. Reprinted with permission from Popular Electronics, 12/90, p. 23. © Copyright Gernsback Publications, Inc., 1990.

Fig. 20-11. Elektor Electronics, 7/89, p. S36.

Fig. 20-12. Reprinted from EDN, Design Ideas Special Issue, Vol. IV, 7/20/89, p. 29. © 1991 Cahners Publishing Company, a division of Reed Publishing USA.

Fig. 20-13. Reprinted from EDN, 1/90, p. 140. © 1991 Cahners Publishing Company, a division of Reed Publishing USA.

Fig. 20-14. Reprinted from EDN, 11/90, p. 22. © 1991 Cahners Publishing Company, a division of Reed Publishing USA.

Fig. 20-15. Harris, Analog Product Data Book, 1988, p.10–23.

Fig. 20-16. Reprinted with permission from Popular Electronics, 1/90, p. 24. © Copyright Gernsback Publications, Inc., 1990.

Chapter 21

Fig. 21-1. Reprinted from EDN, 7/90, p. 192. © 1991 Cahners Publishing Company, a division of Reed Publishing USA.

Fig. 21-2. Reprinted from EDN, 7/89, p. 151. © 1991 Cahners Publishing Company, a division of Reed Publishing USA.

Fig. 21-3. Elektor Electronics, 12/90, p. 513–514.

Fig. 21-4. Elektor Electronics, 12/90, p. 56.

Fig. 21-5. Reprinted with permission from Electronic Design, 6/91, p. 110. Copyright 1991, Penton Publishing.

Fig. 21-6. Elektor Electronics, 12/90, p. 530–531.

Fig. 21-7. Reprinted from EDN, 3/90, p. 140. © 1991 Cahners Publishing Company, a division of Reed Publishing USA.

Fig. 21-8. Linear Technology, Design Note No. 2.

Fig. 21-9. Reprinted with permission from Electronic Design, 9/90, p. 93.

Fig. 21-10. Linear Technology, 6/27/91, Advertisement.

Fig. 21-11. Linear Technology, 6/27/91, Advertisement.

Fig. 21-12. Maxim, 1986 Power Supply Circuits, p. 26.

Fig. 21-13. Reprinted with permission from Radio-Electronics, 6/86, p. 59. © Copyright Gernsback Publications, Inc., 1986.

Fig. 21-14. Reprinted with permission from Radio-Electronics, 9/89, p. 60. © Copyright Gernsback Publications, Inc., 1989.

Chapter 22

Fig. 22-1. Reprinted from EDN, 10/90, p. 132. © 1991 Cahners Publishing Company, a division of Reed Publishing USA.

Fig. 22-2. RF Design, 7/90, p. 63.

Fig. 22-3. RF Design, 7/90, p. 63.

Fig. 22-4. Reprinted with permission from Popular Electronics, 11/91, p. 68. © Copyright Gernsback Publications, Inc., 1991.

Fig. 22-5. Reprinted with permission from Popular Electronics, 7/91, p. 58. © Copyright Gernsback Publications, Inc., 1991.

Fig. 22-6. 73 Amateur Radio, 10/90, p. 65.

Fig. 22-7. Valpey-Fisher Corp., A User's Guide to Quartz Crystal Oscillators.

Fig. 22-8. RF Design, 7/90, p. 64.

Fig. 22-9. Valpey-Fisher Corp., A User's Guide to Quartz Crystal Oscillators.

Fig. 22-10. Valpey-Fisher Corp., A User's Guide to Quartz Crystal Oscillators.

Fig. 22-11. Valpey-Fisher Corp., A User's Guide to Quartz Crystal Oscillators.

Fig. 22-12. Valpey-Fisher Corp., A User's Guide to Quartz Crystal Oscillators.

Fig. 22-13. Valpey-Fisher Corp., A User's Guide to Quartz Crystal Oscillators.

Fig. 22-14. Reprinted with permission from Popular Electronics, 11/90, p. 69. © Copyright Gernsback Publications, Inc., 1990.

Fig. 22-15. Valpey-Fisher Corp., A User's Guide to Quartz Crystal Oscillators.

Fig. 22-16. Reprinted from EDN, 11/90, p. 234. © 1991 Cahners Publishing Company, a division of Reed Publishing USA.

Fig. 22-17. Elektor Electronics, 302 Circuits, p. 128–129.

Fig. 22-18. William Sheets.

Chapter 23

Fig. 23-1. Reprinted from EDN, 12/89, p. 262. © 1991 Cahners Publishing Company, a division of Reed Publishing USA.

Fig. 23-2. Harris, Analog Product Data Book, 1988, p. 10–49.

Fig. 23-3. Harris, Analog Product Data Book, 1988, p. 2–105.

Fig. 23-4. Reprinted with permission from Radio-Electronics, 12/89, p. 58. © Copyright Gernsback Publications, Inc., 1989.

Fig. 23-5. Reprinted with permission from Electronic Design, 6/90, p. 102–103. Copyright 1991, Penton Publishing.

Chapter 24

Fig. 24-1. Elektor Electronics, 12/90, p. 520.

Fig. 24-2. Reprinted from EDN, 12/89, p. 260–261. © 1991 Cahners Publishing Company, a division of Reed Publishing USA.

Fig. 24-3. Reprinted with permission from Popular Electronics, 8/91, p. 73. © Copyright Gernsback Publications, Inc., 1991.

Fig. 24-4. Reprinted with permission from Popular Electronics, 3/91, p. 92. © Copyright Gernsback Publications, Inc., 1991.

Fig. 24-5. Reprinted from EDN, 5/90, p. 173. Copyright 1991, Penton Publishing.

Fig. 24-6. Reprinted with permission from Radio-Electronics, 12/88, p. 72. © Copyright Gernsback Publications, Inc., 1988.

Fig. 24-7. ARRL Handbook, 1991, p. 18–13.

Fig. 24-8. Reprinted with permission from Popular Electronics, 9/91, p. 75. © Copyright Gernsback Publications, Inc., 1991.

Fig. 24-9. RCA, ICAN #6629, Design Guide for Fire Detection Systems, p. 6.

Fig. 24-10. Reprinted with permission from Popular Electronics, 5/91, p. 63. © Copyright Gernsback Publications, Inc., 1991.

Fig. 24-11. RF Design, 4/90, p. 35–36.

Fig. 24-12. Reprinted with permission from Electronic Design, 12/90, p. 61. Copyright 1991, Penton Publishing.

Fig. 24-13. Reprinted with permission from Popular Electronics, 3/90, p. 46. © Copyright Gernsback Publications, Inc., 1990.

Fig. 24-14. ARRL Handbook, 1991, p. 18–14.

Fig. 24-15. Harris, Analog Product Data Book, 1988, p. 10–84.

Fig. 24-16. Reprinted with permission from Popular Electronics, 9/91, p. 74. © Copyright Gernsback Publications, Inc., 1991.

Fig. 24-17. Reprinted with permission from Popular Electronics, 9/90, p. 25. © Copyright Gernsback Publications, Inc., 1990.

Fig. 24-18. Reprinted with permission from Radio-Electronics, 9/89, p. 59. © Copyright Gernsback Publications, Inc., 1989.

Chapter 25

Fig. 25-1. Reprinted from EDN, 12/90, p. 228–230. © 1991 Cahners Publishing Company, a division of Reed Publishing USA.

Fig. 25-2. 73 Amateur Radio, 7/90, p. 9–11.

Chapter 26

Fig. 26-1. Reprinted with permission from Electronic Design, 6/91, p. 109–114. Copyright 1991, Penton Publishing.

Fig. 26-2. Reprinted with permission from Electronic Design, 3/91, p. 87–88. Copyright 1991, Penton Publishing.

Fig. 26-3. Reprinted from EDN, 6/89, p. 207. © 1991 Cahners Publishing Company, a division of Reed Publishing USA.

Fig. 26-4. 73 Amateur Radio, 3/91, p. 74.

Fig. 26-5. Reprinted with permission from Electronic Design, 11/89, p. 111. Copyright 1989, Penton Publishing.

Fig. 26-6. Reprinted from EDN, 6/91, p. 174. © 1991 Cahners Publishing Company, a division of Reed Publishing USA.

Chapter 27

Fig. 27-1. Elektor Electronics, 12/90, p. 526–527.

Fig. 27-2. Reprinted with permission from Electronic Design, 12/88, p. 104. Copyright 1991, Penton Publishing.

Fig. 27-3. Reprinted with permission from Radio-Electronics, 9/87, p. 65. © Copyright Gernsback Publications, Inc., 1987.

Chapter 28

Fig. 28-1. Reprinted with permission from Popular Electronics, 11/90, p. 61–62. © Copyright Gernsback Publications, Inc., 1990

Fig. 28-2. Reprinted with permission from Popular Electronics, 12/90, p. 24. © Copyright Gernsback Publications, Inc., 1990.

Chapter 29

Fig. 29-1. 73 Amateur Radio, 2/91, p. 24.

Fig. 29-2. Reprinted with permission from Popular Electronics, 10/89, p. 26. © Copyright Gernsback Publications, Inc., 1989.

Fig. 29-3. Elektor Electronics, 302 Circuits, p. 134.

Chapter 30

Fig. 30-1. Electronic Engineering, 2/90, p. 30.

Fig. 30-2. Reprinted with permission from Popular Electronics, 11/89, p. 101. © Copyright Gernsback Publications, Inc., 1989.

Fig. 30-3. Elektor Electronics, 12/90, p. 530.

Fig. 30-4. Elektor Electronics, 7/89, p. S37.

Fig. 30-5. ARRL Handbook, 1991, p. 16 – 6.

Fig. 30-6. Reprinted from EDN, 2/91, p. 186 – 188. © 1991 Cahners Publishing Company, a division of Reed Publishing USA.

Fig. 30-7. Reprinted with permission from Radio-Electronics, 2/90, p. 58. © Copyright Gernsback Publications, Inc., 1990.

Fig. 30-8. Reprinted with permission from Radio-Electronics, 5/89, p. 58. © Copyright Gernsback Publications, Inc., 1989.

Fig. 30-9. Reprinted with permission from Radio-Electronics, 5/89, p. 58. © Copyright Gernsback Publications, Inc., 1989.

Fig. 30-10. Reprinted with permission from Wels' Think Tank, Gernsback Publications Inc., p. 52.

Fig. 30-11. Reprinted with permission from Radio-Electronics, 5/89, p. 58. © Copyright Gernsback Publications, Inc., 1989.

Fig. 30-12. Reprinted with permission from Radio-Electronics, 2/90, p. 58. © Copyright Gernsback Publications, Inc., 1990.

Fig. 30-13. Reprinted with permission from Radio-Electronics, 2/90, p. 58. © Copyright Gernsback Publications, Inc., 1990.

Fig. 30-14. Electronics Today International, 4/79, p. 76.

Fig. 30-15. Reprinted with permission from Radio-Electronics, 10/90, p. 78. © Copyright Gernsback Publications, Inc., 1990.

Chapter 31

Fig. 31-1. Reprinted from EDN, 10/88, p. 327. © 1991 Cahners Publishing Company, a division of Reed Publishing USA.

Fig. 31-2. Reprinted from EDN, 6/84, p. 242. © 1991 Cahners Publishing Company, a division of Reed Publishing USA.

Fig. 31-3. Reprinted with permission from Radio-Electronics, 2/91, p. 33 – 34. © Copyright Gernsback Publications, Inc., 1991.

Fig. 31-4. Reprinted with permission from Radio-Electronics, 2/87, p. 36. © Copyright Gernsback Publications, Inc., 1987.

Fig. 31-5. Reprinted from EDN, 7/73, p. 87. © 1991 Cahners Publishing Company, a division of Reed Publishing USA.

Fig. 31-6. Reprinted with permission from Popular Electronics, 1/91, p. 25. © Copyright Gernsback Publications, Inc., 1991.

Fig. 31-7. Reprinted from EDN, 7/73, p. 87. © 1991 Cahners Publishing Company, a division of Reed Publishing USA.

Fig. 31-8. Elektor Electronics, 12/90 Supplement, p. 43.

Fig. 31-9. Reprinted with permission from Popular Electronics, 6/89, p. 22. © Copyright Gernsback Publications, Inc., 1989.

Chapter 32

Fig. 32-1. Reprinted with permission from Popular Electronics, 3/91, p. 24. © Copyright Gernsback Publications, Inc., 1991.

Fig. 32-2. Reprinted from EDN, 12/90, p. 217. © 1991 Cahners Publishing Company, a division of Reed Publishing USA.

Fig. 32-3. Elektor Electronics, 301 Circuits, p. 131.

Fig. 32-4. Elektor Electronics, 12/90, p. 55 – 56.

Fig. 32-5. Reprinted with permission from Popular Electronics, 9/91, p. 76. © Copyright Gernsback Publications, Inc., 1991.

Fig. 32-6. Reprinted with permission from Radio-Electronics, 7/90, p. 59. © Copyright Gernsback Publications, Inc., 1990.

Fig. 32-7. Reprinted with permission from Popular Electronics, Fact Card No. 107. © Copyright Gernsback Publications, Inc.

Fig. 32-8. Reprinted with permission from Popular Electronics, 3/90, p. 25. © Copyright Gernsback Publications, Inc., 1990.

Fig. 32-9. Reprinted with permission from Popular Electronics, 11/90, p. 22. © Copyright Gernsback Publications, Inc., 1990.

Fig. 32-10. Reprinted with permission from Popular Electronics, 1/90, p. 43. © Copyright Gernsback Publications, Inc., 1990.

Fig. 32-11. Reprinted with permission from Popular Electronics, 1/90, p. 24. © Copyright Gernsback Publications, Inc., 1990.

Chapter 33

Fig. 33-1. Reprinted with permission from Radio-Electronics, 6/91, p. 56. © Copyright Gernsback Publications, Inc., 1991.

Fig. 33-2. Reprinted with permission from Popular Electronics, Fact Card No. 134. © Copyright Gernsback Publications, Inc.

Fig. 33-3. Elektor Electronics, 12/90, p. 510 – 511.

Fig. 33-4. Reprinted with permission from Popular Electronics, Fact Card No. 98. © Copyright Gernsback Publications, Inc.

Fig. 33-5. Reprinted with permission from Radio-Electronics, 6/91, p. 58 – 59. © Copyright Gernsback Publications, Inc., 1991.

Chapter 34

Fig. 34-1. Hands-On Electronics, 11/86, p. 33.

Fig. 34-2. Reprinted with permission from Electronic Design, 5/91, p. 95. Copyright 1991, Penton Publishing.

Fig. 34-3. Reprinted with permission from Radio-Electronics, 2/91, p. 23. © Copyright Gernsback Publications, Inc., 1991.

Fig. 34-4. Reprinted with permission from Radio-Electronics, 6/91, p. 56 – 57. © Copyright Gernsback Publications, Inc., 1991.

Fig. 34-5. Reprinted with permission from Radio-Electronics, 12/89, p. 57. © Copyright Gernsback Publications, Inc., 1989.

Fig. 34-6. Reprinted with permission from Electronic Design, 8/90, p. 94 – 96. Copyright 1991, Penton Publishing.

Fig. 34-7. Hands-On Electronics, 8/87, p. 93.

Fig. 34-8. Reprinted with permission from Electronic Design, 6/90, p. 100. Copyright 1991, Penton Publishing.

Fig. 34-9. Reprinted with permission from Electronic Design, 9/90, p. 89. Copyright 1991, Penton Publishing.

Chapter 35

Fig. 35-1. Reprinted with permission from Radio-Electronics, 4/90, p. 33 – 36. © Copyright Gernsback Publications, Inc., 1990.

Fig. 35-2. Reprinted with permission from Popular Electronics, 6/89, p. 90. © Copyright Gernsback Publications, Inc., 1989.

Fig. 35-3. Reprinted with permission from Radio-Electronics, 12/89, p. 57. © Copyright Gernsback Publications, Inc., 1989.

Fig. 35-4. Reprinted with permission from Popular Electronics, 10/77, p. 82. © Copyright Gernsback Publications, Inc., 1977.

Fig. 35-5. Reprinted with permission from Popular Electronics, 1/90, p. 70. © Copyright Gernsback Publications, Inc., 1990.

Fig. 35-6. Reprinted with permission from Popular Electronics, 12/89, p. 24 – 25. © Copyright Gernsback Publications, Inc., 1989.

Chapter 36

Fig. 36-1. Reprinted from EDN, 10/90, p. 235. © 1991 Cahners Publishing Company, a division of Reed Publishing USA.

Chapter 37

Fig. 37-1. Electronic Engineering, 4/90, p. 42.

Fig. 37-2. Elektor Electronics, 12/90, p. 532 – 533.

Fig. 37-3. Reprinted with permission from Popular Electronics, 9/91, p. 22. © Copyright Gernsback Publications, Inc., 1991.

Fig. 37-4. 73 Amateur Radio, 10/90, p. 46.

Fig. 37-5. Reprinted with permission from Popular Electronics, 3/91, p. 25. © Copyright Gernsback Publications, Inc., 1991.

Fig. 37-6. Reprinted with permission from Radio-Electronics, 2/91, p. 83. © Copyright Gernsback Publications, Inc., 1991.

Fig. 37-7. Elektor Electronics, 7/89, p. S44.

Fig. 37-8. Elektor Electronics, 7/89, p. 40 – 41.

Fig. 37-9. Elektor Electronics, 7/89, p. S41.

Fig. 37-10. Reprinted with permission from Radio-Electronics, 7/90, p. 59. © Copyright Gernsback Publications, Inc., 1990.

Fig. 37-11. Ham Radio, 4/90, p. 66.

Chapter 38

Fig. 38-1. Reprinted with permission from R-E Experimenter's Handbook, 1987, p. 59. © Copyright Gernsback Publications, Inc., 1987.

Fig. 38-2. Reprinted with permission from Popular Electronics, 4/90, p. 91. © Copyright Gernsback Publications, Inc., 1990.

Fig. 38-3. Reprinted with permission from Radio-Electronics, 4/89, p. 47. © Copyright Gernsback Publications, Inc., 1989.

Fig. 38-4. Reprinted with permission from Popular Electronics, 9/90, p. 23. © Copyright Gernsback Publications, Inc., 1990.

Fig. 38-5. Reprinted with permission from R-E Experimenter's Handbook, 1987, p. 59. © Copyright Gernsback Publications, Inc., 1987.

Fig. 38-6. Reprinted with permission from Popular Electronics, 1/90, p. 27. © Copyright Gernsback Publications, Inc., 1990.

Fig. 38-7. Reprinted with permission from Radio-Electronics, 4/91, p. 73. © Copyright Gernsback Publications, Inc., 1991.

Fig. 38-8. Reprinted with permission from Popular Electronics, 2/90, p. 24. © Copyright Gernsback Publications, Inc., 1990.

Fig. 38-9. Reprinted with permission from Popular Electronics/Hands-On Electronics, 4/89, p. 83. © Copyright Gernsback Publications, Inc., 1989.

Fig. 38-10. Reprinted with permission from Popular Electronics, 4/90, p. 90 – 91. © Copyright Gernsback Publications, Inc., 1990.

Fig. 38-11. Reprinted with permission from Popular Electronics, 3/90, p. 28. © Copyright Gernsback Publications, Inc., 1990.

Fig. 38-12. Reprinted with permission from Popular Electronics, 12/90, p. 27. © Copyright Gernsback Publications, Inc., 1990.

Fig. 38-13. Reprinted with permission from Popular Electronics, 2/82, p. 99. © Copyright Gernsback Publications, Inc., 1982.

Chapter 39

Fig. 39-1. Elektor Electronics, 12/90, p. 529.

Fig. 39-2. Reprinted from EDN, 2/91, p. 105. © 1991 Cahners Publishing Company, a division of Reed Publishing USA.

Fig. 39-3. DATEL, Data Conversion Components, p. 4 – 17.

Fig. 39-4. Reprinted with permission from Electronic Design, 12/88, p. 67. Copyright 1988, Penton Publishing.

Chapter 40

Fig. 40-1. Elektor Electronics, 12/90, p. 515 – 516.

Chapter 41

Fig. 41-1. Reprinted with permission from Electronic Design, 8/90, p. 94. Copyright 1990, Penton Publishing.

Fig. 41-2. Elektor Electronics, 12/90, p. 512.

Fig. 41-3. Reprinted with permission from Electronic Design, 4/91, p. 102. Copyright 1991, Penton Publishing.

Fig. 41-4. Harris, Analog Product Data Book, 1988, p. 2 – 105.

Chapter 42

Fig. 42-1. 73 Amateur Radio, 5/91, p. 14 – 15.

Fig. 42-2. QST, 6/91, p. 45.

Fig. 42-3. Reprinted by permission of Texas Instruments, Linear and Interface Circuits Applications, 1985, Vol. 1, p. 4 – 3.

Chapter 43

Fig. 43-1. Reprinted with permission from Popular Electronics, 8/90, p. 27. © Copyright Gernsback Publications, Inc., 1990.

Fig. 43-2. Reprinted with permission from Popular Electronics/Hands-On Electronics, 1/89, p. 44. © Copyright Gernsback Publications, Inc., 1989.

Fig. 43-3. Elektor Electronics, 7/89, p. 38.

Fig. 43-4. Reprinted with permission from Popular Electronics, 8/89, p. 23. © Copyright Gernsback Publications, Inc., 1989.

Fig. 43-5. Elektor Electronics, 301 Circuits, p. 8.

Fig. 43-6. Reprinted with permission from Popular Electronics, 8/90, p. 25. © Copyright Gernsback Publications, Inc., 1990.

Fig. 43-7. Reprinted with permission from Popular Electronics, 8/90, p. 23. © Copyright Gernsback Publications, Inc., 1990.

Fig. 43-8. Reprinted with permission from Popular Electronics, 11/90, p. 27. © Copyright Gernsback Publications, Inc., 1990.

Fig. 43-9. Reprinted with permission from Popular Electronics, 9/90, p. 26. © Copyright Gernsback Publications, Inc., 1990.

Fig. 43-10. Reprinted with permission from Popular Electronics, 9/90, p. 26. © Copyright Gernsback Publications, Inc., 1990.

Fig. 43-11. Elektor Electronics, 12/90, p. 58 – 59.

Fig. 43-12. Reprinted with permission from Popular Electronics, 9/91, p. 19. © Copyright Gernsback Publications, Inc., 1991.

Fig. 43-13. Reprinted with permission from Popular Electronics, 9/91, p. 18. © Copyright Gernsback Publications, Inc., 1991.

Fig. 43-14. Reprinted with permission from Radio-Electronics, 9/87, p. 65. © Copyright Gernsback Publications, Inc., 1987.

Chapter 44

Fig. 44-1. Reprinted with permission from Electronic Design, 5/91, p. 117. Copyright 1991, Penton Publishing.

Fig. 44-2. Reprinted with permission from Electronic Design, 12/89, p. 71. Copyright 1991, Penton Publishing.

Chapter 45

Fig. 45-1. Intersil, Applications Handbook, 1988, p. 5–8.

Fig. 45-2. Electronic Engineering, 4/90, p. 27.

Fig. 45-3. Reprinted from EDN, 10/25/90, p. 79. © 1991 Cahners Publishing Company, a division of Reed Publishing USA.

Chapter 46

Fig. 46-1. Reprinted with permission from Electronic Design, 2/91, p. 81. Copyright 1991, Penton Publishing.

Fig. 46-2. Reprinted with permission from Popular Electronics, 8/89, p. 26–27. © Copyright Gernsback Publications, Inc., 1989.

Fig. 46-3. Reprinted with permission from Radio-Electronics, 4/91, p. 33–42. © Copyright Gernsback Publications, Inc., 1991.

Fig. 46-4. Reprinted with permission from Popular Electronics, 7/91, p. 59–61. © Copyright Gernsback Publications, Inc., 1991.

Fig. 46-5. Reprinted with permission from R-E Experimenter's Handbook, 1989, p. 45. © Copyright Gernsback Publications, Inc., 1989.

Fig. 46-6. Reprinted with permission from Popular Electronics, 10/77, p. 61. © Copyright Gernsback Publications, Inc., 1977.

Fig. 46-7. Reprinted with permission from Popular Electronics, 4/90, p. 24. © Copyright Gernsback Publications, Inc., 1990.

Fig. 46-8. Reprinted from EDN, 9/3/90, p. 171. © 1991 Cahners Publishing Company, a division of Reed Publishing USA.

Fig. 46-9. Reprinted with permission from Wels' Think Tank, Gernsback Publishing Inc., p. 1.

Fig. 46-10. Reprinted from EDN, 9/90, p. 94. © 1991 Cahners Publishing Company, a division of Reed Publishing USA.

Fig. 46-11. Reprinted with permission from Electronic Design, 5/91, p. 120. Copyright 1991, Penton Publishing.

Fig. 46-12. Reprinted with permission from Radio-Electronics, 8/90, p. 13. © Copyright Gernsback Publications, Inc., 1990.

Fig. 46-13. Reprinted with permission from Popular Electronics, 2/90, p. 86. © Copyright Gernsback Publications, Inc., 1990.

Fig. 46-14. Reprinted with permission from Electronic Design, 11/82, p. 184. Copyright 1982, Penton Publishing.

Fig. 46-15. Ham Radio, 5/89, p. 85–87.

Fig. 46-16. Reprinted with permission from Popular Electronics, 10/90, p. 27. © Copyright Gernsback Publications, Inc., 1990.

Fig. 46-17. Reprinted with permission from Popular Electronics, 2/90, p. 23. © Copyright Gernsback Publications, Inc., 1990.

Fig. 46-18. Reprinted with permission from Popular Electronics, 8/90, p. 82. © Copyright Gernsback Publications, Inc., 1990.

Fig. 46-19. Reprinted with permission from Electronic Design, 4/91, p. 102. Copyright 1991, Penton Publishing.

Fig. 46-20. Reprinted with permission from Popular Electronics, 1/90, p. 22. © Copyright Gernsback Publications, Inc., 1990.

Fig. 46-21. Reprinted with permission from R-E Experimenter's Handbook, 1990, p. 46. © Copyright Gernsback Publications, Inc., 1990.

Fig. 46-22. Reprinted with permission from Electronic Design, 6/91, p. 101. Copyright 1991, Penton Publishing.

Fig. 46-23. Reprinted with permission from R-E Experimenter's Handbook, 1990, p. 43–47. © Copyright Gernsback Publications, Inc., 1990.

Fig. 46-24. Reprinted with permission from Popular Electronics, 2/90, p. 86. © Copyright Gernsback Publications, Inc., 1990.

Fig. 46-25. Reprinted with permission from Popular Electronics, 7/91, p. 71. © Copyright Gernsback Publications, Inc., 1991.

Fig. 46-26. Reprinted with permission from Popular Electronics, 8/89, p. 26. © Copyright Gernsback Publications, Inc., 1989.

Fig. 46-27. Reprinted with permission from Wels' Think Tank, Gernsback Publishing Inc., p. 1.

Fig. 46-28. Reprinted with permission from Popular Electronics, 10/90, p. 23–24. © Copyright Gernsback Publications, Inc., 1990.

Chapter 47

Fig. 47-1. Reprinted from EDN, Special Ideas Issue, Vol. I, 7/88, p. 39. © 1991 Cahners Publishing Company, a division of Reed Publishing USA.

Fig. 47-2. Reprinted from EDN, 5/91, p. 175. © 1991 Cahners Publishing Company, a division of Reed Publishing USA.

Fig. 47-3. Elektor Electronics, 7-8/91, p. 47.

Fig. 47-4. Hands-On Electronics, 9/87, p. 97.

Fig. 47-5. Reprinted with permission from Popular Electronics, 2/91, p. 25. © Copyright Gernsback Publications, Inc., 1991.

Fig. 47-6. Reprinted with permission from Radio-Electronics, 3/89, p. 73. © Copyright Gernsback Publications, Inc., 1989.

Fig. 47-7. Reprinted with permission from Radio-Electronics, 9/89, p. 62. © Copyright Gernsback Publications, Inc., 1989.

Fig. 47-8. Reprinted with permission from Popular Electronics/Hands-On Electronics, 2/89, p. 36. © Copyright Gernsback Publications, Inc., 1989.

Fig. 47-9. Reprinted with permission from Popular Electronics, 11/91, p. 85. © Copyright Gernsback Publications, Inc., 1991.

Fig. 47-10. Reprinted with permission from Radio-Electronics, 9/89, p. 62. © Copyright Gernsback Publications, Inc., 1989.

Fig. 47-11. Reprinted from EDN, 7/91, p. 165. © 1991 Cahners Publishing Company, a division of Reed Publishing USA.

Fig. 47-12. Reprinted with permission from Popular Electronics, 11/91, p. 84. © Copyright Gernsback Publications, Inc., 1991.

Fig. 47-13. Reprinted with permission from Popular Electronics, 3/89, p. 26. © Copyright Gernsback Publications, Inc., 1989.

Fig. 47-14. Reprinted with permission from Radio-Electronics, 9/89, p. 62. © Copyright Gernsback Publications, Inc., 1989.

Fig. 47-15. Elektor Electronics, 7/89, p. 542.

Fig. 47-16. Reprinted with permission from Popular Electronics, 12/90, p. 25. © Copyright Gernsback Publications, Inc., 1990.

Fig. 47-17. Ham Radio, 3/90, p. 82 – 83.

Fig. 47-18. Reprinted with permission from Popular Electronics, 1/91, p. 25. © Copyright Gernsback Publications, Inc., 1991.

Fig. 47-19. Reprinted with permission from Radio-Electronics, 9/89, p. 61. © Copyright Gernsback Publications, Inc., 1989.

Fig. 47-20. Reprinted with permission from Popular Electronics, 9/90, p. 27. © Copyright Gernsback Publications, Inc., 1990.

Chapter 48

Fig. 48-1. QST, 7/90, p. 18 – 19.

Fig. 48-2. 73 Amateur Radio, 2/91, p. 20.

Fig. 48-3. Reprinted with permission from Popular Electronics, 10/89, p. 23. © Copyright Gernsback Publications, Inc., 1989.

Fig. 48-4. Reprinted with permission from Popular Electronics, 7/91, p. 56. © Copyright Gernsback Publications, Inc., 1991.

Fig. 48-5. Reprinted with permission from Radio-Electronics, 12/90, p. 35. © Copyright Gernsback Publications, Inc., 1990.

Fig. 48-6. Reprinted with permission from Popular Electronics, 11/90, p. 70. © Copyright Gernsback Publications, Inc., 1990.

Fig. 48-7. Ham Radio, 6/90, p. 44 – 47.

Fig. 48-8. Reprinted with permission from Popular Electronics, 7/91, p. 58. © Copyright Gernsback Publications, Inc., 1991.

Fig. 48-9. Reprinted with permission from Popular Electronics, 11/90, p. 70. © Copyright Gernsback Publications, Inc., 1990.

Fig. 48-10. Reprinted with permission from Popular Electronics, 7/91, p. 57. © Copyright Gernsback Publications, Inc., 1991.

Fig. 48-11. 73 Amateur Radio, 3/89, p. 66.

Chapter 49

Fig. 49-1. Reprinted with permission from Popular Electronics, 10/89, p. 84. © Copyright Gernsback Publications, Inc., 1989.

Fig. 49-2. Reprinted with permission from Radio-Electronics, 5/85, p.73 – 77. © Copyright Gernsback Publications, Inc., 1985.

Fig. 49-3. Elektor Electronics, 7/89, p. S35.

Fig. 49-4. Elektor Electronics, 7/89, p. S32.

Fig. 49-5. Reprinted with permission from Popular Electronics, 10/90, p. 26. © Copyright Gernsback Publications, Inc., 1990.

Fig. 49-6. Reprinted with permission from Radio-Electronics, 1/91, p. 71. © Copyright Gernsback Publications, Inc., 1991.

Fig. 49-7. Reprinted with permission from Radio-Electronics, 1/91, p. 71. © Copyright Gernsback Publications, Inc., 1991.

Fig. 49-8. Reprinted with permission from Popular Electronics, 9/90, p. 82. © Copyright Gernsback Publications, Inc., 1990.

Fig. 49-9. Reprinted with permission from Popular Elec-

tronics, 9/90, p. 83. © Copyright Gernsback Publications, Inc., 1990.

Chapter 50

Fig. 50-1. Reprinted with permission from Electronic Design, 11/88, p. 144. Copyright 1988, Penton Publishing.

Fig. 50-2. Reprinted with permission from Popular Electronics, 7/89, p. 26. © Copyright Gernsback Publications, Inc., 1989.

Fig. 50-3. Reprinted with permission from Popular Electronics, 9/90, p. 22. © Copyright Gernsback Publications, Inc., 1990.

Fig. 50-4. Reprinted with permission from Popular Electronics/Hands-On Electronics, 12/88, p. 103. © Copyright Gernsback Publications, Inc., 1988.

Fig. 50-5. Reprinted with permission from Popular Electronics, 7/89, p. 26. © Copyright Gernsback Publications, Inc., 1989.

Chapter 51

Fig. 51-1. QST, 5/89, p. 31.
Fig. 51-2. QST, 5/89, p. 31.
Fig. 51-3. QST, 5/89, p. 32.
Fig. 51-4. QST, 5/89, p. 32.
Fig. 51-5. QST, 5/89, p. 33.
Fig. 51-6. QST, 5/89, p. 32.

Chapter 52

Fig. 52-1. Reprinted from EDN, 6/91, p. 161. © 1991 Cahners Publishing Company, a division of Reed Publishing USA.

Fig. 52-2. Reprinted with permission from Electronic Design, 2/91, p. 102. Copyright 1991, Penton Publishing.

Fig. 52-3. ARRL Handbook, 1991, p. 18 – 7.

Fig. 52-4. Reprinted with permission from Popular Electronics, 5/91, p. 25 – 26. © Copyright Gernsback Publications, Inc., 1991.

Fig. 52-5. Ham Radio, 4/89, p. 65.

Fig. 52-6. Reprinted with permission from Electronic Design, 11/90, p. 136. Copyright 1990, Penton Publishing.

Fig. 52-7. Reprinted with permission from Radio-Electronics, 9/89, p. 59. © Copyright Gernsback Publications, Inc., 1989.

Fig. 52-8. Reprinted with permission from Popular Electronics, 2/91, p. 28. © Copyright Gernsback Publications, Inc., 1991.

Fig. 52-9. 73 Amateur Radio, 7/90, p. 70.

Fig. 52-10. Reprinted with permission from Radio-Electronics, 12/88, p. 72. © Copyright Gernsback Publications, Inc., 1988.

Fig. 52-11. Reprinted from EDN, 6/91, p. 182. © 1991 Cahners Publishing Company, a division of Reed Publishing USA.

Fig. 52-12. Maxim, Seminar Applications Book, 1988/89, p. 79.

Fig. 52-13. Reprinted with permission from Radio-Electronics, 9/89, p. 60. © Copyright Gernsback Publications, Inc., 1989.

Fig. 52-14. Electronics Today International, 4/78, p. 31.

Fig. 52-15. Reprinted with permission from Popular Electronics, 9/90, p. 22. © Copyright Gernsback Publications, Inc., 1990.

Chapter 53

Fig. 53-1. Reprinted from EDN, 9/90, p. 172. © 1990 Cahners Publishing Company, a division of Reed Publishing USA.

Fig. 53-2. Reprinted with permission from Popular Electronics, 3/90, p. 24. © Copyright Gernsback Publications, Inc., 1990.

Fig. 53-3. Elektor Electronics, 7/89, p. S38.

Fig. 53-4. Reprinted with permission from Electronic Design, 7/82, p. 194. Copyright 1991, Penton Publishing.

Fig. 53-5. Reprinted with permission from Radio-Electronics, 2/90, p. 57. © Copyright Gernsback Publications, Inc., 1990.

Fig. 53-6. Reprinted with permission from Popular Electronics, 12/77, p. 83. © Copyright Gernsback Publications, Inc., 1977.

Fig. 53-7. ARRL Handbook, 1991, p. 12 – 17.

Fig. 53-8. Reprinted with permission from Radio-Electronics, 4/70, p. 29. © Copyright Gernsback Publications, Inc., 1970.

Chapter 54

Fig. 54-1. Reprinted with permission from Popular Electronics, 12/90, p. 22. © Copyright Gernsback Publications, Inc., 1990.

Fig. 54-2. Reprinted with permission from Popular Electronics, 3/90, p. 70. © Copyright Gernsback Publications, Inc., 1990.

Fig. 54-3. Elektor Electronics, 7/89 Supplement, 545.

Chapter 55

Fig. 55-1. Elektor Electronics, 12/90, p. 54.
Fig. 55-2. Reprinted with permission from Popular Elec-

tronics, 9/90, p. 84. © Copyright Gernsback Publications, Inc., 1990.

Fig. 55-3. RF Design, 7/88, p. 43.

Fig. 55-4. Reprinted with permission from Popular Electronics, 11/90, p. 75. © Copyright Gernsback Publications, Inc., 1990.

Fig. 55-5. Reprinted with permission from Radio-Electronics, 7/90, p. 59. © Copyright Gernsback Publications, Inc., 1990.

Fig. 55-6. Reprinted with permission from Popular Electronics, 5/91, p. 76. © Copyright Gernsback Publications, Inc., 1991.

Chapter 56

Fig. 56-1. Elektor Electronics, 7/89 Supplement, p. 39–40.

Fig. 56-2. Reprinted from EDN, Design Ideas Special Issue, 7/20/89, Vol. IV, p. 36. © 1991 Cahners Publishing Company, a division of Reed Publishing USA.

Fig. 56-3. Reprinted from EDN, 12/89, p. 260. © 1991 Cahners Publishing Company, a division of Reed Publishing USA.

Fig. 56-4. Reprinted from EDN, 12/89, p. 257. © 1991 Cahners Publishing Company, a division of Reed Publishing USA.

Fig. 56-5. Reprinted from EDN, 5/91, p. 121. © 1991 Cahners Publishing Company, a division of Reed Publishing USA.

Fig. 56-6. Reprinted with permission from Popular Electronics, 8/91, p. 35–36. © Copyright Gernsback Publications, Inc., 1991.

Fig. 56-7. Harris, Analog Product Data Book, 1988, p. 10–162.

Fig. 56-8. Reprinted with permission from Radio-Electronics, 1/91, p. 9–10. © Copyright Gernsback Publications, Inc., 1991.

Chapter 57

Fig. 57-1. ARRL Handbook, 1991, p. 12–31.

Fig. 57-2. ARRL Handbook, 1991, p. 12–31.

Fig. 57-3. ARRL Handbook, 1991, p. 12–31.

Chapter 58

Fig. 58-1. Reprinted from EDN, 2/91, p. 183. © 1991 Cahners Publishing Company, a division of Reed Publishing USA.

Fig. 58-2. Reprinted with permission from Electronic Design, 12/90, p. 63. Copyright 1990, Penton Publishing.

Fig. 58-3. Reprinted from EDN, 7/91, p. 165. © 1991

Cahners Publishing Company, a division of Reed Publishing USA.

Fig. 58-4. Reprinted with permission from Popular Electronics, 11/90, p. 76. © Copyright Gernsback Publications, Inc., 1990.

Fig. 58-5. Reprinted with permission from Radio-Electronics, 1/89, p. 69. © Copyright Gernsback Publications, Inc., 1989.

Fig. 58-6. Reprinted with permission from Electronic Design, 5/91, p. 121. Copyright 1991, Penton Publishing.

Fig. 58-7. Reprinted with permission from Electronic Design, 12/88, p. 69. Copyright 1988, Penton Publishing.

Fig. 58-8. Reprinted with permission from Radio-Electronics, 12/88, p. 72. © Copyright Gernsback Publications, Inc., 1988.

Fig. 58-9. Maxim, Seminar Applications Book, 1988/89, p. 58.

Fig. 58-10. Reprinted with permission from Radio-Electronics, 1/89, p. 69. © Copyright Gernsback Publications, Inc., 1989.

Chapter 59

Fig. 59-1. Reprinted from EDN, 1/90, p. 138. © 1991 Cahners Publishing Company, a division of Reed Publishing USA.

Fig. 59-2. Elektor Electronics, 6/91, p. 43–45.

Fig. 59-3. Ham Radio, 4/90, p. 169.

Fig. 59-4. Elektor Electronics, 6/91, p. 43–45.

Fig. 59-5. QST, 10/90, p. 25.

Fig. 59-6. Hands-On Electronics, 11/87, p. 93.

Fig. 59-7. Reprinted with permission from Popular Electronics, 1/91, p. 27. © Copyright Gernsback Publications, Inc., 1991.

Fig. 59-8. Reprinted with permission from Popular Electronics, 10/89, p. 22. © Copyright Gernsback Publications, Inc., 1989.

Chapter 60

Fig. 60-1. Reprinted with permission from Popular Electronics, 7/91, p. 71–72. © Copyright Gernsback Publications, Inc., 1991.

Fig. 60-2. Reprinted from EDN, 11/90, p. 234. © 1991 Cahners Publishing Company, a division of Reed Publishing USA.

Fig. 60-3. QST, 6/91, p. 27–29.

Fig. 60-4. Reprinted with permission from Wels' Think Tank, Gernsback Publishing Inc., p. 37.

Fig. 60-5. Reprinted with permission from Electronic

Design, 12/88, p. 103. Copyright 1988, Penton Publishing.

Fig. 60-6. Reprinted with permission from Popular Electronics/Hands-On Electronics, 5/89, p. 86. © Copyright Gernsback Publications, Inc., 1989.

Fig. 60-7. Reprinted with permission from Hands-On Electronics/Popular Electronics, 1/89, p. 26.

Fig. 60-8. Reprinted with permission from Radio-Electronics, 8/86, p. 83. © Copyright Gernsback Publications, Inc., 1986.

Fig. 60-9. Reprinted with permission from Wels' Think Tank, Gernsback Publications Inc., p. 2.

Fig. 60-10. Reprinted with permission from Radio-Electronics, 1/89, p. 71. © Copyright Gernsback Publications, Inc., 1989.

Fig. 60-11. Harris, Analog Product Data Book, 1988, p. 10-147.

Chapter 61

Fig. 61-1. Reprinted with permission from Electronic Design, 9/89, p. 93–94. Copyright 1989, Penton Publishing.

Fig. 61-2. Reprinted with permission from Popular Electronics, Fact Card No. 65. © Copyright Gernsback Publications, Inc.

Fig. 61-3. Reprinted with permission from Popular Electronics, 3/91, p. 40–42. © Copyright Gernsback Publications, Inc., 1991.

Fig. 61-4. Reprinted with permission from Popular Electronics, 1/91, p. 28–29. © Copyright Gernsback Publications, Inc., 1991.

Fig. 61-5. Reprinted with permission from Wels' Think Tank, Gernsback Publications Inc., p. 39.

Fig. 61-6. Reprinted with permission from Popular Electronics, Fact Card No. 170. © Copyright Gernsback Publications, Inc.

Fig. 61-7. Reprinted with permission from Popular Electronics/Hands-On Electronics, 12/88, p. 58. © Copyright Gernsback Publications, Inc., 1988.

Chapter 62

Fig. 62-1. Reprinted from EDN, 1/91, p. 154–156. © 1991 Cahners Publishing Company, a division of Reed Publishing USA.

Fig. 62-2. Hands-On Electronics, Fact Card No. 57.

Fig. 62-3. Reprinted from EDN, Design Ideas Special Issue, Vol. IV, 7/20/89, p. 12. © 1991 Cahners Publishing Company, a division of Reed Publishing USA.

Fig. 62-4. Reprinted from EDN, 2/91, p. 106. © 1991 Cahners Publishing Company, a division of Reed Publishing USA.

Fig. 62-5. Reprinted with permission from Popular Electronics, Fact Card No. 65. © Copyright Gernsback Publications, Inc.

Fig. 62-6. Reprinted with permission from Popular Electronics, 8/90, p. 20. © Copyright Gernsback Publications, Inc., 1990.

Fig. 62-7. Reprinted with permission from Radio-Electronics, 9/87, p. 64. © Copyright Gernsback Publications, Inc., 1987.

Fig. 62-8. Reprinted with permission from Electronic Design, 6/22/89, p. 105. Copyright 1989, Penton Publishing.

Fig. 62-9. Hands-On Electronics, Fact Card No. 40.

Chapter 63

Fig. 63-1. Reprinted from EDN, Design Ideas Special Issue, Vol. IV, 7/20/89, p. 14. © 1991 Cahners Publishing Company, a division of Reed Publishing USA.

Fig. 63-2. Reprinted from EDN, 1/90, p. 218. © 1991 Cahners Publishing Company, a division of Reed Publishing USA.

Fig. 63-3. Electronic Engineering, 9/88, p. 31.

Fig. 63-4. Reprinted from EDN, 6/91, p. 162. © 1991 Cahners Publishing Company, a division of Reed Publishing USA.

Fig. 63-5. Reprinted with permission from Electronic Design, 11/90, p. 136. Copyright 1990, Penton Publishing.

Fig. 63-6. Reprinted with permission from Popular Electronics, 6/89, p. 25. © Copyright Gernsback Publications, Inc., 1989.

Fig. 63-7. Reprinted with permission from Electronic Design, 4/91, p. 104. Copyright 1991, Penton Publishing.

Fig. 63-8. Reprinted with permission from Wels' Think Tank, Gernsback Publications Inc., p. 52.

Fig. 63-9. Elektor Electronics, 12/90 Supplement, p. 41.

Fig. 63-10. Reprinted with permission from Radio-Electronics, 10/90, p. 14. © Copyright Gernsback Publications, Inc., 1990.

Fig. 63-11. Elektor Electronics, 12/90 Supplement, p. 536.

Fig. 63-12. Reprinted with permission from Radio-Electronics, 3/89, p. 53. © Copyright Gernsback Publications, Inc., 1989.

Fig. 63-13. Reprinted with permission from Electronic Design, 2/91, p. 104. Copyright 1991, Penton Publishing.

Fig. 63-14. Reprinted with permission from Electronic Design, 3/91, p. 86. Copyright 1991, Penton Publishing.

Fig. 63-15. 73 Amateur Radio, 11/90, p. 67.

Fig. 63-16. 73 Amateur Radio, 7/90, p. 68.

Fig. 63-17. 73 Amateur Radio, 7/90, p. 69.

Fig. 63-18. 73 Amateur Radio, 7/90, p. 77.

Fig. 63-19. Reprinted with permission from Electronic Design, 12/90, p. 77. Copyright 1990, Penton Publishing.

Fig. 63-20. Reprinted with permission from Radio-Electronics, 12/88, p. 72. © Copyright Gernsback Publications, Inc., 1988.

Fig. 63-21. Linear Technology, Design Note #47.

Fig. 63-22. Linear Technology, Design Note #47.

Fig. 63-23. Reprinted with permission from Radio-Electronics, 5/91, p. 49 – 55. © Copyright Gernsback Publications, Inc., 1991.

Fig. 63-24. Elektor Electronics, 7-8/89, p. 28.

Fig. 63-25. Ham Radio, 4/89, p. 67.

Fig. 63-26. Linear Technology, Linear Databook, 1986, p. 2 – 112.

Fig. 63-27. Reprinted from EDN, 1/91, p. 152. © 1991 Cahners Publishing Company, a division of Reed Publishing USA.

Fig. 63-28. Reprinted from EDN, 8/2/90, p. 96. © 1991 Cahners Publishing Company, a division of Reed Publishing USA.

Fig. 63-29. Reprinted with permission from Radio-Electronics, 3/89, p. 54 – 55. © Copyright Gernsback Publications, Inc., 1989.

Fig. 63-30. Reprinted with permission from Electronic Design, 6/91, p. 109. Copyright 1991, Penton Publishing.

Fig. 63-31. Reprinted with permission from Electronic Design, 3/91, p. 117. Copyright 1991, Penton Publishing.

Fig. 63-32. Reprinted with permission from Popular Electronics, Fact Card No. 134. © Copyright Gernsback Publications, Inc.

Chapter 64

Fig. 64-1. Reprinted with permission from Radio-Electronics, 8/91, p. 42. © Copyright Gernsback Publications, Inc., 1991.

Fig. 64-2. Linear Technology, Application Note 45.

Fig. 64-3. Reprinted with permission from R-E Experimenter's Handbook, 1990, p. 61. © Copyright Gernsback Publications, Inc., 1990.

Fig. 64-4. Reprinted with permission from Popular Electronics, 6/89, p. 88. © Copyright Gernsback Publications, Inc., 1989.

Fig. 64-5. Reprinted with permission from Radio-Electronics, 2/91, p. 34. © Copyright Gernsback Publications, Inc., 1991.

Chapter 65

Fig. 65-1. QST, 9/89, p. 23.

Fig. 65-2. Reprinted with permission from Radio-Electronics, 3/90, p. 31 – 34. © Copyright Gernsback Publications, Inc., 1990.

Fig. 65-3. Reprinted with permission from Popular Electronics, 12/90, p. 22. © Copyright Gernsback Publications, Inc., 1990.

Fig. 65-4. Hands-On Electronics, Fact Card No. 40.

Fig. 65-5. Reprinted with permission from Popular Electronics, 9/90, p. 27. © Copyright Gernsback Publications, Inc., 1990.

Fig. 65-6. Reprinted with permission from Popular Electronics, 4/90, p. 45 – 46. © Copyright Gernsback Publications, Inc., 1990.

Fig. 65-7. Linear Technology, 10/90, Design Notes #40.

Fig. 65-8. Reprinted with permission from Popular Electronics, 11/90, p. 24. © Copyright Gernsback Publications, Inc., 1990.

Fig. 65-9. Reprinted with permission from Radio-Electronics, 1/89, p. 70. © Copyright Gernsback Publications, Inc., 1989.

Chapter 66

Fig. 66-1. Reprinted from EDN, 11/90, p. 292. © 1991 Cahners Publishing Company, a division of Reed Publishing USA.

Fig. 66-2. Reprinted with permission from Popular Electronics, 6/89, p. 26. © Copyright Gernsback Publications, Inc., 1989.

Fig. 66-3. Reprinted with permission from Electronic Design, 5/90, p. 80. Copyright 1990, Penton Publishing.

Fig. 66-4. Ham Radio, 4/90, p. 73 – 78.

Fig. 66-5. Reprinted from EDN, 1/91, p. 170. © 1991 Cahners Publishing Company, a division of Reed Publishing USA.

Fig. 66-6. Reprinted with permission from Electronic

Fig. 66-8. Hands-On Electronics, Fact Card No. 37.

Fig. 66-9. Reprinted from EDN, 4/91, p. 188. © 1991 Cahners Publishing Company, a division of Reed Publishing USA.

Design, 9/89, p. 94–95. Copyright 1989, Penton Publishing.

Fig. 66-7. Reprinted from EDN, 12/90, p. 218. © 1991 Cahners Publishing Company, a division of Reed Publishing USA.

Chapter 67

Fig. 67-1. Reprinted with permission from Popular Electronics, 10/89, p. 104. © Copyright Gernsback Publications, Inc., 1989.

Fig. 67-2. Elektor Electronics, 12/90 Supplement, p. 42.

Fig. 67-3. Elektor Electronics, 7-8/89, p. 67.

Fig. 67-4. Reprinted with permission from Popular Electronics, 3/91, p. 55–57. © Copyright Gernsback Publications, Inc., 1991.

Fig. 67-5. Reprinted with permission from Popular Electronics, 7/91, p. 57. © Copyright Gernsback Publications, Inc., 1991.

Fig. 67-6. Reprinted with permission from Radio-Electronics, 12/75, p. 42. © Copyright Gernsback Publications, Inc., 1975.

Fig. 67-7. Reprinted with permission from Popular Electronics/Hands-On Electronics, 5/89, p. 28. © Copyright Gernsback Publications, Inc., 1989.

Fig. 67-8. TAB Books, Third Book of Electronic Projects, p. 39.

Chapter 68

Fig. 68-1. Reprinted with permission from Radio-Electronics, 5/91, p. 40–42. © Copyright Gernsback Publications, Inc., 1991.

Fig. 68-2. Reprinted with permission from Popular Electronics, 8/90, p. 80. © Copyright Gernsback Publications, Inc., 1990.

Fig. 68-3. GE, Application Note 90.16, p. 29.

Fig. 68-4. Reprinted with permission from Popular Electronics, 5/91, p. 74. © Copyright Gernsback Publications, Inc., 1991.

Fig. 68-5. Reprinted with permission from Radio-Electronics, 12/88, p. 76. © Copyright Gernsback Publications, Inc., 1988.

Fig. 68-6. Elektor Electronics, 302 Circuits, p. 203.

Fig. 68-7. Reprinted with permission from Popular Electronics, 8/90, p. 80. © Copyright Gernsback Publications, Inc., 1990.

Fig. 68-8. Elektor Electronics, 302 Circuits, p. 165–166.

Fig. 68-9. Reprinted from EDN, 3/75, p. 73. © 1991 Cahners Publishing Company, a division of Reed Publishing USA.

Fig. 68-10. Reprinted from EDN, 10/90, p. 238. © 1991 Cahners Publishing Company, a division of Reed Publishing USA.

Fig. 68-11. Hands-On Electronics, Fact Card No. 86.

Chapter 69

Fig. 69-1. Reprinted with permission from R-E Experimenter's Handbook, 1987, p. 122. © Copyright Gernsback Publications, Inc., 1987.

Fig. 69-2. Reprinted with permission from R-E Experimenter's Handbook, 1987, p. 122. © Copyright Gernsback Publications, Inc., 1987.

Chapter 70

Fig. 70-1. Reprinted with permission from Popular Electronics, 6/89, p. 59–62. © Copyright Gernsback Publications, Inc., 1989.

Fig. 70-2. Reprinted with permission from Radio-Electronics, 12/90, p. 81. © Copyright Gernsback Publications, Inc., 1990.

Fig. 70-3. Reprinted from EDN, 3/79, p. 130. © 1991 Cahners Publishing Company, a division of Reed Publishing USA.

Fig. 70-4. Reprinted with permission from Radio-Electronics, 12/90, p. 82. © Copyright Gernsback Publications, Inc., 1990.

Fig. 70-5. Reprinted with permission from Popular Electronics, 11/90, p. 77. © Copyright Gernsback Publications, Inc., 1990.

Chapter 71

Fig. 71-1. 73 Amateur Radio, 11/90, p. 34.

Fig. 71-2. 73 Amateur Radio, 4/91, p. 26.

Fig. 71-3. Reprinted with permission from Popular Electronics, 8/91, p. 72. © Copyright Gernsback Publications, Inc., 1991.

Fig. 71-4. Reprinted with permission from Popular Electronics, 5/90, p. 85 and 96. © Copyright Gernsback Publications, Inc., 1990.

Fig. 71-5. Reprinted with permission from Popular Electronics, 10/89, p. 24. © Copyright Gernsback Publications, Inc., 1989.

Fig. 71-6. Reprinted with permission from Popular Electronics, 3/90, p. 96. © Copyright Gernsback Publications, Inc., 1990.

Fig. 71-7. Reprinted with permission from Popular Electronics, 2/82, p. 99. © Copyright Gernsback Publications, Inc., 1982.

Fig. 71-8. Reprinted with permission from Radio-Elec-

tronics, 1/91, p. 56. © Copyright Gernsback Publications, Inc., 1991.

Fig. 71-9. Reprinted with permission from Popular Electronics, 10/89, p. 22. © Copyright Gernsback Publications, Inc., 1989.

Chapter 72

Fig. 72-1. QST, 8/90, p. 59.

Fig. 72-2. ARRL Handbook, 1991, p. 12 – 30.

Fig. 72-3. Reprinted with permission from Popular Electronics, 3/90, p. 48. © Copyright Gernsback Publications, Inc., 1990.

Fig. 72-4. ARRL Handbook, 1991, p. 12 – 26.

Fig. 72-5. 73 Amateur Radio, 3/90, p. 13.

Fig. 72-6. ARRL Handbook, 1991, p. 14 – 6.

Chapter 73

Fig. 73-1. Elektor Electronics, 7/89 Supplement, p. S33.

Fig. 73-2. Reprinted from EDN, Design Ideas Special Issue, Vol. IV, 7/20/89, p. 28 – 29. © 1991 Cahners Publishing Company, a division of Reed Publishing USA.

Fig. 73-3. Reprinted with permission from Electronic Design, 12/88, p. 105. Copyright 1988, Penton Publishing.

Fig. 73-4. Reprinted from EDN, 4/91, p. 179. © 1991 Cahners Publishing Company, a division of Reed Publishing USA.

Fig. 73-5. Reprinted with permission from Electronic Design, 2/91, p. 104. Copyright 1991, Penton Publishing.

Fig. 73-6. Reprinted from EDN, 6/90, p. 266. © 1991 Cahners Publishing Company, a division of Reed Publishing USA.

Fig. 73-7. Reprinted with permission from Radio-Electronics, 12/88, p. 72. © Copyright Gernsback Publications, Inc., 1988.

Chapter 74

Fig. 74-1. Elektor Electronics, 12/90, p. 512 – 513.

Fig. 74-2. Intersil, Applications Handbook, 1988, p. 6 – 20.

Fig. 74-3. Reprinted with permission from Radio-Electronics, 12/88, p. 72. © Copyright Gernsback Publications, Inc., 1988.

Chapter 75

Fig. 75-1. GE, Optoelectronics, Third Edition, p. 139.

Fig. 75-2. Reprinted from EDN, 12/89, p. 258. © 1991 Cahners Publishing Company, a division of Reed Publishing USA.

Fig. 75-3. Elektor Electronics, 12/90, p. 525.

Fig. 75-4. Reprinted with permission from Popular Electronics, 5/91, p. 86. © Copyright Gernsback Publications, Inc., 1991.

Fig. 75-5. Reprinted from EDN, 7/90, p. 185. © 1991 Cahners Publishing Company, a division of Reed Publishing USA.

Fig. 75-6. Reprinted from EDN, 9/89, p. 170. © 1991 Cahners Publishing Company, a division of Reed Publishing USA.

Chapter 76

Fig. 76-1. QST, 7/90, p. 29 – 31.

Fig. 76-2. CQ, 7/90, p. 61 – 65.

Fig. 76-3. Elektor Electronics, 301 Circuits, p. 38 – 39.

Fig. 76-4. QST, 10/90, p. 18 – 21.

Fig. 76-5. North Country Radio, P.O. Box 53, Wykagyl Station, New Rochelle, NY 10804.

Fig. 76-6. Elektor Electronics, 12/90, p. 57 – 58.

Fig. 76-7. Reprinted with permission from Popular Electronics, 1/91, p. 58. © Copyright Gernsback Publications, Inc., 1991.

Fig. 76-8. QST, 7/90, p. 28 – 30.

Fig. 76-9. Reprinted with permission from Popular Electronics, 1/91, p. 58. © Copyright Gernsback Publications, Inc., 1991.

Fig. 76-10. Ham Radio, 11/88, p. 60 – 62.

Fig. 76-11. Reprinted with permission from Popular Electronics, 12/89, p. 23. © Copyright Gernsback Publications, Inc., 1989.

Fig. 76-12. Reprinted with permission from Popular Electronics, 1/91, p. 59. © Copyright Gernsback Publications, Inc., 1991.

Fig. 76-13. Reprinted with permission from Popular Electronics, 1/91, p. 59. © Copyright Gernsback Publications, Inc., 1991.

Fig. 76-14. Reprinted with permission from Popular Electronics, 1/91, p. 59. © Copyright Gernsback Publications, Inc., 1991.

Fig. 76-15. Reprinted with permission from Popular Electronics, 1/90, p. 98. © Copyright Gernsback Publications, Inc., 1990.

Fig. 76-16. Reprinted with permission from Popular Electronics, 11/90, p. 69. © Copyright Gernsback Publications, Inc., 1990.

Fig. 76-17. Reprinted with permission from Popular Electronics, 1/90, p. 98. © Copyright Gernsback Publications, Inc., 1990.

Fig. 76-18. Reprinted with permission from Popular Electronics, 1/90, p. 74. © Copyright Gernsback Publications, Inc., 1990.

Fig. 76-19. Copyright of Motorola, MOSPOWER Design Catalog, 1/83, p. 5 – 6. Used by permission.

Fig. 76-20. Reprinted with permission from Popular Electronics, 1/90, p. 74. © Copyright Gernsback Publications, Inc., 1990.

Fig. 76-21. Reprinted with permission from Popular Electronics, 1/90, p. 74. © Copyright Gernsback Publications, Inc., 1990.

Fig. 76-22. Reprinted with permission from Popular Electronics, 1/90, p. 74. © Copyright Gernsback Publications, Inc., 1990.

Chapter 77

Fig. 77-1. Elektor Electronics, 7/89, p. S31.

Fig. 77-2. North Country Radio, P.O. Box 53, Wykagyl Station, New Rochelle, NY 10804.

Fig. 77-3. Reprinted with permission from Popular Electronics, 11/90, p. 85. © Copyright Gernsback Publications, Inc., 1990.

Fig. 77-4. Elektor Electronics, 3/91, p. 37 – 41.

Fig. 77-5. Reprinted with permission from Popular Electronics, 3/90, p. 48. © Copyright Gernsback Publications, Inc., 1990.

Fig. 77-6. Reprinted with permission from Popular Electronics, 8/90, p. 88. © Copyright Gernsback Publications, Inc., 1990.

Fig. 77-7. Reprinted with permission from Radio-Electronics, 10/89, p. 43. © Copyright Gernsback Publications, Inc., 1989.

Fig. 77-8. Ham Radio, 4/90, p. 21.

Fig. 77-9. Reprinted with permission from Radio-Electronics, 11/90, p. 25. © Copyright Gernsback Publications, Inc., 1990.

Chapter 78

Fig. 78-1. Reprinted from EDN, 12/90, p. 153. © 1991 Cahners Publishing Company, a division of Reed Publishing USA.

Chapter 79

Fig. 79-1. Elektor Electronics, 12/90, p. 523.

Fig. 79-2. Reprinted from EDN, 10/90, p. 206. © 1991 Cahners Publishing Company, a division of Reed Publishing USA.

Fig. 79-3. Elektor Electronics, 7/89, p. 542.

Fig. 79-4. Micro Linear Advertisement

Fig. 79-5. Reprinted with permission from Popular Electronics, 3/91, p. 92. © Copyright Gernsback Publications, Inc., 1991.

Fig. 79-6. Reprinted with permission from Popular Electronics, 1/90, p. 85. © Copyright Gernsback Publications, Inc., 1990.

Fig. 79-7. Reprinted with permission from Electronic Design, 4/91, p. 99. Copyright 1991, Penton Publishing.

Fig. 79-8. Reprinted with permission from Popular Electronics, 1/90, p. 84. © Copyright Gernsback Publications, Inc., 1990.

Fig. 79-9. Reprinted with permission from Radio-Electronics, 7/89, p. 52. © Copyright Gernsback Publications, Inc., 1989.

Fig. 79-10. Reprinted with permission from Popular Electronics, 1/90, p. 101. © Copyright Gernsback Publications, Inc., 1990.

Fig. 79-11. Reprinted from EDN, 11/90, p. 236. © 1991 Cahners Publishing Company, a division of Reed Publishing USA.

Fig. 79-12. GE/RCA, BiMOS Operational Amplifier Circuit Ideas, 1987, p. 7.

Fig. 79-13. 73 Amateur Radio, 12/76, p. 97 – 99.

Fig. 79-14. Reprinted with permission from Popular Electronics, 3/91, p. 88. © Copyright Gernsback Publications, Inc., 1991.

Fig. 79-15. Reprinted from EDN, 11/90, p. 23. © 1991 Cahners Publishing Company, a division of Reed Publishing USA.

Chapter 80

Fig. 80-1. Hands-On Electronics, 1-2/86, p. 67.

Fig. 80-2. Reprinted with permission from Radio-Electronics, 7/90, p. 58. © Copyright Gernsback Publications, Inc., 1990.

Fig. 80-3. Reprinted with permission from Radio-Electronics, 7/90, p. 57. © Copyright Gernsback Publications, Inc., 1990.

Fig. 80-4. Reprinted with permission from National Semiconductor Corp., CMOS Databook, 1981, p. 8 – 46.

Fig. 80-5. Reprinted with permission from Popular Electronics, 7/89, p. 22. © Copyright Gernsback Publications, Inc., 1989.

Fig. 80-6. Reprinted with permission from Radio-Electronics, 12/89, p. 56. © Copyright Gernsback Publications, Inc., 1989.

Chapter 81

Fig. 81-1. Elektor Electronics, 7/89 Supplement, p. 41–42.

Fig. 81-2. Elektor Electronics, 7/89, p. 537.

Fig. 81-3. Reprinted with permission from Popular Electronics, 10/89, p. 85. © Copyright Gernsback Publications, Inc., 1989.

Fig. 81-4. Reprinted with permission from Popular Electronics, 1/91, p. 27. © Copyright Gernsback Publications, Inc., 1991.

Fig. 81-5. Reprinted with permission from Popular Electronics, 9/90, p. 63–66. © Copyright Gernsback Publications, Inc., 1990.

Fig. 81-6. Elektor Electronics, 301 Circuits, p. 130.

Fig. 81-7. Reprinted with permission from Radio-Electronics, 12/89, p. 56. © Copyright Gernsback Publications, Inc., 1989.

Fig. 81-8. Reprinted from EDN, 8/2/91, p. 112. © 1991 Cahners Publishing Company, a division of Reed Publishing USA.

Chapter 82

Fig. 82-1. Reprinted with permission from Radio-Electronics, 4/87, p. 30. © Copyright Gernsback Publications, Inc., 1987.

Fig. 82-2. Hands-On Electronics, 11/86, p. 32.

Fig. 82-3. Reprinted from EDN, 6/88, p. 175. © 1991 Cahners Publishing Company, a division of Reed Publishing USA.

Fig. 82-4. Reprinted with permission from Electronic Design, 10/89, p. 85. Copyright 1991, Penton Publishing.

Chapter 83

Fig. 83-1. Elektor Electronics, 301 Circuits, p. 32–35.

Fig. 83-2. Reprinted from EDN, 11/90, p. 232. © 1991 Cahners Publishing Company, a division of Reed Publishing USA.

Fig. 83-3. Reprinted by permission of Texas Instruments, Linear and Interface Circuit Applications, 1985, p. 7–15.

Fig. 83-4. Reprinted with permission from Radio-Electronics, 12/88, p. 76. © Copyright Gernsback Publications, Inc., 1988.

Fig. 83-5. Reprinted with permission from Radio-Electronics, 7/89, p. 53. © Copyright Gernsback Publications, Inc., 1989.

Fig. 83-6. Reprinted with permission from Popular Electronics, Fact Card No. 98. © Copyright Gernsback Publications, Inc.

Fig. 83-7. Reprinted with permission from Radio-Electronics, 7/89, p. 53. © Copyright Gernsback Publications, Inc., 1989.

Fig. 83-8. Reprinted with permission from Radio-Electronics, 12/89, p. 56. © Copyright Gernsback Publications, Inc., 1989.

Fig. 83-9. Reprinted by permission of Texas Instruments, Linear and Interface Circuit Applications, 1985, Vol. 1, p. 3–16.

Fig. 83-10. Reprinted with permission from Radio-Electronics, 12/88, p. 76. © Copyright Gernsback Publications, Inc., 1988.

Chapter 84

Fig. 84-1. Elektor Electronics, 12/90, p. 531–532.

Fig. 84-2. Reprinted from EDN, 7/85, p. 282. © 1991 Cahners Publishing Company, a division of Reed Publishing USA.

Fig. 84-3. Reprinted with permission from Popular Electronics, 12/89, p. 29–34. © Copyright Gernsback Publications, Inc., 1989.

Fig. 84-4. Reprinted with permission from Radio-Electronics, 12/89, p. 58. © Copyright Gernsback Publications, Inc., 1989.

Fig. 84-5. QST, 6/91, p. 45.

Fig. 84-6. Reprinted with permission from Popular Electronics, 11/90, p. 25. © Copyright Gernsback Publications, Inc., 1990.

Fig. 84-7. Reprinted with permission from Popular Electronics, 11/90, p. 25. © Copyright Gernsback Publications, Inc., 1990.

Fig. 84-8. Reprinted from EDN, 10/90, p. 132. © 1991 Cahners Publishing Company, a division of Reed Publishing USA.

Fig. 84-9. RF Design, 9/90, p. 76.

Fig. 84-10. Elektor Electronics, 12/90, p. 532.

Fig. 84-11. Elektor Electronics, 12/90 Supplement, p. 43.

Fig. 84-12. Electronics Today International, 4/78, p. 31.

Chapter 85

Fig. 85-1. 73 Amateur Radio, 12/90, p. 54.

Fig. 85-2. Hands-On Electronics, 8/87, p. 92.

Chapter 86

Fig. 86-1. Reprinted with permission from Popular Electronics, 6/91, p. 53–57. © Copyright Gernsback Publications, Inc., 1991.

Fig. 86-2. Reprinted with permission from Popular Elec-

tronics, 8/90, p. 33–36. © Copyright Gernsback Publications, Inc., 1990.

Fig. 86-3. Reprinted with permission from Popular Electronics, 8/90, p. 29–32. © Copyright Gernsback Publications, Inc., 1990.

Fig. 86-4. Reprinted with permission from Popular Electronics, 6/89, p. 33. © Copyright Gernsback Publications, Inc., 1989.

Fig. 86-5. Elektor Electronics, 7/89 Supplement, p. 526.

Fig. 86-6. Ham Radio, 12/89, p. 32.

Fig. 86-7. Reprinted with permission from Popular Electronics, 1/91, p. 81. © Copyright Gernsback Publications, Inc., 1991.

Fig. 86-8. Reprinted with permission from Radio-Electronics, 7/90, p. 8. © Copyright Gernsback Publications, Inc., 1990.

Fig. 86-9. Reprinted from EDN, 12/90, p. 153–154. © 1991 Cahners Publishing Company, a division of Reed Publishing USA.

Fig. 86-10. Reprinted with permission from Wels' Think Tank, Gernsback Publications Inc., p. 48.

Fig. 86-11. Reprinted with permission from Popular Electronics, 9/91, p. 21. © Copyright Gernsback Publications, Inc., 1991.

Fig. 86-12. Hands-On Electronics, 11/87, p. 85.

Fig. 86-13. TAB Books, The Build-It Book of Electronic Projects, p. 24.

Fig. 86-14. Reprinted with permission from Popular Electronics, 5/91, p. 86. © Copyright Gernsback Publications, Inc., 1991.

Fig. 86-15. Reprinted with permission from Popular Electronics, 8/90, p. 80. © Copyright Gernsback Publications, Inc., 1990.

Fig. 86-16. Reprinted with permission from Popular Electronics, 7/91, p. 22. © Copyright Gernsback Publications, Inc., 1991.

Fig. 86-17. Reprinted with permission from Popular Electronics, Fact Card No. 170. © Copyright Gernsback Publications, Inc.

Fig. 86-18. Hands-On Electronics, Winter 1985, p. 65.

Fig. 86-19. Reprinted with permission from Popular Electronics, 8/90, p. 81. © Copyright Gernsback Publications, Inc., 1990.

Fig. 86-20. Elektor Electronics, 7/89, p. 540.

Fig. 86-21. Reprinted with permission from Popular Electronics, 9/91, p. 22. © Copyright Gernsback Publications, Inc., 1991.

Fig. 86-22. Reprinted with permission from Popular Electronics, 2/91, p. 25–26. © Copyright Gernsback Publications, Inc., 1991.

Fig. 86-23. Reprinted with permission from Popular Electronics, 11/90, p. 26. © Copyright Gernsback Publications, Inc., 1990.

Fig. 86-24. Reprinted with permission from General Electric Semiconductor Department, Optoelectronics, Second Edition, p. 118.

Chapter 87

Fig. 87-1. Reprinted from EDN, 9/89, p. 152. © 1991 Cahners Publishing Company, a division of Reed Publishing USA.

Fig. 87-2. GE/RCA, BiMOS Operational Amplifiers Circuit Ideas, 1987, p. 25.

Fig. 87-3. Reprinted with permission from Popular Electronics, 1/91, p. 22. © Copyright Gernsback Publications, Inc., 1991.

Chapter 88

Fig. 88-1. Elektor Electronics, 12/90, p. 533.

Fig. 88-2. Elektor Electronics, 7/89, p. 528.

Fig. 88-3. Reprinted with permission from Radio-Electronics, 12/88, p. 72. © Copyright Gernsback Publications, Inc., 1988.

Fig. 88-4. Reprinted with permission from Radio-Electronics, 6/91, p. 58. © Copyright Gernsback Publications, Inc., 1991.

Fig. 88-5. Reprinted with permission from Radio-Electronics, 12/88, p. 71. © Copyright Gernsback Publications, Inc., 1988.

Fig. 88-6. Reprinted from EDN, 4/91, p. 180. © 1991 Cahners Publishing Company, a division of Reed Publishing USA.

Chapter 89

Fig. 89-1. Electronic Engineering, 12/90, p. 28.

Fig. 89-2. Reprinted with permission from R-E Experimenter's Handbook, 1990, p. 22–23. © Copyright Gernsback Publications, Inc., 1990.

Fig. 89-3. Reprinted with permission from Electronic Design, 4/90, p. 85–88. Copyright 1991, Penton Publishing.

Fig. 89-4. 42 New Ideas, Gernsback Publications Inc., 1984, p. 13.

Fig. 89-5. Linear Technology, 10/90, Design Notes #40.

Chapter 90

Fig. 90-1. Elektor Electronics, 7/89, p. 536–537.

Fig. 90-2. 73 Amateur Radio, 8/90, p. 82.

Fig. 90-3. Reprinted with permission from Popular Electronics, 12/89, p. 82–83. © Copyright Gernsback Publications, Inc., 1989.

Fig. 90-4. Reprinted with permission from Popular Electronics, 12/89, p. 82. © Copyright Gernsback Publications, Inc., 1989.

Fig. 90-5. Reprinted with permission from Popular Electronics, 12/89, p. 27. © Copyright Gernsback Publications, Inc., 1989.

Fig. 90-6. Hands-On Electronics, Fact Card No. 49.

Fig. 90-7. Reprinted from EDN, 8/2/90, p. 86. © 1991 Cahners Publishing Company, a division of Reed Publishing USA.

Fig. 90-8. 73 Amateur Radio, 2/91, p. 78.

Fig. 90-9. Hands-On Electronics, 1-2/86, p. 52.

Fig. 90-10. Copyright of Motorola, Thyristor Device Data, Series A 1985, p. 1-6-51. Used by permission.

Fig. 90-11. Reprinted with permission from Popular Electronics, 1/91, p. 80. © Copyright Gernsback Publications, Inc., 1991.

Chapter 91

Fig. 91-1. GE/RCA, BiMOS Operational Amplifier Circuit Ideas, 1987, p. 22.

Fig. 91-2. Reprinted with permission from Radio-Electronics, 2/90, p. 58. © Copyright Gernsback Publications, Inc., 1990.

Fig. 91-3. Hands-On Electronics, 8/87, p. 91.

Chapter 92

Fig. 92-1. Elektor Electronics, 7/89, p. 538-539.

Fig. 92-2. Reprinted with permission from Popular Electronics, 2/91, p. 23-24. © Copyright Gernsback Publications, Inc., 1991.

Fig. 92-3. GE/RCA, BiMOS Operational Amplifier Circuit Ideas, 1987, p. 28.

Fig. 92-4. Reprinted with permission from Wels' Think Tank, Gernsback Publications Inc., p. 39.

Fig. 92-5. Reprinted with permission from Radio-Electronics, 7/90, p. 58. © Copyright Gernsback Publications, Inc., 1990.

Fig. 92-6. Hands-On Electronics, 9/87, p. 89.

Chapter 93

Fig. 93-1. 73 Amateur Radio, 6/89, p. 65.

Fig. 93-2. 73 Amateur Radio, 11/90, p. 78.

Fig. 93-3. Reprinted with permission from Radio-Electronics, 6-7/89, p. 45-50. © Copyright Gernsback Publications, Inc., 1989.

Fig. 93-4. Elektor Electronics, 7/89, p. 521.

Fig. 93-5. 73 Amateur Radio, 4/91, p. 76.

Fig. 93-6. 73 Amateur Radio, 4/90, p. 46-47.

Fig. 93-7. 73 Amateur Radio, 7/90, p. 16.

Chapter 94

Fig. 94-1. Reprinted with permission from R-E Experimenter's Handbook, 1987, p. 93-94. © Copyright Gernsback Publications, Inc., 1987.

Fig. 94-2. Reprinted with permission from Radio-Electronics, 7/85, p. 61. © Copyright Gernsback Publications, Inc., 1985.

Fig. 94-3. Reprinted with permission from R-E Experimenter's Handbook, 1987, p. 93. © Copyright Gernsback Publications, Inc., 1987.

Chapter 95

Fig. 95-1. Reprinted from EDN, 12/90, p. 224-226. © 1991 Cahners Publishing Company, a division of Reed Publishing USA.

Fig. 95-2. Elektor Electronics, 302 Circuits, p. 180.

Fig. 95-3. Elektor Electronics, 6/91, p. 29.

Fig. 95-4. Reprinted with permission from Radio-Electronics, 12/89, p. 81. © Copyright Gernsback Publications, Inc., 1989.

Fig. 95-5. Reprinted from EDN, 2/90, p. 185. © 1991 Cahners Publishing Company, a division of Reed Publishing USA.

Fig. 95-6. Reprinted with permission from Radio-Electronics, 10/90, p. 59-60. © Copyright Gernsback Publications, Inc., 1990.

Fig. 95-7. Reprinted with permission from Radio-Electronics, 7/89, p. 41. © Copyright Gernsback Publications, Inc., 1989.

Fig. 95-8. Electronic Products, 4/88, p. 36.

Fig. 95-9. Reprinted from EDN, 12/90, p. 222. © 1991 Cahners Publishing Company, a division of Reed Publishing USA.

Fig. 95-10. Elektor Electronics, 302 Circuits, p. 78-79.

Fig. 95-11. Reprinted with permission from Radio-Electronics, 7/91, p. 73. © Copyright Gernsback Publications, Inc., 1991.

Chapter 96

Fig. 96-1. Reprinted with permission from Popular Electronics, 3/90, p. 28. © Copyright Gernsback Publications, Inc., 1990.

Fig. 96-2. Reprinted from EDN, 3/91, p. 137-138. © 1991 Cahners Publishing Company, a division of Reed Publishing USA.

Chapter 97

Fig. 97-1. Elektor Electronics, 7/89 Supplement, p. 34.

Fig. 97-2. Reprinted with permission from Popular Electronics, 6/91, p. 39–40. © Copyright Gernsback Publications, Inc., 1991.

Chapter 98

Fig. 98-1. Elektor Electronics, 12/90 Supplement, p. 41–42.

Fig. 98-2. Reprinted with permission from Radio-Electronics, 12/89, p. 56. © Copyright Gernsback Publications, Inc., 1989.

Fig. 98-3. Reprinted with permission from Popular Electronics, 10/89, p. 105. © Copyright Gernsback Publications, Inc., 1989.

Fig. 98-4. Reprinted with permission from Radio-Electronics, 12/89, p. 56. © Copyright Gernsback Publications, Inc., 1989.

Fig. 98-5. Reprinted from EDN, 10/90, p. 222. © 1991 Cahners Publishing Company, a division of Reed Publishing USA.

Fig. 98-6. Reprinted with permission from Popular Electronics, 11/90, p. 102. © Copyright Gernsback Publications, Inc., 1990

Fig. 98-7. Reprinted from EDN, 5/90, p. 174. © 1991 Cahners Publishing Company, a division of Reed Publishing USA.

Fig. 98-8. Reprinted with permission from Radio-Electronics, 12/89, p. 56. © Copyright Gernsback Publications, Inc., 1989.

Fig. 98-9. Reprinted with permission from Radio-Electronics, 12/89, p. 56. © Copyright Gernsback Publications, Inc., 1989.

Fig. 98-10. RF Design, 4/86, p. 41.

Fig. 98-11. Fairchild Camera and Instrument Corp., Linear Databook, 1982, p. 5–24.

Chapter 99

Fig. 99-1. Reprinted from EDN, 6/91, p. 173. © 1991 Cahners Publishing Company, a division of Reed Publishing USA.

Fig. 99-2. Reprinted with permission from Radio-Electronics, 8/91, p. 63. © Copyright Gernsback Publications, Inc., 1991.

Fig. 99-3. Reprinted with permission from Popular Electronics, 10/89, p. 37–38. © Copyright Gernsback Publications, Inc., 1989.

Fig. 99-4. Reprinted with permission from Radio-Electronics, 1/91, p. 41–43. © Copyright Gernsback Publications, Inc., 1991.

Fig. 99-5. Reprinted with permission from Radio-Electronics, 8/91, p. 62. © Copyright Gernsback Publications, Inc., 1991.

Fig. 99-6. Reprinted with permission from Radio-Electronics, 8/91, p. 63. © Copyright Gernsback Publications, Inc., 1991.

Fig. 99-7. Reprinted with permission from Radio-Electronics, 3/89, p. 33–37. © Copyright Gernsback Publications, Inc., 1989.

Fig. 99-8. Reprinted with permission from R-E Experimenter's Handbook, 1990, p. 45. © Copyright Gernsback Publications, Inc., 1990.

Fig. 99-9. Reprinted from EDN, Design Ideas Special Issue, Vol. IV, 7/20/89, p. 16. © 1991 Cahners Publishing Company, a division of Reed Publishing USA.

Chapter 100

Fig. 100-1. Reprinted with permission from Radio-Electronics, 6/91, p. 55. © Copyright Gernsback Publications, Inc., 1991.

Fig. 100-2. Reprinted from EDN, 5/91, p. 178. © 1991 Cahners Publishing Company, a division of Reed Publishing USA.

Fig. 100-3. Reprinted with permission from Radio-Electronics, 6/91, p. 59–60. © Copyright Gernsback Publications, Inc., 1991.

Fig. 100-4. Reprinted with permission from Radio-Electronics, 6/91, p. 56. © Copyright Gernsback Publications, Inc., 1991.

Chapter 101

Fig. 101-1. Elektor Electronics, 12/90 Supplement, p. 40.

Chapter 102

Fig. 102-1. Reprinted from EDN, 6/91, p. 184. © 1991 Cahners Publishing Company, a division of Reed Publishing USA.

Fig. 102-2. Reprinted from EDN, 7/91, p. 166–168. © 1991 Cahners Publishing Company, a division of Reed Publishing USA.

Fig. 102-3. Reprinted with permission from Popular Electronics, 11/89, p. 85. © Copyright Gernsback Publications, Inc., 1989.

Fig. 102-4. Reprinted with permission from Electronic Design, 4/89, p. 77. Copyright 1989, Penton Publishing.

Fig. 102-5. Reprinted with permission from Electronic Design, 11/88, p. 78. Copyright 1988, Penton Publishing.

Fig. 102-6. Reprinted with permission from Electronic Design, 3/91, p. 119–120. Copyright 1991, Penton Publishing.

Fig. 102-7. Reprinted from EDN, 1/91, p. 168. © 1991 Cahners Publishing Company, a division of Reed Publishing USA.

Chapter 103

Fig. 103-1. Reprinted with permission from R-E Experimeter's Handbook, 1990, p. 155. © Copyright Gernsback Publications, Inc., 1990.

Fig. 103-2. Reprinted with permission from Popular Electronics, 8/89, p. 27. © Copyright Gernsback Publications, Inc., 1989.

Chapter 104

Fig. 104-1. Reprinted with permission from Electronic Design, 7/91, p. 136. Copyright 1991, Penton Publishing.

Fig. 104-2. Reprinted with permission from Popular Electronics, 1/90, p. 26. © Copyright Gernsback Publications, Inc., 1990.

Fig. 104-3. Reprinted with permission from Radio-Electronics, 6/86, p. 60. © Copyright Gernsback Publications, Inc., 1986.

Fig. 104-4. Harris, Analog Product Data Book, 1988, p. 2–106.

Fig. 104-5. Reprinted with permission from Radio-Electronics, 6/86, p. 58. © Copyright Gernsback Publications, Inc., 1986.

Index

buzzers (*cont.*)
  continuous tone 2kHZ, I-11
  gated 2kHz, I-12

C

cable bootstrapping, I-34
cable tester, III-539
calibrated circuit, DVM auto, I-714
calibrated tachometer, III-598
calibration standard, precision, I-406
calibrators
  crystal, 100 kHz, I-185
  electrolytic-capacitor reforming
    circuit, IV-276
  ESR measurer, IV-279
  oscilloscope, II-433, III-436
  portable, I-644
  square-wave, 5 V, I-423
  tester, IV-265
  wave-shaping circuits, high slew
    rates, IV-650
cameras (*see* photography-related
  circuits; television-related circuits;
  video circuits)
canceller, central image, III-358
capacitance buffers
  low-input, III-498
  low-input, stabilized, III-502
capacitance meters, I-400, II-91-94,
  III-75-77
  A/D, three-and-a-half digit, III-76
  capacitance-to-voltage, II-92
  digital, II-94
capacitance multiplier, I-416, II-200
capacitance relay, I-130
capacitance switched light, I-132
capacitance-to-pulse width converter,
  II-126
capacitance-to-voltage meter, II-92
capacitor discharge
  high-voltage generator, III-485
  ignition system, II-103
capacity tester, battery, III-66
car port, automatic light controller, II-
  308
cars (*see* automotive circuits)
carrier-current circuits, III-78-82, IV-
  91-93
  AM receiver, III-81
  audio transmitter, III-79
  data receiver, IV-93
  data transmitter, IV-92
  FM receiver, III-80
  intercom, I-146
  power-line modem, III-82
  receiver, I-143
  receiver, single transistor, I-145
  receiver, IC, I-146

remote control, I-146
transmitter, I-144
transmitter, integrated circuit, I-145
carrier-operated relay (COR), IV-461
carrier system receiver, I-141
carrier transmitter with on/off 200kHz
  line, I-142
cascaded amplifier, III-13
cassette bias oscillator, II-426
cassette interface, telephone, III-618
cassette-recorders (*see* tape-recorder
  circuits)
centigrade thermometer, I-655, II-648,
  II-662
central image canceller, III-358
charge pool power supply, III-469
charge pumps
  positive input/negative output, I-418,
    III-360
  regulated for fixed power supplies,
    IV-396
chargers (*see* battery charger)
chase circuit, I-326, III-197
Chebyshev filters (*see also* filter circuits)
  bandpass, fourth-order, III-191
  fifth-order multiple feedback low-
    pass, II-219
  high-pass, fourth-order, III-191
chime circuit, low-cost, II-33
chopper amplifier, I-350, II-7, III-12
checkers
  buzz box continuity and coil, I-551
  car battery condition, I-108
  crystal, I-178, I-186
  zener diode, I-406
chroma demodulator with RGB matrix,
  III-716
chug-chug sound generator, III-576
circuit breakers (*see also* protection
  circuits)
  12ns, II-97
  ac, III-512
  high-speed electronic, II-96
  trip circuit, IV-423
circuit protection (*see* protection circuits)
clamp-on-current probe compensator,
  II-501
clamp-limiting amplifiers, active, III-15
clamping circuits
  video signal, III-726
  video summing amplifier and, III-710
class-D power amplifier, III-453
clippers, II-394, IV-648
  audio-powered noise, II-396
  audio clipper/limiter, IV-355
  zener-design, fast and symmetrical,
    IV-329

clock circuits, II-100-102, III-83-85
  60Hz clock pulse generator, II-102
  adjustable TTL, I-614
  comparator, I-156
  crystal oscillators, micropower
    design, IV-122
  digital, with alarm, III-84
  gas discharge displays, 12-hour, I-253
  oscillator/clock generator, III-85
  phase lock, 20-Mhz to Nubus, III-
    105
  run-down clock for games, IV-205
  sensor touch switch and clock, IV-
    591
  single op amp, III-85
  source, clock source, I-729
  three-phase from reference, II-101
  TTL, wide-frequency, III-85
  Z80 computer, II-121
clock generators
  oscillator, I-615, III-85
  precision, I-193
  pulse generator, 60 Hz, II-102
clock radio, I-542
  AM/FM, I-543
CMOS circuits
  555 astable true rail to rail square
    wave generator, II-596
  9-bit, III-167
  coupler, optical, III-414
  crystal oscillator, III-134
  data acquisition system, II-117
  flasher, III-199
  inverter, linear amplifier from, II-11
  mixer, I-57
  optical coupler, III-414
  oscillator, III-429, III-430
  short-pulse generator, III-523
  timer, programmable, precision, III-
    652
  touch switch, I-137
  universal logic probe, III-499
coaxial cable, five-transistor pulse
  booster, II-191
Cockcroft-Walton cascaded voltage
  doubler, IV-635
code-practice oscillator, I-15, I-20, I-
  22, II-428-431, IV-373, IV-375, IV-
  376
coil drivers, current-limiting, III-173
coin flipper circuit, III-244
color amplifier, video, III-724
color-bar generator, IV-614
color organ, II-583, II-584
color video amplifier, I-34
Colpitts crystal oscillator, I-194, I-572,
  II-147
  1-to-20 MHz, IV-123
  frequency checker, IV-301

698

ignition circuit, electronic, automotive, IV-65

ignition cut-off circuit, automotive, IV-53

ignition substitute, automotive, III-41

ignition system, capacitor discharger, I-103

ignition timing light, II-60

ignitor, III-362

illumination stabilizer, machine vision, II-306

image canceller, III-358

immobilizer, II-50

impedance converter, high to low, I-41

incandescent light flasher, III-198

indicators (see also alarms; control circuits; detectors; monitors; sensors), III-268-270, IV-210-218

ac-current indicator, IV-290

ac-power indicator, LED display, IV-214

ac/dc indicator, IV-214

alarm and, I-337

automotive-temperature indicator, PTC thermistor, II-56

balance indicator, IV-215

bar-graph driver, transistorized, IV-213

battery charge/discharge, I-122

battery condition, I-121

battery level, I-124

battery threshold, I-124

battery voltage, solid-state, I-120

beat frequency, I-336

CW offset indicator, IV-213

dial pulse, III-613

field-strength (see field-strength meters)

in-use indicator, telephone, II-629

infrared detector, low-noise, II-289

lamp driver, optically coupled, III-413

level, three-step, I-336

low-battery, I-124

low-voltage, III-769

mains-failure indicator, IV-216

On indicator, IV-217

on-the-air, III-270

overspeed, I-108

overvoltage/undervoltage, I-150

peak level, I-402

phase sequence, I-476

receiver-signal alarm, III-270

rf output, IV-299

rf-actuated relay, III-270

simulated, I-417

sound sensor, IV-218

stereo-reception, III-269

SWR warning, I-22

telephone, in-use indicator, II-629, IV-560, IV-563

telephone, off-hook, I-633

temperature indicator, IV-570

transmitter-output indicator, IV-218

undervoltage, battery operated equipment, I-123

visual modulation, I-430

visual level, III-269

voltage, III-758-772

voltage, visible, I-338, III-772

voltage-level, I-718, III-759

voltage-level, five step, I-337

voltage-level, ten-step, I-335

volume indicator, audio amplifier, IV-212

VU meter, LED display, IV-211

zero center, FM receivers, I-338

in-use indicator, telephone, II-629

induction heater, ultrasonic, 120-KHz 500-W, III-704

inductors

active, I-417

simulated, II-199

infrared circuits, II-288-292, III-271-277, IV-219-228

data link, I-341

detector, III-276, IV-224

detector, low-noise, II-289

emitter drive, pulsed, II-292

fan controller, IV-226

laser rifle, invisible pulsed, II-291

loudspeaker link, remote, I-343

low-noise detector for, II-289

object detector, long-range, III-273

people-detector, IV-225

proximity switch, infrared-activated, IV-345

receivers, I-342, II-292, III-274, IV-220-221

receivers, remote-control, I-342

remote controller, IV-224

remote-control tester, IV-228

remote-extender, IV-227

transmitter, I-343, II-289, II-290, III-274, III-276, III-277, IV-226-227

transmitter, digital, III-275

transmitter, remote-control, I-342

transmitter, voice-modulated pulse FM, IV-228

wireless speaker system, III-272

injectors

three-in-one set: logic probe, signal tracer, injector, IV-429

injector-tracers, I-522

single, II-500

signal, I-521

input selectors, audio, low distortion, II-38

input/output buffer, analog multiplexers, III-11

instrumentation amplifiers, I-346, I-348, I-349, I-352, II-293-295, III-278-284, IV-229-234

± 100 V common mode range, III-294

current collector head amplifier, II-295

differential, I-347, I-354, III-283

differential, biomedical, III-282

differential, high-gain, I-353

differential, input, I-354

differential, variable gain, I-349

extended common-mode design, IV-234

high-impedance low drift, I-355

high-speed, I-354

low-drift/low-noise dc amplifier, IV-232

low-signal level/high-impedance, I-350

low-power, III-284

meter driver, II-296

preamp, oscilloscope, IV-230-231

re-amp, thermocouple, III-283

precision FET input, I-355

saturated standard cell amplifier, II-296

strain gauge, III-280

triple op amp, I-347

ultra-precision, III-279

variable gain, differential input, I-349

very high-impedance, I-354

wideband, III-281

instrumentation meter driver, II-296

integrators, II-297-300, III-285-286

active, inverting buffer, II-299

JFET ac coupled, II-200

gamma ray pulse, I-536

long time, II-300

low drift, I-423

noninverting, improved, II-298

photocurrent, II-326

programmable reset level, III-286

ramp generator, initial condition reset, III-327

resettable, III-286

intercoms, I-415, II-301-303, III-287-292

bidirectional, III-290

carrier current, I-146

hands-off, III-291

party-line, II-303

pocket pager, III-288

telephone-intercoms, IV-557

two-way, III-292

two-wire design, IV-235-237

interfaces (see also computer circuits), IV-238-242

680x, 650x, 8080 families, III-98

timers, I-666, I-668, II-671-681, III-650-655, IV-578-586
0.1 to 90 second, I-663
741 timer, I-667
adjustable, IV-585
adjustable ac .2 to 10 seconds, II-681
alarm, II-674
appliance-cutoff timer, IV-583
CMOS, programmable precision, III-652
circuit, II-675
darkroom, I-480
elapsed time/counter timer, II-680
electronic egg, I-665
IC, crystal-stabilized, II-151
interval, programmable, II-678
interval, programmable, thumbwheel, I-660
long-delay, PUT, I-219
long-duration, PUT, II-675
long-duration, time delay, IV-585
long-interval, programmable, IV-581, IV-582
long-interval, RC, I-667
long-term electronic, II-672
long-time, III-653
mains-powered, IV-579
one-shot, III-654
photographic, I-485
photographic, darkroom enlarger, III-445
photographic, photo-event timer, IV-379
reaction timer, game circuit, IV-204
SCR design, IV-583
sequential, I-661-662, I-663, III-651
sequential UJT, I-662
slide-show, III-444
slides, photographic, III-448
solid-state, industrial applications, I-664
ten-minute ID timer, IV-584
three-minute, III-654
thumbwheel-type, programmable interval, I-660
time-out circuit, IV-586
transmit-time limiter, IV-580
triangle-wave generator, linear, III-222
variable duty-cycle output, III-240
voltage-controlled, programmable, II-676
washer, I-668
watchdog timer/alarm, IV-584
timing light, ignition, II-60
timing threshold and load driver, III-648
tone alert decoder, I-213
tone annunciator, transformerless, III-27-28

tone burst generators, I-604, II-90
European repeaters, III-74
tone controls (*see also* sound generators), I-677, II-682-689, III-656-660, IV-587-589
active bass and treble, with buffer, I-674
active control, IV-588
audio amplifier, II-686
Baxandall tone-control audio amplifier, IV-588
equalizer, ten-band octave, III-658
equalizer, ten-band graphic, active filter, II-684
guitar treble booster, II-683
high-quality, I-675
high-z input, hi fi, I-676
microphone preamp, I-675, II-687
mixer preamp, I-58
passive circuit, II-689
preamplifier, high-level, II-688
preamplifier, IC, I-673, III-657
preamplifier, microphone, I-675, II-687
preamplifier, mixer, I-58
rumble/scratch filter, III-660
three-band active, I-676, III-658
three-channel, I-672
tremolo circuit, IV-589
Wien-bridge filter, III-659
tone decoders, I-231, III-143
dual time constant, II-166
24 percent bandwidth, I-215
relay output, I-213
tone-dial decoder, I-631
tone detectors, 500-Hz, III-154
tone-dial decoder, I-630, I-631
tone-dial encoder, I-629
tone-dial generator, I-629
tone-dialing telephone, III-607
tone encoder, I-67
subaudible, I-23
tone-dial encoder, I-629
two-wire, II-364
tone generators (*see* sound generators)
tone probe, digital IC testing with, II-504
tone ringer, telephone, II-630, II-631
totem-pole driver, bootstrapping, III-175
touch circuit, I-137
touch switches, I-131, I-135-136, II-690-693, III-661-665, IV-590-594
CMOS, I-137
bistable multivibrator, touch-triggered, I-133
double-button latching, I-138
hum-detecting touch sensor, IV-594
lamp control, three-way, IV-247

low-current, I-132
On/Off, II-691, III-663, IV-593
line-hum, III-664
momentary operation, I-133
negative-triggered, III-662
positive-triggered, III-662
sensor switch and clock, IV-591
time-on touch switch, IV-594
touchomatic, II-693
two-terminal, III-663
Touchtone generator, telephone, III-609
toxic gas detector, II-280
toy siren, II-575
TR circuit, II-532
tracers
audio reference signal, probe, I-527
bug, III-358
closed-loop, III-356
receiver, III-357
track-and-hold circuits, III-667
sample-and-hold circuit, III-549, III-552
signal, III-668
tracking circuits, III-666-668
positive/negative voltage reference, III-667
preregulator, III-492
track-and-hold, III-667
track-and-hold, signal, III-668
train chuffer sound effect, II-588
transceivers (*see also* receivers; transmitters), IV-595-603
CE, 20-m, IV-596-598
CW, 5 W, 80-meter, IV-602
hand-held, dc adapter, III-461
hand-held, speaker amplifiers, III-39
HF transceiver/mixer, IV-457
ultrasonic, III-702, III-704
transducer amplifiers, III-669-673
flat-response, tape, III-673
NAB preamp, record, III-673
NAB preamp, two-pole, III-673
photodiode amplifier, III-672
preamp, magnetic phono, III-671, III-673
tape playback, III-672
voltage, differential-to-single-ended, III-670
transducers, I-86
bridge type, amplifier, II-84, III-71
detector, magnetic transducer, I-233
sonar, switch and, III-703
temperature, remote sensor, I-649
transistors and transistorized circuits
flashers, II-236, III-200
frequency tripler, nonselective, saturated, II-252
headphone amplifier, II-43

on/off switch for op amp, IV-546
pulse generator, IV-437
sorter, I-401
tester, I-401, IV-281
transmission indicator, II-211
transmitters (see also receivers; transceivers), III-674-691, IV-595-603
2-meter, IV-600-601
acoustic-sound transmitter, IV-311
amateur radio, 80-M, III-675
amateur TV, IV-599
beacon, III-683, IV-603
broadcast, 1-to-2 MHz, I-680
carrier current, I-144, I-145, III-79
computer circuit, 1-of-8 channel, III-
100
CW, 1 W, III-678
CW, 10 W, one-tube, I-681
CW, 40 M, III-684
CW, 902 MHz, III-686
CW, HF low-power, IV-601
CW, QRP, III-690
fiber optic, III-177
FM, I-681
FM, infrared, voice-modulated pulse,
IV-228
FM, multiplex, III-688
FM, one-transistor, III-687
FM, (PRM) optical, I-367
FM, snooper, III-680
FM, voice, III-678
FM, wireless microphone, III-682,
III-685, III-691
half-duplex information transmission
link, low-cost, III-679
HF, low-power, IV-598
infrared, I-343, II-289, II-290, III-
277, IV-226-227
infrared, digital, III-275
infrared, FM, voice-modulated pulse,
IV-228
infrared, remote control with
receiver, I-342
line-carrier, with on/off, 200 kHz, I-
142
low-frequency, III-682
multiplexed, 1-of-8 channel, III-395
negative key-line keyer, IV-244
optical, I-363, IV-368
optical, FM, 50 kHz center frequency, II-417
optical, receiver for, II-418
oscillator and, 27 and 49 MHz, I-680
output indicator, IV-218
remote sensors, loop-type, III-70
television, III-676
ultrasonic, 40 kHz, I-685
VHF, modulator, III-684
VHF, tone, III-681

treasure locator, lo-parts, I-409
treble booster, guitar, II-683
tremolo circuits, I-59, III-692-695, IV-
589
voltage-controlled amplifier, I-598
triac circuits
ac-voltage controller, IV-426
contact protection, II-531
dimmer switch, II-310, III-303
dimmer switch, 800W, I-375
drive interface, direct dc, I-266
microprocessor array, II-410
relay-contact protection with, II-531
switch, inductive load, IV-253
trigger, I-421
voltage doubler, III-468
zero point switch, II-311
zero voltage, I-623
triangle-to-sine converter, II-127
triangle/square wave oscillator, II-422
triangle-wave generators, III-234
square-wave, III-225, III-239
square-wave, precision, III-242
square-wave, wide-range, III-242
timer, linear, III-222
trickle charger, 12 V battery, I-117
triggers
50-MHz, III-364
camera alarm, III-444
flash, photography, xenon flash, III-
447
optical Schmitt, I-362
oscilloscope-triggered sweep, III-438
remote flash, I-484
SCR series, optically coupled, III-411
sound/light flash, I-482
triac, I-421
triggered sweep, add-on, I-472
tripler, nonselective, transistor saturation, II-252
trouble tone alert, II-3
TTL circuits
clock, wide-frequency, III-85
coupler, optical, III-416
gates, siren using, II-576
Morse code keyer, II-25
square wave to triangle wave converter, II-125
TTL to MOS logic converter, II-125
TTL oscillators, I-179, I-613
1MHz to 10MHz, I-178
television display using, II-372
crystal, I-197
sine/square wave oscillator, IV-512
tube amplifier, high-voltage isolation,
IV-426
tuners
antenna tuner, 1-to-30 MHz, IV-14
FM, I-231

guitar and bass, II-362
turbo circuits, glitch free, III-186
twang-twang circuit, II-592
twilight-triggered circuit, II-322
twin-T notch filters, III-403
two-state siren, III-567
two-tone generator, II-570
two-tone siren, III-562
two-way intercom, III-292
two's complement, D/A conversion
system, binary, 12-bit, III-166

U

UA2240 staircase generator, III-587
UHF transmissions
field-strength meters, IV-165
rf amplifiers, UHF TV-line amplifier,
IV-482, IV-483
source dipper, IV-299
TV preamplifier, III-546
VHF/UHF rf diode switch, IV-544
wideband amplifier, high performance
FETs, III-264
UJT circuits
battery chargers, III-56
metronome, II-355
monostable circuit, bias voltage
change insensitive, II-268
ultrasonic circuits (see also sound-
operated circuits), III-696-707, IV-
604-606
arc welding inverter, 20 KHz, III-700
induction heater, 120-KHz 500-W,
III-704
pest-controller, III-706, III-707
pest-repeller, I-684, II-685, III-699,
IV-605-606
ranging system, III-697
receiver, III-698, III-705
sonar transducer/switch, III-703
sound source, IV-605
switch, I-683
transceiver, III-702, III-704
transmitter, I-685
undervoltage detector, IV-138
undervoltage monitor, III-762
uninterruptible power supply, II-462
+5V, III-477
unity-gain amplifiers
inverting, I-80
inverting, wideband, I-35
ultra high Z, ac, II-7
unity-gain buffer
stable, with good speed and high-
input impedance, II-6
unity-gain follower, I-27
universal counters
10 MHz, II-139

voltage regulators (*cont.*)

switching, 5 V, 6 A 25kHz, separate ultrastable reference, I-497

switching, 6 A, variable output, I-513

switching, 200 kHz, I-491

switching, multiple output, for use with MPU, I-513

switching, step down, I-493

switching, high-current inductorless, III-476

switching, low-power, III-490

variable, III-491, IV-468-470

variable, current source, III-490

zener design, programmable, IV-470

voltage sources

millivolt, zenerless, I-696

programmable, I-694

voltage splitter, III-738

voltage-to-current converter, I-166, II-124, III-110, IV-118

power, I-163

zero IB error, III-120

voltage-to-frequency converters, I-707, III-749-757, IV-638-642

1 Hz-to-10MHz, III-754

1 Hz-to-30 MHz, III-750

1Hz-to-1.25 MHz, III-755

5 KHz-to-2MHz, III-752

10Hz to 10 kHz, I-706, III-110

accurate, III-756

differential-input, III-750

function generators, potentiometer-position, IV-200

low-cost, III-751

low-frequency converter, IV-641

negative input, I-708

optocoupler, IV-642

positive input, I-707

precision, II-131

preserved input, III-753

ultraprecision, I-708

wide-range, III-751, III-752

voltage-to-pulse duration converter, II-124

voltmeters

3 1/2 digit, I-710

3 1/2 digital true rms ac, I-713

5-digit, III-760

ac, III-765

ac, wide-range, III-772

add-on thermometer for, III-640

bar-graph, I-99, II-54

dc, III-763

dc, high-input resistance, III-762

digital, III-4

digital, 3.5-digit, full-scale, four-decade, III-761

digital, LED readout, IV-286

FET, I-714, III-765, III-770

high-input resistance, III-768

millivoltmeters (*see* millivoltmeters)

rf, I-405, III-766

wide-band ac, I-716

voltohmmeter, phase meter, digital readout, IV-277

volume amplifier, II-46

volume control circuits, IV-643-645

telephone, II-623

volume indicator, audio amplifier, IV-212

VOR signal simulator, IV-273

vox box, II-582, IV-623

Vpp generator, EPROM, II-114

VU meters

extended range, II-487, I-715

LED display, IV-211

W

waa-waa circuit, II-590

wailers (*see* alarms; sirens)

wake-up call, electronic, II-324

walkman amplifier, II-456

warblers (*see* alarms; sirens)

warning devices

auto lights-on warning, II-55

high-level, I-387

high-speed, I-101

light, III-317

light, battery-powered, II-320

low-level, audio output, I-391

speed, I-96

varying-frequency alarm, II-579

water-level sensors (*see* fluid and moisture detectors)

water-temperature gauge, automotive, IV-44

wattmeter, I-17

wave-shaping circuits (*see also* wave-form generators), IV-646-651

capacitor for high slew rates, IV-650

clipper, glitch-free, IV-648

flip-flop, S/R, IV-651

harmonic generator, IV-649

phase shifter, IV-647

rectifier, full-wave, IV-650

signal conditioner, IV-649

waveform generators (*see also* burst generators; function generators; sound generators; square-wave generators; wave-shaping circuits), II-269, II-272

audio, precision, III-230

four-output, III-223

harmonic generator, IV-649

high-speed generator, I-723

precise, II-274

ramp generators, IV-443-447

sawtooth generator, digital, IV-444, IV-446

sine-wave, IV-505, IV-506

sine-wave, 60 Hz, IV-507

sine-wave, audio, II-564

sine-wave, LC, IV-507

sine-wave, LF, IV-512

sine-wave oscillator, audio, III-559

staircase generators, IV-443-447

staircase generator/frequency divider, I-730

stepped waveforms, IV-447

triangle and square waveform, I-726

VCO and, III-737

wavemeter, tuned RF, IV-302

weather-alert decoder, IV-140

weight scale, digital, II-398

Wheel-of-Fortune game, IV-206

whistle, steam locomotive, II-589, III-568

who's first game circuit, III-244

wide-range oscillators, I-69, III-425

variable, I-730

wide-range peak detectors, III-152

hybrid, 500 kHz-1 GHz, III-265

instrumentation, III-281

miniature, III-265

UHF amplifiers, high-performance FETs, III-264

wideband amplifiers

low-noise/low drift, I-38

two-stage, I-689

rf, IV-489, IV-490, IV-491

rf, HF, IV-492

rf, JFET, IV-493

rf, MOSFET, IV-492

rf, two-CA3100 op amp design, IV-491

unity gain inverting, I-35

wideband signal splitter, III-582

wideband two-pole high pass filter, II-215

Wien-bridge filter, III-659

notch filter, II-402

Wien-bridge oscillators, I-62-63, I-70, III-429, IV-371, IV-377, IV-511

CMOS chip in, II-568

low-distortion, thermally stable, III-557

low-voltage, III-432

sine wave, I-66, I-70, II-566

sine-wave, three-decade, IV-510

sine-wave, very-low distortion, IV-513

single-supply, III-558

variable, III-424

wind-powered battery charger, II-70

windicator, I-330

window circuits, II-106, III-90, III-776-

Other Bestsellers of Related Interest

ENCYCLOPEDIA OF ELECTRONIC CIRCUITS
Vol. 1—Rudolf F. Graf

". . . schematics that encompass virtually the entire spectrum of electronics technology . . . This is a well worthwhile book to have handy." —Modern Electronics

Discover hundreds of the most versatile electronic and integrated circuit designs, all available at at the turn of a page. You'll find circuit diagrams and schematics for a wide variety of practical applications. Many entries also include clear, concise explanations of the circuit configurations and functions. 768 pages, 1,762 illustrations. Book No. 1938, $29.95 paperback, $60.00 hardcover

THE ILLUSTRATED DICTIONARY OF ELECTRONICS—5th Edition
—Rufus P. Turner and Stan Gibilisco

This completely revised and updated edition defines more than 27,000 practical electronics terms, acronyms, and abbreviations. Find up-to-date information on basic electronics, computers, mathematics, electricity, communications, and state-of-the-art applications—all discussed in a nontechnical style. The author also includes 360 new definitions and 125 illustrations and diagrams. 736 pages, 650 illustrations. Book No. 3345, $26.95 paperback, $39.95 hardcover

ELECTRONIC CONVERSIONS:
Symbols and Formulas—2nd Edition
—Rufus P. Turner and Stan Gibilisco

This revised and updated edition supplies all the formulas, symbols, tables, and conversion factors commonly used in electronics. Exceptionally easy to use, the material is organized by subject matter. Its format is ideal and you can save time by directly accessing specific information. Topics cover only the most-needed facts about the most often used conversions, symbols, formulas, and tables. 280 pages, 94 illustrations. Book No. 2865, $14.95 paperback only

ELECTRONIC DATABOOK—4th Edition
—Rudolf F. Graf

If it's electronic, it's here—current, detailed, and comprehensive! Use this book to broaden your electronics information base. Revised and expanded to include all up-to-date information, this fourth edition makes any electronic job easier and less time-consuming. You'll find information that will aid in the design of local area networks, computer interfacing structure, and more! 528 pages, 131 illustrations. Book No. 2958, $24.95 paperback only

BUILD YOUR OWN TEST EQUIPMENT
—Homer L. Davidson

Build more than 30 common electronic testing devices, ranging from simple continuity and polarity testers to signal injectors and power supplies. Also learn how test instruments work, how they are used, and how to save money. Each project includes a complete parts list with exact part numbers. 300 pages, 324 illustrations. Book No. 3475, $17.95 paperback, $27.95 hardcover

ELECTRONIC SIGNALS AND SYSTEMS:
Television, Stereo, Satellite TV, and
Automotive—Stan Prentiss

Study signal analysis as it applies to the operation and signal-generating capabilities of today's most advanced electronic devices with this handbook. It explains the composition and use of a wide variety of test instruments, transmission media, satellite systems, stereo broadcast and reception facilities, antennas, television equipment, and even automotive electrical systems. You'll find coverage of C- and Ku-band video, satellite master TV systems, high-definition television, C-QUAM® AM stereo transmission and reception, and more. 328 pages, 186 illustrations. Book No. 3557, $19.95 paperback, $29.95 hardcover

MASTERING ELECTRONICS MATH
—2nd Edition—R. Jesse Phagan

A self-paced text for hobbyists and a practical tool-box reference for technicians, this book guides you through the practical calculations needed to design and troubleshoot circuits and electronics components. Clear explanations and sample problems illustrate each concept, including how each is used in common electronics applications. If you want to gain a strong understanding of electronics math and stay on top of your profession, this book will be a valuable tool for you! 344 pages, 270 illustrations. Book No. 3589, $17.95 paperback, $27.95 hardcover

BOB GROSSBLATT'S GUIDE TO CREATIVE CIRCUIT DESIGN—Robert Grossblatt

Robert Grossblatt, *Radio Electronics'* popular columnist brings his unique circuit design philosophy and style to this hands-on guide. Emphasizing the importance of scientific method over technical knowledge, it walks you through the circuit design process—from brainwork to paperwork to boardwork—and suggests ways for making your bench time as efficient as possible. 248 pages, 129 illustrations. Book No. 3610, $17.95 paperback, $28.95 hardcover

SECRETS OF RF CIRCUIT DESIGN
—Joseph J. Carr

This book explains in clear, nontechnical language what RF is, how it works, and how it differs from other electromagnetic frequencies. You'll learn the basics of receiver operation, the proper use and repair components in RF circuits, and principles of radio signal propagation from low frequencies to microwave. You'll enjoy experiments that explore such problems as electromagnetic interface. 416 pages, 411 illustrations. Book No. 3710, $19.95 paperback, $32.95 hardcover

INTERNATIONAL ENCYCLOPEDIA OF INTEGRATED CIRCUITS—2nd Edition
—Stan Gibilisco

The most thorough coverage of foreign and domestic integrated circuits is available today in this giant resource. Seven separate sections detail thousands of ICs and their applications, including all relevant information, charts, and tables. This second edition of a unique all-in-one reference tells what each IC is, what it does, how it does it, and what its relationship is to other ICs and their applications. 1,168 pages, 4,605 illustrations. Book No. 3802, $84.95 hardcover only

ELECTRONIC POWER CONTROL: Circuits, Devices & Techniques—Irving M. Gottlieb

This guide focuses on the specific digital circuits used in electronic power applications. It presents state-of-the-art approaches to analysis, troubleshooting, and implementation of new solid-state devices. Gottlieb shows you how to adapt various power-control techniques to your individual needs. He uses descriptive analysis and real-world applications wherever possible, employing mathematical theory only when relevant. 272 pages, 197 illustrations. Book No. 3837, $17.95 paperback, $27.95 hardcover

ENCYCLOPEDIA OF ELECTRONICS
—2nd Edition—Stan Gibilisco and Neil Sclater, Co-Editors-in-Chief
Praise for the first edition:
". . . a fine one-volume source of detailed information for the whole breadth of electronics."

—Modern Electronics

The second edition brings you more than 950 pages of listings and cover virtually every electronics concept and component imaginable. From basic electronics terms to state-of-the-art applications, this is the most complete and comprehensive electronics reference available! 976 pages, 1,400 illustrations. Book No. 3389, $69.50 hardcover only

MASTER HANDBOOK OF ELECTRONIC TABLES AND FORMULAS—5th Edition
—Martin Clifford
". . . a source of quick, accurate, and easy-to-use solutions to electronics problems . . . to be used as a reference book for hobbyists as well as professionals."

Electronics for You, on a previous edition
It's a completely revised and updated edition of the classic reference used by thousands of hobbyists, technicians, and engineers. Reflecting the latest developments in the field, you'll never again have to stop and make pencil and paper calculations, hunt up your calculator for simple conversions, or search through reference books trying to find a specific formula. Everything you need is right here—logically organized and fully indexed. Martin Clifford includes the most current information on everything from resistance formulas, sine waves, capacitance, impedance vectors, decibels, and more. 544 pages, 490 illustrations. Book No. 3739, $22.95 paperback, $39.95 hardcover